Lecture Notes in Computer Science   12573

More information about this subseries at http://www.springer.com/series/7409

Jakub Lokoč · Tomáš Skopal ·
Klaus Schoeffmann · Vasileios Mezaris ·
Xirong Li · Stefanos Vrochidis ·
Ioannis Patras (Eds.)

MultiMedia Modeling

27th International Conference, MMM 2021
Prague, Czech Republic, June 22–24, 2021
Proceedings, Part II

Springer

Editors
Jakub Lokoč 🔟
Charles University
Prague, Czech Republic

Klaus Schoeffmann 🔟
Klagenfurt University
Klagenfurt, Austria

Xirong Li 🔟
Renmin University of China
Beijing, China

Ioannis Patras 🔟
Queen Mary University of London
London, UK

Tomáš Skopal 🔟
Charles University
Prague, Czech Republic

Vasileios Mezaris 🔟
CERTH-ITI
Thessaloniki, Greece

Stefanos Vrochidis 🔟
CERTH-ITI
Thessaloniki, Greece

ISSN 0302-9743   ISSN 1611-3349   (electronic)
Lecture Notes in Computer Science
ISBN 978-3-030-67834-0   ISBN 978-3-030-67835-7   (eBook)
https://doi.org/10.1007/978-3-030-67835-7

LNCS Sublibrary: SL3 – Information Systems and Applications, incl. Internet/Web, and HCI

This Springer imprint is published by the registered company Springer Nature Switzerland AG
The registered company address is: Gewerbestrasse 11, 6330 Cham, Switzerland

Preface

These two-volume proceedings contain the papers accepted at MMM 2021, the 27th International Conference on MultiMedia Modeling.

Organized for more than 25 years, MMM has become a respected and well-established international conference bringing together excellent researchers from academic and industrial areas. During the conference, novel research works from MMM-related areas (especially multimedia content analysis; multimedia signal processing and communications; and multimedia applications and services) are shared along with practical experiences, results, and exciting demonstrations. The 27th instance of the conference was organized in Prague, Czech Republic on June 22–24, 2021. Due to the COVID-19 pandemic, the conference date was shifted by five months, however the Proceedings were published in January in accordance with the original plan. Despite the pandemic, MMM 2021 received a large number of submissions organized in different tracks.

Specifically, 211 papers were submitted to seven MMM 2021 tracks. Each paper was reviewed by at least two reviewers (but mostly three) from the Program Committee, while the TPC chairs and special event organizers acted as meta-reviewers. Out of 166 regular papers, 73 were accepted for the proceedings. In particular, 40 papers were accepted for oral presentation and 33 papers for poster presentation. Regarding the remaining tracks, 16 special session papers were accepted as well as 2 papers for a demo presentation and 17 papers for participation at the Video Browser Showdown 2021. Overall, the MMM 2021 program comprised 108 papers from the seven tracks with the following acceptance rates:

Tracks	#Papers	ACCEPTANCE rates
Full papers (oral)	40	24%
Full papers (oral + poster)	73	44%
Demos	2	67%
SS1: MAPTA	4	50%
SS2: MDRE	5	71%
SS3: MMARSat	3	100%
SS4: MULTIMED	4	67%
Video Browser Showdown	17	94%

The special sessions are traditionally organized to extend the program with novel challenging problems and directions. The MMM 2021 program included four special sessions:

- SS1: Multimedia Analytics: Perspectives, Tools, and Applications (MAPTA)
- SS2: Multimedia Datasets for Repeatable Experimentation (MDRE)
- SS3: Multimodal Analysis and Retrieval of Satellite Images (MMARSat)
- SS4: Multimedia and Multimodal Analytics in the Medical Domain and Pervasive Environments (MULTIMED)

Besides the four special sessions, the anniversary 10th Video Browser Showdown represented an important highlight of MMM 2021 with a record number of 17 participating systems in this exciting (and challenging!) competition. In addition, two highly respected speakers were invited to MMM 2021 to present their impressive talks and results in multimedia-related topics. Specifically, we would like to thank Cees Snoek from the University of Amsterdam, and Pavel Zezula from Masaryk University.

Last but not least, we would like to thank all members of the MMM community who contributed to the MMM 2021 event. We also thank all authors of submitted papers, all reviewers, and all members of the MMM 2021 organization team for their great work and support. They all helped MMM 2021 to be an exciting and inspiring international event for all participants!

January 2021

Jakub Lokoč
Tomáš Skopal
Klaus Schoeffmann
Vasileios Mezaris
Xirong Li
Stefanos Vrochidis
Ioannis Patras

Organization

Organizing Committee

General Chairs

Jakub Lokoč	Charles University, Prague
Tomáš Skopal	Charles University, Prague

Program Chairs

Klaus Schoeffmann	Klagenfurt University
Vasileios Mezaris	CERTH-ITI, Thessaloniki
Xirong Li	Renmin University of China

Special Session and Tutorial Chairs

Werner Bailer	Joanneum Research
Marta Mrak	BBC Research & Development

Panel Chairs

Giuseppe Amato	ISTI-CNR, Pisa
Fabrizio Falchi	ISTI-CNR, Pisa

Demo Chairs

Cathal Gurrin	Dublin City University
Jan Zahálka	Czech Technical University in Prague

Video Browser Showdown Chairs

Klaus Schoeffmann	Klagenfurt University
Werner Bailer	Joanneum Research
Jakub Lokoč	Charles University, Prague
Cathal Gurrin	Dublin City University

Publicity Chairs

Phoebe Chen	La Trobe University
Chong-Wah Ngo	City University of Hong Kong
Bing-Kun Bao	Nanjing University of Posts and Telecommunications

Publication Chairs

Stefanos Vrochidis	CERTH-ITI, Thessaloniki
Ioannis Patras	Queen Mary University of London

Steering Committee

Phoebe Chen	La Trobe University
Tat-Seng Chua	National University of Singapore
Kiyoharu Aizawa	University of Tokyo
Cathal Gurrin	Dublin City University
Benoit Huet	Eurecom
Klaus Schoeffmann	Klagenfurt University
Richang Hong	Hefei University of Technology
Björn Þór Jónsson	IT University of Copenhagen
Guo-Jun Qi	University of Central Florida
Wen-Huang Cheng	National Chiao Tung University
Peng Cui	Tsinghua University

Web Chair

František Mejzlík   Charles University, Prague

Organizing Agency

Conforg, s.r.o.

Special Session Organizers

Multimedia Datasets for Repeatable Experimentation (MDRE)

Cathal Gurrin	Dublin City University, Ireland
Duc-Tien Dang-Nguyen	University of Bergen, Norway
Björn Þór Jónsson	IT University of Copenhagen, Denmark
Klaus Schoeffmann	Klagenfurt University, Austria

Multimedia Analytics: Perspectives, Tools and Applications (MAPTA)

Björn Þór Jónsson	IT University of Copenhagen, Denmark
Stevan Rudinac	University of Amsterdam, The Netherlands
Xirong Li	Renmin University of China, China
Cathal Gurrin	Dublin City University, Ireland
Laurent Amsaleg	CNRS-IRISA, France

Multimodal Analysis and Retrieval of Satellite Images

Ilias Gialampoukidis	Centre for Research and Technology Hellas, Information Technologies Institute, Greece
Stefanos Vrochidis	Centre for Research and Technology Hellas, Information Technologies Institute, Greece
Ioannis Papoutsis	National Observatory of Athens, Greece

Guido Vingione   Serco Italy, Italy
Ioannis Kompatsiaris   Centre for Research and Technology Hellas,
                              Information Technologies Institute, Greece

MULTIMED: Multimedia and Multimodal Analytics in the Medical Domain and Pervasive Environments

Georgios Meditskos   Centre for Research and Technology Hellas,
                              Information Technologies Institute, Greece
Klaus Schoeffmann   Klagenfurt University, Austria
Leo Wanner   ICREA – Universitat Pompeu Fabra, Spain
Stefanos Vrochidis   Centre for Research and Technology Hellas,
                              Information Technologies Institute, Greece
Athanasios Tzioufas   Medical School of the National and Kapodistrian
                              University of Athens, Greece

MMM 2021 Program Committees and Reviewers Regular and Special Sessions
Program Committee

Olfa Ben Ahmed   EURECOM
Laurent Amsaleg   CNRS-IRISA
Evlampios Apostolidis   CERTH ITI
Ognjen Arandjelović   University of St Andrews
Devanshu Arya   University of Amsterdam
Nathalie Aussenac   IRIT CNRS
Esra Açar   Middle East Technical University
Werner Bailer   JOANNEUM RESEARCH
Bing-Kun Bao   Nanjing University of Posts and Telecommunications
Ilaria Bartolini   University of Bologna
Christian Beecks   University of Munster
Jenny Benois-Pineau   LaBRI, UMR CNRS 5800 CNRS,
                              University of Bordeaux
Roberto Di Bernardo   Engineering Ingegneria Informatica S.p.A.
Antonis Bikakis   University College London
Josep Blat   Universitat Pompeu Fabra
Richard Burns   West Chester University
Benjamin Bustos   University of Chile
K. Selçuk Candan   Arizona State University
Ying Cao   City University of Hong Kong
Annalina Caputo   University College Dublin
Savvas Chatzichristofis   Neapolis University Pafos
Angelos Chatzimichail   Centre for Research and Technology Hellas
Edgar Chavez   CICESE
Mulin Chen   Northwestern Polytechnical University
Zhineng Chen   Institute of Automation, Chinese Academy of Sciences
Zhiyong Cheng   Qilu University of Technology
Wei-Ta Chu   National Cheng Kung University

Andrea Ciapetti	Innovation Engineering
Kathy Clawson	University of Sunderland
Claudiu Cobarzan	Klagenfurt University
Rossana Damiano	Università di Torino
Mariana Damova	Mozaika
Minh-Son Dao	National Institute of Information and Communications Technology
Petros Daras	Information Technologies Institute
Mihai Datcu	DLR
Mathieu Delalandre	Université de Tours
Begum Demir	Technische Universität Berlin
Francois Destelle	Dublin City University
Cem Direkoğlu	Middle East Technical University – Northern Cyprus Campus
Jianfeng Dong	Zhejiang Gongshang University
Shaoyi Du	Xi'an Jiaotong University
Athanasios Efthymiou	University of Amsterdam
Lianli Gao	University of Science and Technology of China
Dimos Georgiou	Catalink EU
Negin Ghamsarian	Klagenfurt University
Ilias Gialampoukidis	CERTH ITI
Nikolaos Gkalelis	CERTH ITI
Nuno Grosso	
Ziyu Guan	Northwest University of China
Gylfi Gudmundsson	Reykjavik University
Silvio Guimaraes	Pontifícia Universidade Católica de Minas Gerais
Cathal Gurrin	Dublin City University
Pål Halvorsen	SimulaMet
Graham Healy	Dublin City University
Shintami Chusnul Hidayati	Institute of Technology Sepuluh Nopember
Dennis Hoppe	High Performance Computing Center Stuttgart
Jun-Wei Hsieh	National Taiwan Ocean University
Min-Chun Hu	National Tsing Hua University
Zhenzhen Hu	Nanyang Technological University
Jen-Wei Huang	National Cheng Kung University
Lei Huang	Ocean University of China
Ichiro Ide	Nagoya University
Konstantinos Ioannidis	CERTH ITI
Bogdan Ionescu	University Politehnica of Bucharest
Adam Jatowt	Kyoto University
Peiguang Jing	Tianjin University
Hyun Woo Jo	Korea University
Björn Þór Jónsson	IT-University of Copenhagen
Yong Ju Jung	Gachon University
Anastasios Karakostas	Aristotle University of Thessaloniki
Ari Karppinen	Finnish Meteorological Institute

Jiro Katto	Waseda University
Junmo Kim	Korea Advanced Institute of Science and Technology
Sabrina Kletz	Klagenfurt University
Ioannis Kompatsiaris	CERTH ITI
Haris Kontoes	National Observatory of Athens
Efstratios Kontopoulos	Elsevier Technology
Markus Koskela	CSC – IT Center for Science Ltd.
Yu-Kun Lai	Cardiff University
Woo Kyun Lee	Korea University
Jochen Laubrock	University of Potsdam
Khiem Tu Le	Dublin City University
Andreas Leibetseder	Klagenfurt University
Teng Li	Anhui University
Xirong Li	Renmin University of China
Yingbo Li	Eurecom
Wu Liu	JD AI Research of JD.com
Xueting Liu	The Chinese University of Hong Kong
Jakub Lokoč	Charles University
José Lorenzo	Atos
Mathias Lux	Klagenfurt University
Ioannis Manakos	CERTH ITI
José M. Martinez	Universidad Autònoma de Madrid
Stephane Marchand-Maillet	Viper Group – University of Geneva
Ernesto La Mattina	Engineering Ingegneria Informatica S.p.A.
Thanassis Mavropoulos	CERTH ITI
Kevin McGuinness	Dublin City University
Georgios Meditskos	CERTH ITI
Robert Mertens	HSW University of Applied Sciences
Vasileios Mezaris	CERTH ITI
Weiqing Min	ICT
Wolfgang Minker	University of Ulm
Marta Mrak	BBC
Phivos Mylonas	National Technical University of Athens
Henning Muller	HES-SO
Duc Tien Dang Nguyen	University of Bergen
Liqiang Nie	Shandong University
Tu Van Ninh	Dublin City University
Naoko Nitta	Osaka University
Noel E. O'Connor	Dublin City University
Neil O'Hare	Yahoo Research
Jean-Marc Ogier	University of La Rochelle
Vincent Oria	NJIT
Tse-Yu Pan	National Cheng Kung University
Ioannis Papoutsis	National Observatory of Athens
Cecilia Pasquini	Universität Innsbruck
Ladislav Peška	Charles University

Zheng Wang	National Institute of Informatics
Leo Wanner	ICREA/UPF
Wolfgang Weiss	JOANNEUM RESEARCH
Lai-Kuan Wong	Multimedia University
Tien-Tsin Wong	The Chinese University of Hong Kong
Marcel Worring	University of Amsterdam
Xiao Wu	Southwest Jiaotong University
Sen Xiang	Wuhan University of Science and Technology
Ying-Qing Xu	Tsinghua University
Toshihiko Yamasaki	The University of Tokyo
Keiji Yanai	The University of Electro-Communications
Gang Yang	Renmin University of China
Yang Yang	University of Science and Technology of China
You Yang	Huazhong University of Science and Technology
Zhaoquan Yuan	Southwest Jiaotong University
Jan Zahálka	Czech Technical University in Prague
Hanwang Zhang	Nanyang Technological University
Sicheng Zhao	University of California, Berkeley
Lei Zhu	Huazhong University of Science and Technology

Additional Reviewers

Hadi Amirpour	Hanyuan Liu
Eric Arazo	Katrinna Macfarlane
Gibran Benitez-Garcia	Danila Mamontov
Adam Blažek	Thanassis Mavropoulos
Manliang Cao	Anastasia Moumtzidou
Ekrem Çetinkaya	Vangelis Oikonomou
Long Chen	Jesus Perez-Martin
Přemysl Čech	Zhaobo Qi
Julia Dietlmeier	Tomas Soucek
Denis Dresvyanskiy	Vajira Thambawita
Negin Ghamsarian	Athina Tsanousa
Panagiotis Giannakeris	Chenglei Wu
Socratis Gkelios	Menghan Xia
Tomáš Grošup	Minshan Xie
Steven Hicks	Cai Xu
Milan Hladik	Gang Yang
Wenbo Hu	Yaming Yang
Debesh Jha	Jiang Zhou
Omar Shahbaz Khan	Haichao Zhu
Chengze Li	Zirui Zhu

Contents – Part II

Contents – Part I

MSCANet: Adaptive Multi-scale Context Aggregation Network for Congested Crowd Counting

Yani Zhang[1,2], Huailin Zhao[1(✉)], Fangbo Zhou[1], Qing Zhang[2], Yanjiao Shi[2], and Lanjun Liang[1]

[1] School of Electrical and Electronic Engineering, Shanghai Institute of Technology, Shanghai, China
zhao_huailin@yahoo.com
[2] School of Computer Science and Information Engineering, Shanghai Institute of Technology, Shanghai, China

Abstract. Crowd counting has achieved significant progress with deep convolutional neural networks. However, most of the existing methods don't fully utilize spatial context information, and it is difficult for them to count the congested crowd accurately. To this end, we propose a novel Adaptive Multi-scale Context Aggregation Network (MSCANet), in which a Multi-scale Context Aggregation module (MSCA) is designed to adaptively extract and aggregate the contextual information from different scales of the crowd. More specifically, for each input, we first extract multi-scale context features via atrous convolution layers. Then, the multi-scale context features are progressively aggregated via a channel attention to enrich the crowd representations in different scales. Finally, a 1×1 convolution layer is applied to regress the crowd density. We perform extensive experiments on three public datasets: ShanghaiTech Part_A, UCF_CC_50 and UCF-QNRF, and the experimental results demonstrate the superiority of our method compared to current the state-of-the-art methods.

Keywords: Crowd counting · Adaptive multi-scale context aggregation · Crowd density estimation

1 Introduction

Crowd counting is a fundamental task for computer vision-based crowd analysis, aiming at automatically detecting the crowd congestion. It has wide-ranging practical applications in video surveillance, metropolis security and human behavior analysis, etc. However, the task often suffers from several challenging factors that frequently appear in crowd scenes, such as severe occlusion, scale variations, diverse crowd distributions, etc. These factors make it difficult to estimate the crowdedness, especially in highly congested scenes.

© Springer Nature Switzerland AG 2021
J. Lokoč et al. (Eds.): MMM 2021, LNCS 12573, pp. 1–12, 2021.
https://doi.org/10.1007/978-3-030-67835-7_1

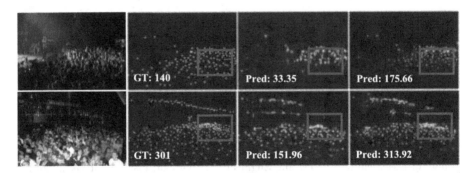

Fig. 1. Representative examples in the UCF-QNRF dataset [9]. From left to right: input images, ground-truth, results of CSRNet [12], results of MSCANet. Compared to CSRNet, MSCANet can effectively handle the ambiguity of appearance between crowd and background objects.

Many deep learning-based models have been proposed to solve this problem and have greatly increased the accuracy of crowd counting. However, it is still difficult for these networks to accurately count the congested crowd scenes, especially in the case of foreground crowd and background object sharing visually similar appearance, as shown in the first column of Fig. 1. Most existing models often miss counting the correct crowd regions and produce inaccurate estimation. For example, single-scale crowd counting networks [12,21,26] have the fixed receptive field so that they cannot efficiently handle the scale variations of people's head. Multi-scale crowd counting networks [3,13,15,23,24,28] are carefully designed to represent different scales of people efficiently. However, most of them don't fully consider the utilization of the spatial context information. Liu et al. [14] and Chen et al. [2] apply different methods for modeling scale-aware context features, but they aggregate different context features directly, not adaptively, which hinders the performance of the counting network. The approach of multi-scale context aggregation has some improved space because not all features from a specific scale are useful for final crowd counting. We argue that different scales of spatial contextual information should be adaptively aggregated.

Therefore, in this work, we propose a novel Adaptive Multi-scale Context Aggregation Network (MSCANet) for congested crowd counting. The core of the network is a Multi-scale Context Aggregation module (MSCA) that adaptively learns a multi-scale context representation. MSCA adopts a multi-branch structure that applies atrous convolution with different dilation rates to capture multi-scale context features. Then, the extracted features are progressively aggregated in neighboring branches via channel attention mechanism [7] to obtain a richer global scene representation. MSCANet consists of multiple MSCAs connected in a cascaded manner, where each MSCA is followed by an up-sampling layer to transform the multi-scale features into higher-resolution representations. The features from the last MSCA are further processed by a 1×1 convolution layer to regress the density of the crowd. MSCANet can be easily instantiated with various network backbones and optimized in an end-to-end manner. We evaluate it on three congested crowd counting datasets: ShanghaiTech Part_A [28], UCF_CC_50 [8] and UCF-QNRF [9]. Experiment results show the superiority of our model. In conclusion, the contributions of this paper are as follows:

- We develop a MSCA to aggregate small-scale context information with larger-scale context information in a cascade and adaptive manner, which can generate more compact context features for crowd representations in different scales.
- Based on multiple MSCAs, we propose a MSCANet to output multi-scale context features with different resolutions. It can efficiently handle the ambiguous appearance challenge, especially under congested crowd scenes with complex background.
- MSCANet achieves promising results to the other state-of-the-art methods on ShanghaiTech Part_A, UCF_CC_50, and UCF-QNRF datasets. More remarkably, on congested crowd scenes, the performance of MSCANet significantly outperforms other context-based crowd counting networks like [2,12,14], which demonstrates the effectiveness of our method.

2   Related Work

Multi-scale Methods. Many crowd counting networks were carefully designed to address the scale variation of people. For instance, Zhang et al. [28] designed a multi-column crowd counting network (MCNN) to recognize different scales of people. Onoro-Rubio et al. [15] established the input image pyramid to learn a multi-scale non-linear crowd density regression model. Liu et al. [13] adopted modified inception modules and deformable convolutional layers to model scale variations of crowd. Chen et al. [2] and Cao et al. [1] fed extracted features to multi-branch atrous convolution layers with different dilated rates. Similar with [1,2], we adopt different atrous convolution layers to model people's scale. However, we aggregate different scales of feature in an adaptive and cascade way, which can generate more compact context vectors for the final crowd counting.

Context-Based Methods. Efficiently acquiring and utilizing contextual information can effectively improve the performance of many computer vision tasks, such as video object tracking [31,33] and segmentation [30,32,34], temporal action localization [25], human-object interaction detection [35], and so on. A common way of encoding context information for crowd counting is that classifies each input image into one of crowd density-level label. For example, Sam et al. [17] and Wang et al. [22] applied a crowd density classifier to give each input image a density-level label. Then the density-level label guided the input image to be processed by its corresponding density-level estimation branch. Sindagi et al. [19] utilized *global context estimator* and *local context estimator* to give the crowd density-level labels to images and image patches, respectively. The outputs of the above two estimators and *density map estimator* were fed into

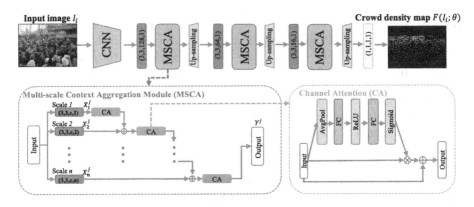

Fig. 2. Overview of our proposed network. 1) Each image I_i is firstly fed to a backbone network (CNN) for feature extraction. 2) The extracted features are progressively processed by multiple Multi-scale Context Aggregation modules to extract and aggregate multi-scale context information Y^j. 3) Multi-scale context features are fed into a 1×1 convolution layer to predict the crowd density map $F(I_i; \theta)$.

fusion-CNN to predict high-quality crowd density map. Gao et al. [5] also proposed a *Random High-level Density Classification* to give each ROI a density level label for extracting context information. Different from the above works, we directly extract and fuse multi-scale context information with different atrous convolution layers rather than with the crowd density classification network.

3   Proposed Method

Similar with [12,28], we formulate crowd counting as the pixel-wise regression problem. Specially, each head location x_j in image I_i is convolved with a 2-D normalized Gaussian kernel G_σ:

$$F_i(x) = \sum_{j=1}^{M} \delta(\boldsymbol{x} - \boldsymbol{x}_j) \times G_\sigma(\boldsymbol{x}), \tag{1}$$

where $\delta(\cdot)$ represents the Dirac delta function, σ stands for standard deviation, M is the total crowd number of I_i and is equal to the sum of all pixel values in crowd density map F_i. The crowd counting network establishes the non-linear transformation between the input image I_i and its corresponding crowd density map F_i. The crowd density map F_i is generated by Eq. 1. L_2 loss is chosen as the network loss function:

$$L(\Theta) = \frac{1}{2N} \sum_{i=1}^{N} \|F(I_i; \Theta) - F_i\|_2^2, \tag{2}$$

where Θ represents the learning parameters of crowd counting network and the $F(I_i; \Theta)$ denotes the output of crowd counting network. Specially, in this paper, we explore a new multi-scale contextual information aggregation method called MSCA. The details are introduced in the next subsection.

3.1   Multi-scale Context Aggregation Module

Aggregating different scales of context information is an effective way to handle the scale variation of people. However, small-scale context information is usually not reliable due to limited receptive fields. Directly aggregating the small-scale context information with large-scale context information introduces irrelevant and useless features, which hinders the counting performance. Thus, we need a selection mechanism to adaptively choose reliable small-scale context features for aggregating them with the large-scale context features. Based on this consideration, we propose a MSCA module, and its detailed structure is shown in Fig. 2.

MSCA module includes multi-branch atrous convolution layers with different dilated rates. We denote the ith scale context feature in resolution size $j \in \{\frac{1}{2^{r-1}} \cdots \frac{1}{4}, \frac{1}{2}, 1\}$ as $X_i^j \in \mathbb{R}^{jW \times jH \times C}$, where i is equal to the dilate rate and r denotes the reduction ratio decided by the backbone network. Before X_i^j aggregating with X_{i+1}^j, we adopt a selection function f to adaptively choose reliable features of X_i^j. The final aggregated context features $Y^j \in \mathbb{R}^{jW \times jH \times C}$ are computed as follows:

$$Y^j = f(\cdots f(f(X_1^j) \oplus X_2^j) \oplus X_3^j) \oplus \cdots \oplus X_n^j), \tag{3}$$

where $\oplus$ represents the element-wise summation.

We apply a channel attention [7] to implement the selecting function f without extra supervised information. As shown in Fig. 2, each feature is firstly sent to a global spatial average pooling (F_{avg}). Then the features are processed by a bottleneck structure, which consists of two fully connected layers. Finally, the output features are normalized into (0,1) by a sigmoid function. The detail process is as follows:

$$\alpha_i = W_2^{fc}(W_1^{fc}(F_{avg}(X_i^j))), \tag{4}$$

where $\alpha_i \in \mathbb{R}^{jW \times jH \times C}$ denotes the adaptive output coefficient. W_1^{fc} and W_2^{fc} represent weights of fully connected layers respectively and W_1^{fc} is followed by a ReLU function.

Besides, we add residual connections between the input and output of CA for better optimization. The residual formulation is as follows:

$$f(X_i^j) = X_i^j + \alpha_i X_i^j, \quad i = 1 \cdots n. \tag{5}$$

3.2   Multi-scale Context Aggregation Network

Based on MSCA, we propose a MSCANet for congested crowd counting, as shown in Fig. 2. Given an input image I_i, we firstly deploy CNN to extract features. Then, the extracted features are fed into multiple MSCA modules in a cascaded manner. Each MSCA is followed by an up-sampling layer to transform the multi-scale context features into higher-resolution representations. Finally, the generated multi-scale context features are processed by one 1×1 convolution layer for predicting the crowd density map.

Fig. 3. Different structures of multi-scale context module. (a) multi-scale context aggregation module (MSCA) w/o channel attention (CA); (b) cascade context pyramid module (CCPM); (c) scale pyramid module (SPM); (d) scale-aware context module (SACM).

3.3   Compared to Other Context Modules

We compare MSCA with another three context modules from [2,14,27], as shown in Fig. 3. To generate a compact context feature, Cascade Context Pyramid Module (CCPM) [27] progressively aggregates large-scale contextual information with small-scale contextual information, as shown in Fig. 3(b). The process of CCPM for extracting aggregated context features $Y^j \in \mathbb{R}^{jW \times jH \times C}$ is as follows:

$$Y^j = g(\cdots g(g(X_n^j \oplus X_{n-1}^j) \oplus X_{n-2}^j) \oplus \cdots \oplus X_1^j), \qquad (6)$$

where $g(\cdot)$ represents the residual block (res) from [6]. Unlike CCPM, we fuse contextual information from small to large in an adaptive way.

Spatial pyramid module (SPM) [2] directly processes input features $U \in \mathbb{R}^{W \times H \times C}$ with different dilate convolution branches (diaconv) and the output of each branch is equally processed by element-wise sum operation, as shown in Fig. 3(c). The process of SPM is as follows:

$$Y^j = \sum_{i=1}^{n} X_i^j = \sum_{i=1}^{n} W_i^{diaconv}(U). \qquad (7)$$

where $W_i^{diaconv}$ denotes weights of dilated convolution layers. Different from SPM, MSCA module adaptively emphasizes reliable information from different scales of context information.

Liu et al. [14] apply spatial pyramid pooling [29] to model multi-scale context features from local features, and then extract contrast features from the difference between local features and multi-scale context features for enhancing the crowd representations in different scales. We modify the above method and design a Scale-aware Context Module (SACM) for crowd counting, as shown in Fig. 3(d). The details of SACM is as follows:

$$Y^j = \sum_{i=1}^{n} X_i^j = \sum_{i=1}^{n} U_p(W_i^{conv}(P_{ave_i}(U))), \qquad (8)$$

where $P_{ave_i}(\cdot)$ represents the adaptive average pooling layer which averages the input feature U into $i \times i$ blocks, W_i^{conv} represents the weights of convolution layers and U_p represents the bilinear interpolation operation for upsampling. Compared to SACM, we apply a different approach to extract scale-aware context features. Experiments in the next section demonstrate the priority of our MSCA module.

4   Experiments

4.1   Datasets

We carry out the experiments on three challenging datasets with congested crowd scene, *i.e.*, ShanghaiTech Part_A [28], UCF_CC_50 [8] and UCF-QNRF [9]:

- **ShanghaiTech Part_A** includes 482 images from the Internet (300 images for training and 182 images for testing). The number of people in this dataset varies from 33 to 3139, which poses a challenge for the network to handle large variations in crowd.
- **UCF_CC_50** only contains 50 images from the Internet, and the maximum number of people is equal to 4543. Limited train images and large variations in the number of people are significant challenges for crowd counting methods. We follow the standard setting in [8] to conduct the five fold cross-validation.
- **UCF-QNRF** is a new proposed dataset that includes 1535 high-quality crowd images in total. There are 1201 images for training and 334 images for testing. The number of people in the UCF-QNRF dataset varies from 49 to 12865.

4.2   Implementation Details

We take the first ten layers of VGG-16 pre-trained on ImageNet as the feature extractor. The initial learning rate is set to 1×10^{-5}. The SGD with momentum is chosen as the optimizer. All experiments are implemented in C^3 Framework [4, 20] on PC with a single RTX 2080 Ti GPU card and an Intel(R) Core(TM) i7-8700 CPU. The data pre-processing settings and data augmentation strategies of the above three datasets all follow C^3 Framework. During the training, the batch size is set to 4 on UCF_CC_50 and set to 1 on the other datasets.

4.3   Evaluation Metrics

Mean absolute error (MAE) and mean squared error (MSE) are chosen as our evaluation metric:

$$MAE = \frac{1}{N} \sum_{1}^{N} |z_i - \hat{z}_i|, \tag{9}$$

$$MSE = \sqrt{\frac{1}{N} \sum_{1}^{N} (z_i - \hat{z}_i)^2}, \tag{10}$$

where z_i denotes the ground truth number of people in image I_i and $\hat{z}_i$ denotes the predicted number of people in image I_i.

Table 1. Comparisons of the different state-of-the-art methods on Shang-haiTech_PartA, UCF_CC_50 and UCF-QNRF datasets respectively.

Method	ShanghaiPartA		UCF_CC_50		UCF-QNRF	
	MAE	MSE	MAE	MSE	MAE	MSE
Lempitsky et al. [11]	–	–	493.4	487.1	–	–
Zhang et al. [26,28]	181.8	277.7	467.0	498.5	–	–
Idrees et al. [8,9]	–	–	419.5	541.6	315	508
MCNN [9,28]	110.2	173.2	377.6	509.1	277	–
Switching CNN [9,17]	90.4	135.0	318.1	439.2	228	445
CL [9]	–	–	–	–	132	191
CP-CNN [19]	73.6	106.4	298.8	**320.9**	–	–
CSRNet(baseline) [12]	68.2	115.0	266.1	397.5	–	–
ic-CNN(one stage) [16]	69.8	117.3	–	–	–	–
ic-CNN(two stage) [16]	68.5	116.2	–	–	–	–
CFF [18]	65.2	109.4	–	–	–	–
TEDNet [10]	**64.2**	109.1	249.4	354.5	113	188
Our method	66.5	**102.1**	**242.84**	329.82	**104.1**	**183.8**

Fig. 4. Impacts of different pyramid scale settings on UCF-QNRF. From left to right: input image, ground truth, result of PS = {1}, result of PS = {1, 2}, result of PS = {1, 2, 3}, result of PS = {1, 2, 3, 4}.

4.4   Comparison with State-of-the-Arts

We compare the proposed model with state-of-the-arts [8–12, 16–19, 26, 28] on three datasets, and the comparison results are shown in Table 1.

ShanghaiTech Part_A. Compared to CSRNet, MSCANet achieves 1.7 and 12.9 improvement on MAE and MSE, respectively. Moreover, MSCANet also achieves comparable results against other state-of-the-art methods, which demonstrates the effectiveness of MSCANet.

UCF_CC_50. Although there are limited images for training, MSCANet also achieves comparable results to the other state-of-the-arts. More remarkably, MSCANet surpasses the performance of TEDNet [10] 6.56 and 24.68 on MAE and MSE respectively.

UCF-QNRF. Compared to the-state-of-the art methods, like TEDNet [10], our method achieves 8.9 and 4.2 improvement on MAE and MSE metric respectively. The above improvements own to the effect of MSCA, which generates multi-scale context features used for crowd counting.

Fig. 5. Impacts of CA on UCF-QNRF. From left to right: input image, ground-truth, result of MSCA w/o CA, result of MSCA.

4.5 Ablation Study

Multi-scale Context Aggregation Module. We first study the impacts of different pyramid scale settings. Pyramid scale setting (PS) represents what dilated convolution branches are used in MSCA module, and the value of PS denotes the dilated rate of each branch. We investigate different PS settings to find the suitable combination. As shown in Table 2, the performances of MSCANet are gradually improved with the increment number of PS, reaching the best results at PS $= \{1, 2, 3\}$. Continually increasing the number of PS doesn't bring extra performance improvement because a larger receptive field brings extra unrelated information, which hinders the performance of MSCANet. As shown in Fig. 4, we visualize the output of MSCANet with different pyramid scale settings. The estimated result of PS $= \{1, 2, 3\}$ is very close to the ground truth. Therefore, we set PS $= \{1, 2, 3\}$ in the following experiments.

Then, we investigate the architecture changes of MSCA. Our baseline network is MSCANet which removes all MSCA modules (Decoder). Besides, to evaluate the effectiveness of CA for feature aggregation, we replace the channel attention modules (CA) into the residual blocks (res) (MSCA w/o CA). Table 3 shows comparison results of the above changes. We find that MSCA outperforms MSCA w/o CA and Decoder in terms of MAE. Figure 5 shows the visual results of the impacts of CA. We find that MSCA w/o CA performs worse than MSCA, which also confirms the importance of CA in MSCANet.

Table 2. Comparisons of our proposed method with different pyramid scales setting (PS) on UCF-QNRF dataset. The value of PS is the dilated rate of each dilation convolution branch from MSCA.

PS	MAE	MSE
{1}	110.9	197.2
{1, 2}	105.2	184.6
{1, 2, 3}	**104.1**	**183.8**
{1, 2, 3, 4}	104.8	186.1

Table 3. Comparisons of our proposed method with different architecture changes on UCF-QNRF dataset.

Configuration	MAE	MSE
Decoder (baseline)	111.3	**182.0**
MSCA w/o CA	105.7	186.9
MSCA	**104.1**	183.8
CSRNet (our reimplementation)	118.8	204.4
CAN [14]	107	183
CCPM	111.9	182.3
SPM	108.1	187.2
SACM	116.2	211.2

Multi-scale Context Modules. We first compare MSCANet with the other context-based crowd counting networks, such as *Congested Scene Recognition Network* (CSRNet) [12] and *Context-aware Network* (CAN) [14], which also adopt the first 10 layers of VGG-16 pre-trained on ImageNet to extract features. The detail results are shown in Table 3. We see that MSCANet surpasses the performance of CSRNet and CAN on MAE and MSE metrics, respectively. Then, we explore the performance of MSCA, CCPM, SPM, and SACM. For fair comparisons, all of them have three branch structures and the feature extractor is the same as MSCANet. The comparison results are displayed in Table 3.

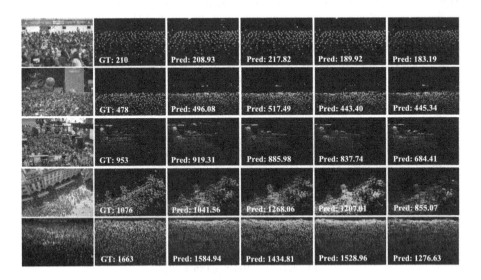

Fig. 6. Visual comparision of different multi-scale context modules on UCF-QNRF. From left to right: input images, ground-truth, results of our method, results of CCPM, results of SPM, results of SACM.

We find that MSCA achieves the best MAE metric on the UCF-QNRF dataset. Figure 6 displays the estimated results of representative images with different crowd density levels. We see that MSCA can handle congested cases better than the other three context modules. Qualitative and quantitative results demonstrate the priority of MSCA.

5   Conclusion

In this paper, we propose a MSCANet for congested crowd counting, which presents a new approach for extracting context information of different people's scale. MSCANet is implemented by multiple MSCAs, which consists of different atrous convolution layers and channel attention modules. Atrous convolution layers with different dilated rates extract multi-scale contextual features and channel attention modules adaptively aggregate the multi-scale context features in a cascade manner. Experiments are performed on three congested crowd datasets, and our MSCANet achieves favorable results against the other state-of-the-art methods.

Acknowledgements. This work is supported by Natural Science Foundation of Shanghai under Grant No. 19ZR1455300, and National Natural Science Foundation of China under Grant No. 61806126.

References

1. Cao, X., Wang, Z., Zhao, Y., Su, F.: Scale aggregation network for accurate and efficient crowd counting. In: ECCV (2018)
2. Chen, X., Bin, Y., Sang, N., Gao, C.: Scale pyramid network for crowd counting. In: WACV (2019)
3. Deb, D., Ventura, J.: An aggregated multicolumn dilated convolution network for perspective-free counting. In: CVPR Workshop (2018)
4. Gao, J., Lin, W., Zhao, B., Wang, D., Gao, C., Wen, J.: C^3 framework: an open-source pytorch code for crowd counting. arXiv preprint arXiv:1907.02724 (2019)
5. Gao, J., Wang, Q., Li, X.: PCC net: perspective crowd counting via spatial convolutional network. IEEE TCSVT 1 (2019)
6. He, K., Zhang, X., Ren, S., Sun, J.: Deep residual learning for image recognition. In: CVPR (2016)
7. Hu, J., Shen, L., Sun, G.: Squeeze-and-excitation networks. In: CVPR (2018)
8. Idrees, H., Saleemi, I., Seibert, C., Shah, M.: Multi-source multi-scale counting in extremely dense crowd images. In: CVPR (2013)
9. Idrees, H., et al.: Composition loss for counting, density map estimation and localization in dense crowds. In: ECCV (2018)
10. Jiang, X., et al.: Crowd counting and density estimation by trellis encoder-decoder networks. In: CVPR (2019)
11. Lempitsky, V., Zisserman, A.: Learning to count objects in images. In: NeurIPS (2010)
12. Li, Y., Zhang, X., Chen, D.: CSRNet: dilated convolutional neural networks for understanding the highly congested scenes. In: CVPR (2018)

13. Liu, N., Long, Y., Zou, C., Niu, Q., Pan, L., Wu, H.: ADCrowdNet: an attention-injective deformable convolutional network for crowd understanding. In: CVPR (2019)
14. Liu, W., Salzmann, M., Fua, P.: Context-aware crowd counting. In: CVPR (2019)
15. Oñoro-Rubio, D., López-Sastre, R.J.: Towards perspective-free object counting with deep learning. In: Leibe, B., Matas, J., Sebe, N., Welling, M. (eds.) ECCV 2016. LNCS, vol. 9911, pp. 615–629. Springer, Cham (2016). https://doi.org/10.1007/978-3-319-46478-7_38
16. Ranjan, V., Le, H., Hoai, M.: Iterative crowd counting. In: ECCV (2018)
17. Sam, D.B., Surya, S., Babu, R.V.: Switching convolutional neural network for crowd counting. In: CVPR (2017)
18. Shi, Z., Mettes, P., Snoek, C.G.M.: Counting with focus for free. In: ICCV (2019)
19. Sindagi, V.A., Patel, V.M.: Generating high-quality crowd density maps using contextual pyramid CNNs. In: ICCV (2017)
20. Wang, Q., Gao, J., Lin, W., Yuan, Y.: Learning from synthetic data for crowd counting in the wild. In: CVPR (2019)
21. Wang, S., Lu, Y., Zhou, T., Di, H., Lu, L., Zhang, L.: SCLNet: spatial context learning network for congested crowd counting. Neurocomputing **404**, 227–239 (2020)
22. Wang, S., Zhao, H., Wang, W., Di, H., Shu, X.: Improving deep crowd density estimation via pre-classification of density. In: Liu, D., Xie, S., Li, Y., Zhao, D., El-Alfy, E.S. (eds.) ICONIP 2017. LNCS, vol. 10636, pp. 260–269. Springer, Cham (2017). https://doi.org/10.1007/978-3-319-70090-8_27
23. Wang, Z., Xiao, Z., Xie, K., Qiu, Q., Zhen, X., Cao, X.: In defense of single-column networks for crowd counting. In: BMVC (2018)
24. Xie, Y., Lu, Y., Wang, S.: RSANet: deep recurrent scale-aware network for crowd counting. In: ICIP (2020)
25. Yang, L., Peng, H., Zhang, D., Fu, J., Han, J.: Revisiting anchor mechanisms for temporal action localization. IEEE TIP **29**, 8535–8548 (2020)
26. Zhang, C., Li, H., Wang, X., Yang, X.: Cross-scene crowd counting via deep convolutional neural networks. In: CVPR (2015)
27. Zhang, P., Liu, W., Lei, Y., Lu, H., Yang, X.: Cascaded context pyramid for full-resolution 3D semantic scene completion. arXiv preprint arXiv:1908.00382 (2019)
28. Zhang, Y., Zhou, D., Chen, S., Gao, S., Ma, Y.: Single-image crowd counting via multi-column convolutional neural network. In: CVPR (2016)
29. Zhao, H., Shi, J., Qi, X., Wang, X., Jia, J.: Pyramid scene parsing network. In: CVPR (2017)
30. Zhou, T., Li, J., Wang, S., Tao, R., Shen, J.: MATNet: motion-attentive transition network for zero-shot video object segmentation. IEEE TIP **29**, 8326–8338 (2020)
31. Zhou, T., Lu, Y., Di, H.: Locality-constrained collaborative model for robust visual tracking. IEEE TCSVT **27**(2), 313–325 (2015)
32. Zhou, T., Lu, Y., Di, H., Zhang, J.: Video object segmentation aggregation. In: ICME (2016)
33. Zhou, T., Lu, Y., Lv, F., Di, H., Zhao, Q., Zhang, J.: Abrupt motion tracking via nearest neighbor field driven stochastic sampling. Neurocomputing **165**, 350–360 (2015)
34. Zhou, T., Wang, S., Zhou, Y., Yao, Y., Li, J., Shao, L.: Motion-attentive transition for zero-shot video object segmentation. In: AAAI (2020)
35. Zhou, T., Wang, W., Qi, S., Ling, H., Shen, J.: Cascaded human-object interaction recognition. In: CVPR (2020)

Tropical Cyclones Tracking Based on Satellite Cloud Images: Database and Comprehensive Study

Cheng Huang[1], Sixian Chan[1,2], Cong Bai[1,2(✉)], Weilong Ding[1,2], and Jinglin Zhang[3(✉)]

[1] College of Computer Science, Zhejiang University of Technology, Hangzhou, China
congbai@zjut.edu.cn
[2] Key Laboratory of Visual Media Intelligent Processing Technology of Zhejiang Province, Hangzhou, China
[3] College of Atmospheric Sciences, Nanjing University of Information Science and Technology, Nanjing, China
jinglin.zhang@nuist.edu.cn

Abstract. The tropical cyclone is one of disaster weather that cause serious damages for human community. It is necessary to forecast the tropical cyclone efficiently and accurately for reducing the loss caused by tropical cyclones. With the development of computer vision and satellite technology, high quality meteorological data can be got and advanced technologies have been proposed in visual tracking domain. This makes it possible to develop algorithms to do the automatic tropical cyclone tracking which plays a critical role in tropical cyclone forecast. In this paper, we present a novel database for **T**ypical **C**yclone **T**racking based on **S**atellite **C**loud **I**mage, called **TCTSCI**. To the best of our knowledge, TCTSCI is the first satellite cloud image database of tropical cyclone tracking. It consists of 28 video sequences and totally 3,432 frames with 6001×6001 pixels. It includes tropical cyclones of five different intensities distributing in 2019. Each frame is scientifically inspected and labeled with the authoritative tropical cyclone data. Besides, to encourage and facilitate research of multimodal methods for tropical cyclone tracking, TCTSCI provides not only visual bounding box annotations but multimodal meteorological data of tropical cyclones. We evaluate 11 state-of-the-art and widely used trackers by using OPE and EAO metrics and analyze the challenges on TCTSCI for these trackers.

Keywords: Typical cyclone tracking · Satellite cloud images · Database · Evaluation

1 Introduction

The tropical cyclone is one of the common weather systems in the tropics, subtropics, and temperate zones. It can bring abundant rainfall and change the

© Springer Nature Switzerland AG 2021
J. Lokoč et al. (Eds.): MMM 2021, LNCS 12573, pp. 13–25, 2021.
https://doi.org/10.1007/978-3-030-67835-7_2

temperature dramatically [25] and will bring the meteorological disasters to coastal cities. For example, there are twenty-nine tropical cyclones generated in the region of the northwest pacific and south China sea in 2019. Twenty-one of them landed on the nearby countries and cause a lot of damages and casualties. Hence, to acquire more meteorological data and understand the weather system better, a great number of meteorological satellites are launched into space. They can obtain valuable meteorological data, including the satellite cloud image and other professional weather data [2]. The resolution of cloud images from satellites become higher and higher as well as the meteorological data from satellites become more accurate and diverse. These images and data can help people to make more accurate short-term and long-term weather forecasting include tropical cyclone forecasting and a good automatic tracking is an important prerequisite for accurate tropical cyclone forecast [20].

With the rapid development of visual tracking, many famous tracking databases and trackers are proposed [17]. If tropical cyclones can be tracked automatically and efficiently, which means that disasters can be predicted and reduced. Therefore, it is necessary to develop better algorithms for tracking them. The basis for developing a robust algorithm is to have a large number of valid experimental data. In addition, databases play a critical role in the development of trackers and provide platforms to evaluate trackers. Combining the high-quality meteorological data with the state-of-the-art visual tracking algorithms, it is possible to track the tropical cyclone automatically and accurately. Unfortunately, there is no such kind of database for tropical cyclones tracking.

In this paper, we propose a novel database named **T**ropical **C**yclone **T**racking based on **S**atellite **C**loud **I**mage (TCTSCI) which includes not only the satellite cloud images but the rich meteorological data of tropical cyclones. As a start-up version, it collects twenty-eight tropical cyclones generated in the region of the northwest pacific and south China sea in 2019. Then we evaluate some state-of-the-art and popular trackers on TCTSCI and analyze the results of these trackers comprehensively. To sum up, the contributions of our work are three parts:

- TCTSCI is the first satellite cloud image database for tropical cyclone tracking. Each frame is carefully inspected and labeled with the authoritative tropical cyclone data from China Meteorological Administration and Shanghai Typhoon Institute. By releasing TCTSCI, we aimed to provide a platform for developing more robust algorithms of tropical cyclone tracking by deep learning technologies. We will release TCTSCI at https://github.com/zjut-cvb315/TCTSCI.
- Different from most existing visual tracking databases, TCTSCI provides not only visual bounding box annotations but abundant meteorological data of tropical cyclones. Those data offer a further aspect to improve the algorithm. By collecting these data, we aim to encourage and facilitate the research of tropical cyclone tracking based on multi-modal data.
- Comprehensive experiments are executed on the proposed TCTSCI. The classical and deep-learning based trackers are evaluated using two different visual tracking metrics. And challenges of TCTSCI are analyzed.

2   The Proposed TCTSCI Database

2.1   Data Preprocessing

Data Source of Satellite Cloud Image. Japan Aerospace Exploration Agency (JAXA) P-Tree System, receiving the data from geostationary Himawari-8 [2] satellite equipped with the most advanced new generation of geostationary meteorological observation remote sensor. It is also the first satellite that can take color images in the world. Compared with previous remote sensors, it achieves higher spatial resolution and more observation wavelength types, so there are various types of data provided by Himawari-8. In this paper, we select the Himawari $L1 - level$ data as raw data, whose format is NetCDF. The spatial resolution is $2.22\,$km and temporal resolution is $10\,$min. The sixteen observation bands with 6001×6001 pixels cover the region ($60°S$–$60°N$, $80°E$–$160°W$).

Data Visualization. To better observe and analyze tropical cyclones, it is necessary to visualize the original data. We select tbb_8, tbb_10, tbb_11 and tbb_12 to synthesize the RGB images, corresponding take tbb_8 minus tbb_10 as the red channel, tbb_12 minus tbb_13 as the green channel, and tbb_8 as the blue channel separately. The reasons are as follows:

- Different bands have different physical properties respectively. The tropical cyclone belongs to air mass and jet stream. According to the properties of the Himawari-8, tbb_8, tbb_10, tbb_12 and tbb_13 are appropriate for observing and analyzing air mass and jet stream [11]. In the channel R(tbb_8 - tbb_10), we can observe the cloud area, cloudless, dry area, and humid area. In channel G(tbb_12 - tbb_13), we can observe the atmospheric ozone contents which are beneficial to analyze the jet stream. In channel B(tbb_8), we can observe water vapor of the tropospheric upper/middle layer contents which are also beneficial to analyze the jet stream. So the RGB images synthesized by such a method can provide appropriate information for tropical cyclone tracking.
- Tropical cyclone tracking is a continuous process. This process can take several days whatever day or night. So we should avoid using visible light bands that will lose at night. The tbb_8, tbb_10, tbb_11 and tbb_12 are infrared bands. They can capture data 24 h per day. Consequently, we choose them to synthesize the RGB images.

Metadata of Tropical Cyclone. Original meteorological data of tropical cyclones is captured from China Meteorological Administration [6] and Shanghai Typhoon Institute [24]. It includes all the data on the formation and development of tropical cyclones, like the ID of the tropical cyclone, name, date of occurrence, grade, geographical coordinates, maximum sustained winds, moving direction, central pressure, the radius of storm axis and the predicted path by China Meteorological Administration. These data plays a critical role in the annotation of TCTSCI.

Table 1. Tropical Cyclone in TCTSCI. MaxV is the max velocity. TS (tropical storm), STS (severe tropical storm), TY (typhoon), STY (severe typhoon), and SuperTY (super typhoon) are the intensity of tropical cyclone.

TropicalCyclone	Frames	MaxV(m/s)	Intensity
BAILU	109	30	STS
BUALOI	142	58	SuperTY
DANAS	148	23	TS
FAXAI	121	50	STY
FENGSHEN	127	45	STY
FRANCISCO	109	42	STY
FUNGWONG	73	30	STS
HAGIBIS	172	65	SuperTY
HALONG	127	65	SuperTY
KALMAEGI	172	38	TY
KAMMURI	217	55	SuperTY
KROSA	244	45	STY
LEKIMA	214	62	SuperTY
LINGLING	159	55	SuperTY
MATMO	55	25	STS
MITAG	127	40	TY
MUN	67	18	TS
NAKRI	154	33	TY
NARI	34	18	TS
NEOGURI	79	42	STY
PABUK	112	28	STS
PEIPAH	22	18	TS
PHANFONE	148	42	STY
PODUL	70	25	STS
SEPAT	19	20	TS
TAPAH	106	33	TY
WIPHA	100	23	TS
WUTIP	205	55	SuperTY

Completion of Missing Meteorological Data. Original meteorological data is not enough for annotating every satellite cloud image we get. Because sometimes China Meteorological Administration updates the meteorological data of tropical cyclones every three or six hours. However, we select the satellite cloud images every hour during the tropical cyclone. In order to annotate the rest of the images without meteorological data, we need to complete the missing data by

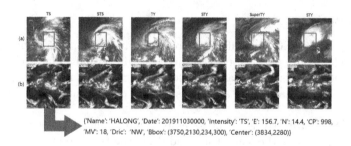

('Name': 'HALONG', 'Date': 201911030000, 'Intensity': 'TS', 'E': 156.7, 'N': 14.4, 'CP': 998, 'MV': 18, 'Dric': 'NW', 'Bbox': (3750,2130,234,300), 'Center': (3834,2280))

Fig. 1. A sequence of tropical cyclone HALONG and its annotation. The images in row (b) are the close-up of the region containing the tropical cyclone in row (a). To see them clearly, we zoom the corresponding area by 36 times. The bottom part is the meteorological data of the #0001 frame in tropical cyclone HALONG. Best viewed in color.

analyzing the meteorological data, such as moving direction, center coordinate and the radius of storm axis, of the adjacent frames with complete data. We use the interpolation method to complete the center coordinate of tropical cyclones for every image. There is a sequence images: $\{t'_1, t_2, t_3, t'_4, ..., t'_n\}$, t'_i is the image with meteorological data, t_i is the image without meteorological data. The center coordinate of tropical cyclone in t_i can be interpolated as follows:

$$X_{t_i} = \frac{X_{t'_j} - X_{t'_k}}{k - j} \times (i - j) \tag{1}$$

$$Y_{t_i} = \frac{Y_{t'_j} - Y_{t'_k}}{k - j} \times (i - j) \tag{2}$$

X_{t_i} is the X-coordinate of the tropical cyclone center. Y_{t_i} is the Y-coordinate of the tropical cyclone center. t'_j and $t'_k (k > j)$ are the adjacent frames which have complete data, like t'_1 and t'_4.

Summary of TCTSCI. Although 29 tropical cyclones are distributing in the region ($0°$–$55°N$, $105°E$–$180°E$) in 2019, we do not find enough meteorological data of KAJIKI to annotate corresponding satellite images. So finally, TCTSCI consists of 28 sequences of tropical cyclones with 3,432 frames totally. It includes tropical cyclones of five different intensities distributing in the region ($0°$–$55°N$, $105°E$–$180°$) in year 2019. We obtain all the data, including satellite cloud images and corresponding meteorological data. Table 1 shows detailed information of tropical cyclones in TCTSCI. The bottom of Fig. 1 shows the more detailed meteorological data of one satellite cloud image. The meaning of abbreviations: 'E' - longitude, 'N' - northern latitude, 'CP' - center pressure (hPa), 'MV' - maximum sustained winds (m/s), 'Dric' - moving direction, 'Bbox' - bounding box, and 'Center' - tropical cyclone center. In addition, these data are rich enough to describe tropical cyclones independently.

2.2   Annotation

Annotation in TCTSCI is different from the annotations in other tracking databases which are only labeled subjectively in visual. The annotation in TCTSCI needs to associate satellite cloud images with meteorological data. Therefore, it is necessary to establish a standard of annotation.

Annotation Standard. We load the satellite cloud image and the corresponding meteorological data at the same time and draw the tropical cyclone center and the bounding box on satellite cloud image, which is determined by the radius of storm axis. In this way, we can check the annotation intuitively. To improve the quantity of annotation better, we check the annotation results twice after labeled a tropical cyclone. The first time, the accuracy of the bounding box on images is checked by the software specifically developed for this purpose. The second time, the meteorology-related experts are invited to check the accuracy and consistency of the bounding box. This task is not complete until all of us are satisfied. After overcoming the difficulties above, a database is obtained finally with both meteorological and visual annotation, with some examples shown in Fig. 1.

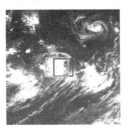

Fig. 2. The different bounding boxes of different standards. The red bounding box is annotated by the meteorological data of the Non-eye tropical cyclone definition. The green one is the visual bounding box. Best viewed in color.

Difference Between Eye Tropical Cyclone and Non-eye Tropical Cyclone. Tropical cyclones are generally divided into two categories: Eye tropical cyclone and Non-eye tropical cyclone [26]. Eye tropical cyclones are usually mature tropical cyclones. Their centers are in the eye region or the geometric center in most cases. As for Non-eye tropical cyclones, they are usually forming or subdued tropical cyclones and their centers are not easy to determine only by visual observation. As shown in Fig. 2. The green bounding box is annotated by the visual standard. This bounding box contains the entire region of the air mass. As for the red one, it seems to deviate from the air mass but it is the right bounding box that annotated with the meteorological data. The reason why the red one seems to deviate from the air mass is that the tropical cyclone in Fig. 2 is a Non-eye tropical cyclone and its meteorological center is not same as the geometric center that we observed. To determine the Non-eye tropical cyclone center, the related meteorological knowledge is required. Above all, the goal we propose TCTSCI is to promote research in tropical cyclone tracking. It is necessary to annotate TCTSCI with a meteorological standard. And looking at the red bounding box, the tropical cyclone center is not the geometric center of the rectangle. It is because the radius of storm axis of different directions (east, south, west, and north) is not always the same. Those characteristics mentioned above are challenges for traditional trackers and deep-learning based trackers, which are not adapt to TCTSCI.

2.3   Attributes

Different from existing tracking databases of nature objects, we do not set some special and challenging attributes to discommode those trackers deliberately. We only display the unique charm of tropical cyclones. Of course, some attributes of TCTSCI are also challenges for the trackers. TCTSCI have six attributes, including difference from visual bounding box (DFVBB), hue variation (HV), deformation (DEF), rotation (ROT), scale variable (SV), and background clutter (BC). These attributes are defined in Table 2.

Table 2. Description of 6 different attributes in TCTSCI

Attr	Description
DFVBB	Different From Visual Bounding Box: The annotation flows meteorological standard
HV	Hue Variable: The hue changes in some frames because of the satellite problem
DEF	Deformation: The tropical cyclone is deformable during tracking
ROT	Rotation: The tropical cyclone rotates in images
SV	Scale Variable: The size of tropical cyclone changes during tracking
BC	Background Clutter: The background is complex with other airmass and jet stream

3   Evaluation

3.1   Evaluation Metric

Following famous and authoritative protocols (VOT2019 [15] and OTB-2015 [22]), we perform an **One-Pass** **Evaluation** (**OPE** [23]) to evaluate trackers with the metric of **precision** and **success** and perform an **Expected Average Overlap** (**EAO** [14]) to evaluate different trackers with the metric of **accuracy,** **robustness** and **EAO**.

One-Pass Evaluation. One-Pass Evaluation is a famous visual tracking metric proposed by OTB [22]. The main indexes in OPE are precision and success. Precision measures how close that the distance between the centers of a tracker bounding box ($bbox_{tr}$) and the corresponding ground truth bounding box ($bbox_{gt}$). Success measures the intersection over union of pixels in $bbox_{tr}$ and those in $bbox_{gt}$.

Table 3. Evaluated Trackers.

Tracker name	Representation	Search	Venue
SiamBAN [5]	Deep	DS	CVPR'20
SiamMask [21]	Deep	DS	CVPR'19
DIMP [3]	Deep	DS	ICCV'19
MDNet [19]	Deep	RS	CVPR'16
CSRT [16]	HOG, CN, PI	DS	IJCV'18
KCF [9]	HOG, GK	DS	PAMI'15
TDL [13]	BP	RS	PAMI'11
MEDIANFLOW [12]	BP	RS	ICPR'10
MOSSE [4]	PI	DS	CVPR'10
MIL [1]	H	DS	CVPR'09
BOOSTING [8]	BP, H, HOG	RS	BMVC'06

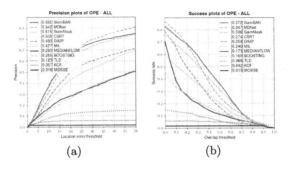

(a)   (b)

Fig. 3. Evaluation results on TCTSCI using precision and success. Best viewed in color.

Expected Average Overlap. Expected Average Overlap is another visual tracking metric that widely used to evaluate trackers. It is proposed by VOT [14] and contains three main indexes: accuracy, robustness and EAO. The accuracy is calculated by the average overlap that the bounding box predicted by the tracker overlaps with the ground truth. As for robustness, it is calculated by the times that the tracker loses the target during tracking. A failure is indicated when the overlap measure becomes zero. And EAO is an index that calculated base on accuracy and robustness.

3.2   Evaluated Trackers

To determine the overall performance of different trackers in TCTSCI, we evaluate 11 state-of-the-art and popular trackers on TCTSCI, including deep-learning based trackers: SiamBan [5], SiamMask [21], DIPM [3], MDNet [19] and classical trackers based on hand-crafted features: CSRT [16], KCF [9], TDL [13], MEDIANFLOW [12], MOSSE [4], MIL [1], BOOSTING [8]. Table 3 summarizes

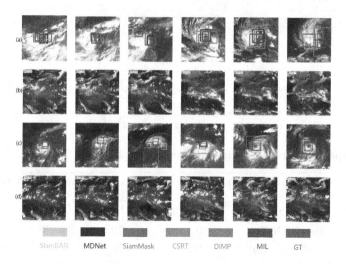

Fig. 4. Qualitative evaluation in three different tropical cyclone intensities: Tropical Storm (MUN), Severe Tropical Storm (PABUK), Typhoon (MITAG), and Severe Typhoon (KROSA). The images in row (a), (c) are the close-up of the region containing tropical cyclone in row (b), (d) respectively. To see them clearly, we zoom the corresponding area by about 10 36 times. Best viewed in color.

these trackers with their representation schemes, search strategies, and other details. The meaning of Representation abbreviations are as follows: PI - Pixel Intensity, HOG - Histogram of Oriented Gradients, CN - Color Names, CH - Color Histogram, GK - Gaussian Kernel, BP - Binary Pattern, H - Haar, Deep - Deep Features. Search: PF - Particle Filter, RS - Random Sampling, DS - Dense Sampling.

3.3 Evaluation Results with OPE

We evaluate these trackers mentioned above on TCTSCI without any changes. To report the evaluation results intuitively, we draw plot of precision and success, as shown in Fig. 3. SiamBAN obtains the best performance. It means SiamBAN can better locate the position and determine the size of tropical cyclones. As for MDNet, it is a method of update its model during tracking. So it gets a good score with the cost of expansive computation. The best non-deep-learning based tracker is CSRT. It even gets a better score than DIMP, a deep-learning based tracker. What is more, we show the qualitative results of six popular trackers, including SiamBan, SiamMask, DIPM, MDNet, CSRT and MIL, in four classes of tropical cyclones containing Tropical Storm, Severe Tropical Storm, Typhoon and Severe Typhoon in Fig. 4. Unfortunately, the results on TCTSCI of these trackers are far worse than the results on natural image tracking. It still has room for improvement.

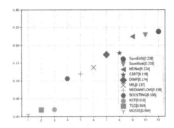

Fig. 5. Expected averaged overlap performance on TCTSCI. Best viewed in color.

Table 4. Evaluation results on TCTSCI using EAO, Accuracy and Robustness.

Tracker name	Accuracy↑	Robustness↓	EAO↑
SiamBAN [5]	0.454	0.702	0.238
SiamMask [21]	0.446	0.761	0.228
MDNet [19]	0.392	0.702	0.224
CSRT [16]	0.413	1.083	0.178
DIMP [3]	0.322	0.936	0.174
MIL [1]	0.384	1.404	0.137
MEDIAN-FLOW [12]	0.138	0.702	0.119
BOOSTING [8]	0.372	1.989	0.106
KCF [9]	–	11.586	0.019
TLD [13]	–	9.977	0.019
MOSSE [4]	–	15.243	0.004

3.4 Evaluation Results with EAO

As for EAO, we also evaluate the trackers mentioned above on TCTSCI. We show the results on accuracy, robustness and EAO through a plot and a table, as shown in Fig. 5 and Table 4. SiamBAN achieves the best result again. It means that SiamBAN loses fewer times during tracking the tropical cyclone than others and has the best Composite performance index of EAO. Different form OPE, SiamMask gets a better score than MDNet with 0.228 EAO score. And tracker CSRT obtains the best performance once again in non-deep-leaning based trackers. KCF, TLD and MOSSE do not work for TCTSCI, as they fail too many times during tracking to calculate the scores of accuracy. So we using '−' to replace their scores. Unsurprisingly, the EAO results on TCTSCI are also worse than the results on natural image tracking.

3.5 Analysis

Deep-Learning Based Trackers Perform Better. Whatever in OPE or EAO, the top three trackers are always the deep-learning based trackers. This phenomenon is similar to the experimental results on other famous tracking databases: TrackingNet [18], LaSOT [7] and GOT-10K [10]. In TCTSCI, the reasons of some non-deep leaning trackers do not have good results even do not work are the features extraction of targets, most of the area is air mass, in TCTSCI are too insufficient. The deep neural network can extract richer features than the hand-crafted features.

Multi-modal Data Based Trackers are the Future. Comparing to the classical tracking performances on OTB or VOT, the results on TCTSCI are worse. On the one hand, the features of tropical cyclones in TCTSCI are so insufficient that trackers can not extract rich features efficiently. On the other hand, the annotation strategy, designed for the meteorological tropical cyclone database, of TCTSCI is different from these typical tracking databases. It is a challenge for these trackers which are tailor-made for typical tracking databases. However human experts use more than just images information to track and predict tropical cyclone. For example, the grade of the tropical cyclone can limit the size of the bounding box and the direction of the tropical cyclone can limit the search region of the next frame. So we hope to simulate this behavior of human experts. There is no doubt that multi-modal data based tracker is the better option. If there is a multi-modal based tracker that can utilize both the satellite cloud images data and the meteorological data of tropical cyclone efficiently, the results will be better. However, there is no such appropriate multi-modal data based tracker so far.

4 Conclusion

In this paper, We propose TCTSCI, a novel visual tracking database with fully annotated sequences from Himawari-8 satellite and rich meteorological data of tropical cyclone from China Meteorological Administration and Shanghai Typhoon Institute. To the best of our knowledge, TCTSCI is the first database of tropical cyclone tracking with rich meteorological information. An extended baseline for 11 state-of-the-art and popular trackers have been comprehensively studied on TCTSCI. From the experimental results, we have found that the deep-learning based trackers get the best performance and the results on TCTSCI have been far worse than the results on classical tracking database. On one hand, more efficient tracker for satellite images of tropical cyclone should be developed. On the other hand, multi-modal data based trackers that may track tropical cyclones successfully. By proposing TCTSCI, we hope to break the barrier between the field of computer vision and meteorology. It will attract more attentions from cross-disciplinary researchers and promote the development of tropical cyclone tracking.

Acknowledgments. This work is partially supported by National Key Research and Development Program of China (No. 2018YFE0126100), National Natural Science Foundation of China under Grant (No. 61906168, 61976192, 41775008 and 61702275).

References

1. Babenko, B., Yang, M.H., Belongie, S.: Visual tracking with online multiple instance learning. In: 2009 IEEE Conference on Computer Vision and Pattern Recognition, pp. 983–990. IEEE (2009)
2. Bessho, K., et al.: An introduction to Himawari-8/9–Japan's new-generation geostationary meteorological satellites. J. Meteorol. Soc. Jpn. Ser. II **94**(2), 151–183 (2016)
3. Bhat, G., Danelljan, M., Gool, L.V., Timofte, R.: Learning discriminative model prediction for tracking. In: Proceedings of the IEEE International Conference on Computer Vision, pp. 6182–6191 (2019)
4. Bolme, D.S., Beveridge, J.R., Draper, B.A., Lui, Y.M.: Visual object tracking using adaptive correlation filters. In: 2010 IEEE Computer Society Conference on Computer Vision and Pattern Recognition, pp. 2544–2550. IEEE (2010)
5. Chen, Z., Zhong, B., Li, G., Zhang, S., Ji, R.: Siamese box adaptive network for visual tracking. In: Proceedings of the IEEE/CVF Conference on Computer Vision and Pattern Recognition, pp. 6668–6677 (2020)
6. CMA: Typhoon network of central meteorological station. http://typhoon.nmc.cn/web.html (2015)
7. Fan, H., et al.: LaSOT: a high-quality benchmark for large-scale single object tracking. In: Proceedings of the IEEE Conference on Computer Vision and Pattern Recognition, pp. 5374–5383 (2019)
8. Grabner, H., Grabner, M., Bischof, H.: Real-time tracking via on-line boosting. In: Proceedings of British Machine Vision Conference (BMVC), vol. 1, pp. 47–56 (2006). https://doi.org/10.5244/C.20.6
9. Henriques, J.F., Caseiro, R., Martins, P., Batista, J.: High-speed tracking with kernelized correlation filters. IEEE Trans. Pattern Anal. Mach. Intell. **37**(3), 583–596 (2014)
10. Huang, L., Zhao, X., Huang, K.: Got-10k: a large high-diversity benchmark for generic object tracking in the wild. IEEE Trans. Pattern Anal. Mach. Intell. (2019)
11. JMA: Airmass RGB analysis of air mass and jet stream (2015). https://www.jma.go.jp/jma/jma-eng/satellite/VLab/RGB-Airmass.pdf
12. Kalal, Z., Mikolajczyk, K., Matas, J.: Forward-backward error: automatic detection of tracking failures. In: 2010 20th International Conference on Pattern Recognition, pp. 2756–2759. IEEE (2010)
13. Kalal, Z., Mikolajczyk, K., Matas, J.: Tracking-learning-detection. IEEE Trans. Pattern Anal. Mach. Intell. **34**(7), 1409–1422 (2011)
14. Kristan, M., et al.: The visual object tracking vot2015 challenge results. In: Proceedings of the IEEE International Conference on Computer Vision Workshops, pp. 1–23 (2015)
15. Kristan, M., et al.: The seventh visual object tracking vot2019 challenge results. In: Proceedings of the IEEE International Conference on Computer Vision Workshops (2019)
16. Lukežič, A., Vojíř, T., Zajc, L.Č., Matas, J., Kristan, M.: Discriminative correlation filter tracker with channel and spatial reliability. Int. J. Comput. Vis. **7**(126), 671–688 (2018)
17. Marvasti-Zadeh, S.M., Cheng, L., Ghanei-Yakhdan, H., Kasaei, S.: Deep learning for visual tracking: a comprehensive survey. arXiv preprint arXiv:1912.00535 (2019)
18. Muller, M., Bibi, A., Giancola, S., Alsubaihi, S., Ghanem, B.: TrackingNet: a large-scale dataset and benchmark for object tracking in the wild. In: Proceedings of the European Conference on Computer Vision (ECCV), pp. 300–317 (2018)

19. Nam, H., Han, B.: Learning multi-domain convolutional neural networks for visual tracking. In: Proceedings of the IEEE Conference on Computer Vision and Pattern Recognition, pp. 4293–4302 (2016)
20. Reichstein, M., Camps-Valls, G., Stevens, B., Jung, M., Denzler, J., Carvalhais, N., et al.: Deep learning and process understanding for data-driven earth system science. Nature **566**(7743), 195–204 (2019)
21. Wang, Q., Zhang, L., Bertinetto, L., Hu, W., Torr, P.H.: Fast online object tracking and segmentation: a unifying approach. In: Proceedings of the IEEE Conference on Computer Vision and Pattern Recognition, pp. 1328–1338 (2019)
22. Wu, Y., Lim, J., Yang, M.: Object tracking benchmark. IEEE Trans. Pattern Anal. Mach. Intell. **37**(9), 1834–1848 (2015)
23. Wu, Y., Lim, J., Yang, M.H.: Online object tracking: a benchmark. In: Proceedings of the IEEE Conference on Computer Vision and Pattern Recognition, pp. 2411–2418 (2013)
24. Ying, M., et al.: An overview of the China meteorological administration tropical cyclone database. J. Atmos. Oceanic Technol. **31**(2), 287–301 (2014)
25. Yong, C., Jing, J.: Typical modes of tropical cyclone rainfalls over China during the typhoon seasons and significant impact facts. J. Nanjing Univ. (Nat. Sci.) **1** (2011)
26. Zhang, Q., Lai, L., Wei, H., Zong, X.: Non-clear typhoon eye tracking by artificial ant colony. In: 2006 International Conference on Machine Learning and Cybernetics, pp. 4063–4068. IEEE (2006)

Image Registration Improved by Generative Adversarial Networks

Shiyan Jiang[1], Ci Wang[2(✉)], and Chang Huang[2]

[1] School of Computer Science and Technology, East China Normal University, Shanghai, China
[2] School of Communication and Electronic Engineering, East China Normal University, Shanghai, China
cwang@cs.ecnu.edu.cn

Abstract. The performances of most image registrations will decrease if the quality of the image to be registered is poor, especially contaminated with heavy distortions such as noise, blur, and uneven degradation. To solve this problem, a generative adversarial networks (GANs) based approach and the specified loss functions are proposed to improve image quality for better registration. Specifically, given the paired images, the generator network enhances the distorted image and the discriminator network compares the enhanced image with the ideal image. To efficiently discriminate the enhanced image, the loss function is designed to describe the perceptual loss and the adversarial loss, where the former measures the image similarity and the latter pushes the enhanced solution to natural image manifold. After enhancement, image features are more accurate and the registrations between feature point pairs will be more consistent.

Keywords: Image registration · GANs · Image enhancement

1 Introduction

In the past decades, more and more images are taken by mobile phones for their convenience. However, their images are often contaminated with different distortions such as noise, blur, and uneven degradation, and sometimes their qualities are not high enough for registration [1]. For example, mobile phones use digital zoom instead of the optical zoom for imaging. The optical zoom changes the image resolution through adjusting the lens, object, and focus position, and often results in a clear captured image. On the contrary, digital zoom adjusts image resolution through interpolation, which produces an image with few details and some unexpected distortions. In addition, the lens of the mobile phone is poor and will introduce some uneven image deformation. Finally, mobile phone manufacturers use their own algorithms for image enhancement, and these algorithms may exacerbate the registration uncertainty.

Traditional image registration can be classified into three categories. Template-based registration directly uses correlation to find the best matching position, such as Mean Absolute Differences (MAD), Sequential Similarity Detection Algorithm (SSDA) [2], etc. These algorithms have high time complexity and are sensitive to light as well as

© Springer Nature Switzerland AG 2021
J. Lokoč et al. (Eds.): MMM 2021, LNCS 12573, pp. 26–35, 2021.
https://doi.org/10.1007/978-3-030-67835-7_3

image scale. Domain-based registration transforms images to other domains, such as Fourier [3], wavelet domain before registration. These algorithms are more flexible, but their calculation speed is slower. Different from the domain-based registration, the feature-based registration extracts image features and matches the pixels by their feature similarity, such as SIFT [4], SURF [5], etc. Inspired by its excellent performance, image features can also be extracted by Convolutional Neural Networks (CNN). On the unlabeled data, Alexey et al. [6] train a CNN to generate robust features. Yang et al. [7] use the pretrained VGG network [8] to generate multi-scale feature descriptors. Although these features are superior to SIFT, they are still insufficient for the degraded images registration. Alternatively, CNN is directly used for image registration. For example, DeTone et al. [9] propose the HomographyNet structure for the supervised learning, which includes a regression network and a classification network. The regression network estimates the homography parameters and the classification network produces a quantized homography distribution. Based on the greedy supervised algorithm, Liao et al. [10] use reinforcement learning for end-to-end training as well as image registration. As for the unsupervised learning, Nguyen et al. [11] train a deep CNN to estimate planar homographies. All these registration algorithms are designed for high-quality images, and their performances are significantly reduced on the images with strong distortions. Therefore, it is necessary to enhance the image before registration.

Image enhancement can be done in spatial-domain [12], frequency-domain [13], or hybrid-domain [14]. Recently, some algorithms based on GANs [15] have produced better performance. For example, Andrey et al. [16] propose an end-to-end GANs model with color loss, content loss, texture loss to map mobile phone photos to the Digital Single Lens Reflex (DSLR) photos. Based on the framework of two-way GANs, Chen et al. [17] propose an unpaired learning method to minimize mapping loss, cycle consistency loss, and adversarial loss. Huang et al. [18] propose a range scaling global U-Net model for image enhancement on small mobile devices. In addition, denoising and brightness recovery by GANs can enhance the image either [19–22]. However, all these methods fail to produce the satisfactory images due to some reasons. For example, U-net structure is poor at enhancing the blurred images, but the images to be processed are often with blur and other distortions simultaneously.

For mobile phone applications, an end-to-end model with specific loss functions is proposed to restore images. Considering the huge impact of pixel distortion on registration performance, a pixel-wise loss function is introduced to measure the pixel differences between the enhanced image and the ideal image. If mean square error (MSE) is used to measure the image content differences, the GANs generated image is too smooth to be registered. Hence, a pretrained loss network is used to measure image feature similarities, rather than MSE. With these improvements, our method provides the images with better quality in terms of registration performance.

The rest of the paper is organized as follows: Sect. 2 introduces the background and gives the design of network architecture and loss functions; Sect. 3 shows implementation details and experimental results; Sect. 4 concludes the paper.

2 Proposed Method

The performance of registration is poor when the images to be registered are with distortions. For better registration, images should be enhanced before registration. Hence, the enhancement to registration is formulated as an instance of image-to-image translation solved by GANs.

2.1 Background

Although traditional GANs have good performance, they still suffer from training instability. For this, Wasserstein GAN(WGAN) uses the earth mover distance [23] to measure the difference between the data distribution and the model distribution to improve training stability. However, WGAN will not converge if the parameters of its discriminator are not bounded during training. To further improve WGAN training, the gradient penalty is introduced in WGAN-GP [24] to restrict gradient and improve training speed.

2.2 Proposed Network Structure

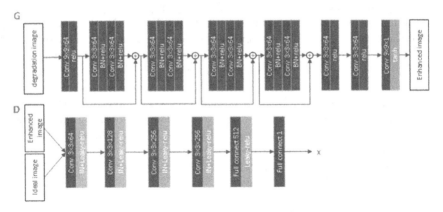

Fig. 1. Architecture of our model

Generator network and discriminator network are designed to achieve the competitive balance. The generator network tries to enhance an image to cheat the discriminator network, and the discriminator network tries to distinguish the enhanced image from the ideal image. The input of the generator network is the grayscale image and the output of the generator network is the enhanced grayscale image. The enhanced image and the ideal image are fed into the discriminator network for judgement and a value representing their difference is its output. The generator network starts with a 9×9 kernel size convolutional layer to transform the image into feature space. After that are four residual blocks, which are consisted of two 3×3 kernel size convolutional layers alternated with batch-normalization (BN) layers. Finally, two 3×3 kernel size convolutional layers and a 9×9 kernel size convolutional layer are used to transform features back to the image. All layers in the generator network are followed by a ReLU activation function, except for the last one, where the tanh function is applied. The discriminate network consists of four 3×3 convolutional layers, each of them followed by an instance normalization (IN) layer and a LeakyReLU activation function. Then two full connection layers are introduced to measure the distance between the ideal image and the enhanced image. The first full connection layer contains 512 neurons followed by LeakyReLU activation function, and the second full connection layer contains only one neuron. The whole network structures are shown in Fig. 1.

2.3 Loss Function

Besides network architecture, loss function is another key work in network design. A synthesized loss function is proposed in this paper, and each loss function is defined as follows.

Pixel-Wise Loss. Pixel-wise loss pays attention to the pixel difference in the enhanced image and ideal image. To measure the pixel difference, the loss function is defined as follow:

$$L_p = \left\| G(I_d) - I_p \right\|_1 \tag{1}$$

where $\|\cdot\|_1$ denotes L1 norm and G is our generator network. I_d denotes the degraded image and I_p is its corresponding ideal image.

Content Loss. Being the image classification network, the pretrained VGG19 [25] is widely used as the perceptual loss function. The definition of our content loss is based on it. Let φ_i be the feature map convoluted by the i-th layer of VGG-19, the content loss is L2 norm of feature difference between the enhanced and ideal images:

$$L_c = \frac{1}{C_i H_i W_i} \left\| \varphi_i(G(I_d)) - \varphi_i(I_p) \right\|_2 \tag{2}$$

where C_i, H_i and W_i denotes the number, height, and width of the feature maps, respectively.

Adversarial Loss. The adversarial loss measures the distribution distance. Therefore, minimizing the adversarial loss leads to the improved perceptual quality of the enhanced image. Adversarial loss is defined as follow:

$$L_a = -D(G(I_d)) \tag{3}$$

Total Variation Loss. Total variation (TV) loss [26] enforces the spatial smoothness for the enhanced images to avoid the potential noise, and keep the high-frequency components simultaneously. TV loss is defined as:

$$L_{tv} = \frac{1}{CHW} \left\| \nabla_X(G(I_d)) - \nabla_x(G(I_d)) \right\|_1 \tag{4}$$

where C, H, W are the dimensions of the enhanced image, respectively.

Total Loss. The final loss is defined as a weighted sum of the previous four losses with the following coefficients:

$$L_{total} = L_a + 2 \cdot L_c + 500 \cdot L_{tv} + 10 \cdot L_p \tag{5}$$

The weights are chosen based on preliminary experiments on the training data.

3 Experiments

In this section, the details of experimental settings are presented, and the proposed method is compared with state-of-the-art methods [16, 30, 31] on both synthetic and real-world datasets.

3.1 Dataset

In the experiments, the tea cakes are used as the analysis object. There are many pit-like structures on the tea cake. When shooting at close distance, it is difficult to select the focus plane for the mobile phone, so that the captured image often has partial unclearness. When lighting condition or shooting angle changed, the quality difference causes serious feature mismatch between the image pair, so we use the tea cake image to verify the effect of image enhancement. There are sixty tea cakes taken by four types of mobile phones, including Huawei Mate20X, Mix3, iPhone X, Oppo R17. The images captured by the handheld mobile phone at different moments may have different shooting angle and focus, so the captured images are difficult to be aligned and enhanced with the GANs. To solve this problem, we simulate the degradation process of the camera by injecting degradation into the high-quality image to form the synthetic dataset for GANs training. Then, the trained networks are used for the real-world dataset to verify the algorithm robustness. For different images of the same tea cake, the enhanced image should have more matching feature points, but for different tea cakes, the number of matching feature points of the enhanced image should not increase significantly.

Synthetic Dataset. In the experiments, we map the low quality of the inferior mobile phone image to the level of the image of the high-quality mobile phone through the enhancement algorithm. For instance, iPhone X mobile phone captures the ideal image. To simulate severely degraded image, some kinds of degradations are injected into ideal images to produce the degraded images. The ideal images and the corresponding degraded images constitute the synthetic pairs.

Real-World Dataset. Also, iPhone X mobile phone captures ideal images, and Mix3, Huawei Mate20X, and Oppo R17 are used to capture the degraded images in a relatively dark environment by autofocus. Under this arrangement, the images captured by Huawei Mate20X and Oppo R17 are with extra distortions such as out of focus and dark noise. The ideal images and the corresponding degraded images constitute the real-world pairs, which are shown in Fig. 2.

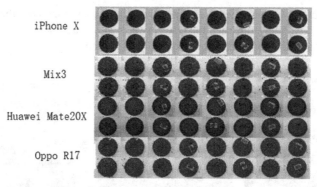

Fig. 2. The images captured by iPhone X as the ideal images. The images captured by Mix3, Huawei Mate20X and Oppo R17 as the severely distorted images.

3.2 Implementation Details

The 256×256 image patch pairs are cropped from synthetic image pairs to produce over 30k training and 3k test samples. Besides, there are more than 1k images from the real-world dataset for tests. The generator network and discriminator network are both trained with the training images. The parameters of the networks are optimized using Adam [28] with a learning rate $\beta = 10^{-5}$. After training, the feature-based registration methods are used for evaluating the performance of the enhancements. To eliminate the false matching, the bi-directional matching method is adopted and its implementation steps are given as:

(1) Forward match: for the feature point $p_{i,j}$ in an image I_1, two feature points $p_{i,j}^1, p_{i,j}^2$ are found in I_2 which are closest to and second closest to $p_{i,j}$ in terms of feature similarity measured by L1 norm. Given the threshold T, $p_{i,j}^1$ is the matching point of $p_{i,j}$ only if the ratio of the smallest distance to the next smallest distance is lower than T. The feature point pair are represented by $\left(p_{i,j}, p_{i,j}^1\right)$. The smaller the T, the more accurate the registration results will be.

(2) Reverse match: the same approach in forward match is adopted to find the corresponding matching feature points $q_{j,i}^1$ on I_1 for feature points $q_{j,i}$ on I_2, which are represented by $\left(q_{j,i}^1, q_{j,i}\right)$.

If the matching point is correct in both forward and reverse results, it is a reasonable match. After bi-directional matching, the Random Sample Consensus (RANSAC) algorithm [27] is performed to further cull mismatch points.

3.3 Results

Enhancement Performance. Limited by the space, only one example is represented in Fig. 3 to visually demonstrate the improvements of the proposed methods. As shown in Fig. 3, our method exhibits impressive enhancement performance, generating results with rich and credible image textures. For other comparison methods, they still leave some blur or noise. Only our method restores the clear image details in the red box, while the results of the competing methods still have some distortions.

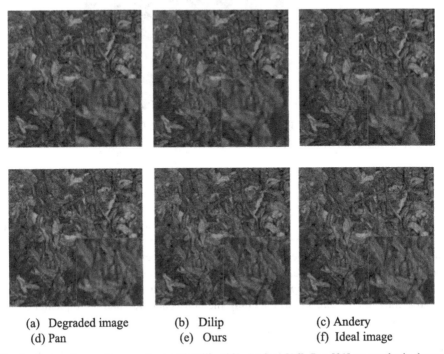

(a) Degraded image	(b) Dilip	(c) Andery
(d) Pan	(e) Ours	(f) Ideal image

Fig. 3. Comparisons of our method with Dilip [30], Andrey [16], Pan [31] on synthetic dataset. (Color figure online)

Registration Performance. Distorted images have fewer matching points than clear images when registered with the same image, so the enhancement effect can be judged by the number of matching points. SIFT [3], SURF [4] feature, and the bi-directional matching are used to quantitatively evaluate registration performance. The enhanced images shown in Fig. 3 are registered with the image patch from the same tea cake or another tea cake. Table 1 shows the number of matching points between the enhanced images and the ideal image of the same tea cake, and the best results as well as their percentage gains are in bold. Our method gets the most matching point pairs, indicating that our enhanced images are closer to ideal images from the objective metric view. Because there are many similar structures in tea cakes, for high-resolution images, even for different tea cake pairs, there still are a small number of feature point pairs that can be matched. But for image patches with a smaller size, such as 256×256, the number of mismatched point pairs is usually zero. The number of matching points is still zero even after image is enhanced, which indicates that our method will not cause feature confusion. For further testings, ten images are randomly chosen from real-world images to register after enhancement. The average numbers of matching points are shown in

Table 2. As shown in Table 2, our model considerably outperforms other methods on the real-world dataset. Meanwhile, these ten images are registered with each other and the number of their matching points is still zero, which proves the robustness of our enhanced method again.

Table 1. Number of the correct matching points and percentage gains

	Degraded	Dilip [30]	Andrey [16]	Pan [31]	Ours	Gain
Sift	372	394	515	406	**523**	41%
Surf	192	206	339	205	**331**	72%

Table 2. Average number of the correct matching points and percentage gains

	Degraded	Dilip [30]	Andrey [16]	Pan [31]	Ours	Gain
Sift	58	85	83	79	**88**	52%
Surf	16	28	32	32	**36**	125%

4 Conclusion

This paper presents an end to end deep learning architecture to enhance images for better registration. Considering the complexity of distortions, the perceptual loss of the proposed GAN architecture has two terms, i.e. the content loss and pixel-wise loss, to produce the images with higher objective quality. The experiments on both real-world and synthetic datasets confirm that the enhanced images will have much more stable feature points. This improvement is beneficial to improve the image registration performance, which has practical applications in the image registration of mobile phones.

References

1. Zitováand, B., Flusser, J.: Image registration methods: a survey. Image Vis. Comput. **21**, 977–1000 (2003)
2. Barnea, D.I., Silverman, H.F.: A class of algorithms for fast digital image registration. IEEE Trans. Comput. **21**, 179–186 (1972)
3. Guo, X., Xu, Z., Lu, Y., Pang, Y.: An application of Fourier–Mellin transform in image registration. In: Proceedings of the 5th International Conference on Computer and Information Technology (CIT), Shanghai, China, September 2005, pp. 619–623 (2005)
4. Lowe, D.: Distinctive image features from scale-invariant keypoints. IJCV **60**(2), 91–110 (2004)
5. Bay, H., Tuytelaars, T., Van Gool, L.: SURF: speeded up robust features. In: Leonardis, A., Bischof, H., Pinz, A. (eds.) ECCV 2006. LNCS, vol. 3951, pp. 404–417. Springer, Heidelberg (2006). https://doi.org/10.1007/11744023_32

6. Dosovitskiy, A., Fischer, P., Springenberg, J.T., Riedmiller, M., Brox, T.: Discriminative unsupervised feature learning with exemplar convolutional neural networks. IEEE Trans. Pattern Anal. Mach. Intell. **38**(9), 1734–1747 (2016)

7. Yang, Z., Dan, T., Yang, Y.: Multi-temporal remote sensing image registration using deep convolutional features. IEEE Access **6**, 38544–38555 (2018)

8. Simonyan, K., Zisserman, A.: Very deep convolutional networks for large-scale image recognition. arXiv preprint arXiv:1409.1556 (2014)

9. DeTone, D., Malisiewicz, T., Rabinovich, A.: Deep image homography estimation. arXiv preprint arXiv:1606.03798 (2016)

10. Liao, R., et al.: An artificial agent for robust image registration. In: AAAI, pp. 4168–4175 (2017)

11. Nguyen, T., Chen, S.W., Shivakumar, S.S., Taylor, C.J., Kumar, V.: Unsupervised deep homography: a fast and robust homography estimation model. arXiv preprint arXiv:1709.03966 (2017)

12. Divya, K.A., Roshna, K.I.: A survey on various image enhancement algorithms for naturalness preservation. Int. J. Comput. Sci. Inf. Technol. **6**, 2043–2045 (2015)

13. Bedi, S., Khandelwal, R.: Various image enhancement techniques—a critical review. Int. J. Adv. Res. Comput. Commun. Eng. **2**(3), 1605–1609 (2013)

14. Yang, F., Wu, J.: An improved image contrast enhancement in multiple-peak images based on histogram equalization. In: 2010 International Conference on Computer Design and Applications, vol. 1, no. 4, pp. 346–349. IEEE (2010)

15. Goodfellow, I., et al.: Generative adversarial nets. In: Advances in Neural Information Processing Systems, pp. 2672–2680 (2014)

16. Ignatov, A., Kobyshev, N., Vanhoey, K., Timofte, R., Van Gool, L.: DSLR-quality photos on mobile devices with deep convolutional networks. In: Proceedings of IEEE International Conference on Computer Vision (ICCV), October 2017

17. Chen, Y.-S., Wang, Y.-C., Kao, M.-H., Chuang, Y.-Y.: Deep photo enhancer: unpaired learning for image enhancement from photographs with GANs. In: Proceedings of the IEEE Conference on Computer Vision Pattern Recognition, pp. 6306–6314 (2018)

18. Huang, J., et al.: Range scaling global U-net for perceptual image enhancement on mobile devices. In: Leal-Taixé, L., Roth, S. (eds.) ECCV 2018. LNCS, vol. 11133, pp. 230–242. Springer, Cham (2019). https://doi.org/10.1007/978-3-030-11021-5_15

19. Tripathi, S., Lipton, Z.C., Nguyen, T.Q.: Correction by projection: denoising images with generative adversarial networks. arXiv:1803.04477 (2018)

20. Chen, J., Chen, J., Chao, H., Yang, M.: Image blind denoising with generative adversarial network-based noise modeling. In: Proceedings of the IEEE International Conference Computer Vision Pattern Recognition, pp. 3155–3164, June 2018

21. Meng, Y., Kong, D., Zhu, Z., Zhao, Y.: From night to day: GANs based low quality image enhancement. Neural Process. Lett. **50**(1), 799–814 (2019). https://doi.org/10.1007/s11063-018-09968-2

22. Jiang, Y., et al.: Enlightengan: deep light enhancement without paired supervision. arXiv: 1906.06972 (2019)

23. Arjovsky, M., Chintala, S., Bottou, L.: Wasserstein GAN. arXiv preprint arXiv:1701.07875 (2017)

24. Gulrajani, I., Ahmed, F., Arjovsky, M., Dumoulin, V., Courville, A.: Improved training of Wasserstein GANs. arXiv e-prints arXiv:1704.00028 (2017). Advances in Neural Information Processing Systems 31 (NIPS 2017)

25. Simonyan, K., Zisserman, A.: Very deep convolutional networks for large-scale image recognition. In: International Conference on Learning Representations (ICLR) (2015)

26. Aly, H.A., Dubois, E.: Image up-sampling using total variation regularization with a new observation model. IEEE Trans. Image Process. **14**(10), 1647–1659 (2005)

27. Vedaldi, A., Fulkerson, B.: VLFeat: an open and portable library of computer vision algorithms (2008)
28. Kingma, D.P., Ba, J.: Adam: a method for stochastic optimization. CoRR, abs/1412.6980 (2014)
29. Wang, Z., Bovik, A.C., Sheikh, H.R., Simoncelli, E.P.: Image quality assessment: from error visibility to structural similarity. IEEE Trans. Image Process. 13(4), 600–612 (2004)
30. Krishnan, D., Tay, T., Fergus, R.: Blind deconvolution using a normalized sparsity measure. In: CVPR, pp. 233–240 (2011)
31. Pan, J., Hu, Z., Su, Z., Yang, M.-H.: Deblurring text imagesvia L0-regularized intensity and gradient prior. In: CVPR, pp. 2901–2908 (2014)
32. Rublee, E., Rabaud, V., Konolige, K., Bradski, G.: ORB: an efficient alternative to SIFT or SURF. In: Proceedings of the IEEE International Conference on Computer Vision, Barcelona, Spain, November 2011, pp. 2564–2571 (2011)
33. Calonder, M., Lepetit, V., Özuysal, M., Trzinski, T., Strecha, C., Fua, P.: BRIEF: computing a local binary descriptor very fast. IEEE Trans. Pattern Anal. Machine Intell. 34(7), 1281–1298 (2011)

Deep 3D Modeling of Human Bodies from Freehand Sketching

Kaizhi Yang, Jintao Lu, Siyu Hu, and Xuejin Chen[✉] [iD]

University of Science and Technology of China, Hefei 230026, China
{ykz0923,ljt123,sy891228}@mail.ustc.edu.cn, xjchen99@ustc.edu.cn

Abstract. Creating high-quality 3D human body models by freehand sketching is challenging because of the sparsity and ambiguity of hand-drawn strokes. In this paper, we present a sketch-based modeling system for human bodies using deep neural networks. Considering the large variety of human body shapes and poses, we adopt the widely-used parametric representation, SMPL, to produce high-quality models that are compatible with many further applications, such as telepresence, game production, and so on. However, precisely mapping hand-drawn sketches to the SMPL parameters is non-trivial due to the non-linearity and dependency between articulated body parts. In order to solve the huge ambiguity in mapping sketches onto the manifold of human bodies, we introduce the skeleton as the intermediate representation. Our skeleton-aware modeling network first interprets sparse joints from coarse sketches and then predicts the SMPL parameters based on joint-wise features. This skeleton-aware intermediate representation effectively reduces the ambiguity and complexity between the two high-dimensional spaces. Based on our light-weight interpretation network, our system supports interactive creation and editing of 3D human body models by freehand sketching.

Keywords: Freehand sketch · Human body · 3D modeling · Skeleton joint · SMPL

1 Introduction

The human body is one of the commonest subjects in virtual reality, telepresence, animation production, etc. Many studies have been conducted in order to produce high-quality meshes for human bodies in recent years. However, 3D modeling is a tedious job since most interaction systems are based on 2D displays. Sketching, which is a natural and flexible tool to depict desired shapes, especially at the early stage of a design process, has been attracting a great deal of research works for easy 3D modeling [14,21,24]. In this paper, we propose a powerful sketch-based system for high-quality 3D modeling of human bodies.

Creating high-quality body models is non-trivial because of the non-rigidness of human bodies and articulation between body parts. A positive thing is that human bodies have a uniform structure with joint articulations. Many parametric

© Springer Nature Switzerland AG 2021
J. Lokoč et al. (Eds.): MMM 2021, LNCS 12573, pp. 36–48, 2021.
https://doi.org/10.1007/978-3-030-67835-7_4

Fig. 1. Our sketch-based body modeling system allows common users to create high-quality 3D body models in diverse shapes and poses easily from hand-drawn sketches.

representations have been proposed to restrict the complicated body models to a manifold [13,15]. Taking advantage of deep neural networks (DNNs), a large number of techniques have been developed to reconstruct human bodies from 2D images [10,12,25–28].

However, hand-drawn sketches show different characteristics compared to natural images. The inevitable distortions in sketch lead to much more ambiguity in the ill-posed problem of 3D interpretation from a single view image. Moreover, hand-drawn sketches are typically sparse and cannot carry as rich texture information as real images which makes sketch-based modeling become more challenging. To address the ambiguity in mapping coarse and sparse sketches to high-quality body meshes, we propose a skeleton-aware interpretation neural network, taking advantage of both non-parametric skeleton construction and parametric regression for more natural-looking meshes. In the non-parametric interpretation stage, we employ a 2D CNN to encode the input sketch into a high-dimensional vector and then predict the 3D positions of joints on the body skeleton. In the parametric interpretation stage, we estimate the body shape and pose parameters from the 2D sketch features and the 3D joints, taking advantage of the SMPL [13] representation in producing high-quality meshes. The disjunctive network branches for pose and shape respectively along with the skeleton-first and mesh-second interpretation architecture efficiently facilitate the quality and accuracy of sketch-based body modeling.

While existing supervised 3D body reconstruction required paired 2D input and 3D models, we build a large-scale database that contains pairs of synthesized sketches and 3D models. Our proposed method achieves the highest reconstruction accuracy compared with state-of-the-art image-based reconstruction approaches. Furthermore, with the proposed light-weight deep neural network, our sketch-based body modeling system supports interactive creation and editing of high-quality body models of diverse shapes and poses, as Fig. 1 shows.

2 Related Work

Sketch-Based Content Creation. Sketch-based content creation is an essential problem in multimedia and human-computer interaction. Traditional methods make 3D interpretation from line drawings based on reasoning geometric constraints, such as convexity, parallelism, orthogonality to produce 3D

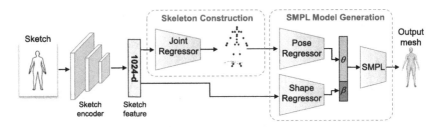

Fig. 2. Overview of our approach with the skeleton as an intermediate representation for model generation. A skeleton-aware representation is introduced before pose parameters regression.

shapes [16,24]. Recently, deep learning has significantly promoted the performance of sketch-based systems, such as sketch-based 3D shape retrieval [20], image synthesis, and editing [11]. Using a volumetric representation, [8] designs two networks to respectively predict a start model from the first view sketches and update the model according to the drawn strokes in other views. [14] predicts depth and normal maps under multiple views and fuses them to get a 3D mesh. An unsupervised method is proposed by searching similar 3D models with the input sketch and integrating them to generate the final shape [21]. Due to the limited capability of shape representation, the generated shapes by these approaches are very rough in low resolutions. Moreover, precise control of the 3D shapes with hand-drawn sketches is not supported.

Object Reconstruction from Single-View Image. With the advance of large-scale 3D shape datasets, DNN-based approaches have made great progress in object reconstruction from single images [1,3,4,6,22,23]. [1] predicts an occupancy grid from images and incrementally fusing the information of multiple views through a 3D recurrent reconstruction network. In [3], by approximating the features after image encoder and voxel encoder, a 3D CNN is used to decode the feature into 3D shapes. Also, many efforts have been devoted to mesh generation from single-view images [4,22]. Starting from a template mesh, they interpret the vertex positions for the target surface through MLPs or graph-based networks. A differentiable mesh sampling operator [23] is proposed to sample point clouds on meshes and predict the vertex offsets for mesh deformation. In comparison, we combine non-parametric joint regression with a parametric representation to generate high-quality meshes for human bodies from sketches.

Human Body Reconstruction. Based on the unified structure of human bodies, parametric representation is usually used in body modeling. Based on Skinned Multi-Person Linear Model (SMPL) [13], many DNN-based techniques extract global features from an image and directly regress the model parameters [9,18,26–28]. However, a recent study [10] shows that direct estimation of vertex positions outperforms parameter regression because of the highly nonlinear correlation between SMPL parameters and body shapes. They employ a graph-CNN to regress vertex positions and attach a network for SMPL param-

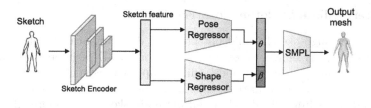

Fig. 3. A plain network (MLP-Vanilla) to regress SMPL parameters from sketches for body mesh generation.

eter regression to improve the mesh quality. Except for global features that are directly extracted from image, semantic part segmentation can also be involved to help reconstruct the body models [17,19,26,27]. Besides single images, multi-view images [12,25] or point clouds [7] can be other options to reconstruct body models. However, different from natural images with rich texture information or point clouds with accurate 3D information, hand-drawn sketches are much more sparse and coarse without sufficient details and precision, making the body modeling task more challenging.

3 Our Method

Hand-drawn sketches are typically coarse and only roughly depict body shapes and poses, leaving details about fingers or expressions unexpressed. In order to produce naturally-looking and high-quality meshes, we employ the parametric body representation, SMPL [13] to supplement shape details. Therefore, the sketch-based modeling task becomes a regression task of predicting SMPL parameters. The SMPL parameters consist of shape parameters $\beta \in \mathbb{R}^{10}$ and pose parameters θ for the rotations associated with $N_j = 24$ joints. Similar to [10], we regress the 3×3 matrix representation for each joint rotation, i.e. $\theta \in \mathbb{R}^{216}$, to avoid the challenge of regressing the quaternions in the axis-angle representation.

When mapping sketches into SMPL parameters, a straightforward method is to encode the input sketch image into a high-dimensional feature vector and then regress the SMPL parameters from the encoded sketch feature, as Fig. 3 shows. A K_s dimensional feature vector $\mathbf{f}_{sketch}$ can be extracted from the input sketch S using a CNN encoder. From the sketch feature $\mathbf{f}_{sketch}$, the shape parameters β and the pose parameters θ are then separately regressed using MLPs.

However, due to the non-linearity and complicated dependency between the pose parameters, it is non-trivial to find a direct mapping from the sketch feature space to the SMPL parameter space. We bring the skeleton as an intermediate expression, which unentangles the global pose and the significantly nonlinear articulated rotations. Figure 2 shows the pipeline of our skeleton-aware model generation from sketches.

3.1   Intermediate Skeleton Construction

Although the rough sketch could not precisely describe local details of human bodies, it effectively delivers posture information. A 3D skeleton can efficiently describe the pose of a human body, while shape details such as fingers, facial features are not considered in this stage. Based on the underlying common sparsity, the mapping from 2D sketches and 3D body skeletons is more effective to interpret. We use the skeleton structure that is composed of $N_j = 24$ ordered joints with fixed connections, as defined in [13]. In order to construct the 3D body skeleton based on the input sketch, the 3D locations of the N_j joints can be regressed using multi-layer perceptrons from the extracted sketch feature f_{sketch}, as shown in Fig. 4. By concatenating the N_j joints' 3D coordinates, the pose parameters θ can be regressed from the $N_j \times 3$ vector using MLPs.

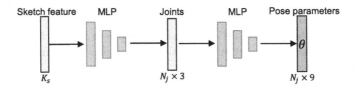

Fig. 4. MLP-Joint Network. With skeleton as an intermediate representation, we first regress the positions of N_j joints on the skeleton with three MLPs and then regress the SMPL pose parameters with another three MLPs.

3.2   Joint-Wise Pose Regression

Direct regression of the non-linear pose parameters from the joint positions inevitably abandons information in the original sketch. On the other hand, naively regression the pose parameters from the globally embedded sketch feature ignores the local articulation between joints. Since pose parameters represent the articulated rotation associated with the joints under a unified structure, we propose an effective joint-wise pose regression strategy to regress the highly non-linear rotations. First, we replace the global regression of joint positions with deformation from a template skeleton. Second, we regress the pose parameters from the embedded joint features rather than the joint positions. Figure 5 shows the detailed architecture of our pose regression network.

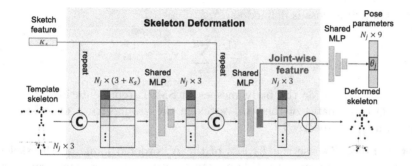

Fig. 5. Detailed architecture of our joint-wise pose regression in the Deform+JF network. We first regress the local offset of each joint from the template skeleton. The pose parameters are regressed from the last layer feature for each joint rather than the deformed joint position using shared MLPs.

Skeleton Deformation. Different from general objects, despite the complex shapes and high dimensional meshes, the human bodies have a unified structure, so a template skeleton could efficiently embed the structure and all the pose of human bodies could be concisely explained as the joints movement in skeleton instead of the deformation of all the vertices in the mesh. We adopt a two-level deformation network for skeleton construction. Each joint coordinate (x, y, z) on the template skeleton is concatenated with the K_s dimensional global sketch feature $\mathbf{f}_{sketch}$ to go through a shared MLP to estimate (d_x^1, d_y^1, d_z^1) for the first step deformation. Another deformation step is performed in a similar manner to get joint-wise deformation (d_x^2, d_y^2, d_z^2). Adding the predicted joint offset to the template skeleton, we get a deformed body skeleton.

Joint-Wise Pose Regression. Benefited from the shared MLP in the skeleton deformation network, we further present joint-wise pose regression based on the high-dimensional vector as joint features (the green vector in Fig. 5). In practice, as all the joints will go through the same MLPs in the deformation stage, this high dimensional representation contains sufficient information which not only from spatial location but also from the original sketch. The joint feature is then fed into another shared MLPs to predict the rotation matrix of each joint.

3.3   Loss

Based on the common pipeline with a skeleton as the intermediary, the above-mentioned variants of our sketch-based body modeling network can all be trained end-to-end. The loss function includes a vertex loss, SMPL parameter loss for the final output, as well as a skeleton loss for the intermediate skeleton construction.

Vertex Loss. To measure the difference between the output body mesh and the ground truth model, we apply a vertex loss L_{vertex} on the final output SMPL models. The vertex loss is defined as

$$L_{vertex} = \frac{1}{N_{ver}} \sum_{i=1}^{N_{ver}} \|\hat{\mathbf{v}}_i - \mathbf{v}_i\|_1, \tag{1}$$

where $N_{ver} = 6890$ for the body meshes in SMPL representation. $\hat{\mathbf{v}}_i$ and $\mathbf{v}_i$ are the 3D coordinates of the i-th vertex in the ground-truth model and the predicted model respectively.

Skeleton Loss. We use the L_2 loss of the N_j ordered joints on the skeleton to constrain the intermediate skeleton construction with the joint regressor or skeleton deformation network.

$$L_{skeleton} = \frac{1}{N_j} \sum_{i=1}^{N_j} \|\hat{\mathbf{p}}_i - \mathbf{p}_i\|_2, \tag{2}$$

where $\hat{\mathbf{p}}$ is the ground truth location of each joint, and $\mathbf{p}_i$ is the predicted joint location.

SMPL Regression Loss. For the SMPL regressor, we use the L_2 losses on the SMPL parameters θ and β. The entire network is trained end-to-end with the three loss items as:

$$L_{model} = L_{vertex} + L_{skeleton} + L_\theta + \lambda L_\beta, \tag{3}$$

where $\lambda = 0.1$ to balance the parameter scales.

4   Experiments and Discussion

Our system is the first one that uses DNNs for sketch-based body modeling. There is no existing public dataset for training DNNs and make a comprehensive evaluation. In this section, we first introduce our sketch-model datasets. Then we explain more implementation details and training settings. We conduct both quantitative evaluation on multiple variants and qualitative test with freehand sketching to demonstrate the effectiveness of our method.

4.1   Dataset

To evaluate our network, we build a dataset with paired sketches and 3D models. The 3D models come from the large-scale dataset AMASS [15] which contains a huge variety of body shapes and poses with fully rigged meshes. We select 25 subjects from a large subset BMLmovi in AMASS dataset to test our algorithm for its large diversity of body shapes and poses. For each subject, we pick one model every 50 frames. In total, we collect 7097 body models with varied shapes and poses. For each subject, we randomly select 60% for training, 20% for validation, and 20% for the test. To generated paired sketch and model for training and quantitative evaluation, we render sketches for each body model using suggestive contours [2] to imitate hand-drawn sketches. For each model, we render a sketch-style image from one view. Figure 6 shows two examples of synthesized sketches from 3D models and hand-drawn sketches by a non-expert user. We can see that the synthesized sketch images look similar to hand-drawn sketches. By rendering a large number of synthesized sketches, the cost of collecting sketch and model pairs is significantly reduced and quantitative evaluation is allowed. Our experiments demonstrate that freehand sketching is well supported in our sketch-based body modeling system without performance degradation.

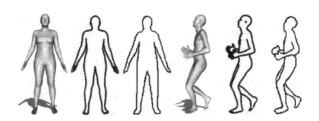

Fig. 6. Two examples of synthesized sketches (middle) from a 3D model (left) and hand-drawn sketches (right).

4.2   Network Details and Training Settings

We use ResNet-34 [5] as our sketch encoder and set the sketch feature dimension $K_s = 1024$ in our method. We introduce three variants of network to map the sketch feature $\mathbf{f}_{sketch} \in \mathbb{R}^{1024}$ into SMPL parameters $\beta \in \mathbb{R}^{10}$ and $\theta \in \mathbb{R}^{216}$. The first, called MLP-Vanilla, as shown in Fig. 3, consists MLP of (512, 256, 10) for the shape regressor and (512, 512, 256, 216) for the pose regressor. The second, call MLP-Joint, employs an MLP of (512, 256, 72) for the joint regressor and an MLP of (256, 256, 216) for pose regressor as Fig. 4 shows. The third one is our full model (named Deform+JF) involving skeleton deformation and joint-wise features for skeleton construction and pose regression, as Fig. 5 shows. In the skeleton deformation part, the two-level shared-MLP of (512, 256, 3) is used to map the $3 + 1024$ input to the 3D vertex offset for each joint. For pose regression, we use the shared-MLP of (128, 64, 9) to regress the 9 elements of a rotation matrix from the 256-D joint-wise feature for each joint.

The networks are trained end-to-end. Each network was trained for 400 epochs with a batch size of 32. Adam optimizer is used with $\beta_1 = 0.9$, $\beta_2 = 0.999$. The learning rate is set as 3e-4 without learning rate decay.

4.3   Results and Discussion

Comparison with Leading Methods. In order to evaluate the effectiveness of our method, we compare our method with two state-of-the-art reconstruction

Table 1. Quantitative evaluation of body reconstruction performance (errors in mm) on the test set. The number of network parameters and GFLOPs are also reported.

Method	RE	MPJPE	N_{para} (M)	GFLOPS
3DN [23]	119.06	95.19	151.77	190.58
CMR [10]	43.22	37.45	40.58	4.046
MLP-Vanilla	42.49	36.42	23.43	3.594
MLP-Joint	39.75	34.10	23.27	3.594
Deform+JF	**38.09**	**32.38**	23.82	3.625

techniques, training by replacing images with synthesized sketches. We select the convolutional mesh regression (CMR) [10] for single-image human body reconstruction and the 3DN model [23] which is a general modeling framework based on mesh deformation. We follow their default training settings except removing the 2D projected joint loss in CMR since it's non-trivial to estimate the perspective projection matrix from hand-drawn sketches.

Two error metrics, the reconstruction error (RE) and the mean per joint position error (MPJPE), are used to evaluate the performance of sketch-based modeling. 'RE' is the mean Euclidean distance between corresponding vertexes in the ground truth and predicted models. The 'MPJPE' reports the mean per joint position error after aligning the root joint. Table 1 lists the reconstruction performance of the three variants of our method and two SOTA methods. The three variants of our network all outperform existing image-based methods when trained for sketch-based body modeling. Taking advantage of the intermediate skeleton representation, our full model effectively extracts features from sketches and produces precise poses (32.38 mm of MPJPE) and shapes (38.09 mm of RE). Moreover, three variants of our method only have about 23M parameters (about half of CMR [10], 0.15 of 3DN [23]) and the least computational complexity.

Ablation Study. With regard to the three variants of our method, the MLP-Vanilla version directly regresses the shape pose parameters from global features extracted from the input sketch. As the existing image-based method may not suitable for sparse sketches, we set the vanilla version as our baseline. Table 1 shows that this plain network is effective in mapping sparse sketches on the parametric body space. It works slightly better than the CMR model. Bringing our main concept of using the skeleton as the intermediate representation, our MLP-Joint network reduces the reconstruction error 2.74 mm compared with MLP-Vanilla. Furthermore, by introducing joint features into the pose regressor, our Deform+JF network improves the modeling accuracy by 1.66 mm.

Qualitative Comparison. Figure 7 shows a group of 3D body models interpreted from synthesized sketches by different methods. As 3DN [23] does not use any prior body structure, it has difficulty to regress a naturally-looking body

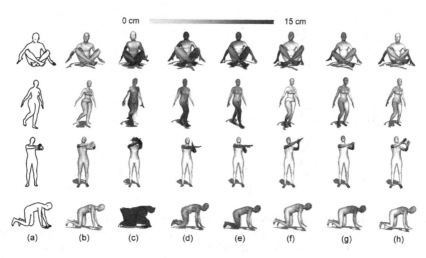

Fig. 7. 3D Meshes generated from synthesized sketches using different methods. The vertex errors are color-coded. From left to right: (a) input sketches; (b) ground truth; (c) results of 3DN [23]; (d) deformation results of CMR [10]; (e) SMPL regression results of CMR [10]; (f) results of MLP-Vanilla; (g) results of our MLP-Joint network; (h) results of our Deform+JF network which regresses parameters from joint-wise features.

model in a huge space. As one of the state-of-the-art method of image-based body reconstruction, CMR's [10] dense deformation by graph convolutions for regressing 3D locations of a large number of vertices rely on rich image features, its performance degrades in sparse and rough sketches. In comparison, as shown in the last three columns in Fig. 7, our method achieves high-quality body models from sketches. From Fig. 7 (f) to (h), the generated body models get better as the reconstruction errors get lower, especially for the head and limbs. And our Deform+JF network shows the most satisfactory visual results.

Skeleton Interpretation. We also compared the joint errors that are directly regressed from the sketch features and after SMPL regression in our network. Table 2 lists the errors of skeleton joints that directly regressed from sketch features and those calculated from the output SMPL models. We can see that our Deform+JF with SMPL could achieve the lowest MPJPE with 32.38 mm. On the other hand, although without strict geometric constraints on joints in the SMPL model, our MLP-Joint and Deform+JF's skeleton regressor still get high-quality skeletons, as shown in Fig. 8. Moreover, by introducing joint features, the Deform+JF network not only obtains more precise final SMPL models but also achieves better skeleton estimation accuracy (46.99 mm) than the MLP-Joint network (62.49 mm), which proves the effectiveness of joint-wise features.

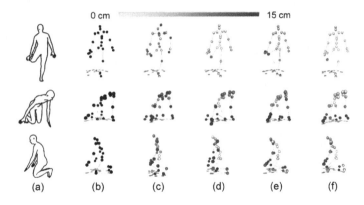

Fig. 8. Reconstructed body skeletons from sketches using different methods. From left to right: (a) input sketch; (b) ground truth; (c) regression result from MLP-Joint; (d) skeleton from the output SMPL model by MLP-Joint; (e) regression result from Deform+JF; (f) skeleton from the output SMPL model by Deform+JF.

Table 2. Comparison MPJPE (in mm) of skeletons generated by different ways.

Method	MLP-Joint		Deform+JF	
	Regressed	SMPL	Regressed	SMPL
MPJPE	62.49	34.10	46.99	32.38

Fig. 9. Interactive editing by sketching. The user modifies some strokes as highlighted to change body poses.

4.4   Body Modeling by Freehand Sketching

Though training with synthetic sketches, our model is also able to interpret 3D body shapes accurately from freehand sketching. We asked several graduate students to test our sketch-based body modeling system. Figure 1 and Fig. 9 show a group of results in diverse shapes and poses. Though the sketches drawn by non-expert users are very coarse with severe distortions, our system is able to capture the shape and pose features and produce visually pleasing 3D models. As shown in Fig. 9, based on our lightweight model, users could modify the sketch in part and get a modified 3D model in real-time rate which allows non-expert users to easily create high-quality 3D body models with fine-grained control on desired shapes and poses by freehand sketching and interactive editing.

5   Conclusions

In this paper, we introduce a sketch-based human body modeling system. In order to precisely mapping the coarse sketches into the manifold of human bodies, we propose to involve the body skeleton as an intermediary for regression of shape and pose parameters in the SMPL representation which allow effective encoding of the global structure prior of human bodies. Our method outperforms state-of-the-art techniques in the 3D interpretation of human bodies from hand-drawn sketches. The underlying sparsity of sketches and shape parameters allows us to use a lightweight network to achieve accurate modeling in real-time rates. Using our sketch-based modeling system, common users can easily create visually pleasing 3D models in a large variety of shapes and poses.

Acknowledgements. This work was supported by the National Key Research & Development Plan of China under Grant 2016YFB1001402, the National Natural Science Foundation of China (NSFC) under Grant 61632006, as well as the Fundamental Research Funds for the Central Universities under Grant WK3490000003.

References

1. Choy, C.B., Xu, D., Gwak, J.Y., Chen, K., Savarese, S.: 3D-R2N2: a unified approach for single and multi-view 3D object reconstruction. In: Leibe, B., Matas, J., Sebe, N., Welling, M. (eds.) ECCV 2016. LNCS, vol. 9912, pp. 628–644. Springer, Cham (2016). https://doi.org/10.1007/978-3-319-46484-8_38
2. DeCarlo, D., et al.: Suggestive contours for conveying shape. In: ACM SIGGRAPH, pp. 848–855 (2003)
3. Girdhar, R., Fouhey, D.F., Rodriguez, M., Gupta, A.: Learning a predictable and generative vector representation for objects. In: Leibe, B., Matas, J., Sebe, N., Welling, M. (eds.) ECCV 2016. LNCS, vol. 9910, pp. 484–499. Springer, Cham (2016). https://doi.org/10.1007/978-3-319-46466-4_29
4. Groueix, T., et al.: A papier-mâché approach to learning 3D surface generation. In: Proceedings of the IEEE Conference on Computer Vision and Pattern Recognition, pp. 216–224 (2018)
5. He, K., et al.: Deep residual learning for image recognition. In: Proceedings of the IEEE Conference on Computer Vision and Pattern Recognition, pp. 770–778 (2016)
6. Wu, J., et al.: Learning shape priors for single-view 3D completion and reconstruction. In: Proceedings of the European Conference on Computer Vision, pp. 646–662 (2018)
7. Jiang, H., et al.: Skeleton-aware 3D human shape reconstruction from point clouds. In: Proceedings of the IEEE International Conference on Computer Vision, pp. 5431–5441 (2019)
8. Delanoy, J., et al.: 3D sketching using multi-view deep volumetric prediction. Proc. ACM Comput. Graph. Interact. Tech. 1(1), 1–22 (2018)
9. Kanazawa, A., et al.: End-to-end recovery of human shape and pose. In: Proceedings of the IEEE Conference on Computer Vision and Pattern Recognition, pp. 7122–7131 (2018)

10. Kolotouros, N., et al.: Convolutional mesh regression for single-image human shape reconstruction. In: Proceedings of the IEEE Conference on Computer Vision and Pattern Recognition, pp. 4501–4510 (2019)

11. Li, Y., et al.: LinesToFacePhoto: face photo generation from lines with conditional self-attention generative adversarial networks. In: Proceedings of the 27th ACM International Conference on Multimedia, pp. 2323–2331 (2019)

12. Liang, J., et al.: Shape-aware human pose and shape reconstruction using multi-view images. In: Proceedings of the IEEE International Conference on Computer Vision, pp. 4352–4362 (2019)

13. Loper, M., et al.: SMPL: a skinned multi-person linear model. ACM Trans. Graph. **34**(6), 1–16 (2015)

14. Lun, Z., et al.: 3D shape reconstruction from sketches via multi-view convolutional networks. In: International Conference on 3D Vision, pp. 67–77 (2017)

15. Mahmood, N., et al.: AMASS: archive of motion capture as surface shapes. In: Proceedings of the IEEE International Conference on Computer Vision, pp. 5442–5451 (2019)

16. Olsen, L., et al.: Sketch-based modeling: a survey. Comput. Graph. **33**(1), 85–103 (2009)

17. Omran, M., et al.: Neural body fitting: unifying deep learning and model based human pose and shape estimation. In: International Conference on 3D Vision, pp. 484–494 (2018)

18. Tan, J.K.V., et al.: Indirect deep structured learning for 3d human body shape and pose prediction (2017)

19. Venkat, A., et al.: HumanMeshNet: polygonal mesh recovery of humans. In: Proceedings of the IEEE International Conference on Computer Vision Workshops (2019)

20. Wang, F., et al.: Sketch-based 3D shape retrieval using convolutional neural networks. In: Proceedings of the IEEE Conference on Computer Vision and Pattern Recognition, pp. 1875–1883 (2015)

21. Wang, L., et al.: Unsupervised learning of 3D model reconstruction from hand-drawn sketches. In: Proceedings of the 26th ACM International Conference on Multimedia, pp. 1820–1828 (2018)

22. Wang, N., et al.: Pixel2Mesh: generating 3D mesh models from single RGB images. In: Proceedings of the European Conference on Computer Vision, pp. 52–67 (2018)

23. Wang, W., et al.: 3DN: 3D deformation network. In: Proceedings of the IEEE Conference on Computer Vision and Pattern Recognition, pp. 1038–1046 (2019)

24. Zeleznik, R.C., et al.: SKETCH: an interface for sketching 3D scenes. In: ACM SIGGRAPH Courses, pp. 9-es (2006)

25. Zhang, H., et al.: DaNet: decompose-and-aggregate network for 3D human shape and pose estimation. In: Proceedings of the 27th ACM International Conference on Multimedia, pp. 935–944 (2019)

26. Zheng, Z., et al.: DeepHuman: 3D human reconstruction from a single image. In: Proceedings of the IEEE International Conference on Computer Vision, pp. 7739–7749 (2019)

27. Xu, Y., et al.: DenseRaC: joint 3D pose and shape estimation by dense render-and-compare. In: Proceedings of the IEEE International Conference on Computer Vision, pp. 7760–7770 (2019)

28. Kolotouros, N., et al.: Learning to reconstruct 3D human pose and shape via model-fitting in the loop. In: Proceedings of the IEEE International Conference on Computer Vision, pp. 2252–2261 (2019)

Two-Stage Real-Time Multi-object Tracking with Candidate Selection

Fan Wang, Lei Luo, and En Zhu$^{(\boxtimes)}$

School of Computer, National University of Defense Technology, Changsha, China
{wangfan10,l.luo,enzhu}@nudt.edu.cn

Abstract. In recent years, multi-object tracking is usually treated as a data association problem based on detection results, also known as tracking-by-detection. Such methods are often difficult to adapt to the requirements of time-critical video analysis applications which consider detection and tracking together. In this paper, we propose to accomplish object detection and appearance embedding via a two-stage network. On the one hand, we accelerate network inference process by sharing a set of low-level features and introducing a Position-Sensitive RoI pooling layer to better estimate the classification probability. On the other hand, to handle unreliable detection results produced by the two-stage network, we select candidates from outputs of both detection and tracking based on a novel scoring function which considers classification probability and tracking confidence together. In this way, we can achieve an effective trade-off between multi-object tracking accuracy and speed. Moreover, we conduct a cascade data association based on the selected candidates to form object trajectories. Extensive experiments show that each component of the tracking framework is effective and our real-time tracker can achieve state-of-the-art performance.

Keywords: Two-stage · Real-time · Multi-object tracking · Candidate selection

1 Introduction

Multi-object tracking (MOT) remains one of the big challenges of computer vision, which underpins critical application significance ranging from visual surveillance to autonomous driving. In order to meet the requirements of real-time applications, MOT is not only a problem of data association, but also needs to take object detection and appearance embedding into consideration.

F. Wang and L. Luo—Equal contribution.
This work was supported by the Natural Science Foundation of China under contracts 61872377.

J. Lokoč et al. (Eds.): MMM 2021, LNCS 12573, pp. 49–61, 2021.
https://doi.org/10.1007/978-3-030-67835-7_5

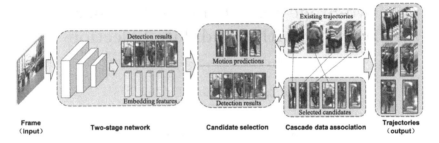

Fig. 1. The proposed two-stage real-time multi-object tracking framework. The two-stage network takes current frame as input, and produces corresponding detection results and appearance embedding features. Then, these two outputs will feed to the candidate selection part to get more accurate object positions for subsequent cascade data association. Finally, The matched candidates will be connected to the existing trajectories, and the remaining unmatched candidates from detection will be viewed as new trajectories.

The problem of tracking multiple objects in a video sequence poses several challenging tasks including object detection, appearance embedding, motion prediction and dealing with occlusions. Usually, object detection and appearance embedding are the core parts of MOT. As discussed in [31], recent progresses on MOT can be primarily divided into three categories: (i) Separate Detection and Embedding (SDE) methods [7,32,34] use two separate models to accomplish object detection and appearance embedding, respectively; (ii) Two-stage methods [30,33] share the Region Proposal Network (RPN) [25] feature map in the detection stage to accelerate the learning of appearance embedding feature; (iii) Joint Detection and Embedding (JDE) methods [31,36] use single-shot deep network to achieve object detection and embedding simultaneously. SDE methods belonging to the typical *tracking-by-detection* paradigm can easily achieve state-of-the-art performance which benefit a lot from the advance in both object detection and embedding. However, these two separate compute-intensive models bring critical challenges for SDE methods to build a real-time MOT system. On the other hand, JDE methods have advantage in tracking speed, but often difficult to train a single-shot deep network which can satisfy both object detection and appearance embedding. The experimental results in [31] show that the performance of object detection is still far from the SDE methods. In order to achieve a compromise between tracking speed and detection accuracy, Two-stage methods adopt the Faster R-CNN framework [25] to conduct detector and embedding jointly by sharing the same set of low-level features. Our method belongs to this category.

We propose a two-stage deep network with candidate selection for joint detection and tracking. As shown in Fig. 1, the proposed two-stage deep network is a CNN-based model that not only produces object candidate bounding boxes, but also goes further to extract features of these boxes to identify them. We follow region-based fully convolutional neural network (R-FCN) [8] to design a two-branch multi-task deep network. We employ a spatial transformer layer (STL) [14] follows RPN to handle the spatial variance problem and get better shared features. In object detection branch, we adopt the Position-Sensitive RoI pooling layer from R-FCN to obtain strong discriminative ability for object classification. Usually, multi-task object detection network produces many unreliable detection results which will greatly affect the tracking performance. In order to deal with the impact of unreliable detection results, we borrow the candidate selection idea from [19] to further improve the tracking performance. We filter out low-confidence object candidate boxes (detection and tracking boxes) by constructing a unified scoring function. Finally, we use a cascade data association method based on the selected boxes to obtain the trajectories of objects in the video sequence.

Our contribution is three fold. First, we design a two-stage multi-task deep network to accomplish object detection and appearance embedding simultaneously. Second, we introduce a candidate selection method to mitigate the impact of unreliable object detection results generated by the two-stage network on the tracking performance. Third, we present a cascade data association strategy, which utilizes spatial information and deeply learned person re-identification (ReID) features.

2 Related Work

In this section, we briefly review the related works on MOT by classify them into the tracking-by-detection methods and simultaneous detection and tracking methods. We discuss the pros and cons of these methods and explain the relationship with our approach.

2.1 Tracking-by-Detection Methods

Tracking-by-detection methods breaks MOT task down to two steps: (i) the detection step, which localizes all objects of interest in a single video frame by a series of boxes; and (ii) the associate step, where detected objects are assigned and connected to existing trajectories based on spatial and appearance information. Such methods can easily achieve state-of-the-art performance because they can use the most suitable model in each step respectively. Many works [15,17,34,37] are devoted to improving the overall tracking performance by enhancing the detection quality. It is also a research hotspot to learn better ReID features for data association [5,6,23,29]. Data association usually uses the Kalman Filter and Hungarian algorithm to accomplish the linking task. A small number of works [12,20,40] use more complicated association strategies such as group models and RNNs.

The obvious disadvantage of tracking-by-detection methods is that it contains two compute-intensive components: a object detection model and an appearance embedding model. Therefore, MOT systems based on this kind of method are

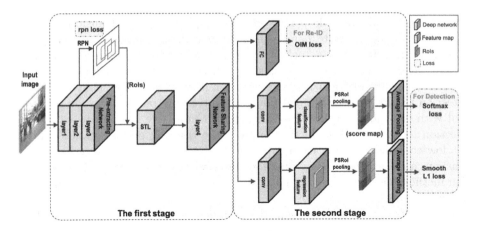

Fig. 2. The structure of our two-stage network. In the first stage, RoIs are generated by RPN and shared features are extracted. STL is employed into backbone network to deal with the spatial variance problem. In the second stage, two branches are appended to the shared feature map to accomplish object detection and appearance embedding. The Position-Sensitive RoI pooling layer is introduced into object detection branch to obtain strong discriminative ability for object classification.

very slow, and it is difficult to meet the real-time requirements in many applications. The two-stage multi-task deep network we proposed completes object detection and appearance embedding via partial feature layer sharing, which saves computation to a certain extent and accelerates the speed of multi-object tracking.

2.2 Simultaneous Detection and Tracking Methods

Simultaneous detection and tracking methods aim to use a single-shot deep network to accomplish object detection and appearance embedding simultaneously. Tracktor [3] removes the box association by directly propagating identities of region proposals using bounding box regression. JDE [31] is introduced on the top of YOLOv3 [24] framework to learn object detection and embedding in a shared model. FairMOT [36] treats the MOT problem as a pix-wise keypoint (object center) estimation and identity classification problem on top of high-resolution feature map. CenterTrack [39] present a point-based framework for joint detection and tracking.

Obviously, these methods have a great advantage in tracking speed, while the tracking accuracy is usually lower than that of tracking-by-detection methods. We deeply investigate the reasons and find that the one-shot detector in shared network produces an amount of unreliable detection results which causes severe ambiguities. To address this problem, we introduce a candidate selection approach to filter out those poor detection results which improves tracking accuracy a lot.

3   Proposed Method

In this section, we present the details for the backbone network, the object detection and appearance embedding branch, the candidate selection process and the cascade data association process, respectively.

3.1   Backbone Network

ResNet-50 [13] is chosen as our backbone in order to strike a good balance between accuracy and speed. We use the first three layers (*layer1, layer2 and layer3*) of ResNet50 as a Pre-extracting Network to extract base feature for follow-up tasks. Next, there is a RPN [25] with a STL [14] to generate region proposals of objects. These generated regions of interest (RoIs) will be directly used for object detection and appearance embedding. The *layer4* of ResNet50 works as a Features Sharing Network to extract shared features. Denote the size of the input image as $H_{image} \times W_{image}$, then the output feature has the shape of $C \times H \times W$ where $H = H_{image}/32$ and $W = W_{image}/32$.

3.2   Two Branches

As shown in Fig. 2, three parallel heads are appended to the shared feature map extracted by the backbone network to accomplish object detection (including object classification and box regression) and appearance embedding tasks. In object detection branch, we employ the Position-Sensitive RoI pooling layer to explicitly encode spatial information into score maps. The object classification probability is estimated from k^2 Position-Sensitive score maps z. We define each RoI as $x = (x_0, y_0, w, h)$, where (x_0, y_0) denotes the top-left point and w, h represent the width and height of the region. Refer to the method in [19], we split a RoI into $k \times k$ bins which represent different spatial locations of the object. We extract responses of $k \times k$ bins from k^2 score maps and finally can get the classification probability as follows:

$$s_{det} = p(y|z, x) = \sigma(\frac{1}{wh} \sum_{i=1}^{k^2} \sum_{(x,y) \in bin_i} z_i(x,y)), \tag{1}$$

where $\sigma(x) = \frac{1}{1+e^{-x}}$ is the sigmoid function, and z_i denotes the i-th score map. Appearance embedding branch aims to generate features that can distinguish different instances of the same class. Ideally, the distance between different instances should be larger than that between the same instance. To achieve this goal, we apply a fully connected layer with 256 neurons on top of the shared feature map to extract appearance embedding feature for each location.

The learning objective of each parallel head in the two-stage network can be modeled as a multi-task learning problem. The joint objective can be written as a linear sum of losses from every component,

$$\mathcal{L}_{total} = \mathcal{L}_{rpn} + \mathcal{L}_{cls} + \mathcal{L}_{reg} + \mathcal{L}_{emb}, \tag{2}$$

where $\mathcal{L}_{rpn}$ denotes loss function of RPN, $\mathcal{L}_{cls}$ and $\mathcal{L}_{reg}$ are the learning objectives of object classification and box regression, respectively. $\mathcal{L}_{emb}$ represents the loss function of appearance embedding. In object detection branch, we choose softmax loss and smooth-L1 loss for $\mathcal{L}_{cls}$ and $\mathcal{L}_{reg}$ respectively. In appearance embedding branch, we adopt Online Instance Matching (OIM) loss [33] for $\mathcal{L}_{emb}$. Because in the network training phase, OIM loss not only can use the information of unlabeled objects, but also is more easily to converge.

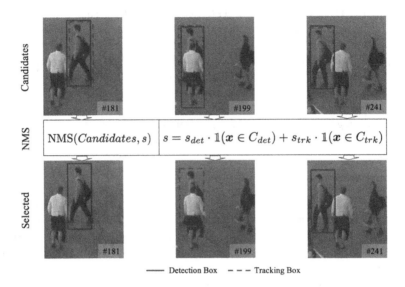

Fig. 3. Candidate selection based on unified scores. Candidates from detection and tracking with higher confidence will be selected out for data association.

3.3   Candidate Selection

Simultaneous detection and tracking methods usually produce an amount of unreliable detection results which will affect tracking performance according to the analysis in Sect. 1. A feasible idea is to adopt a motion model to handle unreliable detection caused by varying visual properties of objects and occlusion in crowded scenes. Thus, we use Kalman filter to estimate new locations of objects based on existing trajectories, and define these predictions as tracking boxes. As shown in Fig. 3, we try to merge the candidates including detection and tracking boxes through a unified score function which is defined as follows:

$$s = s_{det} \cdot \mathbb{1}(\boldsymbol{x} \in C_{det}) + s_{trk} \cdot \mathbb{1}(\boldsymbol{x} \in C_{trk}), \tag{3}$$

where $\boldsymbol{x}$ denotes a certain candidate box. $\mathbb{1}(\cdot)$ is the indicator function that equals 1 if the input is *ture*, otherwise equals 0. C_{det} and C_{trk} denote the candidates from detection and tracking respectively, and s_{det} and s_{trk} represent the corresponding scores. Non-Maximal Suppression (NMS) is performed with the scored candidates for subsequent data association. Ideally, we can filter out unreliable detection results with lower scores, and replace poor detection results with high-score tracking boxes to locate objects. The score of detection box has been explained in Sect. 3.2. Here we focus on the scoring method of tracking box.

During multi-object tracking, a complete trajectory may be splited into a set of tracklets since a trajectory can be interrupted and retrieved for times during its lifetime. We define tracklet confidence as follows to measure the accuracy of the Kalman filter using temporal information:

$$s_{trk} = \max(1 - \log(1 + \alpha \cdot L_{trk}), 0) \cdot \mathbb{1}(L_{det} > 2), \qquad (4)$$

where L_{det} denotes the number of detection results associated to the tracklet, and L_{trk} represents the number of tracking boxes after the last detection is associated, α is a coefficient which ensures s_{det} and s_{trk} are on the same order of magnitude. Equation (4) shows that the more the Kalman filter is used for prediction, the lower the confidence of the tracklet. Because the accuracy of the Kalman filter could decrease if it is not update by detection over a long time. Every time a trajectory is retrieved from lost state, the Kalman filter will be re-initialized. Therefore, we only consider the information of the last tracklet to formulate tracklet confidence.

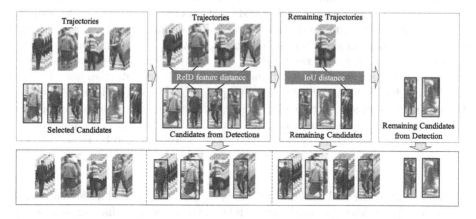

Fig. 4. Illustration of cascade data association. The upper part (yellow box) describes the data association process, and the lower part (red box) describes the trajectory update during data association. Objects in blue boxes represent the selected candidates from detection, and objects in red boxes represent the selected candidates from tracking. After the candidates and the existing trajectories are matched by appearance information and spatial information, the remaining unmatched detection results will be viewed as new object trajectories. (Color figure online)

3.4   Cascade Data Association

As shown in Fig. 4, the selected candidates are the basis of our cascade data association. Specifically, we first apply data association on candidates from detection, using appearance embedding feature with a threshold τ_d for the maximum distance. Each trajectory stores an appearance representation of the object by saving embedding feature from the associated detection. Then, we associate the remaining candidates with unassociated trajectories based on Intersection over Union (IoU) with a threshold τ_{iou}. Finally, the remaining detection results are initialized as new trajectories. Within the cascade data associate, all detection results and appearance embedding features are extracted by a two-stage network. Moreover, the candidate selection process requires very little computation, which means our MOT algorithm can meet the real-time requirement for many applications.

4   Experiments

4.1   Datasets and Metrics

Datasets. To evaluate the performance of our proposed tracker, we conducted sufficient experiments on the MOT16 [21] dataset which is a widely used benchmark for multi-object tracking. A total of 14 videos are included in MOT16 dataset (7 for training and 7 for testing), with public detections obtained using the Deformable Part-based Model (DPM) [22]. In order to achieve better object detection and embedding performance, we build a large-scale dataset by combining the training images from seven public datasets for person detection and search. In particular, we use the ETH [10] and the CityPerson [35] datasets which only contain bounding box annotations to train the object detection branch, and the CalTech [9], 2DMOT2015 [18], MOT17 [21], CUHK-SYSU [33] and PRW [38] datasets which contain both bounding box and identity annotations are used to train both of the object detection and embedding branches. Because part of the videos overlaps in the three datasets 2DMOT2015, MOT17 and ETH, we remove them from the training dataset for fair evaluation.

Metrics. Although multiple Object Tracking Accuracy (MOTA) [4] has been used as the main metric for evaluating MOT in recent years, it has exposed some obvious flaws. In particular, MOTA can not adequately illustrate ReID performance which will mark a tracker performs well even though it has a large number of ID switches. On the contrary, Identity-F1 [26] shows a certain advantage in measuring long consistent trajectories without switches. In this paper, We comprehensively focus and emphasize MOTA and IDF1 to evaluate the tracking performance. Specifically, MOTA is defined as follows:

$$MOTA = 1 - \frac{(FN + FP + IDSW)}{GT} \in (-\infty, 1], \tag{5}$$

where GT is the number of ground truth boxes, FP is the number of false positives in the whole video, FN is the number of false negatives in the whole video, and $IDSW$ is the total number of ID switches. And IDF1 is defined as follows:

$$IDF1 = \frac{2}{\frac{1}{IDP} + \frac{1}{IDR}}, \tag{6}$$

where IDP is the identification precision and IDR is the identification recall.

4.2   Implementation Details

We employ ResNet-50 as the backbone of two-stage network for object detection and appearance embedding. In appearance embedding branch, we try to obtain object appearance feature with 256 dimensions. In object detection branch, we set $k = 7$ for Position-Sensitive score maps. To train the two-stage network on a GeForce TITAN Xp GPU, we use SGD optimizer with learning rate of 1e-4 and the batch size of 4 for 20 epochs. Refer to [19], we set $\tau_{nms} = 0.3$ for NMS, minimum unified score $\tau_s = 0.4$ and coefficient $\alpha = 0.05$ for candidate selection to filter out low-quality candidates. At the same time, we set $\tau_d = 0.4$ and $\tau_{iou} = 0.3$ for cascade data association.

4.3   Experimental Results

Evaluation on Test Set. We validated our proposed method on the MOT16 test set. As shown in Table 1, we compare our tracker with other offline and online trackers. In particular, all the compared trackers follows the "tracking-by-detection" paradigm, which means they pay more attention to data association rather than detection in the tracking process. The calculation of tracking speed by these methods mainly refers to data association, and does not consider the time consumption of object detection. However, our proposed method takes into account the time consumption of object detection, object embedding, and data association. Experimental results show that our method has great advantages in the two metrics of MOTA and IDF1, even is better than some offline methods. It should be noted that although our tracker still has a gap in performance compared to MOTDT tracker [19], we have a great advantage in tracking speed.

Ablation Studies. In order to demonstrate the effectiveness of our method, we ablate the two main components: candidate selection and embedding feature on the MOT16 training set. The baseline method uses the detection results produced by two-stage network to form object trajectories based on IoU. Based on the baseline method, we realize the cascade data association by adding embedded features which improves MOTA by 1% and IDF1 by 5.1%. If we only introduce the candidate selection mechanism, there will only be an improvement in MOTA. An intuitive explanation is that the candidate selection process eliminates unreliable detection results and improves MOTA by effectively reducing the number of false negatives. From the results of ablation experiment, it can be seen that

Table 1. Evaluation results on the MOT16 test set. The arrow after each metric indicates that the higher (↑) or lower (↓) value is better.

Tracker	Mode	MOTA↑	IDF1↑	MT(%)↑	ML(%)↓	FP↓	FN↓	IDSW↓	FPS↑
LINF1 [11]	batch	41.0	45.7	11.6	51.3	7896	99224	**430**	**4.2**
MHT_bLSTM [16]	batch	42.1	47.8	14.9	44.4	11637	93672	753	1.8
JMC [28]	batch	46.3	46.3	15.5	**39.7**	**6375**	90914	657	0.8
LMP [29]	batch	**48.8**	**51.3**	**18.2**	40.1	6654	**86245**	481	0.5
HISP [2]	online	37.4	30.5	7.6	50.9	**3222**	108865	2101	3.3
EAMTT [27]	online	38.8	42.4	7.9	49.1	8114	102452	965	11.8
GMPHD_ReID [1]	online	40.4	49.7	11.2	43.3	6572	101266	792	31.6
MOTDT [19]	online	47.6	**50.9**	15.2	38.3	9253	**85431**	792	**20.6**
Ours*	online	42.0	48.5	**17.3**	**33.5**	19056	85855	921	12

the candidate selection component focuses on improving MOTA metric, while the ReID component focuses on improving IDF1 metric. By combining these two components, our proposed method can ultimately improve MOTA by 3.8% and IDF1 by 5.8%, and almost has the best results for all metrics.

Table 2. Evaluation results on the MOT16 training set in terms of different components used. **CS:** candidate selection, **ReID:** appearance embedding feature. The arrow after each metric indicates that the higher (↑) or lower (↓) value is better.

Method	CS	ReID	MOTA↑	MOTP↑	IDF1↑	IDSW ↓
Baseline			45.9	85.3	50.2	895
		✓	46.9	85.2	55.3	592
	✓		48.4	86.2	47.5	922
Proposed	✓	✓	**49.7**	86.1	**56.0**	**441**

5   Conclusion

In this paper, we propose an end-to-end online multi-object detection and tracking framework. Our method trains a two-stage network on a large scale dataset for object detection and appearance embedding. The candidate selection based on the score map extracted by two-stage network is introduced to reduce the impact of unreliable detection results on multi-object tracking. Our two-stage tracker associates objects greedily through time and runs in nearly real time. At the same time, the experimental results show the competitiveness in metrics of MOTA and IDF1.

References

1. Baisa, N.L.: Online multi-object visual tracking using a GM-PHD filter with deep appearance learning. In: 2019 22th International Conference on Information Fusion (FUSION), pp. 1–8. IEEE (2019)
2. Baisa, N.L.: Robust online multi-target visual tracking using a hisp filter with discriminative deep appearance learning. arXiv preprint arXiv:1908.03945 (2019)
3. Bergmann, P., Meinhardt, T., Leal-Taixe, L.: Tracking without bells and whistles. In: Proceedings of the IEEE International Conference on Computer Vision, pp. 941–951 (2019)
4. Bernardin, K., Stiefelhagen, R.: Evaluating multiple object tracking performance: the clear mot metrics. EURASIP J. Image Video Process. 1–10 (2008)
5. Bewley, A., Ge, Z., Ott, L., Ramos, F., Upcroft, B.: Simple online and realtime tracking. In: 2016 IEEE International Conference on Image Processing (ICIP), pp. 3464–3468. IEEE (2016)
6. Chen, J., Sheng, H., Zhang, Y., Xiong, Z.: Enhancing detection model for multiple hypothesis tracking. In: Proceedings of the IEEE Conference on Computer Vision and Pattern Recognition Workshops, pp. 18–27 (2017)
7. Choi, W.: Near-online multi-target tracking with aggregated local flow descriptor. In: Proceedings of the IEEE International Conference on Computer Vision, pp. 3029–3037 (2015)
8. Dai, J., Li, Y., He, K., Sun, J.: R-FCN: object detection via region-based fully convolutional networks. In: Advances in Neural Information Processing Systems, pp. 379–387 (2016)
9. Dollár, P., Wojek, C., Schiele, B., Perona, P.: Pedestrian detection: a benchmark. In: 2009 IEEE Conference on Computer Vision and Pattern Recognition, pp. 304–311. IEEE (2009)
10. Ess, A., Leibe, B., Schindler, K., Van Gool, L.: Robust multiperson tracking from a mobile platform. IEEE Trans. Pattern Anal. Mach. Intell. 1831–1846 (2009)
11. Fagot-Bouquet, L., Audigier, R., Dhome, Y., Lerasle, F.: Improving multi-frame data association with sparse representations for robust near-online multi-object tracking. In: Leibe, B., Matas, J., Sebe, N., Welling, M. (eds.) ECCV 2016. LNCS, vol. 9912, pp. 774–790. Springer, Cham (2016). https://doi.org/10.1007/978-3-319-46484-8_47
12. Fang, K., Xiang, Y., Li, X., Savarese, S.: Recurrent autoregressive networks for online multi-object tracking. In: 2018 IEEE Winter Conference on Applications of Computer Vision (WACV), pp. 466–475. IEEE (2018)
13. He, K., Zhang, X., Ren, S., Sun, J.: Deep residual learning for image recognition, pp. 770–778 (2016)
14. Jaderberg, M., Simonyan, K., Zisserman, A., et al.: Spatial transformer networks. In: Advances in Neural Information Processing Systems, pp. 2017–2025 (2015)
15. Kieritz, H., Hubner, W., Arens, M.: Joint detection and online multi-object tracking. In: Proceedings of the IEEE Conference on Computer Vision and Pattern Recognition Workshops, pp. 1459–1467 (2018)
16. Kim, C., Li, F., Rehg, J.M.: Multi-object tracking with neural gating using bilinear LSTM. In: Ferrari, V., Hebert, M., Sminchisescu, C., Weiss, Y. (eds.) ECCV 2018. LNCS, vol. 11212, pp. 208–224. Springer, Cham (2018). https://doi.org/10.1007/978-3-030-01237-3_13
17. Kim, S.J., Nam, J.Y., Ko, B.C.: Online tracker optimization for multi-pedestrian tracking using a moving vehicle camera. IEEE Access 48675–48687 (2018)

18. Leal-Taixé, L., Milan, A., Reid, I., Roth, S., Schindler, K.: Motchallenge 2015: Towards a benchmark for multi-target tracking. arXiv preprint arXiv:1504.01942 (2015)
19. Long, C., Haizhou, A., Zijie, Z., Chong, S.: Real-time multiple people tracking with deeply learned candidate selection and person re-identification. In: ICME (2018)
20. Mahmoudi, N., Ahadi, S.M., Rahmati, M.: Multi-target tracking using CNN-based features: CNNMTT. Multimedia Tools Appl. **78**, 7077–7096 (2019)
21. Milan, A., Leal-Taixé, L., Reid, I., Roth, S., Schindler, K.: Mot16: A benchmark for multi-object tracking. arXiv preprint arXiv:1603.00831 (2016)
22. Ott, P., Everingham, M.: Shared parts for deformable part-based models. In: CVPR 2011, pp. 1513–1520. IEEE (2011)
23. Ran, N., Kong, L., Wang, Y., Liu, Q.: A robust multi-athlete tracking algorithm by exploiting discriminant features and long-term dependencies. In: Kompatsiaris, I., Huet, B., Mezaris, V., Gurrin, C., Cheng, W.-H., Vrochidis, S. (eds.) MMM 2019. LNCS, vol. 11295, pp. 411–423. Springer, Cham (2019). https://doi.org/10.1007/978-3-030-05710-7_34
24. Redmon, J., Farhadi, A.: Yolov3: An incremental improvement. arXiv preprint arXiv:1804.02767 (2018)
25. Ren, S., He, K., Girshick, R., Sun, J.: Faster R-CNN: towards real-time object detection with region proposal networks. In: Advances in Neural Information Processing Systems, pp. 91–99 (2015)
26. Ristani, E., Solera, F., Zou, R., Cucchiara, R., Tomasi, C.: Performance measures and a data set for multi-target, multi-camera tracking. In: Hua, G., Jégou, H. (eds.) ECCV 2016. LNCS, vol. 9914, pp. 17–35. Springer, Cham (2016). https://doi.org/10.1007/978-3-319-48881-3_2
27. Sanchez-Matilla, R., Poiesi, F., Cavallaro, A.: Online multi-target tracking with strong and weak detections. In: Hua, G., Jégou, H. (eds.) ECCV 2016. LNCS, vol. 9914, pp. 84–99. Springer, Cham (2016). https://doi.org/10.1007/978-3-319-48881-3_7
28. Tang, S., Andres, B., Andriluka, M., Schiele, B.: Multi-person tracking by multicut and deep matching. In: Hua, G., Jégou, H. (eds.) ECCV 2016. LNCS, vol. 9914, pp. 100–111. Springer, Cham (2016). https://doi.org/10.1007/978-3-319-48881-3_8
29. Tang, S., Andriluka, M., Andres, B., Schiele, B.: Multiple people tracking by lifted multicut and person re-identification. In: Proceedings of the IEEE Conference on Computer Vision and Pattern Recognition, pp. 3539–3548 (2017)
30. Voigtlaender, P., et al.: Mots: multi-object tracking and segmentation. In: Proceedings of the IEEE Conference on Computer Vision and Pattern Recognition, pp. 7942–7951 (2019)
31. Wang, Z., Zheng, L., Liu, Y., Wang, S.: Towards real-time multi-object tracking. arXiv preprint arXiv:1909.12605 (2019)
32. Wojke, N., Bewley, A., Paulus, D.: Simple online and realtime tracking with a deep association metric. In: 2017 IEEE international conference on image processing (ICIP), pp. 3645–3649. IEEE (2017)
33. Xiao, T., Li, S., Wang, B., Lin, L., Wang, X.: Joint detection and identification feature learning for person search. In: Proceedings of the IEEE Conference on Computer Vision and Pattern Recognition, pp. 3415–3424 (2017)
34. Yu, F., Li, W., Li, Q., Liu, Yu., Shi, X., Yan, J.: POI: multiple object tracking with high performance detection and appearance feature. In: Hua, G., Jégou, H. (eds.) ECCV 2016. LNCS, vol. 9914, pp. 36–42. Springer, Cham (2016). https://doi.org/10.1007/978-3-319-48881-3_3

35. Zhang, S., Benenson, R., Schiele, B.: Citypersons: a diverse dataset for pedestrian detection. In: Proceedings of the IEEE Conference on Computer Vision and Pattern Recognition, pp. 3213–3221 (2017)
36. Zhang, Y., Wang, C., Wang, X., Zeng, W., Liu, W.: A simple baseline for multi-object tracking. arXiv preprint arXiv:2004.01888 (2020)
37. Zhao, D., Fu, H., Xiao, L., Wu, T., Dai, B.: Multi-object tracking with correlation filter for autonomous vehicle. Sensors 2004 (2018)
38. Zheng, L., Zhang, H., Sun, S., Chandraker, M., Yang, Y., Tian, Q.: Person re-identification in the wild. In: Proceedings of the IEEE Conference on Computer Vision and Pattern Recognition, pp. 1367–1376 (2017)
39. Zhou, X., Koltun, V., Krähenbühl, P.: Tracking objects as points. arXiv:2004.01177 (2020)
40. Zhou, Z., Xing, J., Zhang, M., Hu, W.: Online multi-target tracking with tensor-based high-order graph matching. In: 2018 24th International Conference on Pattern Recognition (ICPR), pp. 1809–1814. IEEE (2018)

Tell as You Imagine: Sentence Imageability-Aware Image Captioning

Kazuki Umemura[1(✉)], Marc A. Kastner[2,1], Ichiro Ide[1], Yasutomo Kawanishi[1], Takatsugu Hirayama[1], Keisuke Doman[3,1], Daisuke Deguchi[1], and Hiroshi Murase[1]

[1] Nagoya University, Furo-cho, Chikusa-ku, Nagoya, Aichi 464-8601, Japan
umemurak@murase.is.i.nagoya-u.ac.jp,
{ide,kawanishi,murase}@i.nagoya-u.ac.jp,
{takatsugu.hirayama,ddeguchi}@nagoya-u.jp
[2] National Institute of Informatics, 2-1-2 Hitotsubashi, Chiyoda-ku, Tokyo 101-8430, Japan
mkastner@nii.ac.jp
[3] Chukyo University, 101 Tokodachi, Kaizu-cho, Toyota, Aichi 470-0393, Japan
kdoman@sist.chukyo-u.ac.jp

Abstract. Image captioning as a multimedia task is advancing in terms of performance in generating captions for general purposes. However, it remains difficult to tailor generated captions to different applications. In this paper, we propose a sentence imageability-aware image captioning method to generate captions tailoring to various applications. Sentence imageability describes how easily the caption can be mentally imagined. This concept is applied to the captioning model to obtain a better understanding of the perception of a generated caption. First, we extend an existing image caption dataset by augmenting its captions' diversity. Then, a sentence imageability score for each augmented caption is calculated. A modified image captioning model is trained using this extended dataset to generate captions tailoring to a specified imageability score. Experiments showed promising results in generating imageability-aware captions. Especially, results from a subjective experiment showed that the perception of the generated captions correlates with the specified score.

Keywords: Vision and language · Image captioning · Psycholinguistics

1 Introduction

In recent years, image captioning that automatically generates image descriptions is advancing. State-of-the-art image captioning methods [14,28,29] commonly perform at a visually descriptive level for general purposes, but do not consider the perception of the generated captions. Because of this, it is difficult to tailor the captions to different applications.

J. Lokoč et al. (Eds.): MMM 2021, LNCS 12573, pp. 62–73, 2021.
https://doi.org/10.1007/978-3-030-67835-7_6

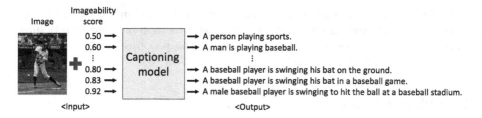

Fig. 1. Example of the proposed imageability-aware captioning. For an input image, the model generates diverse captions with different degrees of visual descriptiveness (i.e. imageability) tailoring to different target applications.

In the context of, e.g., news articles, the visual contents of an image are obvious and not needed to be captioned in detail. Accordingly, a caption should focus on additional information or context, rather than a pure visual description. News article captions also usually include many proper nouns [2]. If proper nouns in a caption are converted into a more abstract word (e.g., replacing *Donald Trump* with *A person*), the description becomes less detailed. Furthermore, such image captions usually include few adjectives [2]. Thus, captions in news articles are visually less descriptive. In contrast, captions targeting at visually impaired people would need a higher degree of visual description to be useful. Similarly, an image description used for image retrieval systems relies on a close connection between the visual contents of an image and the resulting caption.

In this research, we aim to generate captions with different levels of visual descriptiveness for such different applications. For this, we introduce the concept of "sentence imageability." The concept of "imageability" originates from Psycholinguistics [19] and describes how easy it is to mentally imagine the meaning or the content of a word. Extending this idea to a sentence allows us to evaluate the visual descriptiveness of a caption. The proposed method generates diverse captions for an image corresponding to a given imageability score as shown in Fig. 1. Each caption is generated so that it contains a different degree of visual information, making them easier or harder to mentally imagine. This intrinsically tailors them to different target applications.

For this, we first augment the image captions in an existing dataset by replacing the words in them. Next, we propose a method to calculate the sentence imageability score based on word-level imageability scores. Then, we modify an existing image captioning model [29] to generate diverse captions according to sentence imageability scores.

The main contributions of this work are as follows:

- Proposal of a novel captioning method that generates captions tailoring to different applications by incorporating the concept of imageability from Psycholinguistics.
- Evaluation of the generated captions in a crowd-sourced fashion, and showing their imageability scores correlate to the mental image of users.

In Sect. 2 we briefly discuss the related work on Psycholinguistics and image captioning. Next, Sect. 3 introduces the proposed method on image captioning considering sentence imageability. We evaluate the proposed method through experiments in Sect. 4 and conclude the paper in Sect. 5.

2   Related Work

We will briefly introduce related work regarding psycholinguistic word ratings and image captioning.

Psycholinguistics: In 1968, Paivio et al. [19] first proposed the concept of imageability which describes the ease or difficulty with which "words arouse a sensory experience", commonly represented as a word rating on the Lickert scale. Existing dictionaries [6,23,24] used in Psycholinguistics are typically created through labor-intensive experiments, often resulting in rather small corpora. For that reason, researchers have been working towards the estimation of imageability or concreteness using text and image data-mining techniques [10,13,17]. Imageability and similar word ratings have been used in multimodal applications like improving the understanding of text-image relationships [30].

Image Captioning: Image captioning is receiving great attention lately thanks to the advances in both computer vision and natural language processing. State-of-the-art models [14,29] commonly take an attention guided encoder-decoder strategy, in which visual information is extracted from images by deep CNNs and then natural language descriptions are generated with RNNs.

In recent research, the goal to generate captions considering sentimental information, which not only contain the visual description of an image but also tailor to specific styles and sentiments, are receiving an increasing attention. Chen et al. [4] and Guo et al. [8] proposed methods to generate captions for combinations of four kinds of stylized captions: humorous, romantic, positive, and negative styles. Mathews et al. [16] considered the semantics and style of captions separately in order to change the style of captions into, e.g., story-like sentences. Most recently, Shuster et al. [25] proposed a method to better engage image captions to humans by incorporating controllable style and personality traits, such as sweet, dramatic, anxious, and so on. While these works aim to control caption styles, some other works [3,5] aim to adjust the contents of the generated captions. However, although they focus on sentence variety, they do not consider the perception or imageability of the output.

Some methods [2,20] targeting news images have been proposed. Similarly, we expect to be able to generate better image captions for news articles by generating captions with low imageability scores. Furthermore, the proposed method can not only generate captions similar to news image captions when targeting lower imageability scores, but also generate captions for other purposes.

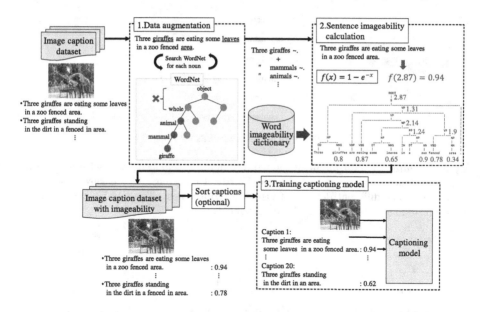

Fig. 2. Training process of the proposed image captioning model.

3 Image Captioning Considering Imageability

In this section, the proposed method is introduced in three steps: (1) Generation of an augmented image caption dataset with a higher variety of captions and different imageability scores for each. (2) Calculation of a sentence imageability score based on word-level imageability scores. (3) Incorporation of a sentence imageability vector into an image captioning model. The training process of the proposed image captioning model is illustrated in Fig. 2.

3.1 Data Augmentation

Existing datasets [12,22] for image captioning commonly provide multiple annotated captions for each image. For example, MSCOCO [12] comes with five distinct captions for each image, differently describing the same contents. However, since our work requires captions with different degrees of visual descriptiveness, the variety of these captions is not sufficient; We require a variety based on different imageability scores, e.g., *human*, *male person*, and *teenage boy* all describe the same object, but result in different degrees of mental clarity if used in a caption. Thus, we augment the sentence variety of existing image-text pairs by replacing words in the original captions. All nouns in a caption are replaced with a selection of hypernyms using the WordNet [18] hierarchy. At most five closest hypernyms are used in order to avoid the word replacements getting too abstract or unrelated to the word from the original caption. Similarly, we do not replace with words too close to the root of the WordNet hierarchy as they

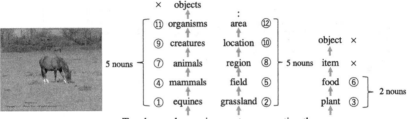

Two brown <u>horses</u> in a <u>pasture</u> are eating the <u>grass</u>.

Fig. 3. Example of data augmentation. Each noun is replaced by a selection of differently abstract words, incorporating the WordNet hierarchy. This ensures a sentence variety with very different degrees of visual descriptiveness. The number next to each hypernym indicates the order of the replacement.

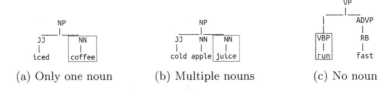

(a) Only one noun   (b) Multiple nouns   (c) No noun

Fig. 4. Three relationship patterns for deciding the most significant words. The square indicates the selected word for each pattern.

get too abstract. By this method, the dataset is augmented to contain a large variety of similar sentences with differently abstract word choices. An example of this method is illustrated in Fig. 3. If there are multiple nouns, the order of replacement occurs as exemplified in the figure. In the experiments, we use this order for sampling a subset of captions in case the augmentation generates too many of them.

3.2 Sentence Imageability Calculation

A sentence imageability score is calculated for each caption in the augmented dataset. Although the concept of word imageability has been part of works such as [19], there are few works that aim to determine the imageability of sentences. To be able to rate the descriptiveness of a caption, we propose to use a sentence imageability score of a caption. As there is no existing method for this, we introduce a way to calculate the sentence imageability score of a caption by using the imageability scores of words composing it. For this method, we assume that the imageability score of a word increases when being modified by other words in the same-level. For example, *coffee* gets less ambiguous when modified by the word *iced* before it (see Fig. 4a).

First, the tree structure of a given sentence is parsed using StanfordCore NLP [15]. We then take a bottom-up approach to calculate the imageability of sub-parts of the sentence based on word imageability scores, normalized to

$[0, 1]$. Along the tree structure, the nodes in the same-level are weighted based on selecting the most significant word (Fig. 4). The significant word is selected according to three relationship patterns: (a) When there is only a single noun at the same depth, it is selected as the significant word for weighting. (b) When there is more than one noun at the same depth, the last one is selected as the significant word. (c) When there is no noun at the same depth, the first word of the sequence is selected as the significant word. Note that stop words and numerals are ignored.

Assuming the imageability score of the most significant word increases by other words modifying it, the imageability score of a sub-tree is calculated with

$$I = x_s \prod_{i=1(\neq s)}^{n} \left(2 - e^{-x_i}\right), \qquad (1)$$

where x_i $(i = 1, \ldots, n \,|\, i \neq s)$ is the score of each word and x_s is the score of the significant word as described above. The resulting score I is used recursively in a bottom-up manner in the parent node. Additionally, when a sub-tree forms a coordinate conjunction, the score I is calculated as the sum of the scores of their nodes. When there is only one node at the same depth of the tree, its score is directly transferred to its parent node. Finally, when reaching the root node of the parsing tree, the score is normalized to the scale of $[0, 1]$ by applying Eq. 2, which represents the imageability score of the sentence.

$$f(x) = 1 - e^{-x} \qquad (2)$$

3.3   Image Captioning

Based on a state-of-the-art captioning model incorporating attention [29], to consider the sentence imageability score of a caption, imageability feature vectors are added. The sentence imageability score for each caption is converted into the same dimensionality as the image feature and caption feature vectors to form an imageability feature vector $\mathbf{A}$. Given a ground-truth caption $\{\mathbf{t}_0, \mathbf{t}_1, \ldots, \mathbf{t}_N\}$, image features $\mathbf{I}_f$ extracted from a pre-trained ResNet network, and an imageability feature $\mathbf{A}$, a caption $c_i := \{\mathbf{w}_0, \mathbf{w}_1, \ldots, \mathbf{w}_N\}$ is generated, where $\mathbf{w}_i$ is a word vector for the i-th word as follows:

$$
\begin{aligned}
\mathbf{x}_t &= W_e \mathbf{w}_{t-1}, \quad t \in \{1, 2, \ldots, N\}, \\
\mathbf{I}_t &= \text{Att}(\mathbf{h}_{t-1}, \mathbf{I}_f), \\
\mathbf{h}_t &= \text{LSTM}(\text{concat}(\mathbf{x}_t, \mathbf{I}_t, \mathbf{A}), \mathbf{h}_{t-1}), \\
\mathbf{w}_t &= \text{softmax}(W_l \mathbf{h}_t),
\end{aligned}
\qquad (3)
$$

where W_e, W_l are learnable parameters, $\mathbf{h}_{t-1}$ is a hidden state of the previous iteration in LSTM. The hidden state $\mathbf{h}_{t-1}$ and the image feature $\mathbf{I}_f$ are input to the Attention Network (Att) and yield the attention weighted vector $\mathbf{I}_t$ of the image. Next, the attention weighted vector $\mathbf{I}_t$, the embedded word vector $\mathbf{x}_t$, and the imageability vector **A** are concatenated. Lastly, the model is trained using the concatenated vector by optimizing Eq. 4, minimizing the cross-entropy loss between the ground-truth word $\mathbf{t}_i$ and the generated word $\mathbf{w}_i$.

$$L = -\sum_{i=0}^{N} q(\mathbf{t}_i) \log p(\mathbf{w}_i), \qquad (4)$$

where p and q are the occurrence probabilities of $\mathbf{w}_i$ and $\mathbf{t}_i$, respectively.

After training, we generate n caption candidates controlled by parameters W_e and W_l, and select the best caption with the smallest Mean Squared Error (MSE) regarding the sentence imageability scores calculated by the method in Sect. 3.2. For generating the image caption, we use the Beam Search which is a search algorithm that stores the top-b sequences (b: beam size) at each time step.

4   Evaluation

We evaluated the proposed method by conducting three experiments. First, the imageability scores of generated captions are evaluated. Second, the generated captions are evaluated by existing image captioning metrics. Lastly, the generated captions are evaluated through a crowd-sourced subjective experiment regarding their actual perception of visual descriptiveness.

4.1   Environment

Word Imageability Dataset: We use the word imageability dictionaries by Scott et al. [24] and Ljubešić et al. [13]. As the latter was predicted through datamining, the former is preferred if a word exists in both of them. Note that captions with a noun not available in the dictionaries are excluded from the datasets.

Image Caption Dataset: We use MSCOCO [12] as the base dataset, which we augment based on the proposed method. Two sampling methods are tested: (1) Sampling based on the order of caption augmentation. It reflects both the order of the nouns in a sentence, and then their respective WordNet hierarchy (*Without Sorting*). (2) Sampling after sorting captions by the calculated sentence imageability scores (*With Sorting*). Here, the augmented captions are sorted by their sentence imageability scores and the captions with the highest and the lowest imageability scores are selected in turn. By this, the model will be trained towards a higher diversity of imageability scores. Note that images with too few captions are excluded from the dataset.

Similarly to prior work, we employ Karpathy-splits [9] resulting in 113,287 images for training and 5,000 images each for validation and testing. After excluding images with an unsufficient number of captions, 109,114 images for training, 4,819 images for validation, and 4,795 images for testing were left.

Table 1. Results of imageability analysis.

Captioning method	Sampling method	Caption Variety ($\uparrow$)	Imageability Range ($\uparrow$)	Average MSE ($\downarrow$)			Average RMSE ($\downarrow$)		
				Low	Mid	High	Low	Mid	High
Proposed	W/o Sorting	**4.68**	0.083	0.405	0.118	**0.011**	0.632	0.334	**0.098**
	With Sorting	4.63	**0.182**	**0.338**	**0.089**	0.014	**0.573**	**0.276**	0.107
Baseline	W/o Sorting	3.50	0.070	0.434	0.131	0.015	0.655	0.354	0.117
	With Sorting	3.26	0.164	0.378	0.103	0.022	0.607	0.300	0.142

Captioning Methods: Using the proposed captioning model, captions for nine levels of imageability in the range of $[0.1, 0.9]$ are generated. We regard the ranges of $[0.1, 0.3]$ as *Low*, $[0.4, 0.6]$ as *Mid*, and $[0.7, 0.9]$ as *High*. For the Beam Search, the beam size is set to $b = 5$. The feature vector has a dimensionality of 512. The proposed method first generates n caption candidates. Next, a candidate with the smallest MSE between the input and the predicted imageability score is chosen. For comparison, a baseline method is prepared, which is a simplified version where a single caption is generated instead of n candidates. We set $n = b$.

4.2 Analysis on the Sentence Imageability Scores

For analyzing the sentence imageability scores, we evaluate the number of unique captions (Caption Variety), the range of imageability scores ($\max - \min$; Imageability Range), as well as the MSE and Root Mean Squared Error (RMSE). The error is calculated between the input and the imageability scores of the generated caption calculated by the method introduced in Sect. 3.2.

The results are shown in Table 1. For all metrics, the results of the proposed method are better than the baseline method. The proposed method generates captions with both a large Caption Variety and a wider Imageability Range. Since the captions generated by Beam Search are all different, there is no identical caption candidate. We select the caption which has the closest imageability score with that of the target, i.e., having the smallest error between the predicted imageability score of the generated caption and the target score. The captions generated like this usually have larger caption variety than the baseline method. Similarly, the proposed method shows smaller MSE and RMSE. For the sampling method, the Imageability Range is wider, and there is smaller error in the low- and the mid-range imageability captions for the *With Sorting* sampling method. On the other hand, the error for high-range imageability captions is smaller for the *Without Sorting* sampling method. As the training data consists of a larger number of long, high-imageability captions, the model favors generating long captions. Due to this, the average error on mid- and high-range imageability scores is lower.

An example of the output of the proposed method is shown in Table 2. While each caption targets a different imageability score, the *Without Sorting* method outputs identical captions. In contrast, the *With Sorting* method produces results resembling the target imageability score.

Table 2. Example of the output of the proposed method corresponding to a given imageability score. The upper half of the generated captions was sampled using the *Without Sorting* method, while the lower half was sampled using the *With Sorting* method in the training phase. For comparison, the caption at the bottom row is generated by a state-of-the-art captioning method [29] not considering imageability.

Image	Imageability	Generated captions
	0.6	A cat laying on top of a device keyboard.
	0.7	A cat laying on top of a device keyboard.
	0.8	A cat laying on top of a device keyboard.
	0.9	A cat laying on top of a device keyboard.
	0.6	A placental is laying on a keyboard on a desk.
	0.7	A vertebrate is laying on a keyboard on a desk.
	0.8	A feline is laying on a keyboard on a desk.
	0.9	A cat is laying on a computer keyboard.
	—	A black and white cat laying next to a keyboard.

Table 3. Results of image captioning metrics.

Captioning method	Sampling method	BLEU-4 ($\uparrow$)			CIDEr ($\uparrow$)			ROUGE ($\uparrow$)			METEOR ($\uparrow$)			SPICE ($\uparrow$)		
		Low	Mid	High	Low	Mid	High	Low	Mid	High	Low	Mid	High	Low	Mid	High
Proposed	W/o Sorting	0.27	0.27	0.26	0.68	0.68	0.68	0.50	0.50	0.50	0.23	0.24	0.24	0.09	0.09	0.09
	With Sorting	0.25	0.27	0.26	0.59	0.50	0.64	0.49	0.50	0.50	0.23	0.23	0.24	0.09	0.09	0.09
Baseline	W/o Sorting	0.28	0.28	0.28	0.71	0.71	0.70	0.51	0.51	0.51	0.24	0.24	0.24	0.09	0.09	0.09
	With Sorting	0.25	0.27	0.28	0.61	0.65	0.65	0.49	0.51	0.51	0.23	0.24	0.24	0.09	0.09	0.09
Comp. [29]	—		0.30			0.91			0.52			0.25			0.18	

4.3 Evaluation of Image Captioning Results

We evaluate the proposed method in the general-purpose image captioning framework. For this, we look at the accuracy of the generated captions through standard metrics for image captioning evaluation, namely BLEU [21], CIDEr [27], ROUGE [11], METEOR [7], and SPICE [1]. For training this model, five captions per image are used, and one caption per image is generated for testing.

The results are shown in Table 3. For comparison, results of a state-of-the-art captioning model which does not consider imageability [29] is shown, which is slightly better than the proposed method. This is because the proposed method focuses on caption diversification. The existing image captioning metrics evaluate the textual similarity to the ground truth, mainly evaluating the linguistic accuracy of the captions. In contrast, the proposed method aims for linguistic diverseness of each caption, intentionally generating different wordings for each caption. Thus, this will naturally decrease the textual similarity, as the generated captions will use different wordings than the ground truth. Following, we aim for a higher diversity of captions (discussed in Sect. 4.2) while maintaining a reasonable captioning quality (discussed here). Therefore, we regard these metrics as not necessarily feasible to evaluate the proposed method. However,

Table 4. Results of subjective evaluation.

(a) Correlation

ρ	#Images
1.0	62 (31%)
0.5	86 (43%)
−0.5	42 (21%)
−1.0	10 (5%)

(b) Examples whose correlation failed

	An organism holding a banana in his hands. An organism holding a banana in his hand. An equipment holding a banana in his hand.
	A structure with a toilet and a sink. An area with a toilet and a sink. A white toilet sitting in a bathroom next to a structure.

the results show similar performance with the general-purpose image captioning method while still considering an additional factor; the imageability of a caption.

4.4 Subjective Evaluation

In order to evaluate the actual perception of the generated captions, we evaluate three captions each for 200 randomly chosen images in a crowd-sourced subjective experiment on the Amazon Mechanical Turk platform[1]. As the majority of generated imageability scores is above 0.5, we focus on the upper half range of imageability scores. To compare how differently generated captions are perceived, we thus uniformly sample three generated captions, resulting in imageability scores of 0.5, 0.7, and 0.9. For each caption pair, we ask 15 English-speaking participants from the US to judge which of the presented two captions has a higher sentence imageability. Note that we do not present the participants the image itself, but rather let them judge the imageability solely based on the textual contents of a caption. Based on the judgments, we rank the three captions for each image using Thurstone's paired comparison method [26].

We calculate the Spearman's rank correlation ρ between the target imageability scores and the actually perceived order obtained by asking participants of the crowd-sourced survey. The average correlation for all images was 0.37, which confirms that the perceived imageability of captions matches relationship between the target values to some extent. To further understand the results, we look into the distribution of the correlation shown in Table 4a. We found that the number of "correctly" selected responses in line with the imageability scores was very high (approx. 65.8%). However, there are a few outliers with strong negative correlations. These results bring down the overall average performance for all images. Table 4b shows outliers whose captions have strong negative correlations. We found that in these cases, the generated captions were not describing the image contents correctly. In other cases, similar captions seem to have prevented the participants to decide which one had higher imageability.

---

[1] https://www.mturk.com/.

5   Conclusion

In this paper we proposed and evaluated an adaptive image captioning method considering imageability. By this method, we aim to control the degree of visual descriptiveness of a caption. For future work, we expect to generate captions with larger variety in terms of imageability. For that, we will try to augment captions in terms of their length as training data.

Acknowledgment. Parts of this research were supported by JSPS KAKENHI 16H02846 and MSR-CORE16 program.

References

1. Anderson, P., Fernando, B., Johnson, M., Gould, S.: SPICE: semantic propositional image caption evaluation. In: Leibe, B., Matas, J., Sebe, N., Welling, M. (eds.) ECCV 2016. LNCS, vol. 9909, pp. 382–398. Springer, Cham (2016). https://doi.org/10.1007/978-3-319-46454-1_24

2. Biten, A.F., Gomez, L., Rusinol, M., Karatzas, D.: Good news, everyone! Context driven entity-aware captioning for news images. In: Proceedings of 2019 IEEE Conference Computer Vision Pattern Recognition, pp. 12466–12475 (2019)

3. Chen, S., Jin, Q., Wang, P., Wu, Q.: Say as you wish: fine-grained control of image caption generation with abstract scene graphs. In: Proceedings of 2020 IEEE Conference Computer Vision Pattern Recognition, pp. 9962–9971 (2020)

4. Chen, T., et al.: "Factual" or "Emotional": stylized image captioning with adaptive learning and attention. In: Ferrari, V., Hebert, M., Sminchisescu, C., Weiss, Y. (eds.) ECCV 2018. LNCS, vol. 11214, pp. 527–543. Springer, Cham (2018). https://doi.org/10.1007/978-3-030-01249-6_32

5. Cornia, M., Baraldi, L., Cucchiara, R.: Show, control and tell: a framework for generating controllable and grounded captions. In: Proceedings of 2019 IEEE Conference Computer Vision Pattern Recognition, pp. 8307–8316 (2019)

6. Cortese, M.J., Fugett, A.: Imageability ratings for 3,000 monosyllabic words. Behav. Res. Methods Instrum. Comput. **36**(3), 384–387 (2004)

7. Denkowski, M., Lavie, A.: METEOR universal: language specific translation evaluation for any target language. In: Proceedings of 2014 EACL Workshop on Statistical Machine Translation, pp. 376–380 (2014)

8. Guo, L., Liu, J., Yao, P., Li, J., Lu, H.: MSCap: multi-style image captioning with unpaired stylized text. In: Proceedings of 2019 IEEE Conference Computer Vision Pattern Recognition, pp. 4204–4213 (2019)

9. Karpathy, A., Li, F.F.: Deep visual-semantic alignments for generating image descriptions. In: Proceedings of 2015 IEEE Conference Computer Vision Pattern Recognition, pp. 3128–3137 (2015)

10. Kastner, M.A., et al.: Estimating the imageability of words by mining visual characteristics from crawled image data. Multimed. Tools Appl. **79**(25), 18167–18199 (2020)

11. Lin, C.Y.: ROUGE: a package for automatic evaluation of summaries. In: Proceedings of 2004 ACL Workshop on Text Summarization Branches Out, pp. 74–81 (2004)

12. Lin, T.-Y., et al.: Microsoft COCO: common objects in context. In: Fleet, D., Pajdla, T., Schiele, B., Tuytelaars, T. (eds.) ECCV 2014. LNCS, vol. 8693, pp. 740–755. Springer, Cham (2014). https://doi.org/10.1007/978-3-319-10602-1_48

13. Ljubešić, N., Fišer, D., Peti-Stantić, A.: Predicting concreteness and imageability of words within and across languages via word embeddings. In: Proceedings of 3rd Workshop on Representation Learning for NLP, pp. 217–222 (2018)
14. Lu, J., Yang, J., Batra, D., Parikh, D.: Neural baby talk. In: Proceedings 2018 IEEE Conference Computer Vision Pattern Recognition, pp. 7219–7228 (2018)
15. Manning, C.D., Surdeanu, M., Bauer, J., Finkel, J., Bethard, S.J., McClosky, D.: The Stanford CoreNLP natural language processing toolkit. In: Proceedings of 52nd Annual Meeting Association for Computational Linguistics: System Demonstrations, pp. 55–60 (2014)
16. Mathews, A., Xie, L., He, X.: Semstyle: Learning to generate stylised image captions using unaligned text. In: Proceedings of 2018 IEEE Conference Computer Vision Pattern Recognition, pp. 8591–8600 (2018)
17. Matsuhira, C., et al.: Imageability estimation using visual and language features. In: Proceedings of 2020 ACM International Conference on Multimedia Retrieval, pp. 306–310 (2020)
18. Miller, G.A.: WordNet: a lexical database for English. Commun. ACM **38**(11), 39–41 (1995)
19. Paivio, A., Yuille, J.C., Madigan, S.A.: Concreteness, imagery, and meaningfulness values for 925 nouns. J. Exp. Psycho. **76**(1), 1–25 (1968)
20. Pan, Y., Yao, T., Li, Y., Mei, T.: X-linear attention networks for image captioning. In: Proceeding of 2020 IEEE Conference on Computer Vision and Pattern Recognition, pp. 10971–10980 (2020)
21. Papineni, K., Roukos, S., Ward, T., Zhu, W.J.: BLEU: a method for automatic evaluation of machine translation. In: Proceedings of 40th Annual Meeting Association for Computational Linguistics, pp. 311–318 (2002)
22. Plummer, B.A., Wang, L., Cervantes, C.M., Caicedo, J.C., Hockenmaier, J., Lazebnik, S.: Flickr30k entities: collecting region-to-phrase correspondences for richer image-to-sentence models. In: Proceedings of 15th IEEE International Conference Computer Vision, pp. 2641–2649 (2015)
23. Reilly, J., Kean, J.: Formal distinctiveness of high-and low-imageability nouns: analyses and theoretical implications. Cogn. Sci. **31**(1), 157–168 (2007)
24. Scott, G.G., Keitel, A., Becirspahic, M., Yao, B., Sereno, S.C.: The Glasgow Norms: Ratings of 5,500 Words on Nine Scales. Springer, Heidelberg (2018)
25. Shuster, K., Humeau, S., Hu, H., Bordes, A., Weston, J.: Engaging image captioning via personality. In: Proceedings of 2019 IEEE Conference on Computer Vision and Pattern Recognition, pp. 12516–12526 (2019)
26. Thurstone, L.L.: The method of paired comparisons for social values. J. Abnorm. Psychol. **21**(4), 384–400 (1927)
27. Vedantam, R., Lawrence Zitnick, C., Parikh, D.: CIDEr: consensus-based image description evaluation. In: Proceedings of 2015 IEEE Conference Computer Vision Pattern Recognition, pp. 4566–4575 (2015)
28. Vinyals, O., Toshev, A., Bengio, S., Erhan, D.: Show and tell: a neural image caption generator. In: Proceedings of 2015 IEEE Conference Computer Vision and Pattern Recognition, pp. 3156–3164 (2015)
29. Xu, K., et al.: Show, attend and tell: neural image caption generation with visual attention. In: Proceedings of 32nd International Conference on Machine Learning, pp. 2048–2057 (2015)
30. Zhang, M., Hwa, R., Kovashka, A.: Equal but not the same: understanding the implicit relationship between persuasive images and text. In: Proceedings of 2018 British Machine Vision Conference, No. 8, pp. 1–14 (2018)

Deep Face Swapping via Cross-Identity Adversarial Training

Shuhui Yang, Han Xue, Jun Ling, Li Song, and Rong Xie[✉]

Shanghai Jiao Tong University, Shanghai, China
{louisxiii,xue_han,lingjun,song_li,xierong}@sjtu.edu.cn

Abstract. Generative Adversarial Networks (GANs) have shown promising improvements in face synthesis and image manipulation. However, it remains difficult to swap the faces in videos with a specific target. The most well-known face swapping method, Deepfakes, focuses on reconstructing the face image with auto-encoder while paying less attention to the identity gap between the source and target faces, which causes the swapped face looks like both the source face and the target face. In this work, we propose to incorporate cross-identity adversarial training mechanism for highly photo-realistic face swapping. Specifically, we introduce corresponding discriminator to faithfully try to distinguish the swapped faces, reconstructed faces and real faces in the training process. In addition, attention mechanism is applied to make our network robust to variation of illumination. Comprehensive experiments are conducted to demonstrate the superiority of our method over baseline models in quantitative and qualitative fashion.

Keywords: Face swapping · Auto-encoder · Cross-identity training · Attention mechanism

1 Introduction

In recent years, face swapping [13,18,19,24] has been becoming an upsurging research topic in computer vision, as well as Deepfakes detection [8,14]. The aim of face swapping is to replace the identity of the source face with that of another target face, while retaining identity-irrelevant visual features such as the illumination, expression and pose. Recently, face swapping has attracted widespread attention due to its application prospects in movie industry (e.g., *The Fast and the Furious*), social applications (e.g., *Snapchat, ZAO*), and face replacement for privacy protection [6,21] (in proper use). As one of the most prevalent face swapping method, Deepfakes [24] was made open-sourced on the web in 2017 and aroused heated discussions. Meanwhile, the advent of Deepfakes has attracted widespread attention from researchers and amateurs.

Supported by the Shanghai Key Laboratory of Digital Media Processing and Transmissions, 111 Project (B07022 and Sheitc No. 150633).

J. Lokoč et al. (Eds.): MMM 2021, LNCS 12573, pp. 74–86, 2021.
https://doi.org/10.1007/978-3-030-67835-7_7

Unlike traditional methods which require professional designer to modify the face identity in an image with substantial efforts by hand, Deepfakes is built by deep learning algorithms to automatically replace faces of a person, without too much data annotation cost. The overall network structure of Deepfakes is based on auto-encoder [5], in particular, one encoder and two decoders. The encoder is required to project an input face into latent embeddings, while the two decoders are designed to restore the source face and the target face from these embeddings, respectively.

Albeit its attractiveness, the auto-encoder-based model primarily focuses on image reconstruction. Specifically, the auto-encoder does not perform face swapping during the training process, therefore, there is no extra guidance for the network to encode meaningful identity-invariant information and decode photo-realistic face images with the target identity. As the case stands, we found that the swapped faces preserve the identity information from both source and target face in most results of Deepfakes, which indicates that the model cannot disentangle identity features well. To address this problem, we introduce a cross-identity adversarial training framework to perform identity supervision over the generated images. To perform cross-identity transfer in the training process, we encode the source image into the latent embeddings in common space and then project it to an image with target decoder, which ensures that we can train an identity-gap-robust face swapping network, not only a face reconstruction network.

Moreover, Deepfakes is inefficient to deal with the fusion boundary of swapped faces(see in Fig. 4, second and third rows), which is caused by various illumination and different skin color. To overcome this issue, we further adopt spatial attention mechanism [20] to improve the robustness of the network to illumination change and skin color variation. In addition, by introducing adversarial training strategy, we enforce the generator to synthesize realistic face images accompanied with more details corresponding to the training corpus. In Sec.5, we provide extensive experiments and demonstrate the improvements of the proposed approach.

2   Related Works

Face attribute manipulation, especially face swapping [13,18,19,24] and face reenactment [15,27] have been studied for decades in the field of computer vision and graphics [23]. The advent of Deepfakes promotes the application of face swapping across areas such as movie production, teleconference, and privacy protection. In recent years, there are two main types of methods for face swapping: 3D-based methods and GAN-based methods.

3D-Based Methods. Early methods adopt predefined 3D parametric models to represent human face and manipulate the identity, pose and expression of a face by modifying the corresponding parameters and render it to a 2D image [1, 2]. Face2Face [22] utilizes 3DMM to realize facial movement transfer from source face to target face. DVP [9] estimates the parameters of face by monocular face reconstruction and exchanges parameters of interest to obtain an intermediate rendering, and a deep neural network is further introduced to generate the refined result.

GAN-Based Methods. With the rapid development of generative adversarial networks [4,17], many methods based on GANs have been proposed for face swapping in recent years. Korshunova et al. [12] proposes to consider the pose and expression of the source face as content, and the identity information of the target face as style, then utilizes the style transfer framework integrating a multi-scale architecture to achieve face swapping. FSGAN [18] first transfers the pose and expression of the target face to the source face, and then designs an inpainting network to refine missing areas that may be caused by the occlusion. Finally, a blending network blends the generated face and the target face. FaceShifter [13] establishes a two-stage framework. In the first stage, the authors design a generator with a novel adaptive attentional denormalization mechanism to combine the identity of source face and the attribute of target face in a multi-feature fusion way. In the second stage, a refinement network is trained in self-supervised manner for error fixing, such as occlusion.

3   Our Approach

This section introduces the proposed approach. To achieve high-fidelity face swapping, we establish a symmetrical framework which comprises two branches for source face and target face, respectively. Each branch contains a generator based on the encoder-decoder architecture and a discriminator based on Patch-GAN [7].

3.1   Network Architecture

Figure 1 shows the overall architecture of our symmetrical framework. For the source and target face, we design the corresponding generator G_s/G_t and discriminator D_s/D_t respectively. The generator adopts encoder-decoder structure, and the source and target encoders share weights. Same as Deepfakes, we perform random warp operation on input source image I_s as data augmentation before feeding it to the encoder. For I_s, the source generator G_s reconstructs the source face and the source discriminator D_s distinguishes whether its input image is real or not. On the other hand, the target generator G_t takes I_s as input to perform cross-identity face swapping, and the swapped face is evaluated by the target discriminator D_t. Symmetrically, the similar process is conducted on target image I_t.

For input image I_s, the generator G_s does not directly generate the fake face, but regresses the attention mask $A_{G_s(I_s)}$ and the color mask $C_{G_s(I_s)}$. Mask $A_{G_s(I_s)}$ determines the contribution of each pixel of the color mask $C_{G_s(I_s)}$ to the final generated image, so that the generator does not need to render the

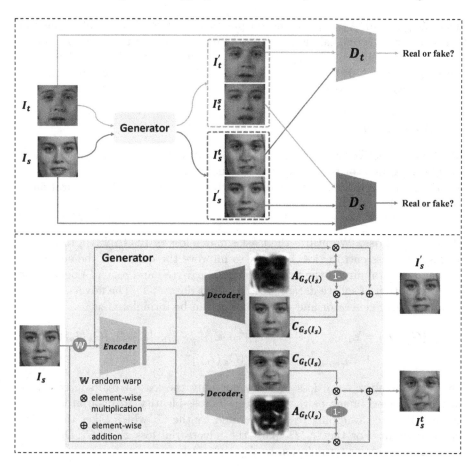

Fig. 1. Overview of the proposed method. Our network contains source and target generator/discriminator pairs, G_s/D_s and G_t/D_t. Note that G_s and G_t share the encoder but have different decoders. The generator regresses the attention mask A and color mask C, and adopts attention mechanism to synthesize the result. The bottom figure shows the training process for source image I_s, G_s tries to reconstruct I_s while G_t is used to generate cross-identity swapped face.

elements in irrelevant region, and only focuses on the pixels we are concerned about, leading to higher quality and more realistic synthesized images. For the input I_s, the reconstructed image generated by G_s can be expressed as:

$$I'_s = G_s(I_s) = A_{G_s(I_s)} \cdot I_{s(w)} + \left(1 - A_{G_s(I_s)}\right) \cdot C_{G_s(I_s)}, \tag{1}$$

where $I_{s(w)}$ represents random warped I_s. Here the discriminator D_s is trained to distinguish between real image I_s and the generated reconstructed image I'_s.

In cross-identity adversarial training branch, we can obtain swapped face by:

$$I^t_s = G_t(I_s) = A_{G_t(I_s)} \cdot I_s + \left(1 - A_{G_t(I_s)}\right) \cdot C_{G_t(I_s)}, \tag{2}$$

where $C_{G_t(I_s)}$ and $A_{G_t(I_s)}$ are the output attention mask and color mask of the generator G_t when taking I_s as input. In order to train an identity-aware face swapping network, the swapped face I_s^t will be evaluated by the target discriminator D_t. This cross-identity adversarial training mechanism is the key to obtain a robust network for identity gap.

3.2   Model Objective

The model objective mainly contains three parts. We design adversarial loss to guide the generator to synthesize high-quality images, reconstruction loss to learn the characteristic of input faces and attention loss to prevent saturation of attention mask.

Adversarial Loss. We utilize the least squares loss of LSGAN [16] to replace the classical cross entropy loss of GAN to improve the quality of the generated image and the stability of the training process. $p_{data}(S)$ and $p_{data}(T)$ denote the data distribution of source dataset S and target dataset T. The adversarial loss of source pair of generator and discriminator can be formulated as:

$$\mathcal{L}_{adv}^s(G_s, D_s) = \mathbb{E}_{I_s \sim p_{data}(S)}\left[D_s(I_s)\right]^2 + \mathbb{E}_{I_s \sim p_{data}(S)}\left[1 - D_s(G_s(I_s))\right]^2 \\ + \mathbb{E}_{I_t \sim p_{data}(T)}\left[1 - D_s(G_s(I_t))\right]^2 \tag{3}$$

Note that our adversarial loss includes not only the loss of real source image and reconstructed image pairs, but also the cross-identity adversarial loss that evaluates the face swapping result generated by the G_s which takes the target face I_t as input. For the G_t/D_t pair, the adversarial loss $\mathcal{L}_{adv}^s(G_t, D_t)$ has a corresponding similar formulation to $\mathcal{L}_{adv}^s(G_t, D_t)$.

Reconstruction Loss. During the training process, G_s/G_t needs to reconstruct the input source/target face. To this end, we use the $L1$ loss as follows:

$$\mathcal{L}_{L1}(G_s, G_t) = \mathbb{E}_{I_s \sim p_{data}(S)}[\|C_{G_s(I_s)} - I_s\|_1] \\ + \mathbb{E}_{I_t \sim p_{data}(T)}\left[\left\|C_{G_t(I_t)} - I_t\right\|_1\right] \tag{4}$$

In order to generate face swapping images that are more in line with the human visual system, we add a structural similarity (SSIM) [26] evaluation to the reconstruction loss. The SSIM loss can be written as:

$$\mathcal{L}_{ssim}(G_s, G_t) = 2 - \mathbb{E}_{I_s \sim p_{data}(S)}[\text{SSIM}(C_{G_s(I_s)}, I_s)] \\ - \mathbb{E}_{I_t \sim p_{data}(T)}[\text{SSIM}(C_{G_t(I_t)}, I_t)] \tag{5}$$

Attention Loss. During training process, we have no direct supervision on the attention mask produced by the generator. Therefore, the attention mask is

(a)   (b)   (c)

Fig. 2. Examples of our dataset. (a) Examples of source face dataset. (b) Examples of target face dataset A. (c) Examples of target face dataset B.

easily saturated to 1, that is, the generator directly uses the input image as the output. To avoid this situation, we regularize the attention mask by:

$$\mathcal{L}_{att}(G_s, G_t) = \mathbb{E}_{I_s \sim p_{data}(S)}\left[\left\|A_{G_s(I_s)}\right\|_2\right] + \mathbb{E}_{I_t \sim p_{data}(T)}\left[\left\|A_{G_t(I_t)}\right\|_2\right] \quad (6)$$

The total loss can be formulated as:

$$\begin{aligned}
\mathcal{L}(G_s, G_t, D_s, D_t) &= \lambda_{L1}\mathcal{L}_{L1}(G_s, G_t) + \lambda_{ssim}\mathcal{L}_{ssim}(G_s, G_t) \\
&+ \lambda_{att}\mathcal{L}_{att}(G_s, G_t) \\
&+ \lambda_{adv}\mathcal{L}_{adv}^s(G_s, D_s) + \lambda_{adv}\mathcal{L}_{adv}^t(G_t, D_t),
\end{aligned} \quad (7)$$

where λ_{L1}, λ_{ssim}, λ_{att} and λ_{adv} are hyper-parameters that control the relative weight of each loss. The training objective can be written as the minmax game:

$$\arg \min_{G_s, G_t} \max_{D_s, D_t} \mathcal{L}(G_s, G_t, D_s, D_t) \quad (8)$$

4   Implementation

Figure 2 shows some samples of the dataset we use. The source images are frames sampled from a music video of Taylor Swift. Meanwhile, we prepare two datasets for the target face, of which dataset A is a photo collection of Scarlett Johnson collected from the Internet, and the other dataset B is image frames sampled from a talking head video. We sample 1/10 from each dataset as the test set, and the remaining images as the train set. In order to demonstrate the robustness of our model to illumination variation, the dataset B of the target person is poorly illuminated and the person's skin appears darker compared to dataset A.

Deepfakes and DeepfaceLab [19] are compared as the baseline model in our experiments. In order to enable a fair comparison, We perform the same pre- and post-processing for different models. In the pre-processing phases, the MTCNN [28] and [25] is applied for face detection and alignment. In the post-processing phase, we use the same color matching and Gaussian blending method to process the raw generated images.

Fig. 3. Comparison of face swapping results on the source dataset with the target dataset A.

In the training phase, we utilize Adam [11] to optimize the network with $\beta_1 = 0.5$ and $\beta_2 = 0.999$. The batch size is set to 8, and all images are resized to 128×128. We train our model for 500 epochs, and perform cross-identity training every 10 iterations. The weight coefficients for the loss terms in Eq. (7) are set to $\lambda_{L1} = 10$, $\lambda_{ssim} = 1$, $\lambda_{att} = 0.01$, $\lambda_{adv} = 0.1$.

5   Experiments and Analysis

In this section, we first compare the proposed model with the state-of-the-art face swapping methods, and verify the superior performance of our method qualitatively and quantitatively. Then we perform ablation study to prove the necessity of our cross-identity training. Finally, we show the experimental results on datasets with larger domain gap, including cross-gender and cross-race scenarios.

5.1   Qualitative Results

Figure 3 shows the face swapping results on the source dataset with the target dataset A. All methods can basically perform face swapping in general. To some extent, the swapped faces all obtain the identity characteristics of the target face, and retain identity-irrelevant information such as the expression and pose of the source face. However, there are obvious artifacts and boundary effects in the images generated by Deepfakes. DeepFaceLab performs better than Deepfakes with sharper edges and more details, but still recovers identity information from source identity, which is inferior to our method. The eyebrow shape of our results is consistent with the target face while other methods incorrectly maintain the eyebrow shape of the source face. Meanwhile, it can be seen that our method can synthesize images with more precise facial details and consistently retain the expression (teeth in second column, closed eyes in fourth column and eye gaze in seventh column) compared to other models.

Fig. 4. Comparison of face swapping results on the source dataset with the target dataset B.

The face swapping results on the source dataset with the target dataset B are illustrated in Fig. 4. It is obvious that the baseline models can hardly handle the illumination and skin color differences between the source and target faces. The post-processing mechanism mentioned above still cannot alleviate the color aberration and boundary effect of swapped faces. Thanks to the attention mechanism, our model can still generate realistic and natural face swapping results, which indicates better adaptability to complex illumination of our model.

5.2   Quantitative Results

For the face swapping task, the most important goal is to preserve the expression of the source face, and obtain the identity of the target face in the swapped result simultaneously. Therefore, we propose two metrics to measure the expression and identity distance between the generated faces and the source/target faces.

Expression Distance. Ekman and Friesen [3] developed the FACS to describe facial expressions in terms of Action Units (AUs), and different expressions of human face can be decomposed into multiple AU combinations. AUs have been applied to facial expression editing task such as GANimation [20]. Here we utilize AU to quantify the facial expressions of the swapped faces and the source faces, and use the Euclidean distance to measure difference between them. We calculate the mean squared error of AU between the source and swapped face as the expression distance ED:

$$ED = \mathbb{E}_{I_s \sim p_{\text{data}}(S_{\text{test}})} \left[\left\| AU_{I_s} - AU_{G_t(I_s)} \right\|_2 \right],\tag{9}$$

where S_{test} denotes test set of source faces, AU_{I_s} and $AU_{G_t(I_s)}$ denote AUs of source face and swapped face respectively.

Table 1. Quantitative comparison results on both target dataset A and B. Baseline models and our model without $\mathcal{L}_{cross}$ are compared, where $\mathcal{L}_{cross}$ denotes the adversarial loss for cross identity training.

Method	$ED \downarrow$		$ID \downarrow$	
	A	B	A	B
Deepfakes	1.0185	2.1055	0.6402	0.5975
DeepFaceLab	0.9762	1.6048	0.6278	0.5807
Ours w/o $\mathcal{L}_{cross}$	0.9279	1.6230	0.6380	0.5811
Ours	**0.9233**	**1.5979**	**0.6136**	**0.5741**

Identity Distance. To measure the identity similarity between the swapped face and target face, we utilize a pretrained face recognition model of DLIB [10]. The pretrained model can embed a face image into an 128-dimensional latent vector, which represent the facial identity features. We compute the average descriptors of target face dataset and the set of swapped faces, then calculate the identity distance ID by:

$$ID = \left\| \mathbb{E}_{t_t \sim p_{\text{data}}(T_{\text{test}})} \left[D_{fr}(I_t) \right] - \mathbb{E}_{I_s \sim p_{\text{data}}(S_{\text{test}})} \left[D_{fr}(G_t(I_s)) \right] \right\|_2,\tag{10}$$

where $D_{fr}(\cdot)$ denotes face recognition descriptor, T_{test} and S_{test} denotes test set of target faces and source faces respectively.

Quantitative comparison results are shown in Table 1. Benefiting from the proposed cross-identity adversarial training strategy, our approach achieves lower ED and ID than baseline models on the source dataset with both target dataset A and B, which proves the superiority of our model in expression preservation and identity transfer.

5.3   Ablation Study

To exploit the importance of cross-identity training, we performed ablation tests on the source dataset with the target dataset A and B with two configurations of our method: without cross-identity training and our full pipeline. The qualitative results are provided in Fig. 5, which shows results of our full model have more consistent identity with the target person. When cross-identity training

Fig. 5. Ablation results. $\mathcal{L}_{cross}$ denotes the adversarial loss for cross identity training. The left and right columns show ablation results on target dataset A and B respectively.

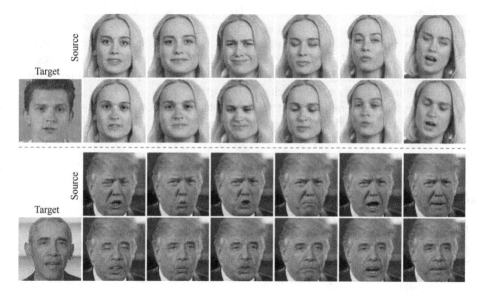

Fig. 6. Face swapping results of difficult cases. The top and bottom rows show cross-gender and cross-race face swapping results respectively.

are not added, the swapped faces on the left are blurry and unnatural, while more unreasonable artifacts appear around the eyebrows and nose regions on the right figure. We also measure the quantitative metric without cross-identity training. According to Table 1, our full model acquires the lower ED and ID, which implies the improvements brought by our proposed cross-identity training mechanism.

5.4 Difficult Cases

In this section, we test our model in more difficult situations including cross-gender (male-female) and cross-race (white-black) face swapping, where the domain gap between the source face and the target face is considerably larger. The face swapping results are shown in Fig. 6. It's easy to find that our model still can generate highly realistic results without noticeable artifacts. The swapped faces maintain the expression of the source face, and identity of the target face is well transferred. Successful swapping under these difficult circumstances demonstrates the robustness of our model to identity diversity.

6 Conclusion

In this paper, we propose a novel face swapping framework which adopt adversarial training for highly photo-realistic face synthesis. To address the identity gap problem of the auto-encoder based method, e.g., Deepfakes, we introduce cross-identity adversarial training strategy and perform identity-aware supervision over the swapped faces by the auxiliary discriminator, leading to identity-aware supervision. In addition, attention mechanism is adopted to improve illumination robustness of our network. Quantitative and qualitative experiments demonstrate that our method outperforms other state-of-the-art models. We further prove the necessity of cross-identity training for bridging large domain gap between source and target faces, and provide highly realistic face swapping results.

References

1. Blanz, V., Scherbaum, K., Vetter, T., Seidel, H.P.: Exchanging faces in images. In: Computer Graphics Forum. vol. 23, pp. 669–676. Wiley Online Library (2004)
2. Dale, K., Sunkavalli, K., Johnson, M.K., Vlasic, D., Matusik, W., Pfister, H.: Video face replacement. In: Proceedings of the 2011 SIGGRAPH Asia Conference, pp. 1–10 (2011)
3. Friesen, E., Ekman, P.: Facial action coding system: a technique for the measurement of facial movement. Palo Alto **3** (1978)
4. Goodfellow, I., et al.: Generative adversarial nets. In: Advances in Neural Information Processing Systems, pp. 2672–2680 (2014)
5. Hinton, G.E., Salakhutdinov, R.R.: Reducing the dimensionality of data with neural networks. Science **313**(5786), 504–507 (2006)
6. Hukkelås, H., Mester, R., Lindseth, F.: DeepPrivacy: a generative adversarial network for face anonymization. In: Bebis, G., et al. (eds.) ISVC 2019. LNCS, vol. 11844, pp. 565–578. Springer, Cham (2019). https://doi.org/10.1007/978-3-030-33720-9_44
7. Isola, P., Zhu, J.Y., Zhou, T., Efros, A.A.: Image-to-image translation with conditional adversarial networks. In: Proceedings of the IEEE Conference on Computer Vision and Pattern Recognition, pp. 1125–1134 (2017)

8. Jiang, L., Li, R., Wu, W., Qian, C., Loy, C.C.: Deeperforensics-1.0: a large-scale dataset for real-world face forgery detection. In: Proceedings of the IEEE/CVF Conference on Computer Vision and Pattern Recognition, pp. 2889–2898 (2020)

9. Kim, H., et al.: Deep video portraits. ACM Trans. Graph. (TOG) **37**(4), 1–14 (2018)

10. King, D.E.: Dlib-ml: a machine learning toolkit. J. Mach. Learn. Res. **10**, 1755–1758 (2009)

11. Kingma, D.P., Ba, J.: Adam: a method for stochastic optimization. arXiv preprint arXiv:1412.6980 (2014)

12. Korshunova, I., Shi, W., Dambre, J., Theis, L.: Fast face-swap using convolutional neural networks. In: Proceedings of the IEEE International Conference on Computer Vision, pp. 3677–3685 (2017)

13. Li, L., Bao, J., Yang, H., Chen, D., Wen, F.: Faceshifter: towards high fidelity and occlusion aware face swapping. arXiv preprint arXiv:1912.13457 (2019)

14. Li, L., et al.: Face x-ray for more general face forgery detection. In: Proceedings of the IEEE/CVF Conference on Computer Vision and Pattern Recognition, pp. 5001–5010 (2020)

15. Ling, J., Xue, H., Song, L., Yang, S., Xie, R., Gu, X.: Toward fine-grained facial expression manipulation. In: Vedaldi, A., Bischof, H., Brox, T., Frahm, J.-M. (eds.) ECCV 2020. LNCS, vol. 12373, pp. 37–53. Springer, Cham (2020). https://doi.org/10.1007/978-3-030-58604-1_3

16. Mao, X., Li, Q., Xie, H., Lau, R.Y., Wang, Z., Paul Smolley, S.: Least squares generative adversarial networks. In: Proceedings of the IEEE International Conference on Computer Vision, pp. 2794–2802 (2017)

17. Mirza, M., Osindero, S.: Conditional generative adversarial nets. arXiv preprint arXiv:1411.1784 (2014)

18. Nirkin, Y., Keller, Y., Hassner, T.: FSGAN: subject agnostic face swapping and reenactment. In: Proceedings of the IEEE International Conference on Computer Vision, pp. 7184–7193 (2019)

19. Petrov, I., et al.: DeepFaceLab: a simple, flexible and extensible face swapping framework. arXiv preprint arXiv:2005.05535 (2020)

20. Pumarola, A., Agudo, A., Martinez, A.M., Sanfeliu, A., Moreno-Noguer, F.: GAN-imation: anatomically-aware facial animation from a single image. In: Ferrari, V., Hebert, M., Sminchisescu, C., Weiss, Y. (eds.) ECCV 2018. LNCS, vol. 11214, pp. 835–851. Springer, Cham (2018). https://doi.org/10.1007/978-3-030-01249-6_50

21. Sun, Q., Tewari, A., Xu, W., Fritz, M., Theobalt, C., Schiele, B.: A hybrid model for identity obfuscation by face replacement. In: Ferrari, V., Hebert, M., Sminchisescu, C., Weiss, Y. (eds.) ECCV 2018. LNCS, vol. 11205, pp. 570–586. Springer, Cham (2018). https://doi.org/10.1007/978-3-030-01246-5_34

22. Thies, J., Zollhofer, M., Stamminger, M., Theobalt, C., Nießner, M.: Face2face: real-time face capture and reenactment of RGB videos. In: Proceedings of the IEEE Conference on Computer Vision and Pattern Recognition, pp. 2387–2395 (2016)

23. Tolosana, R., Vera-Rodriguez, R., Fierrez, J., Morales, A., Ortega-Garcia, J.: Deepfakes and beyond: a survey of face manipulation and fake detection. arXiv preprint arXiv:2001.00179 (2020)

24. Tordzf, Andenixa, K.: Deepfakes/faceswap. https://github.com/deepfakes/faceswap

25. Umeyama, S.: Least-squares estimation of transformation parameters between two point patterns. IEEE Trans. Pattern Anal. Mach. Intell. **4**, 376–380 (1991)

26. Wang, Z., Bovik, A.C., Sheikh, H.R., Simoncelli, E.P.: Image quality assessment: from error visibility to structural similarity. IEEE Trans. Image Process. **13**(4), 600–612 (2004)
27. Xue, H., Ling, J., Song, L., Xie, R., Zhang, W.: Realistic talking face synthesis with geometry-aware feature transformation. In: 2020 IEEE International Conference on Image Processing (ICIP), pp. 1581–1585. IEEE (2020)
28. Zhang, K., Zhang, Z., Li, Z., Qiao, Y.: Joint face detection and alignment using multitask cascaded convolutional networks. IEEE Sig. Process. Lett. **23**(10), 1499–1503 (2016)

Res2-Unet: An Enhanced Network for Generalized Nuclear Segmentation in Pathological Images

Shuai Zhao[1,2], Xuanya Li[3], Zhineng Chen[2], Chang Liu[2],
and Changgen Peng[1(✉)]

[1] Guizhou Key Laboratory of Public Big Data, Guizhou University, Guiyang, China
`cgpeng@gzu.edu.cn`
[2] Institute of Automation, Chinese Academy of Sciences, Beijing, China
{`shuai.zhao,zhineng.chen`}`@ia.ac.cn`
[3] Baidu Inc., Beijing, China
`lixuanya@baidu.com`

Abstract. The morphology of nuclei in a pathological image plays an essential role in deriving high-quality diagnosis to pathologists. Recently, deep learning techniques have pushed forward this field significantly in the generalization ability, i.e., segmenting nuclei from different patients and organs by using the same CNN model. However, it remains challenging to design an effective network that segments nuclei accurately, due to their diverse color and morphological appearances, nuclei touching or overlapping, etc. In this paper, we propose a novel network named Res2-Unet to relief this problem. Res2-Unet inherits the contracting-expansive structure of U-Net. It is featured by employing advanced network modules such as the residual and squeeze-and-excitation (SE) to enhance the segmentation capability. The residual module is utilized in both contracting and expansive paths for comprehensive feature extraction and fusion, respectively. While the SE module enables selective feature propagation between the two paths. We evaluate Res2-Unet on two public nuclei segmentation benchmarks. The experiments show that by equipping the modules individually and jointly, performance gains are consistently observed compared to the baseline and several existing methods.

Keywords: Digital pathology · Nuclear segmentation · U-Net · Convolutional neural network

1 Introduction

With the advances of computing technologies, digital pathology, which scans biopsy samples to high-resolution digital pathological images and uses computational methods to analyze them, is becoming one of the most popular research

---

S. Zhao and X. Li—Have contributed equally to this work.

© Springer Nature Switzerland AG 2021
J. Lokoč et al. (Eds.): MMM 2021, LNCS 12573, pp. 87–98, 2021.
https://doi.org/10.1007/978-3-030-67835-7_8

topics in the community of computer-aided diagnosis. Nuclear segmentation is a primary digital pathology task that aims to extract the contour of every nuclear appearing in a pathological image or image ROI. Recently, the computational method has shown significant advantages in this task compared to the human eye [15]. Compared with other visual tasks, the particularity of cell nucleus segmentation mainly lies in that nuclei are small in size but large in number and with complex spatial context. For example, dye unevenly stained, nuclei touching or overlapping, etc. [3,4]. It makes the quantification of cell density and morphology a challenging task. The segmentation can provide critical clinical parameters that greatly facilitate the cancer diagnosis as well as the following treatment if they were accurately measured.

As far as we know, it is still a complicated task to accurately segment the nuclei across patients and organs by computational methods. The challenges mainly come from three aspects. First, the human-involved staining process introduces variability in color, contrast, shape, etc, which diversifies cell nuclei across patients. Second, touching or overlapping nuclei blurs the contours of themselves, leading to instance boundaries that are hard to distinguish. It also easily results in over- or under-segmentation [6]. Third, tumor heterogeneity makes cells exhibiting differently within and across cell types. It again brings wide and diverse variances in nuclear appearance, magnitude and cell spatial distribution for tissues sampled from different organs [12]. It still remains difficult to find a method that deals with all these challenges well.

Many works for nuclear segmentation have been proposed in the literature. In early years, the methods mainly adopted traditional image analysis techniques such as morphological processing, edge and contour analyses, and their variants along with dedicated pre- and post-processing techniques [1,2,25]. They achieved success in several organs such as prostate and breast but their generalization ability is less considered. In other words, the methods were dataset-specific and need to be modified when migrating to other organs. Machine learning based methods generally gave better generalization capabilities because they could be trained to accommodate nuclear variations, especially for those using deep learning [5]. Deep learning methods automatically learn discriminative features from the raw image to separate background and nuclear foreground. By directly learning to segment the two kinds of pixels, they attained decent performance [7, 26]. Besides, in [23], a deep-based method to detect nuclear centers was proposed. It tried to estimate the probability of a pixel being a nuclear center rather than the nuclear foreground, as only the center-level annotation was available. With a large and diverse training set containing over tens of thousands nuclear boundaries, Kumar et al. [12] formulated the nuclear segmentation as a pixelwise classification problem, being classified as background, foreground or a third class of nuclear boundary. The method attained satisfactory performance and was likely to generalize well among different patients and organs. By leveraging more advanced deep models such as Mask R-CNN [8] and U-Net and its variants [11, 21,27], recently several studies further improved the segmentation performance without loss the generalization ability, given that enhanced nuclear features were extracted [17].

Inspired by previous encouraging works, in this paper we develop a novel U-Net like network, termed as Res2-Unet, that aims at improving the performance of generalized nuclear segmentation. Res2-Unet enjoys the encode-decode structure. It is featured by improving the feature extraction and utilization from three aspects, i.e., encoder, decoder and the skip connection. Specifically, the encoder of Res2-Unet employs advanced residual modules (ERes) to obtain the enhanced features, which robustly extracts generalized nuclear features with rich boundary information. The decoder side is also equipped with a slightly different residual based module (DRes), aiming to perform more comprehensive feature fusion. Meanwhile, instead of flat passing all feature maps from the encoder to decoder side, we adopt an attention based mechanism that employs the SE block to enable a more targeted feature propagation through the skip connection. It has the merits of enhancing the weight of useful feature maps while downweighting redundant feature maps. With these meaningful enhancements, Res2-Unet is deemed to be generalized well in the nuclear segmentation task. To verify our hypothesis, we conduct experiments on two public nuclear segmentation datasets, i.e., MoNuSeg [12] and TN-BC [6], which are both widely used in this specific task. The experimental results show that performance gains are consistently observed when equipping the modules individually and jointly. Res2-Unet also outperforms several popular existing methods.

2   Proposed Method

In the training process, color normalization based pre-processing is performed at first and then image patches are extracted. The patches and their corresponding label masks serve as the network input and output, respectively, from which a dedicated Res2-Unet is trained to segment nuclei. While in the inference process, the same pre-processing is applied and the trained Res2-Unet is employed to perform nuclear segmentation on unseen images. Then, certain post-processing is leveraged to split the nuclear foreground into instances. We will elaborate on these building blocks as follows.

2.1   Pre-processing and Patch Extraction

There are a large variations in image color due to H&E reagents, staining process, sensor response, etc [22,24]. To reduce color differences, we employ the method proposed by Vahadane et al. [24]. It first decomposes the image into sparse and non-negative stain density maps. Then the maps are combined with a pathologist-specified color basis to generate the normalized image. For more details about the method refer to [24]. Several images before and after color normalization are listed in the first and second row of Fig. 3, respectively.

Another practical issue is that the size of a pathological image is too large to directly feed into the GPU for model training. Typically, the image is split into a number of overlapping image patches of several hundred to bypass the memory constraint. The split detail will be explained in the experiments.

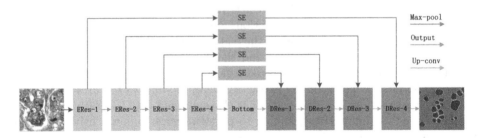

Fig. 1. An overview of the proposed Res2-Unet for nuclear segmentation.

2.2 Res2-Unet

As mentioned above, Res2-Unet improves U-Net from encoder, decoder and the skip connections. Its structure is presented in Fig. 1. Specifically, we use ResNet-50 [9] as the encoder. It is deeper than the convolutional layers used in U-Net, thus beneficial to get richer nuclear feature representation. The encoder is composed of one 7×7 convolutional layer with stride 2 (we omit it in Fig. 1 for simplicity) and four *ERes-N* modules. The four *ERes-N* modules repeat ERes block (see Fig. 2(a)) 3, 4, 6, 3 times, respectively. Each module is followed by a max pooling of stride 2 for downsampling and then a skip connection to connect encoder and decoder. The ERes block consists of a sequence of 1×1, 3×3 and 1×1 convolutional layers, each succeeded by a batch normalization and a ReLU nonlinearity unit. The channels of outputted feature maps is $256 \times N$ for *Res-N*. The encoder shrinks the image size $32\times$ in total.

Res2-Unet appends SE block (see Fig. 2(b)) on the skip connection to generate an attention based feature propagation. As we known, U-Net sent all feature maps to the decoder side in a flat way, where important and less valuable features are equally treated. Inspired by the SE mechanism [10], we add it to drive an attention based channel-wised feature re-weighting. It learns to increase the weight of important feature channels while reducing the redundant channels during the model training. Therefore features are better utilized.

The decoder has an almost symmetric structure as the encoder except for the following differences. First, it enjoys a deconvolution layer for $2\times$ upsampling along adjacent *DRes-N* modules. Second, a DRes block is devised (see Fig. 2(c)). It concatenates features from both the previous layer and lateral skip connection, followed by a block quite similar to ERes except that the last layer undergoes a 3×3 convolution rather than 1×1. A larger filter size is good for feature fusion along the spatial dimension. At last, the last DRes module, i.e., *DRes-4* experiences a deconvolution operation to generate three segmentation maps, corresponding to background, nuclear foreground and boundary, respectively.

2.3 Post-processing

As for the post-processing, we employ the method proposed by [12] to further instantiate the segmentation maps into nuclear instances. The result is shown in the forth row of Fig. 3.

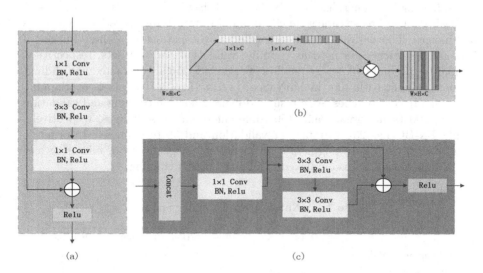

Fig. 2. Details of modules in Res2-Unet, (a) the ERes block, (b) the Squeeze-Excitation (SE) block, (c) the DRes block.

2.4 Loss Function

In deep learning based semantic segmentation, most methods [14,21] use cross entropy (CE) as the loss function. To better distinguish the challenging boundary pixels in our context, we upgrade the CE loss to Focal loss (FL) [13], which increases the loss of hard-to-distinguish pixels. The formula of FL is given by

$$FL = -\sum_{i=1}^{N} \alpha (1 - p_i)^{\gamma} \log(p_i) \tag{1}$$

where α and γ are both hyper-parameters adjusted during the training process. p_i is the predicted probability. Note that when setting α to 1 and γ to 0, FL is equivalent to the CE.

3 Experiments

3.1 Dataset and Implementation Detail

We evaluate Res2-Unet on two public datasets. The first is MoNuSeg collected from The Cancer Genomic Atlas (TCGA) [12,17]. It samples one image per patient to maximize variation from the patient side. It contains 30 H&E stained images of size 1000 × 1000 from 7 different organs (Breast, Liver, Kidney, Prostate, Bladder, Colon, Stomach) with a total of 21,623 annotated nuclei, 13,372 for training and 8,251 for test. The dataset detail is given in Table 1. It has two test sets, i.e., same organ test and different organ test that emphasizes on evaluating the generalization ability. The other is TN-BC [16]. It consists of 33 H&E stained breast cancer images of size 512 × 512 from 7 patients, each has 3–8 images. The dataset is split according to the patient ID, where images of 5, 1 and 1 patients are used as training, validation and test sets, respectively. Similar to [12] and [16], image patches of size 256 × 256 are extracted in a sliding-window manner with 100 (in MoNuSeg) or 128 (in TN-BC) pixels overlapping between adjacent patches. As a result, MoNuSeg has 400 training and validation images, 200 (same organ) and 150 (different organ) test images, respectively. While TN-BC has 234 training, 27 validation and 36 test images. In order to avoid over-fitting, data augmentation including image rotation, mirroring and elastic transformation is employed during the model training.

Table 1. MoNuSeg split for training and test

	Breast	Liver	Kidney	Prostate	Bladder	Colon	Stomach	**Total**
Training and validation	4	4	4	4	0	0	0	16
Same organ test	2	2	2	2	0	0	0	8
Different organ test	0	0	0	0	2	2	2	6
Total	6	6	6	6	2	2	2	30

Res2-Unet is implemented by using TensorFlow 2.0. The training process lasts 200 epochs with a batch-size of 16 images. Adam is used for training and the learning rate is set to 0.001. The experiments are carried out on a workstation with one NVIDIA TITAN Xp GPU. We plan to reproduce Res2-Unet in PaddlePaddle, an open source deep learning framework, in the near future.

3.2   Experiments on MoNuSeg

We compare Res2-Unet with Fiji [20], an open-source software, CNN3 [12] and DIST [17], two deep learning methods that reported encouraging results on MoNuSeg. Fiji is a Java-based software utilizing the watershed algorithm to segment nuclear to instance. CNN3 was a famous method that treated nuclear segmentation as a pixel-level three-class classification task. DIST proposed a new loss function based on CNN3 to better handle the overlapping area of nuclear, and obtained better segmentation results.

As for the evaluation metrics, we adopt Dice and Aggregated Jaccard Index (AJI) [12] to evaluate the performance. Dice is a commonly used pixel-level metric in biomedical image segmentation. While AJI assesses nuclear segmentation algorithms from both target- (nucleus detection) and pixel-level (nucleus shape and size) aspects. It is more in accordance with human perception. Readers can refer to [12] for more details.

Table 2. Performance comparison of different methods on MoNuSeg test data

Organ	Image	Fiji [20]		CNN3 [12]		DIST [17]		Res2-Unet	
		AJI	Dice	AJI	Dice	AJI	Dice	AJI	Dice
Breast	1	0.2772	0.5514	0.4974	0.6885	0.5334	**0.7761**	**0.5490**	0.7685
	2	0.2252	0.5957	0.5769	0.7476	0.5884	0.8380	**0.6301**	**0.8415**
Kidney	1	0.2429	0.5676	0.4792	0.6606	0.5648	0.7805	**0.6315**	**0.8148**
	2	0.3290	0.7089	0.6672	0.7837	**0.5420**	**0.7606**	0.5399	0.7510
Liver	1	0.2539	0.4989	0.5175	0.6726	**0.5466**	**0.7877**	0.4982	0.7788
	2	0.2826	0.5711	0.5148	0.7036	0.4432	0.6684	**0.5162**	**0.7259**
Prostrate	1	0.2356	0.6915	0.4914	**0.8306**	0.6273	0.8030	0.5749	0.7635
	2	0.1592	0.5888	0.3761	0.7537	**0.6294**	**0.7903**	0.6233	0.7884
Bladder	1	0.3730	0.7949	0.5465	**0.9312**	0.6475	0.8623	0.6333	0.8617
	2	0.2149	0.5128	0.4968	0.6304	**0.5467**	0.7768	0.5113	**0.7826**
Stomach	1	0.3757	0.8663	0.4538	**0.8913**	0.6408	0.8547	0.5699	0.8168
	2	0.3586	0.8428	0.4378	**0.8982**	0.6550	0.8520	0.5866	0.8264
Colon	1	0.2295	0.6381	0.4891	0.7679	0.4240	0.7212	**0.5565**	**0.7714**
	2	0.2685	0.6620	**0.5692**	0.7118	0.4484	0.7360	0.4869	**0.7559**
Total		0.2733	0.6493	0.5083	0.7623	0.5598	0.7863	**0.5648**	**0.7890**

Table 2 shows the performance of the refereed methods. Res2-Unet attains the best performance among the compared methods, indicating that it is highly competitive in both instance-level (AJI) and pixel-level (Dice) evaluations. As depicted in Fig. 3, MoNuSeg has complex color appearances and perplexing morphological patterns, even experienced pathologists cannot distinguish the nuclei well. This also highlights the challenge of generalized nuclear segmentation. Encouragingly, it is observed that large performance variant exists even for images of the same organ, e.g., Liver. Moreover, besides those organs appearing in the training data, Res2-Unet also exhibits advantages on those organs that the training set has not involved, i.e., Bladder, Stomach, Colon. It exhibits a good generalization ability. Several segmentation results are shown in the fourth row of Fig. 3.

3.3 Experiments on TN-BC

We then carry out experiments on TN-BC. Existing methods mainly adopt pixel-level metrics including accuracy, F1 score, IoU to evaluate their algorithms in this dataset. In order to compare with the methods, we keep these metrics for

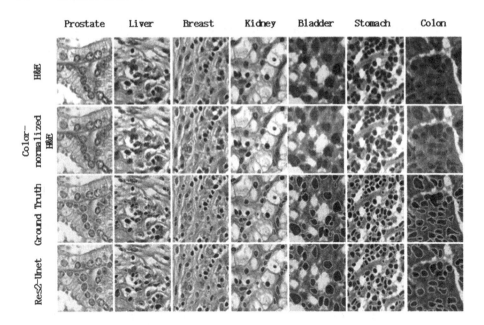

Fig. 3. Segmentation results on MoNuSeg. The first, second, third, fourth rows represents image patches sampled from different organs, images after color normalization, their ground truth, the segmentation results predicted by our Res2-Unet, respectively.

evaluation. Besides, we add AJI as a supplied target-level assessment. The results of Res2-Unet and four existing methods on TN-BC test data are given by Table 3.

As can be seen, despite apparent visual differences appearing between the two datasets, Res2-Unet also get quite competitive results on TN-BC. It attains the highest accuracy and F1 score, while its IoU is also quite well. We also list two illustrative segmentation examples in Fig. 4. The images exhibit different color distribution and spatial context with those on MoNuSeg. However, Res2-Unet is still able to identify them satisfactorily. Note that this is achieved by just using the TN-BC training data to retrain the network without any modification on training strategy and network structure. It again demonstrates the generalization capability of Res2-Unet.

Table 3. Segmentation results on TN-BC

	Accuracy	F1	IoU	AJI
PangNet [19]	0.924	0.676	0.722	–
FCN [14]	0.944	0.763	0.782	–
DeconvNet [18]	0.954	0.805	**0.814**	–
Ensemble [25]	0.944	0.802	0.804	–
Res2-Unet	**0.955**	**0.809**	0.811	**0.602**

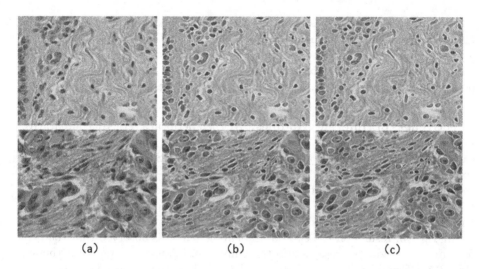

(a)   (b)   (c)

Fig. 4. Illustrative segmentation results on TN-BC. (a) raw image, (b) ground truth, (c) prediction of Res2-Unet.

3.4 Ablation Study

To better understand Res2-Unet, we conduct controlled experiments on both MoNuSeg and TN-BC datasets. All employed experiments are under the same setting except for the specified changes in different comparisons.

Loss Function: FL V.S. CE. To quantify the difference of using FL instead of CE as the loss function, experiments are conducted on the two datasets and Table 4 (Left) gives their AJI scores. As anticipated, FL leads to performance improvement compared to CE in both datasets. It can be explained as that the network learning process is better guided by FL compared to CE. It emphasizes on identifying hard samples, e.g., pixels near to the nuclear boundary, thus beneficial to a more obvious separation to different kinds of pixels.

Table 4. AJI score under different settings. Left: using cross-entropy (CE) and focal loss (FL). Right: with pre-pocessing (w Pre) and without pre-processing (w/o Pre).

	MoNuSeg	TN-BC
CE	0.5613	0.5943
FL	0.5648	0.6020

	MoNuSeg	TN-BC
w/o Pre	0.5562	0.6020
w Pre	0.5648	–

Effectiveness of Pre-processing. We carry out experiments to verify the effectiveness of the color normalization based pre-processing. The results are listed in Table 4 (right). It is seen that the color normalization leads to nearly 1.5% performance improvement on MoNuSeg, attributed to the reduced color diversity. Similar to [16], pre-processing on TN-BC is not applied, as it contains normal epithelial and myoepithelial mammary glands. It would generate quite poor normalization performance for some TN-BC images.

Effectiveness of Different Modules. We then perform a series of comparative experiments to evaluate the effectiveness of the three upgraded modules, i.e., encoder (EN), decoder (DE) and skip connection (SC). Specifically, the following counterparts are taken into account. Res2-EN: U-Net plus with the upgraded encoder. Res2-EN-DE: U-Net plus with the upgraded encoder and decoder. Res2-EN-SC: U-Net plus with the upgraded encoder and skip connection. And Res2-Unet that jointly upgrades all the three modules.

In Table 5, results on the two datasets are given. It is observed that the performance is steadily increased with the number of modules upgraded. Although Res2-EN-DE and Res2-EN-SC behave differently between the two datasets. They reach a consensus on Res2-Unet that the best performance is achieved. We visualize two images in Fig. 5 by using the methods above. It vividly illustrates how the better segmentation result is gradually reached. By step-wisely evaluating the building blocks, we verify that every up-gradation is necessary and contributing to a better segmentation performance.

Table 5. Ablation comparison results on MoNuSeg (left) and TN-BC (right)

Method	AJI	Dice
Res2-EN	0.5236	0.7820
Res2-EN-DE	0.5427	0.7860
Res2-EN-SC	0.5577	**0.7902**
Res2-Unet	**0.5648**	0.7890

Method	Accuracy	F1
Res2-EN	0.9362	0.7415
Res2-EN-DE	0.9448	0.7664
Res2-EN-SC	0.9375	0.7862
Res2-Unet	**0.9558**	**0.809**

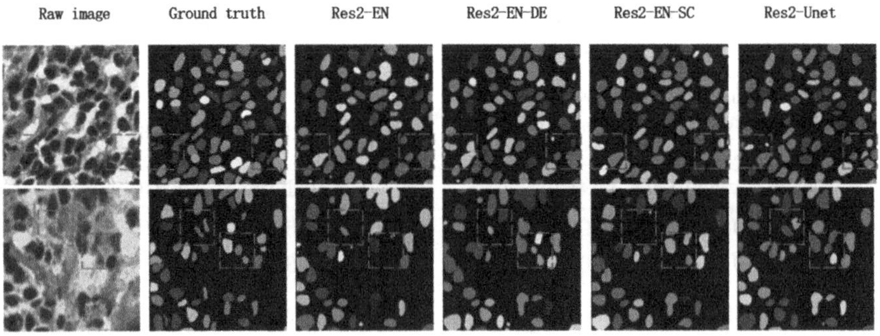

Fig. 5. Visualization of nuclear segmentation results of different methods.

4   Conclusion

We have presented the Res2-Unet to relief the difficulty of generalized nuclear segmentation in pathological images. It leverages the residual and SE blocks to improve the feature extraction and fusion capabilities. The experiments conducted on MoNuSeg and TN-BC basically validate our proposal. Steady performance improvements are observed when evaluating the methods step-by-step. It also outperforms several popular solutions. Our study indicates the generalized nuclear segmentation task can be benefited from advanced network modules.

Acknowledgements. This work was supported by National Key R&D Program of China (No. 2019YFC1710404), the Natural Science Foundation of China (Nos. U1836205, 61772526, 61662009), the Foundation of Guizhou Provincial Key Laboratory of Public Big Data (No. 2019BDKFJJ013), and Baidu Open Research Program.

References

1. Al-Kofahi, Y., Lassoued, W., Lee, W., Roysam, B.: Improved automatic detection and segmentation of cell nuclei in histopathology images. IEEE Trans. Biomed. Eng. **57**(4), 841–852 (2010)
2. Ali, S., Madabhushi, A.: An integrated region-, boundary-, shape-based active contour for multiple object overlap resolution in histological imagery. IEEE Trans. Med. Imaging **31**(7), 1448–1460 (2012)
3. Beck, A.H., et al.: Systematic analysis of breast cancer morphology uncovers stromal features associated with survival. Sci. Transl. Med. **3**(108), 108ra113–108ra113 (2011)
4. Chang, H., Han, J., Borowsky, A., Loss, L., Gray, J.W.: Invariant delineation of nuclear architecture in glioblastoma multiforme for clinical and molecular association. IEEE Trans. Med. Imaging **32**(4), 670–682 (2012)
5. Chen, Z., Ai, S., Jia, C.: Structure-aware deep learning for product image classification. ACM Trans. Multimed. Comput. Commun. Appl. **15**(1s), 4:1–20 (2019)
6. Cheng, J., Rajapakse, J.C., et al.: Segmentation of clustered nuclei with shape markers and marking function. IEEE Trans. Biomed. Eng. **56**(3), 741–748 (2008)
7. Graham, S., et al.: Hover-Net: simultaneous segmentation and classification of nuclei in multi-tissue histology images (2018)
8. He, K., Gkioxari, G., Dollár, P., Girshick, R.: Mask R-CNN. In: Proceedings of the IEEE International Conference on Computer Vision, pp. 2961–2969 (2017)
9. He, K., Zhang, X., Ren, S., Sun, J.: Deep residual learning for image recognition. In: Proceedings of the IEEE Conference on Computer Vision and Pattern Recognition, pp. 770–778 (2016)
10. Hu, J., Shen, L., Sun, G.: Squeeze-and-excitation networks. In: Proceedings of the IEEE Conference on Computer Vision and Pattern Recognition, pp. 7132–7141 (2018)
11. Hu, K., Zhang, Z., Niu, X., Zhang, Y., Cao, C., et al.: Retinal vessel segmentation of color fundus images using multiscale convolutional neural network with an improved cross-entropy loss function. Neurocomputing **309**, 179–191 (2018)
12. Kumar, N., Verma, R., Sharma, S., Bhargava, S., Vahadane, A., Sethi, A.: A dataset and a technique for generalized nuclear segmentation for computational pathology. IEEE Trans. Med. Imaging **36**(7), 1550–1560 (2017)

13. Lin, T.Y., Goyal, P., Girshick, R., He, K., Dollár, P.: Focal loss for dense object detection. In: Proceedings of the IEEE International Conference on Computer Vision, pp. 2980–2988 (2017)
14. Long, J., Shelhamer, E., Darrell, T.: Fully convolutional networks for semantic segmentation. In: Proceedings of the IEEE Conference on Computer Vision and Pattern Recognition, pp. 3431–3440 (2015)
15. Louis, D.N., et al.: Computational pathology: a path ahead. Arch. Pathol. Lab. Med. **140**(1), 41–50 (2015)
16. Naylor, P., Lae, M., Reyal, F., Walter, T.: Nuclei segmentation in histopathology images using deep neural networks. In: 2017 IEEE 14th International Symposium on Biomedical Imaging (ISBI 2017), pp. 933–936. IEEE (2017)
17. Naylor, P., Laé, M., Reyal, F., Walter, T.: Segmentation of nuclei in histopathology images by deep regression of the distance map. IEEE Trans. Med. Imaging **38**(2), 448–459 (2018)
18. Noh, H., Hong, S., Han, B.: Learning deconvolution network for semantic segmentation. In: Proceedings of the IEEE International Conference on Computer Vision, pp. 1520–1528 (2015)
19. Pang, B., Zhang, Y., Chen, Q., Gao, Z., Peng, Q., You, X.: Cell nucleus segmentation in color histopathological imagery using convolutional networks. In: 2010 Chinese Conference on Pattern Recognition (CCPR), pp. 1–5. IEEE (2010)
20. Perez, J.M.M., Pascau, J.: Image Processing with ImageJ. Packt Publishing, Birmingham (2016)
21. Ronneberger, O., Fischer, P., Brox, T.: U-Net: convolutional networks for biomedical image segmentation. In: Navab, N., Hornegger, J., Wells, W.M., Frangi, A.F. (eds.) MICCAI 2015. LNCS, vol. 9351, pp. 234–241. Springer, Cham (2015). https://doi.org/10.1007/978-3-319-24574-4_28
22. Sethi, A., et al.: Empirical comparison of color normalization methods for epithelial-stromal classification in H and E images. J. Pathol. Inform. **7**, 17 (2016)
23. Sirinukunwattana, K., Ahmed Raza, S.E., Tsang, Y.W., Snead, D.R., Cree, I.A., Rajpoot, N.M.: Locality sensitive deep learning for detection and classification of nuclei in routine colon cancer histology images. IEEE Trans. Med. Imaging **35**(5), 1196–1206 (2016)
24. Vahadane, A., et al.: Structure-preserving color normalization and sparse stain separation for histological images. IEEE Trans. Med. Imaging **35**(8), 1962–1971 (2016)
25. Wienert, S., et al.: Detection and segmentation of cell nuclei in virtual microscopy images: a minimum-model approach. Sci. Rep. **2**, 503 (2012)
26. Xing, F., Xie, Y., Yang, L.: An automatic learning-based framework for robust nucleus segmentation. IEEE Trans. Med. Imaging **35**(2), 550–566 (2015)
27. Zhu, Y., Chen, Z., Zhao, S., Xie, H., Guo, W., Zhang, Y.: ACE-Net: biomedical image segmentation with augmented contracting and expansive paths. In: Shen, D., et al. (eds.) MICCAI 2019. LNCS, vol. 11764, pp. 712–720. Springer, Cham (2019). https://doi.org/10.1007/978-3-030-32239-7_79

Automatic Diagnosis of Glaucoma on Color Fundus Images Using Adaptive Mask Deep Network

Gang Yang[1,2], Fan Li[2], Dayong Ding[3], Jun Wu[4(✉)], and Jie Xu[5(✉)]

[1] Key Lab of Data Engineering and Knowledge Engineering, Beijing, China
[2] School of Information, Renmin University of China, Beijing, China
[3] Vistel Inc., Beijing, China
[4] School of Electronics and Information, Northwestern Polytechnical University, Xi'an, China
junwu@nwpu.edu.cn
[5] Beijing Tongren Hospital, Beijing, China
fionahsu920@foxmail.com

Abstract. Glaucoma, a disease characterized by the progressive and irreversible defect of the visual field, requires a lifelong course of treatment once it is confirmed, which highlights the importance of glaucoma early detection. Due to the diversity of glaucoma diagnostic indicators and the diagnostic uncertainty of ophthalmologists, deep learning has been applied to glaucoma diagnosis by automatically extracting characteristics from color fundus images, and that has achieved great performance recently. In this paper, we propose a novel adaptive mask deep network to obtain effective glaucoma diagnosis on retinal fundus images, which fully utilizes the prior knowledge of ophthalmologists on glaucoma diagnosis to synthesize attention masks of color fundus images to locate a reasonable region of interest. Based on the synthesized masks, our method could pay careful attention to the effective visual representation of glaucoma. Experiments on several public and private fundus datasets illustrate that our method could focus on the significant area of glaucoma diagnosis and simultaneously achieve great performance in both academic environments and practical medical applications, which provides a useful contribution to improve the automatic diagnosis of glaucoma.

Keywords: Glaucoma diagnosis · Deep learning · Attention mechanism

This work is supported in part by Beijing Natural Science Foundation (No. 4192029, No. 4202033), Natural Science Basic Research Plan in Shaanxi Province of China (2020JM-129), and Seed Foundation of Innovation and Creation for Postgraduate Students in Northwestern Polytechnical University (CX2020162).

J. Lokoč et al. (Eds.): MMM 2021, LNCS 12573, pp. 99–110, 2021.
https://doi.org/10.1007/978-3-030-67835-7_9

1 Introduction

Glaucoma is one of the leading causes of human vision loss in the world [9]. The symptoms only occur when the glaucoma disease is quite advanced, and it is diagnosed difficultly, so glaucoma is called the silent thief of sight. Although glaucoma cannot be cured, glaucoma detection in an early stage plays a crucial part in helping patients to slow down the disease progression. Therefore, ophthalmologists pay increasing attention to glaucoma early diagnosis. However, by 2020, glaucoma will affect around 80 million people worldwide [13]. As the number of glaucoma patients increases, only relying on ophthalmologists' manual diagnosis and medical devices to detect glaucoma has become unfeasible. Artificial intelligence (AI) applied to medical images provides a promising way to meet the requirement. Retinal fundus images can be gathered widely in hospitals, which offers an acceptable way with lower cost and non-interfering characteristics to patients in glaucoma diagnosis.

The diagnosis of glaucoma focuses on the evaluation of glaucoma optic neuropathy (GON). The content of GON evaluation mainly includes optic nerve head (ONH) and retinal nerve fiber layer (RNFL). As shown in Fig. 1, vertical cup to disc ratio (CDR), retinal nerve fiber layer defects (RNFLD) are frequently used indicators. In addition, the beta zone of parapapillary atrophy (PPA) is closely related to glaucoma, and is often inversely proportional to the disc edge width, and its expansion is related to the progress of glaucoma. In clinical diagnosis, the complex and diverse clinical morphology of the optic papilla, that is, individualized differences, brings many uncertainties to the diagnosis of glaucoma, making it difficult to diagnose glaucoma directly based on specific indicators. Along with AI development, the research of multimedia modeling on medical images used for medical recognition and diagnosis tasks has received increasingly attention [10, 23]. Recently, deep learning methods have been widely researched to diagnose glaucoma on fundus images [1], including convolutional neural network (CNN) [20] and U-Shape convolutional network [15]. However, due to data imbalance and lack of labeled retinal fundus images, they have revealed shortcomings on robustness and generalization. To overcome these problems, prior knowledge has been introduced into deep learning methods, which could obtain higher classification accuracy on glaucoma diagnosis [1, 2, 11]. Similarly, visual attention mechanism could be considered as a kind of powerful prior knowledge to optimize algorithms of pattern recognition, which has been widely adopted and has exposed superior performances in both image captioning and image analysis [7]. Nowadays, attention mechanism is introduced into automatic diagnosis of diseases based on medical images [8]. Sedai et. al proposed a multi-scale attention map, which is generated by a convex combination of class activation maps using the layer-wise weights, to localize the predicted pathology [16]. A Zoom-in-Net, trained with only image-level supervisions, could generate the highlighted suspicious regions with attention maps and predict the disease level accurately based on the high-resolution suspicious patches [22].

In this paper, we introduce an attention mechanism into our adaptive mask deep network, named AMNet, to realize automatic diagnosis of glaucoma on fun-

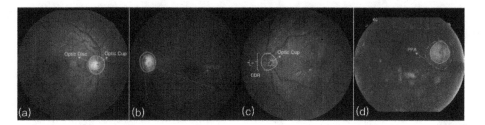

Fig. 1. Retinal fundus images. (a) Health. (b) Glaucoma. (c) Disease, but non-glaucoma. (d) Glaucoma, meanwhile other diseases. Some representations are labeled, including optic disc (OD), optic cup (OC), vertical cup to disc ratio (CDR), retinal nerve fiber layer defects (RNFLD) and parapapillary atrophy (PPA).

dus images. We first present the significant effect of prior knowledge when using deep learning to classify glaucoma on fundus images. Meanwhile, our method reveals a possible effective way to improve AI ability of computer-aided diagnosis. The main contributions of this paper are listed as follows:

1) In our method, an attention mask is generated adaptively, which fully uses the prior knowledge of glaucoma diagnosis about disease representation and symptom localization, importantly, the domains of RNFL, optic disc (OD) and optic cup (OC).
2) A new loss function is introduced to restrict AMNet to pay further attention to the important diagnosis areas, which increases the network robustness to some extent.
3) Experiments on three glaucoma datasets reveal that AMNet could focus on the significant area of glaucoma diagnosis and simultaneously achieve great performance, which demonstrates the vital role prior knowledge plays in glaucoma diagnosis.

2   Method

As a binary classification task, our adaptive mask deep network (AMNet) aims to map an input fundus image $x \in \mathbb{R}^{H \times W \times C}$ (height H, width W and channel C) to a binary category y. Depicted as Fig. 2, swallow mask generator F, attention mask generator $\hat{F}$ and glaucoma classification subnet $\tilde{F}$ constitute AMNet.

Formally, given an input fundus image x, F is designed to generate the swallow mask m that covers the area glaucoma symptoms usually appear. Then $\hat{F}$ takes $\hat{x}$, i.e., element wise multiplication of m and x, as input, and outputs $\hat{m}$ (the weight of each pixel in the $\hat{x}$). Finally, $\tilde{F}$ establishes a formula from $\tilde{x}$, element wise multiplication of $\hat{x}$ and $\hat{m}$, to the final binary category y. Note that F is trained individually, while $\hat{F}$ and $\tilde{F}$ are trained jointly. We will elaborate the detailed information of each phases and the motivation behind them in the following sections. Examples of swallow masks and attention masks are shown in Fig. 3.

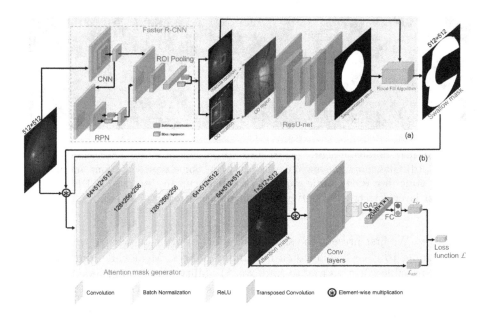

Fig. 2. An overview of the proposed adaptive mask deep network (AMNet): (a) swallow mask generator for swallow mask generation, which consists of three main components: Faster R-CNN, ResU-net and Flood Fill Algorithm; and (b) glaucoma classification subnet accompanied with attention mask generator that is designed for generating attention masks.

Swallow Mask Generation. Unlike other diseases (e.g., diabetic retinopathy), lesion location of glaucoma is relatively fixed. The occurrence of glaucoma is usually accompanied with neuroretinal rim loss (NRL) and RNFLD. Neuroretinal rim (NR) is located between the edge of OD and OC, and RNFLD often appears on the upper and lower vascular arcades (ULVA) of fundus images. So swallow masks are proposed to keep the aforementioned areas crucial for glaucoma diagnosis, i.e., OD and ULVA, and filter the remaining area.

Swallow mask generator F acts as an independent foundation work for subsequent processes. To generate the swallow mask m, two relevant areas, OD and macula, are located by jointing Faster R-CNN [14] and ResU-net [18]. ResU-net is degined for segmentation task, which incorporates residual learning along with skip-connection in an encoder-decoder based architecture. Faster R-CNN [14] serves as OD and macula location module and ResU-net serves as OD segmentation module.

First, given an input fundus image x, Faster R-CNN outputs bounding boxes of OD and macula. Then the cropped OD area based on the predicted bounding box is employed as the input of ResU-net to segment OD with ellipse fitting as post-processing. Finally, the centroid coordinates of both segmented OD area and the bounding box of macula are adopted as the parameters of Flood Fill Algorithm to generate m referred by [19].

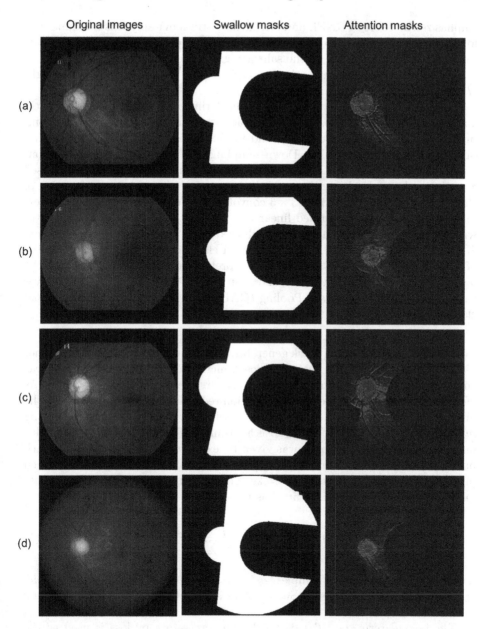

Fig. 3. Illustration of swallow masks and attention masks of our attention mechanism. (a)(b): Non-glaucoma samples; (c)(d): glaucoma samples.

Glaucoma Classification with Attention. After generating the swallow mask m, glaucoma classification subnet $\tilde{F}$ accompanied with attention mask generator $\hat{F}$ is further designed. Although m can effectively make the network emphasize the effect of RNFL and NRL, there still inevitably exists redundancy for glaucoma diagnosis. To tackle this problem, we propose $\hat{F}$ to drive the network pay further attention to the salient region based on m.

$\hat{F}$ is built to yield an attention mask $\hat{m} \in [0,1]^{W \times H}$ (width W and height H) for each corresponding $\hat{x}$ adaptively.

Similar as U-Net [15], $\hat{F}$ consists of a contracting path and an expanding path, and no skip-connection is used. The contraction path is designed to extract comprehensive contextual information, and the expansion path is used for precise pixel-wise weight distribution. Down-sampling is implemented by a 4×4 convolution layer with stride 2, and Up-sampling is implemented by a transposed convolution layer. Each of other convolution mudules is a composite function of three consecutive operations: a 3×3 convolution (Conv) followed by batch normalization (BN) and a rectified linear unit (ReLU). The last convolution layer with kernel size of 1×1 and stride 1 followed by a Sigmoid activation function is used to generate $\hat{m}$. Within the glaucoma classification subnet $\tilde{F}$, layers 1–4 of ResNet50 [5], initialized with ImageNet pretrained weights, is adopted as the feature extractor, i.e, 'Conv layers' shown in Fig. 2. A fully connected (FC) layer that follows a Global Average Pooling (GAP) layer is used to produce the final classification y.

Loss Function. Attention mask generator $\hat{F}$ and glaucoma classification subnet $\tilde{F}$ were trained end to end. A hybrid loss function $\mathcal{L}$ containing two terms was employed during this training process. The first term, i.e., $\mathcal{L}_{att}$, is for $\hat{F}$. As aforementioned, there still exists information redundancy in m. Let ρ denote the proportion of key information in m, the optimization objective of $\hat{F}$ is generating $\hat{m}$ that could keep the critical diagnosis areas in pixel-level by filtering redundant information. The proportion ρ was given by experience and the weight of each pixel was automatically learned through the attention mechanism, which realized pixel-wise screening to focus on those key glaucoma representations. To generate $\hat{m}$, the attention loss $\mathcal{L}_{att}$ is defined as follows:

$$\mathcal{L}_{att}(m, \hat{m}) = \frac{1}{N} \sum_{k=1}^{N} [\frac{\sum (m_{ij}^k \hat{m}_{ij}^k) - \rho \sum m_{ij}^k}{\sum m_{ij}^k}]^2 \quad (1)$$

where N is the number of samples. m^k is the $m \in \{0,1\}$ of sample k, and $\hat{m}^k$ is the $\hat{m} \in [0,1]$ of sample k. Subscript ij stands for pixel coordinate, where $i \in [0,W)$ and $j \in [0,H)$.

The second term, i.e., $\mathcal{L}_{ce}$, is for $\tilde{F}$. The cross-entropy loss is used as the classification loss $\mathcal{L}_{ce}$, which is defined as follows:

$$\mathcal{L}_{ce}(\hat{y}, \tilde{y}) = -\frac{1}{N} \sum_{k=1}^{N} \sum_{t=1}^{T} \hat{y}_t^k \log \frac{e^{\tilde{y}_t^k}}{\sum_p e^{\tilde{y}_p^k}} \quad (2)$$

where $\hat{y}^k$ is the corresponding one-hot label of sample k, and $\tilde{y}^k$ is the predicted vector for sample k. T is the dimension of $\hat{y}^k$ and $\tilde{y}^k$, whose value is 2 (i.e., glaucoma and non-glaucoma categories) in our study. The hybrid loss function $\mathcal{L}$ can be written as

$$\mathcal{L} = \mathcal{L}_{att} + \mathcal{L}_{ce} \tag{3}$$

AMNet utilizes $\mathcal{L}$ to restrict its training processing to generate adaptive attention for glaucoma diagnosis.

Table 1. Dataset partition of our experiments.

Dataset	TRGD			REFUGE			LAG
Purpose	Training	Validation	Test	Training	Validation	Test	Test
G/N[a]	3346/4482	385/596	398/571	40/360	40/360	40/360	629/342

[a]G denotes Glaucoma, N denotes Non-Glaucoma.

3 Experiments

3.1 Experimental Setup

Datasets. To test the practical ability of AMNet, we gathered fundus images from an authoritative ophthalmic hospital and constructed a private glaucoma dataset: Tongren Glaucoma Dataset (TRGD), annotated by 2 ophthalmologists of Beijing Tongren Hospital who have at least 3 years experience on glaucoma diagnosis. Moreover, the public datasets REFUGE[1] [12] and LAG[2] [11] available on the Internet were also adopted. The dataset partition is shown in Table 1. Note that TRGD test set and LAG were just used to evaluate the performance of experimental models that trained on TRGD dataset. Then the test on LAG revealed the model ability of domain adaption. Moreover on REFUGE dataset, we fine-tuned all of the models pre-trained by TRGD dataset and make evaluation.

Evaluation Metrics. We used sensitivity (Sen), specificity (Spe), F1-score (F1) and Area Under Curve (AUC) to report the experimental performance. F1 is defined as harmonic average of Sen and Spe. As Sen and Spe are two trade-off evaluation to some extends, they are not suitable as a basis for choosing an available model in practical application. We chose the model with the highest F1 in validation set to evaluate its performance.

---

[1] https://refuge.grand-challenge.org/.
[2] https://github.com/smilell/AG-CNN.

Table 2. Performance comparison with single-stream methods on the three datasets (indices in %).

Method	TRGD				REFUGE				LAG			
	Sen	Spe	F1	AUC	Sen	Spe	F1	AUC	Sen	Spe	F1	AUC
ResNet50 [5]	**90.2**	89.8	90.0	96.4	85.0	92.8	88.8	95.1	46.2	99.2	63.0	92.8
Inception-ResNet [20]	88.7	89.8	89.3	95.8	80.0	**96.1**	87.3	96.1	39.2	**99.5**	56.2	90.4
Inception [21]	88.7	89.8	89.2	96.4	82.5	89.2	85.7	94.6	46.5	**99.5**	63.4	**93.8**
ABN [4]	89.7	89.8	89.8	96.5	85.0	89.2	87.0	88.8	39.5	99.4	56.5	88.1
MNet [2]	81.4	74.4	77.8	84.5	52.3	91.1	66.6	81.8	**84.8**	57.7	68.7	78.0
AMNet	89.5	**91.4**	**90.4**	**96.6**	**92.5**	88.3	**90.4**	**96.3**	60.5	95.9	**74.2**	91.9

[a]Inception-ResNet denotes the Inception-ResNet-v2 model proposed in [20].
[b]Inception denotes the Inception-v3 model proposed in [21].

Implementation Details. Our experiment was implemented with Pytorch and Scikit library on a Linux server with 4 Titan XP GPUs. For Faster R-CNN training, locating macula and OD used 1892 individual labeled fundus images. We randomly sampled 256 anchors per image to train Region Proposal Network (RPN), including 128 positive samples that had Intersection-over-Union (IoU) overlap higher than 0.7 with any ground-truth box and 128 negative samples with IoU ratio lower than 0.3 for all ground-truth boxes. We randomly sampled 128 regions of interest (RoIs) per image with IoU threshold of 0.5 to train the final classification layer and regression layer, where the sampled positive and negative RoIs had a ratio of up to 1:3. For ResU-net training, the cropped OD regions were resized to 256×256 pixels. Data augmentation in form of random crop, flip, gamma and SamplePairing [6] was employed. For glaucoma classification training, the parameter ρ of the loss function was set to 0.1. Fundus images were all resized to 512×512 pixels. Data augmentation was applied on the training set by random rotation and random horizontal flip. The learning rate was 0.001 and SGD with momentum of 0.9 was used as the optimizer.

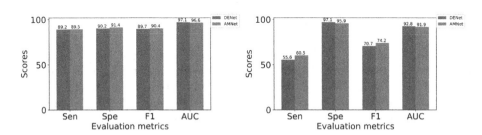

Fig. 4. Performance comparison with DENet (indices in %). Left: performance on TRGD dataset; right: performance on LAG dataset.

Table 3. The ablation study results on two public datasets (indices in %).

Method	REFUGE				LAG			
	Sen	Spe	F1	AUC	Sen	Spe	F1	AUC
AMNet w/o $\hat{F}$	90.0	83.0	86.4	95.6	50.0	97.6	66.1	90.8
AMNet w/o F	90.0	88.3	89.1	96.1	52.9	**98.3**	68.8	91.0
AMNet	**92.5**	**88.3**	**90.4**	**96.3**	**60.5**	95.9	**74.2**	**91.9**

3.2   Results and Discussion

Comparison with Single-Stream Models. We compared AMNet with several advanced single-stream deep neural networks, including ResNet50 [5], Inception-ResNet-v2 [20], Inception-v3 [21], Attention Branch Network (ABN) [4] and MNet [2]. The compared results are shown in Table 2.

As Table 2 shows, compared with other models, AMNet achieves higher F1 and AUC score on the three datasets except for AUC score on LAG dataset. Especially, AMNet can obtain F1 scores of 90.4% and 74.2% on REFUGE and LAG datasets separately, which outperforms the second place algorithm with 1.6% and 5.5% improvement. The results justify the validity of our attention mechanism. This is explicable because compared to these models, AMNet further infuses medical prior knowledge into the attention mechanism.

Comparison with DENet. As a superior **multi-stream** network specially designed for glaucoma screening, DENet [3] was employed for further comparison. The results compared with DENet is listed separately from single-stream models in an extra figure, i.e., Fig. 4. We chose two large datasets, TRGD and LAG, to make this comparison. As the harmonic average of two widely used indicators (Sen and Spe), F1 provides significant reference value for clinical diagnosis. From Fig. 4 we can see, AMNet outperforms DENet in terms of F1. Although our AMNet achieves lower score in AUC compared to DENet, it is reasonable and there still exists room for us to make improvement, considering that DENet is an ensemble network. In general, the performance our AMNet achieves on F1 demonstrates the effectiveness of the adaptive attention mask and the validity of introducing medical prior knowledge into automatic glaucoma diagnosis.

Visualization. In order to be intuitionistic, further visualization was made. We compared the attention maps of AMNet with Gradient-weighted Class Activation Mappings (Grad-CAMs) [17] of ResNet50. Besides, attention maps of ABN was also compared consider that ABN was designed based on attention mechanism without prior knowledge. As shown in Fig. 5, AMNet highlights areas of OD and ULVA in which RNFLD often appears on the original image, while ResNet50 and ABN are incapable of focusing on those areas significant to glaucoma diagnosis. In other word, in AMNet, the adaptive attention mask drives the network

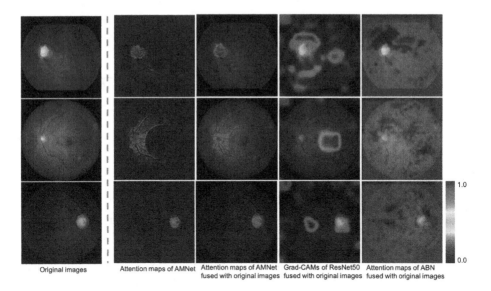

Fig. 5. Visualization comparison with ResNet50 and ABN. From left to right: attention maps of AMNet, attention maps of AMNet fused with original images, Grad-CAMs of ResNet50 fused with original images, and attention maps of ABN fused with original images. For attention maps of AMNet, to prevent confusion caused by color overlays, we only fuse the regions with activation values lower than 0.8 with original images. Top: glaucoma sample that three models all predict correctly. Middle: non-glaucoma sample that three models all predict correctly. Bottom: glaucoma sample that only AMNet predicts correctly.

focus on the mapping computation of diagnosis areas more seriously. Therefore, AMNet has high robustness to apply in practical applications.

Ablation Study. To validate the performance improvement resulted from AMNet, a set of ablation experiments were conducted to evaluate the effect of each component used in attention mechanism, i.e., swallow mask generator F and attention mask generator $\hat{F}$. To validate the effect of F, we removed F from AMNet, that is, the original fundus images were directly feeded into $\hat{F}$ and no element-wise multiplication was conducted with attention mask $\hat{m}$. Similarly, ablation experiment of removing $\hat{F}$ from AMNet was conducted to validate effectiveness of $\hat{F}$. Specifically, the result of element-wise multiplication of the original fundus image x and the swallow mask m was employed as input of 'Conv layers' shown in Fig. 2. As there was no $\hat{m}$, attention loss $\mathcal{L}_{att}$ was removed. Results of ablation experiments are listed in Table 3. Our AMNet obtains the best results in terms of F1 and AUC compared with the methods of removing F and $\hat{F}$ respectively. The ablation study demonstrates the validity of F and $\hat{F}$, which is a further proof of the ability of prior knowledge in automatic glaucoma diagnosis.

4   Conclusion

In this paper, a novel adaptive mask deep network (AMNet) is proposed to infuse reasonable prior knowledge into the classification process, which pays more attention to the areas of neuroretinal rim loss (NRL) and optic disc (OD) and filters redundant pixel information about glaucoma diagnosis. The experiments on three datasets illustrate that our method could focus on the significant area of glaucoma diagnosis and simultaneously obtain great performance in both academic environment and practical medical applications, which reveals the definite effect of prior knowledge on glaucoma classification and provides an effective inspiration to improve the automatic diagnosis of glaucoma.

References

1. Chai, Y., Liu, H., Xu, J.: Glaucoma diagnosis based on both hidden features and domain knowledge through deep learning models. Knowl.-Based Syst. **161**, 147–156 (2018)
2. Fu, H., Cheng, J., Xu, Y., Wong, D.W.K., Liu, J., Cao, X.: Joint optic disc and cup segmentation based on multi-label deep network and polar transformation. IEEE Trans. Med. Imaging **37**(7), 1597–1605 (2018)
3. Fu, H., et al.: Disc-aware ensemble network for glaucoma screening from fundus image. IEEE Trans. Med. Imaging **37**(11), 2493–2501 (2018)
4. Fukui, H., Hirakawa, T., Yamashita, T., Fujiyoshi, H.: Attention branch network: Learning of attention mechanism for visual explanation. In: Proceedings of the IEEE Conference on Computer Vision and Pattern Recognition, pp. 10705–10714 (2019)
5. He, K., Zhang, X., Ren, S., Sun, J.: Deep residual learning for image recognition. In: Proceedings of the IEEE Conference on Computer Vision and Pattern Recognition, pp. 770–778 (2016)
6. Inoue, H.: Data augmentation by pairing samples for images classification. arXiv preprint arXiv:1801.02929 (2018)
7. Jin, J., Fu, K., Cui, R., Sha, F., Zhang, C.: Aligning where to see and what to tell: image caption with region-based attention and scene factorization. arXiv preprint arXiv:1506.06272 (2015)
8. Keel, S., Wu, J., Lee, P.Y., Scheetz, J., He, M.: Visualizing deep learning models for the detection of referable diabetic retinopathy and glaucoma. JAMA Ophthalmol. **137**, 288–292 (2018)
9. Kim, M., Zuallaert, J., De Neve, W.: Few-shot learning using a small-sized dataset of high-resolution fundus images for glaucoma diagnosis. In: Proceedings of the 2nd International Workshop on Multimedia for Personal Health and Health Care, pp. 89–92. ACM (2017)
10. Lai, X., Li, X., Qian, R., Ding, D., Wu, J., Xu, J.: Four models for automatic recognition of left and right eye in fundus images. In: Kompatsiaris, I., Huet, B., Mezaris, V., Gurrin, C., Cheng, W.-H., Vrochidis, S. (eds.) MMM 2019. LNCS, vol. 11295, pp. 507–517. Springer, Cham (2019). https://doi.org/10.1007/978-3-030-05710-7_42
11. Li, L., Xu, M., Wang, X., Jiang, L., Liu, H.: Attention based glaucoma detection: a large-scale database and CNN model. In: Proceedings of the IEEE Conference on Computer Vision and Pattern Recognition, pp. 10571–10580 (2019)

12. Orlando, J.I., et al.: Refuge challenge: a unified framework for evaluating automated methods for glaucoma assessment from fundus photographs. Med. Image Anal. **59**, 101570 (2020)

13. Quigley, H.A., Broman, A.T.: The number of people with glaucoma worldwide in 2010 and 2020. Br. J. Ophthalmol. **90**(3), 262–267 (2006)

14. Ren, S., He, K., Girshick, R., Sun, J.: Faster R-CNN: towards real-time object detection with region proposal networks. In: Advances in Neural Information Processing Systems, pp. 91–99 (2015)

15. Ronneberger, O., Fischer, P., Brox, T.: U-Net: convolutional networks for biomedical image segmentation. In: Navab, N., Hornegger, J., Wells, W.M., Frangi, A.F. (eds.) MICCAI 2015. LNCS, vol. 9351, pp. 234–241. Springer, Cham (2015). https://doi.org/10.1007/978-3-319-24574-4_28

16. Sedai, S., Mahapatra, D., Ge, Z., Chakravorty, R., Garnavi, R.: Deep multiscale convolutional feature learning for weakly supervised localization of chest pathologies in X-ray images. In: Shi, Y., Suk, H.-I., Liu, M. (eds.) MLMI 2018. LNCS, vol. 11046, pp. 267–275. Springer, Cham (2018). https://doi.org/10.1007/978-3-030-00919-9_31

17. Selvaraju, R.R., Cogswell, M., Das, A., Vedantam, R., Parikh, D., Batra, D.: Grad-CAM: visual explanations from deep networks via gradient-based localization. In: Proceedings of the IEEE International Conference on Computer Vision, pp. 618–626 (2017)

18. Shankaranarayana, S.M., Ram, K., Mitra, K., Sivaprakasam, M.: Joint optic disc and cup segmentation using fully convolutional and adversarial networks. In: Cardoso, M.J., et al. (eds.) FIFI/OMIA -2017. LNCS, vol. 10554, pp. 168–176. Springer, Cham (2017). https://doi.org/10.1007/978-3-319-67561-9_19

19. Son, J., Bae, W., Kim, S., Park, S.J., Jung, K.-H.: Classification of findings with localized lesions in fundoscopic images using a regionally guided CNN. In: Stoyanov, D., et al. (eds.) OMIA/COMPAY -2018. LNCS, vol. 11039, pp. 176–184. Springer, Cham (2018). https://doi.org/10.1007/978-3-030-00949-6_21

20. Szegedy, C., Ioffe, S., Vanhoucke, V., Alemi, A.A.: Inception-v4, inception-resnet and the impact of residual connections on learning. In: Thirty-First AAAI Conference on Artificial Intelligence (2017)

21. Szegedy, C., Vanhoucke, V., Ioffe, S., Shlens, J., Wojna, Z.: Rethinking the inception architecture for computer vision. In: Computer Vision & Pattern Recognition (2016)

22. Wang, Z., Yin, Y., Shi, J., Fang, W., Li, H., Wang, X.: Zoom-in-Net: deep mining lesions for diabetic retinopathy detection. In: Descoteaux, M., Maier-Hein, L., Franz, A., Jannin, P., Collins, D.L., Duchesne, S. (eds.) MICCAI 2017. LNCS, vol. 10435, pp. 267–275. Springer, Cham (2017). https://doi.org/10.1007/978-3-319-66179-7_31

23. Wu, J., et al.: AttenNet: deep attention based retinal disease classification in OCT images. In: Ro, Y.M., et al. (eds.) MMM 2020. LNCS, vol. 11962, pp. 565–576. Springer, Cham (2020). https://doi.org/10.1007/978-3-030-37734-2_46

Initialize with Mask: For More Efficient Federated Learning

Zirui Zhu and Lifeng Sun[✉]

Department of Computer Science and Technology, Tsinghua University,
Beijing, China
zhu-zr20@mails.tsinghua.edu.cn, sunlf@tsinghua.edu.cn

Abstract. Federated Learning (FL) is a machine learning framework proposed to utilize the large amount of private data of edge nodes in a distributed system. Data at different edge nodes often shows strong heterogeneity, which makes the convergence speed of federated learning slow and the trained model does not perform well at the edge. In this paper, we propose Federated Mask (FedMask) to address this problem. FedMask uses Fisher Information Matrix (FIM) as a mask when initializing the local model with the global model to retain the most important parameters for the local task in the local model. Meanwhile, FedMask uses Maximum Mean Discrepancy (MMD) constraint to avoid the instability of the training process. In addition, we propose a new general evaluation method for FL. Following experiments on MNIST dataset show that our method outperforms the baseline method. When the edge data is heterogeneous, the convergence speed of our method is 55% faster than that of the baseline method, and the performance is improved by 2%.

Keywords: Federated Learning · Machine learning · Fisher Information Matrix · Maximum Mean Discrepancy

1   Introduction

Along with technical and economical development, more and more intelligent devices come into our lives, such as smartphones, wearable devices, etc. These devices are producing large amount of data everyday, which can significantly improve the user experience if it can be used in machine learning. Unfortunately, because this data is always privacy sensitive and large in quantity, it is not feasible to upload this data to the server for centralized training. To this end, Google proposed the concept of federated learning (FL) [1–3]. It can use the data of each node to complete the training of a shared model without uploading data. FL has many potential applications, such as personalized push, health

This work was supported by the NSFC under Grant 61936011, 61521002, National Key R&D Program of China (No. 2018YFB1003703), and Beijing Key Lab of Networked Multimedia.

J. Lokoč et al. (Eds.): MMM 2021, LNCS 12573, pp. 111–120, 2021.
https://doi.org/10.1007/978-3-030-67835-7_10

prediction, smart city and so on, which makes it become a hot spot in recent research.

FL faces many challenges, one of the biggest challenges is the heterogeneity of data. Recent work [4] argue that, when the data of edge nodes are non-iid, the convergence speed of FL will decrease, and the performance of the model at the edge will also deteriorate. We think that the reason for this problem is that a lot of local knowledge is forgotten in global model which we use to initialize the local model in FL. Therefore, we hope to solve this problem by retaining some local parameters in initialization process, which we also call **partial initialization**. Based on this idea, we propose **Federated Mask** (FedMask). In FedMask, we use Fisher Information Matrix (FIM) as a mask when initializing the local model with the global model to retain the most important parameters for local task. Meanwhile, we use Maximum Mean Discrepancy (MMD) constraint to ensure the stability of the training process. In addition, due to partial initialization, the original evaluation method of FL is no longer applicable, so we propose a new general evaluation method. Following experiments on MNIST dataset show that our method outperforms the baseline method. when the edge data is heterogeneous, the convergence speed of our method is 55% faster than that of the baseline method, and the performance after convergence is improved by 2%.

We make the following contributions: 1) To the best of our knowledge, this is the first paper that introduce the idea of partial initialization in FL. 2) Based on the above idea, we propose FedMask and experiments on MNIST show that its performance is better than the baseline method. 3) What's more, we propose a new general evaluation method for FL.

2   Related Work

FL is a machine learning framework that can use edge data to complete model training while ensuring data privacy. FL is an iterative process and one round of its iteration can be divided into the following steps. Firstly, the server will randomly select several clients to participate in this round of training among all clients. Then the server sends the global model to the selected client. The client uses the global model to initialize the local model and train local model with local data. Finally, the clients upload trained models and a new global is obtained after the server aggregating these models. Such iterations will be repeated for multiple rounds, until the model converges or reaches the time limit. The most representative algorithm in FL is Federated Averaging (FedAvg) [1]. In FedAvg, the server aggregates the models by weighted averaging (the weight is the number of training data held by the client).

In order to solve the problems caused by the heterogeneity of edge data, many works have been done. For example, Yao et al. [6] introduced two-stream model to the local training process, aiming to integrate more global knowledge while adapting the local data. Li et al. [7] proposed FedProx and Shoham et al. [8] proposed FedCurv. They both limit the distance between the local model and the global model during local training to avoid the loss of global knowledge.

The above works can speed up the convergence of FL, but cannot improve the performance of the model at edge nodes.

3   Federated Mask

3.1   Motivation

When FL faces heterogeneous edge data, its convergence speed will decrease, and the performance of the model at the edge will also deteriorate. We think that the reason for this problem is that the global model obtained by model aggregation forgets a lot of knowledge about local tasks. Using such global model to initialize local model will inevitably lead to poor performance of local model on local task, and need extra cost to relearn local knowledge. Therefore, we hope to solve this problem by retaining some important parameters for local task in local model when the local model is initialized by the global model, which we also call **partial initialization**.

According to the research of Yosinski et al. [5], the feature extraction part (shallow layer) of the neural network is general, that is, task-specific knowledge is not learned by the feature extraction part of the network. Hence, we divide the model into feature extractor and classifier. And when we select the parameters that need to be retained, we only need to focus on the classifier. By retaining some of the most important parameters for the local task in the classifier when initializing the local model, we can improve the performance of FL under non-iid edge data.

In addition, the characteristics of FL determine that a client will not participate in each round of training. When the local model is initialized, the classifier of the local model may be out-of-date. In other words, the local classifier cannot use the features extracted by the global feature extractor correctly. Since the selection of training nodes is random, whether this happens or not and the severity of this situation is also random, which will lead to the instability of FL process. In this regard, we can add some constraints to the feature extractor in the local training process to obtain a relatively stable feature extractor to reduce this instability.

3.2   Initialize with Fisher Information Matrix

For determining the importance of parameters, Fisher Information Matrix (FIM) is a good choice. It was originally applied to continual learning [9] to limit the changes of parameters which are important to the previous task.

FIM has many good properties. Firstly, FIM's diagonal element directly represents the importance of the parameter. The larger the diagonal element value, the more important the corresponding parameter is. Secondly, according to the research of Pascanu et al. [10], FIM's diagonal element can be easily approximated by the first-order derivatives of the model. The calculation process is as follows:

$$F = E[(\frac{\partial L(\theta(x_i), y_i)}{\partial \theta})^2] \tag{1}$$

where L is the loss function, θ is the model parameters and (x_i, y_i) is data with label. After getting the diagonal elements of FIM, we can use them to initialize the local model. The process is shown in Algorithm 1, where E_l is local feature extractor, E_g is global feature extractor, C_l is local classifier, C_g is global classifier and β is the hyper-parameter that controls the retention ratio. Since the role of FIM in Algorithm 1 is similar to mask, we call this process Initialize with Mask.

Algorithm 1. Initialize with Mask

1: $E_l \leftarrow E_g$
2: **for** each layer k in classifier **do**
3:     $F_k \leftarrow diag(FIM)$ of local classifier's k-th layer
4:     $n \leftarrow \beta * |C_{l,k}|$
5:     $m \leftarrow$ n-th largest number in F_k
6:     $H \leftarrow F \geq m$ //True is 1 and False is 0
7:     $C_{l,k} \leftarrow H * C_{l,k} + \widetilde{H} * C_{g,k}$
8: **end for**

3.3 Local Training with MMD Constraint

In order to constrain the change of feature extractor in local training, we use a two-stream model on feature extractor (Fig. 1). In our two-stream model, global feature extractor will be freeze, and MMD between the outputs of global and local feature extractor will be added into loss function:

$$Loss = L + \lambda MMD^2(E_g(x_i), E_l(x_i)) \tag{2}$$

where L is the classification loss and λ is the hyper-parameter weighted to MMD loss. And MMD is a measure of the gap between two distributions, it is widely used in domain adaptation [11–13]. For given distributions p and q, the MMD between them can be expressed as:

$$\begin{aligned} MMD^2(p,q) &= \|E[\phi(p)] - E[\phi(q)]\|^2 \\ &= E[K(p,p)] + E[K(q,q)] - 2E[K(p,q)] \end{aligned} \tag{3}$$

where ϕ is a mapping to Reproducing Kernel Hilbert Space (RKHS) and K is the corresponding kernel function. In this paper, we use a standard radial basis function kernel as K.

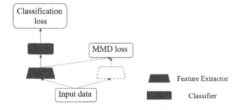

Fig. 1. Local training with MMD loss

3.4   FedMask Implementation

The above are the implementation details of the key steps in FedMask. Now we give the overall implementation of FedMask (Algorithm 2).

Algorithm 2. FedMask

Server:
1: initialize $E_{g,0}$ and $C_{g,0}$
2: **for** each round $t = 1, 2, 3...$ **do**
3:     $m \leftarrow max(C \cdot K, 1)$ $//K$ is number of clients and C is selection ratio
4:     $S_t \leftarrow$ random set of m clients
5:     **for** each client $k \in S_t$ **do**
6:         $(E_{g,t}^k, C_{g,t}^k) \leftarrow Client(k, E_{g,t-1}, C_{g,t-1})$
7:     **end for**
8:     $E_{g,t} \leftarrow \frac{1}{n_{S_t}} \sum_{k \in S_t} n_k \cdot E_{g,t}^k$ $//n$ is data number
9:     $C_{g,t} \leftarrow \frac{1}{n_{S_t}} \sum_{k \in S_t} n_k \cdot C_{g,t}^k$
10: **end for**
Client(k, E_g, C_g)**:**
11: Initialize E_l^k, C_l^k by Algorithm 1
12: minimize $Loss = L + \lambda MMD^2(E_g(x), E_l^k(x))$ for E epochs with batch size B
13: return (E_l^k, C_l^k) to server

For the convenience of comprehension, we also give the corresponding flow chart (Fig. 2) of the Algorithm 2. We retain some important parameters in the classifier when initializing the local model, so that the local model can adapt to the local task better during the FL process. Meanwhile, we also use MMD constraint in local training to avoid the instability of the training process. We can see that our algorithm does not change the framework of FL. This means that our work can be easily deployed to existing FL projects.

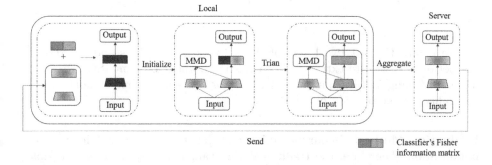

Fig. 2. Flow chart of FedMask

4   Experiments

4.1   Experimental Setup

Dataset. We use MNIST dataset in our experiments. To simulate the FL settings, we use Hsu's [14] method to divide the data into different clients. In this method, the local data distribution is generated by a Dirichlet distribution which is related to the global data distribution as follows:

$$q \sim Dir(\alpha p) \tag{4}$$

where p is the global data distribution, Dir is Dirichlet distribution and α is the parameter controlling the identicalness among clients. A bigger α represents more similar data distributions among clients. According to the above method, we divide the data evenly into 100 clients and select 10 clients for participating in training each round ($C = 0.1$).

Model. We use Convolutional Neural Network (CNN) as the model in our experiments and the network architecture is shown in Table 1.

Table 1. Network architecture

Name	Kernel size	Stride	Output channel	Activation
Conv1	5×5	1	16	Relu
Pool1	2×2	2	-	-
Conv2	5×5	1	32	Relu
Pool2	2×2	2	-	-
Fc1	-	-	512	Relu
Fc2	-	-	256	Relu
Fc3	-	-	128	Relu
Fc4	-	-	10	Softmax

We regard the last two layers of the network as classifiers and the rest of the network as feature extractors. Stochastic Gradient Descent (SGD) is used in local training and learning rate is 10^{-4}. For convenience but without loss of generality, we set $B = 50$, $E = 1$.

Evaluation Method. In our method, the model of each client is different after initialization. Therefore, the traditional FL evaluation method is not feasible and we propose a new general evaluation method for FL. Our evaluation method consists of two parts: Local Test and Global Test.

* **Local Test**   The model will be evaluated by a test set that subject to the local data distribution and the results on each server are averaged to get the final result. This test is more in line with the actual scene of FL, so we take it as the major evaluation.

* **Global Test**   The model will be evaluated by a test set that subject to the global data distribution and the results on each server are averaged to get the final result. The purpose of this test is to determine whether the model has learned global knowledge, so we take it as the minor evaluation.

4.2   Results and Analysis

In this section, we will verify the effectiveness of each design in FedMask, and then evaluate the overall method.

Effectiveness of FIM. In this experiment, we will verify whether it is necessary to use FIM to select important parameters. We choose FedAvg and a special FedMask in which FIM is replaced by a random matrix as the baseline method. In order to eliminate interference from other factors, we set $\beta = 0.1$ and $\lambda = 0$. The results of the experiment are shown in the Fig. 3.

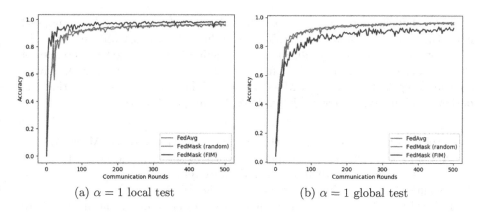

(a) $\alpha = 1$ local test   (b) $\alpha = 1$ global test

Fig. 3. Effectiveness of FIM

In local test with $\alpha = 1$, we can see that performance of FedMask with FIM far exceeds FedAvg and FedMask with random matrix, and the performance of FedAvg and FedMask with random matrix is very similar. In global test with $\alpha = 1$, although the performance of FedMask with FIM is slightly worse, the accuracy reached about 93% after convergence, which is completey acceptable. The above result shows that the important parameters are selected by FIM and helps the local model adapting local task better. However, random matrix has no such function, so the design of using FIM as a mask when initializing the local model is effective.

Effectiveness of MMD. In this experiment, we will verify the effect of MMD constraint in local training. We compare the FedMask with MMD constraint ($\lambda = 0.1$) and without MMD constraint ($\lambda = 0$) and we set $\beta = 0.1$. The results of the experiment are shown in the Fig. 4.

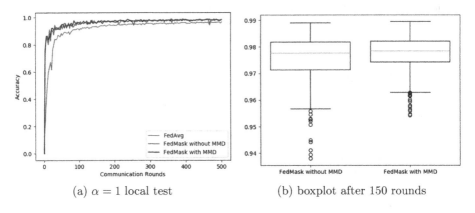

(a) $\alpha = 1$ local test   (b) boxplot after 150 rounds

Fig. 4. Effectiveness of MMD

In the left figure, we can see that the MMD constraint in local training can make the training process more stable. By adding MMD constraint, many peaks in the original training curve disappeared. The figure on the right is a boxplot of the data after 150 rounds in the left figure, from which we can more intuitively see the changes after adding MMD constraint. The above results show that the design of adding MMD constraint in local training is effective.

FedMask. At last, we evaluate the overall performance of FedMask. We set $\beta = 0.1$ and $\lambda = 0.1$. And we experiment in two settings: data distribution is relatively uniform ($\alpha = 10000$) and data is extremely heterogeneous ($\alpha = 1$). The results of the experiments are shown in the Fig. 5.

In local test with $\alpha = 1$, our method needs fewer communication rounds to get to convergence. More concretely, our method reaches the test accuracy of 0.95 by 78 rounds of communication while the FedAvg needs 209 rounds, achieving a reduction of 55%. In addition, our method achieves the accuracy of 98.5% after convergence, while FedAvg is only 96.5%. In global test with $\alpha = 1$, our method converges slower than FedAvg, when the accuracy of FedAvg is 95%, our method has an accuracy of 91.5%. But we should note that in the setting of FL, users are not faced with global data distribution when using the local model, so such results are acceptable.

In local test and global test with $\alpha = 10000$, Our method's performance is basically the same as FedAvg. This shows that our method will not degrade the performance of FL when the data distribution is relatively uniform.

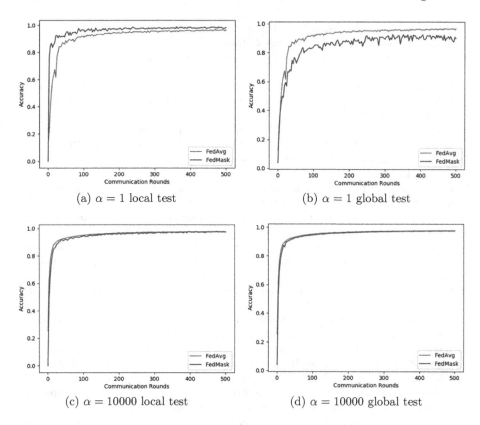

Fig. 5. Performance of FedMask

5   Conclusion

In this paper, we propose FedMask that improves the convergence speed and performance of FL under non-iid edge data, and we introduce a new general evaluation method for FL. Experiments show that the convergence speed of our method is 55% faster than that of the baseline method, and the performance after convergence is improved by 2%.

In future work, we may try to replace FIM with other parameter importance estimation methods, or replace MMD with other constraint methods in FedMask. Of course, combining our work with other FL work is also a worthwhile idea.

References

1. McMahan, H.B., Moore, E., Ramage, D., et al.: Communication-efficient learning of deep networks from decentralized data. In: Proceedings of the 20th International Conference on Artificial Intelligence and Statistics, pp. 1273–1282 (2017)
2. Konečný, J., Mcmahan, B., Ramage, D.: Federated optimization: distributed optimization beyond the data center. Mathematics **2**(1), 115 (2015)

3. Konečný, J., McMahan, H.B., Yu, F.X., et al.: Federated learning: strategies for improving communication efficiency (2016)
4. Zhao, Y., Li, M., Lai, L., et al.: Federated learning with non-IID data. arXiv preprint arXiv:1806.00582 (2018)
5. Yosinski, J., Clune, J., Bengio, Y., et al.: How transferable are features in deep neural networks? In: Advances in Neural Information Processing Systems, vol. 27, pp. 3320–3328 (2014)
6. Yao, X., Huang, C., Sun, L.: Two-stream federated learning: reduce the communication costs [C/OL]. In: IEEE Visual Communications and Image Processing, VCIP 2018, Taichung, Taiwan, 9–12 December 2018, pp. 1–4. IEEE (2018). https://doi.org/10.1109/VCIP.2018.8698609
7. Li, T., Sahu, A.K., Zaheer, M., et al.: Federated optimization in heterogeneous networks. arXiv preprint arXiv:1812.06127 (2018)
8. Shoham, N., Avidor, T., Keren, A., et al.: Overcoming forgetting in federated learning on non-IID data. arXiv preprint arXiv:1910.07796 (2019)
9. Kirkpatrick, J., et al.: Overcoming catastrophic forgetting in neural networks. In: Proceedings of the National Academy of Sciences, vol. 114, no. 13, pp. 3521–3526 (2017)
10. Pascanu, R., Bengio, Y.: Revisiting natural gradient for deep networks. arXiv preprint arXiv:1301.3584 (2013)
11. Long, M., Cao, Y., Wang, J., Jordan, M.I.: Learning transferable features with deep adaptation networks, arXiv preprint arXiv:1502.02791 (2015)
12. Long, M., Wang, J., Ding, G., Sun, J., Philip, S.Y.: Transfer feature learning with joint distribution adaptation. In: 2013 IEEE International Conference on Computer Vision (ICCV), pp. 2200–2207. IEEE (2013)
13. Tzeng, E., Hoffman, J., Zhang, N., Saenko, K., Darrell, T.: Deep domain confusion: maximizing for domain invariance, arXiv preprint arXiv:1412.3474 (2014)
14. Hsu, T.M.H., Qi, H., Brown ,M.: Measuring the effects of non-identical data distribution for federated visual classification (2019)
15. Liang, P.P., Liu, T., Ziyin, L., et al.: Think locally, act globally: federated learning with local and global representations (2020)
16. Dosovitskiy, A., Brox, T.: Inverting visual representations with convolutional networks. In: The IEEE Conference on Computer Vision and Pattern Recognition (CVPR) (2016)
17. Zhuo, J., Wang, S., Zhang, W., Huang, Q.: Deep unsupervised convolutional domain adaptation. In: Proceedings of the 2017 ACM on Multimedia Conference, pp. 261–269. ACM (2017)
18. Gretton, A., et al.: Optimal kernel choice for large-scale two-sample tests. In: Advances in Neural Information Processing Systems, pp. 1205–1213 (2012)
19. Jung, H., Ju, J., Jung, M., Kim, J.: Less-forgetting learning in deep neural networks, arXiv preprint arXiv:1607.00122 (2016)

Unsupervised Gaze: Exploration of Geometric Constraints for 3D Gaze Estimation

Yawen Lu[1], Yuxing Wang[1], Yuan Xin[2], Di Wu[2], and Guoyu Lu[1(✉)]

[1] Intelligent Vision and Sensing Lab, Rochester Institute of Technology,
Rochester, USA
luguoyu@cis.rit.edu
[2] Tecent Deep Sea Lab, Shenzhen, China

Abstract. Eye gaze estimation can provide critical evidence for people attention, which has extensive applications on cognitive science and computer vision areas, such as human behavior analysis and fake user identification. Existing typical methods mostly place the eye-tracking sensors directly in front of the eyeballs, which is hard to be utilized in the wild. And recent learning-based methods require prior ground truth annotations of gaze vector for training. In this paper, we propose an unsupervised learning-based method for estimating the eye gaze in 3D space. Building on top of the existing unsupervised approach to regress shape parameters and initialize the depth, we propose to apply geometric spectral photometric consistency constraint and spatial consistency constraints across multiple views in video sequences to refine the initial depth values on the detected iris landmark. We demonstrate that our method is able to learn gaze vector in the wild scenes more robust without ground truth gaze annotations or 3D supervision, and show our system leads to a competitive performance compared with existing supervised methods.

Keywords: Unsupervised learning · 3D gaze estimation · Geometric constraints

1 Introduction

Eye gaze, as an important cue to indicate human attention, intention and social communication interest, has been investigated in a vast domain, such as educational training [1], human-computer interaction [3,7], and driving attention modelling [4], etc. As a result, gaze estimation is an active research topic to predict the location that a person looks at. Typical methods mostly require intrusive equipment like head-mounted cameras [16,19] to closely track the eyes. These methods could bring accurate results, but are extremely inconvenient to be applied in real-world scenarios, and result in unpleasant user experience (e.g., heavy load on nose and ears, high temperature around eye regions). Recently, deep learning approaches have been explored to contribute to gaze estimation

© Springer Nature Switzerland AG 2021
J. Lokoč et al. (Eds.): MMM 2021, LNCS 12573, pp. 121–133, 2021.
https://doi.org/10.1007/978-3-030-67835-7_11

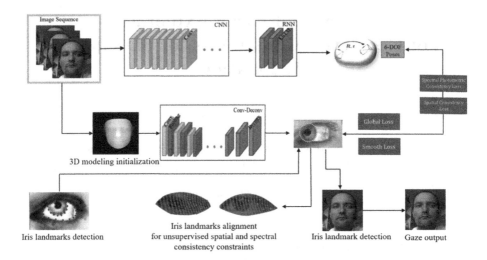

Fig. 1. Overview of the proposed unsupervised deep gaze estimation framework. For training, image sequences extracted from the eye regions are feeded into the network as input. The relative motion and depth estimation are constrained with multi-view geometric and photometric losses. During the inference, only one image is needed to output the gaze.

tasks [12,13,25]. However, for such supervised learning methods, the accuracy and performance largely depend on the availability of the amount of data with ground truth annotations, which is difficult and expensive to collect for gaze estimation tasks. Furthermore, the models trained on one dataset may suffer from error when being applied to unseen scenes.

To address such challenges above and mitigate the dependence on the limited and expensive annotated gaze data, we propose an unsupervised pipeline to learn eye gaze representation. Our scheme simulates the 3D reconstruction and ego-motion estimation approaches to exploit 3D gaze vectors from the refined eye region depth map. After extracting eye regions and iris landmarks, we propose to utilize geometric photometric consistency constraints and spatial loss to guide the learning procedure, thus leading to a more accurate 3D eye shape for pupil curve surface and the gaze vector inference in 3D space from the surface normal direction, as shown in Fig. 1. The experimental results demonstrate the effectiveness of our proposed unsupervised pipeline in predicting eye gazes compared with other most recent supervised approaches. The main contribution of this work can be summarized as follows:

1. We propose a novel unsupervised approach to learn gaze estimation in the wild without gaze annotations. To the best of our knowledge, this is the first work to explore unsupervised learning pipeline on 3D gaze estimation in the wild images; 2. We estimate the depths around the eye regions and iris landmarks to generate 3D eyeball and iris models. From experiments, we further demonstrate the accurate eye region landmark depth estimation is critical for

3D gaze estimation; 3. We explore to use multi-view image sequences as input to constrain the designed network. We introduce photo-metric loss constraints and geometric spatial loss to better refine the depth initialization on the detected iris landmarks.

2   Related Work

Gaze estimation, a measurement defined by the pupil and eyeball center, is difficult due to factors influencing accurate results: head pose, image quality, even gaze in itself. In addition, regressing gaze from natural eye images only is a challenging task, because a key component for regressing gaze, the eyeball center, is unobservable in 2D images. Many current approaches' accuracy deviates from satisfactory gaze measurements. Generally, gaze estimation can be divided into model-based [20] and appearance-based methods [6]. Model-based methods estimate the gaze using a geometric eye model and can be further distinguished as shape-based and corneal-reflection approaches. Several early corneal-reflection studies are proposed according to eye features which can be detected using the reflection principle of an external infrared light source on the outermost layer of the eye and cornea [11,22]. Other works are extended to handle free head poses using multiple devices or light sources [27,28]. Shape-based methods rely on the detected eye shape, such as iris edges, pupil center or eyeball, to estimate the gaze direction [13,17]. Although model-based methods have been applied to practical applications, the accuracy of such methods is still lower since it depends on the accuracy of detected eye shape to a large extent.

In contrast to model-based methods, appearance-based methods do not depend on geometrical features but directly learn a mapping from raw eye images to 3D gaze direction. The appearance-based approaches still exist several challenges due to changes in illumination, eye decorations, occlusions and head motion. Therefore, to overcome these difficulties, appearance-based methods not only need to introduce large and diverse datasets but also typically introduce convolutional neural network architectures. Early works on appearance-based methods are limited to stationary settings and fixed head pose [9,10]. Liang et al. [9] proposed a gaze estimation model based on appearance with a rapid calibration, which first introduced a local eye appearance Center Symmetric Local Binary Pattern (CS-LBP) descriptor for each subregion of eye image to generate an eye appearance feature vector, followed by the spectral clustering to obtain the supervision information of eye manifolds, and then the subject's gaze coordinates are estimated by a sparse semi-supervised Gaussian Process Regression (GPR). More recent appearance-based methods are applied to free head motion in the front of the camera [13,23–25]. Among them Zhang et al. [23,24] adapted the LeNet-5 and VGG-16 structure and head pose angles are connected to the first fully-connected layers. Both methods allow to be used in settings that do not require any user- or device-specific training.

However, there has not been an unsupervised neural network proposed for gaze estimation. And most existing approaches target at gaze estimation from head-mounted cameras, which is not convenient in many real applications. To deal with these issues, in this work, we propose an unsupervised deep neural network method that can be applied to wild image gaze estimation, involving 3D modeling initialization and the exploration of multiple-view geometric spatial and spectral consistency relationship between image sequences.

3   Unsupervised Gaze Estimation Framework

During the training time, to help the network learn and converge better, a pretrained unsupervised spatial transformer network for learning fitted 3D morphable model is utilized to initialize the depth estimation network [2]. Then the proposed unsupervised ego-motion and depth estimation network is to simultaneously update depth values on eye regions and relative gaze motion from monocular input sequences. The problem can be formalized as follows: given a pair of consecutive frames I_{t-1} and I_t, the corresponding depth maps D_{t-1} and D_t are generated from the depth network and the relative pose $P_{t->t-1}$ is predicted from the pose network. In addition, our method designs a geometric spatial loss to overcome depth estimation error caused by the unsupervised 3DMM learning initialization. Furthermore, we generate the synthesized source image $\tilde{I}_s$ by interpolating the target view image I_t with the predicted depth map and camera ego-motion and minimize the photometric error between $\tilde{I}_s$ and I_t, which can optimize the eyeball surface and iris landmark depth estimation. With both spatial and spectral constraints in the learning pipeline, our proposed pipeline is able to continuously refine the depth and gaze estimation, as depicted in Fig. 1.

A face expression in 3D morphable model can be expressed of a linear combination of the deviation from the neutral face, and a face model can be expressed as a sum of a mean shape, a neural face model and a unique offset as:

$$\mathcal{L}_{3DMM} = \bar{\mathcal{L}} + \sum_i \alpha_{i,neural} \mathcal{L}_{i,neural} + \sum_j \alpha_{j,exp} \mathcal{L}_{j,exp} \tag{1}$$

where $\mathcal{L}_{3DMM}$ is the 3DMM expression for an input image, $\bar{\mathcal{L}}$ is the mean shape, and $\mathcal{L}_{i,neural}$ is the i-th eigenvector from the neural face, and $\mathcal{L}_{j,exp}$ is the i-th eigenvector offset from the mean shape. α is the shape parameters for each components used to reconstruct the shape. The main goal of the pretraiend network is to regress the pose and shape parameter θ from each image, and the θ is a combination of both a 2D translation t, an axis-angle 3D rotation R, a log scale factor log and shape parameter α, as $\theta = (R, t, log, \alpha)$.

The initialization network utilizes a Spatial Transformer Network (STN) to regress pose and shape parameters with a purely geometric approach. The multi-view fitting loss, landmark loss and statistical prior shape loss are applied to guide the network learning. The multi-view fitting loss $L_{multi-view}$ is defined to constrain the sampled feature textures on visible pixels in multiple views are consistent and symmetric, benefiting from our multi-view input images. The landmark loss $L_{landmark}$ is introduced to accelerate the convergence by minimizing the distance between the projected 2D landmark points with the ground truth positions with an L2 norm. The statistical prior shape loss L_{prior} is to make sure the shape parameters follow a standard multivariate normal distribution $N(0, I_D)$. The comprehensive objective function for depth initialization is a combination of $L_{multi-view}$, $L_{landmark}$ and L_{prior}. The projected 2D landmarks and the reconstructed 3DMM model are demonstrated in Fig. 2.

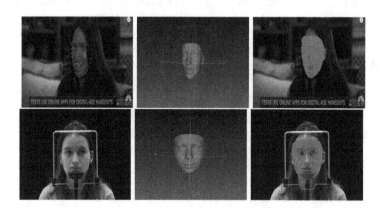

Fig. 2. Demonstration of the learned face model for initialization. Left to right: input image examples with projected landmark points; Learned 3DMM model in 3D space; Corresponding 2D alignment from the reconstructed 3D model.

Building on top of the depth initialization, we introduce our unsupervised gaze estimation network to continuously refine the initial depth and infer the gaze vector from it. The STN-based depth network for regressing face model parameters is mainly to initialize the face depth value. However, though the initialization is able to guide the network to generate a coarse depth prediction, it is not enough to recover fine details in eyeballs and iris regions, resulting in a limited effect on the final gaze output. This is mainly because 3DMM focuses on the entire face, and the model is an average estimation of the face shape. Thus, we propose a photometric consistency with a smoothness term. With the predicted depth D_t and the relative camera motion $P_{t->t-1}$, the pixel on generated synthesized image $\tilde{I}_s(p)$ can be expressed as $KP_{t->t-1}D_tK^{-1}I_t(p)$. The photometric loss is a combination of L_1 loss and Structural Similarity Index Metric (SSIM) [18] loss. L_1 shows good robustness to outliners and overall similarity. Meanwhile, $SSIM$ is better to handle the changes in contrast and illuminance. Therefore, we take the benefits of both loss constraints. The $L1$ loss is given as:

$$L_{photo\_l1} = \sum_p \left\| \tilde{I}_s(p) - I_t(p) \right\|_1 \qquad (2)$$

The corresponding $SSIM$ constraint is expressed as:

$$L_{photo\_ssim} = \frac{1 - SSIM(\tilde{I}_s(p), I_t(p))}{2} \qquad (3)$$

Then the full photo-metric error turns into a linear combination of $L1$ loss and the $SSIM$ term as:

$$L_{photo} = \lambda_1 L_{photo\_l1} + \lambda_2 L_{photo\_ssim} \tag{4}$$

where λ_1 and λ_2 are set as 0.15 and 0.85 respectively following [5,21]. As photometric loss may not be sufficient to prevent small divergent depth values from occluded regions and low-textured areas, we introduce the texture-aware smoothness term as a penalty term to minimize depth-related Laplacian of Gaussian (LoG) filter whose each element is weighted by the corresponding image gradient, described as follows:

$$L_{smooth} = \left\| \frac{|\triangledown^2 (D \otimes G)|}{\|\triangledown I\|} \right\|_1 \times \frac{1}{\|D\|} \tag{5}$$

where $\triangledown$ and $\triangledown^2$ refer to the gradient and Laplacian operator respectively. G represents a 5×5 Gaussian kernel. D and I correspond to the predicted inverse depth map and input image respectively. The motivation to scale the first term by dividing the mean depth value is to normalize the output.

As photometric error with smoothness term has a limitation on repeated and fine regions, we designed a feature-based geometry spatial consistency loss to deal with this issue. More specifically, benefit from 2D face landmark points, we force the warped iris keypoints from the source image and the corresponding key-points on the reference image to share the vertical coordinate. The proposed feature-based spatial consistency loss is introduced as:

$$L_{spatial} = \frac{1}{N} \sum_N (\widetilde{Y}_i - Y_i)^2 \tag{6}$$

where $\widetilde{Y}_i$ is the i th spatial Y-axis coordinate of the warped keypoints from the source image, and the Y_i is the spatial Y-axis coordinate of the correspondence. Their spatial position should be constrained to be close with the predicted depth map and the relative camera poses. The integrated objective function that is used to learn the entire network for gaze estimation is given in the following:

$$L_{total} = \lambda_1 L_{photo} + \lambda_2 L_{smooth} + \lambda_3 L_{spatial} \tag{7}$$

where λ_1, λ_2 and λ_3 are the weights for each constraint introduced above, which are set to be 1, 0.2 and 0.5 respectively. With the refined depth, we model the detected 2D landmarks on iris regions as a 2D ellipse in the input image (Fig. 3), and then project the landmarks into a 3D circle. With the help of the proposed geometric spatial and spectral constraints, we are able to recover the eyeball shape more accurately in fine details.

4   Experiments

To evaluate the effectiveness of our proposed pipeline for eye gaze estimation, we conduct experiments on both MPIIFaceGaze dataset [23] and Columbia Gaze dataset [15] containing various face and gaze orientations. We also compare our proposed pipeline with other recent learning-based gaze estimation methods [25] and [13] to demonstrate the comparable performance and suitability of our unsupervised approaches for gaze estimation.

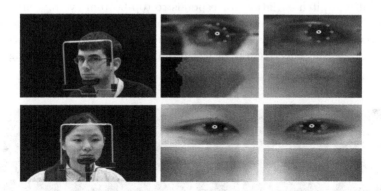

Fig. 3. Demonstration of the zoom-in eye regions with the detected iris landmarks and the predicted depth maps.

4.1 Dataset

Columbia Gaze Dataset [15] comprises 5880 full-face images from 56 individuals (24 female, 32 male). For each person, images were collected on five horizontal head poses ($0°$, $\pm15°$, $\pm30°$), seven horizontal gaze directions ($0°$, $\pm5°$, $\pm10°$, $\pm15°$), and three vertical gaze directions ($0°$, $\pm10°$). We process the neighboring horizontal head poses as adjacent sequences and select 10 males and 10 females for training, and the rest five individuals (3 males and 2 females) for testing.

MPIIFaceGaze dataset [23] was collected to estimate gaze direction angles. It contains 213659 images with the corresponding ground truth gaze position of 15 individuals (six females and nine males). We train the network from the selected 10 individuals (four females and six males) and the rest five for evaluation.

4.2 Training Configuration

The proposed learning framework is implemented with PyTorch library. Similar to [26], we use a snippet of three sequential frames as the input for training. The second image is processed as the target view to compute the photometric consistency, spatial back-projection loss, and global scale depth consistency to minimize their distances. The proposed geometric modeling loss and the smoothness loss are constrained from the predicted depths from the depth network. The training data is augmented with random scaling, cropping and horizontal flips, and we set the input image resolution of 640×384 for training to keep the similar ratio between image height and width). Adam optimizer [8] is applied together with a batch size of 4, while the learning rate is 2e−3. The entire pipeline is trained for 50 epochs.

For validating the accuracy of our proposed unsupervised learning based gaze estimation network, we compare the gaze estimation error of the two eye rotation variables: Gaze yaw and pitch, where the yaw variation corresponds to left-right and pitch variation corresponds to top-bottom eye movements. We compute the mean angular error via cosine similarity in a format of mean errors in degrees on two datasets, compared with other methods. In the meantime, we further validate the effectiveness of the proposed method by providing visual comparison results.

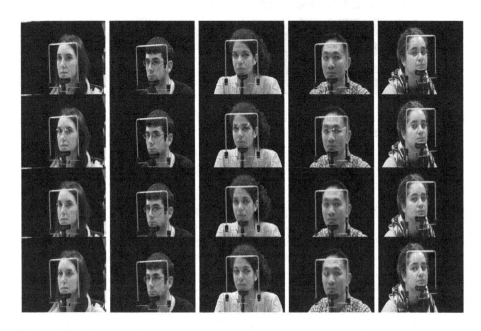

Fig. 4. Visual comparison on gaze estimation between our result and other recent methods on Columbia Gaze dataset. Ground truth vector is repeated on two eyes. First row: raw input image with ground truth gaze vector marked in red arrow; Second row: our result from the proposed full pipeline; Third row: Result from [25] (supervised network); Fifth row: result from [13] (supervised network).

4.3 Qualitative Results and Analysis

We provide various qualitative experiments on diverse datasets to show the effectiveness of our methods. In Fig. 4, we show sample results on test data of Columbia gaze dataset. The predicted gaze vectors in images are drawn in yellow and the ground truth vectors from the dataset are drawn in red. It can be observed that, on Columbia dataset, our method achieves similar performance as [25] and higher accuracy than [13]. On MPIIFaceGaze dataset, as shown in Fig. 5, the result presents similar outputs, when detecting and predicting persons at the same time. Our method tends to predict a slightly larger error compared with the ground truth annotations, which can be explained by that for far and dark objects, the predicted depth has a higher uncertainty because of the increase of the real distance from the camera.

Next, in Fig. 6 we show a depth estimation comparison results, alongside the reconstructed iris landmarks output in 3D space. We observe that though supervised learning 3DMM-based method [14] is able to recover a rough face shape, the details in eye and iris regions are missing, resulting in a relatively large deviation in the later gaze estimation. Compared with the depths and reconstructed 3D landmarks from [14], the results from our full pipeline take advantages of both spatial and spectral consistency constraints, and have a smaller distortion in 3D landmarks.

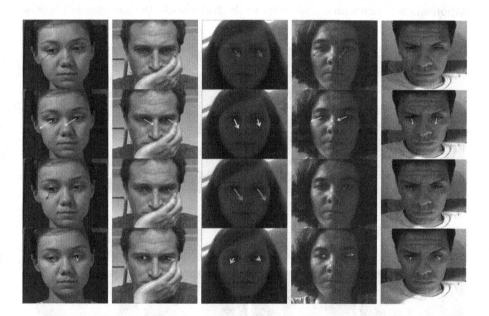

Fig. 5. Visual comparison on gaze estimation between our result and other recent methods on MPIIFaceGaze dataset. First row: raw input image with ground truth gaze vector marked in red arrow; Second row: our result from the proposed full pipeline; Third row and fourth row: result from supervised methods of [25] and [13].

Table 1. Cross-dataset validation of the mean gaze angular error on two datasets.

Test	Train	
	CAVE	MPIIFaceGaze
CAVE	–	8.7 (Ours)/9.1 [25]/8.9 [13]
MPIIFaceGaze	7.8 (Ours)/8.6 [25]/7.6 [13]	–

4.4   Quantitative Results and Analysis

We present the mean angular errors compared with other existing methods on the datasets. In Fig. 7, we separately verify the effectiveness of our methods on Columbia Gaze dataset and MPIIFaceGaze dataset. In the left sub-figure of Fig. 7, our proposed unsupervised method achieves 7.2° in mean angular error, compared with 6.7° for [25] and 5.6 for [13]. Compared with the supervised method trained on the same data, our unsupervised method without using any ground truth annotations is able to generate a comparable result though keeping slightly higher angular errors. Similar results can be observed in the right sub-figure of Fig. 7. When compared with other supervised learning approaches, our proposed unsupervised network presents slightly higher angular errors against [25] by just 0.5° and [13] by 0.6°.

An ablation analysis is also provided in the Fig. 7. We compare different components and determine their roles in the framework. We can clearly observe

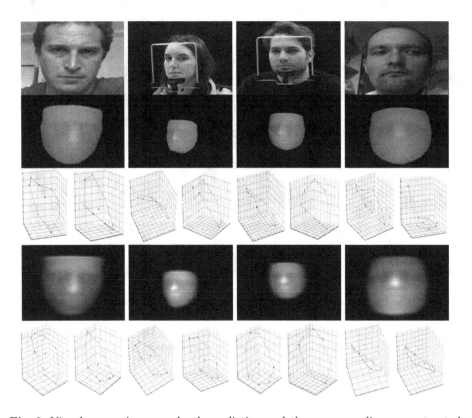

Fig. 6. Visual comparisons on depth prediction and the corresponding reconstructed 3D iris landmarks between our method and recent supervised method [14] on MPI-IFaceGaze and Columbia Gaze dataset. First row: raw input image; Second row: depth predictions from supervised method [14]; Third row: the reconstructed iris landmarks from [14]; Fourth and fifth rows: Above results from our proposed full pipeline.

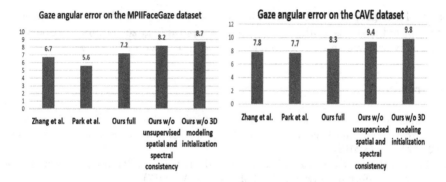

Fig. 7. Left: gaze angular error comparisons on the MPIIFaceGaze dataset: Result from supervised learning method Zhang et al. [25]; and Park et al. [13]; Result from our full pipeline; Our result without the proposed unsupervised spatial and spectral consistency loss; Our result without unsupervised initialization process; Right: 3D gaze angular error comparisons on the CAVE gaze dataset: Same comparison items as above.

that with the help of the proposed spatial and spectral consistency loss, and a 3D modeling initialization process, our full pipeline is able to achieve a 7.2° angular error on the MPIIFace dataset and 8.3° angular error on the CAVE dataset, compared with an obvious decrease in the settings without the proposed unsupervised spatial and spectral consistency and 3D modeling initialization process.

To further validate the effectiveness of our proposed network, we design an evaluation on cross-dataset. As depicted in Table 1, we report the result of our proposed network trained on samples from MPIIFaceGaze dataset and test the performance on Columbia gaze dataset, and the result of our method trained on Columbia gaze dataset and directly test on MPIIFaceGaze dataset. As observed in Table 1, our method achieves an error of 8.7° on CAVE dataset, and an error of 7.8° on MPIIFaceGaze dataset, which is smaller than supervised methods [13,25]. It demonstrates the superior performance of our proposed method that without relying on large amount of labelled gaze data for training, the proposed method performs better on scenarios without training or fine-tuning.

5 Conclusion

In this work, we propose an unsupervised learning-based method to estimate the eye gaze in the wild. Extending the 3D modeling initialization, we further propose photometric consistency constraint and spatial back-projection constraints to optimize the depth and the curvature modeling effect of the eyeball region. With 3D modeling initialization, the refined depth maps with our unsupervised geometrical constraints are able to provide reliable base for gaze estimation. Experimental results on multiple benchmark datasets demonstrate the comparable effectiveness with other supervised methods and excellent generalization ability of our proposed network. With no gaze or 3D ground truth annotations, our proposed network is able to estimate comparable gaze vectors as supervised neural network, which can be widely applied in the wild conditions.

References

1. Atkins, M.S., Tien, G., Khan, R.S., Meneghetti, A., Zheng, B.: What do surgeons see: capturing and synchronizing eye gaze for surgery applications. Surg. Innov. **20**, 241–248 (2013)
2. Bas, A., Huber, P., Smith, W.A., Awais, M., Kittler, J.: 3D morphable models as spatial transformer networks. In: IEEE ICCVW (2017)
3. Drewes, H., De Luca, A., Schmidt, A.: Eye-gaze interaction for mobile phones. In: Proceedings of the Conference on Mobile Technology, Applications, and Systems (2007)
4. Fridman, L., Langhans, P., Lee, J., Reimer, B.: Driver gaze region estimation without use of eye movement. IEEE Intell. Syst. **31**, 49–56 (2016)
5. Godard, C., Mac Aodha, O., Firman, M., Brostow, G.J.: Digging into self-supervised monocular depth estimation. In: IEEE ICCV (2019)
6. Hansen, D.W., Ji, Q.: In the eye of the beholder: a survey of models for eyes and gaze. IEEE TPAMI **32**, 478–500 (2009)
7. Hutchinson, T.E., White, K.P., Martin, W.N., Reichert, K.C., Frey, L.A.: Human-computer interaction using eye-gaze input. IEEE Trans. Syst. Man Cybern. **19**, 1527–1534 (1989)
8. Kingma, D.P., Ba, J.: Adam: a method for stochastic optimization. arXiv preprint arXiv:1412.6980 (2014)
9. Liang, K., Chahir, Y., Molina, M., Tijus, C., Jouen, F.: Appearance-based gaze tracking with spectral clustering and semi-supervised Gaussian process regression. In: Proceedings of Conference on Eye Tracking South Africa (2013)
10. Lu, F., Sugano, Y., Okabe, T., Sato, Y.: Adaptive linear regression for appearance-based gaze estimation. IEEE TPAMI **36**, 2033–2046 (2014)
11. Morimoto, C.H., Amir, A., Flickner, M.: Detecting eye position and gaze from a single camera and 2 light sources. In: Object Recognition Supported by User Interaction for Service Robots, vol. 4, pp. 314–317. IEEE (2002)
12. Park, S., Mello, S.D., Molchanov, P., Iqbal, U., Hilliges, O., Kautz, J.: Few-shot adaptive gaze estimation. In: IEEE ICCV (2019)
13. Park, S., Spurr, A., Hilliges, O.: Deep pictorial gaze estimation. In: ECCV, pp. 721–738 (2018)
14. Sela, M., Richardson, E., Kimmel, R.: Unrestricted facial geometry reconstruction using image-to-image translation. In: IEEE ICCV (2017)
15. Smith, B.A., Yin, Q., Feiner, S.K., Nayar, S.K.: Gaze locking: passive eye contact detection for human-object interaction. In: ACM Symposium on User Interface Software and Technology, pp. 271–280 (2013)
16. Tsukada, A., Shino, M., Devyver, M., Kanade, T.: Illumination-free gaze estimation method for first-person vision wearable device. In: IEEE ICCVW (2011)
17. Valenti, R., Sebe, N., Gevers, T.: Combining head pose and eye location information for gaze estimation. IEEE TIP **21**, 802–815 (2011)
18. Wang, Z., Bovik, A.C., Sheikh, H.R., Simoncelli, E.P.: Image quality assessment: from error visibility to structural similarity. IEEE Trans. Image Process. **13**(4), 600–612 (2004)

19. Whitmire, E., et al.: EyeContact: scleral coil eye tracking for virtual reality. In: Proceedings of ACM International Symposium on Wearable Computers, pp. 184–191 (2016)
20. Wood, E., Baltrusaitis, T., Zhang, X., Sugano, Y., Robinson, P., Bulling, A.: Rendering of eyes for eye-shape registration and gaze estimation. In: IEEE ICCV (2015)
21. Yin, Z., Shi, J.: GeoNet: unsupervised learning of dense depth, optical flow and camera pose. In: IEEE CVPR (2018)
22. Yoo, D.H., Chung, M.J.: A novel non-intrusive eye gaze estimation using cross-ratio under large head motion. CVIU **98**, 25–51 (2005)
23. Zhang, X., Sugano, Y., Fritz, M., Bulling, A.: Appearance-based gaze estimation in the wild. In: IEEE CVPR, pp. 4511–4520 (2015)
24. Zhang, X., Sugano, Y., Fritz, M., Bulling, A.: It's written all over your face: full-face appearance-based gaze estimation. In: IEEE CVPRW (2017)
25. Zhang, X., Sugano, Y., Fritz, M., Bulling, A.: MPIIGaze: real-world dataset and deep appearance-based gaze estimation. IEEE TPAMI **41**, 162–175 (2017)
26. Zhou, T., Brown, M., Snavely, N., Lowe, D.G.: Unsupervised learning of depth and ego-motion from video. In: IEEE CVPR (2017)
27. Zhu, Z., Ji, Q.: Eye gaze tracking under natural head movements. In: IEEE CVPR (2005)
28. Zhu, Z., Ji, Q., Bennett, K.P.: Nonlinear eye gaze mapping function estimation via support vector regression. In: ICPR (2006)

Median-Pooling Grad-CAM: An Efficient Inference Level Visual Explanation for CNN Networks in Remote Sensing Image Classification

Wei Song[1]([✉]) [iD], Shuyuan Dai[1], Dongmei Huang[1,2], Jinling Song[3], and Liotta Antonio[4]

[1] Shanghai Ocean University, Shanghai 201306, China
wsong@shou.edu.cn
[2] Shanghai University of Electric and Power, Shanghai 201306, China
[3] Hebei Normal University of Science and Technology, Qinhuangdao 066000, China
[4] Free University of Bozen-Bolzano, 39100 Bozen-Bolzano, Italy
antonio.liotta@unibz.it

Abstract. Gradient-based visual explanation techniques, such as Grad-CAM and Grad-CAM++ have been used to interpret how convolutional neural networks make decisions. But not all techniques can work properly in the task of remote sensing (RS) image classification. In this paper, after analyzing why Grad-CAM performs worse than Grad-CAM++ for RS images classification from the perspective of weight matrix of gradients, we propose an efficient visual explanation approach dubbed median-pooling Grad-CAM. It uses median pooling to capture the main trend of gradients and approximates the contributions of feature maps with respect to a specific class. We further propose a new evaluation index, confidence drop %, to express the degree of drop of classification accuracy when occluding the important regions that are captured by the visual saliency. Experiments on two RS image datasets and for two CNN models of VGG and ResNet, show our proposed method offers a good tradeoff between interpretability and efficiency of visual explanation for CNN-based models in RS image classification. The low time-complexity median-pooling Grad-CAM could provide a good complement to the gradient-based visual explanation techniques in practice.

Keywords: Visual explanation · Median pooling · CNN networks · Remote sensing images

1 Introduction

Nowadays, deep learning models have excellent performance in various related computer vision tasks and applications, like image recognition [1] and image classification [2, 3]. However, the deep architecture of these models often work as a "black box" and cannot be easily understood by looking at their parameters [4], and thus has lower interpretability than the linear model, decision tree and other simple models. A straightforward visual

© Springer Nature Switzerland AG 2021
J. Lokoč et al. (Eds.): MMM 2021, LNCS 12573, pp. 134–146, 2021.
https://doi.org/10.1007/978-3-030-67835-7_12

explanation of the deep learning models will be useful to explain how the decisions are made to human users. This is particularly necessary in risk-averse applications such as security, health and autonomous navigation, where the reliance of model on the correct features must be ensured.

More interpretable models have been developed to reveal important patterns in the deep learning models. In 2014, a deconvolution approach (DeConvNet) [5] is proposed to better understand what the higher layers represent in a given network. Later, Springenberg et al. [6] extended this work and proposed a guided backpropagation (GBP) which adds an additional guidance signal from the higher layers to usual backpropagation. These works generate high-resolution saliency maps by performing a gradient ascent in pixel space. Another method, Local Interpretable Model-Agnostic Explanations (LIME) [7] works in super-pixel level, generates a set of perturbed samples (super-pixel blocks) close to the original input image, and trains a linear classifier to generate a local explanation. Its explanation performance is affected by the neighborhood of the input and perturbed samples, leading to bias in some datasets [8].

Other methods estimate saliency/attribution distribution at intermediate layer of neural units. Class Activation Map (CAM) [9] computes an attribution map over features from the last convolutional layer. However, it needs to modify the fully-connected layer with a Global Average Pooling (GAP) and retrain a linear classifier for each class. Fukui et al. [10] extended CAM-based model by introducing a branch structure with an attention mechanism. However, the attention branch is not independent from the image classification model. Gradient-weighted CAM (Grad-CAM) [11] uses average gradients to replace the GAP layer to avoid modifying and retraining the original model. What's more, by fusing pixel-space GBP with Grad-CAM, Guided Grad-CAM is able to generate a class-sensitive and high-resolution output to human-interpretability. But Grad-CAM has issues in capturing the entire object and explaining occurrence of multiple objects instances, which is considered to affect performance on recognition task [12]. Grad-CAM++ [12] was therefore proposed to improve the class object capturing, localization and visual appeal. Later, Omeiza et al. proposed Smooth Grad-CAM++ [13] by combining Grad-CAM++ with SmoothGrad [14], which alleviates visual diffusion for saliency maps by averaging over explanations of noisy copies of an input.

All these recent methods aim to produce more visually sharp maps with better localization of objects in the given input images. Their performances were verified on general image datasets that contain objects such as people, animals and plants (e.g., ImageNet and Pascal VOC). However, our previous study [15] found that Grad-CAM could not correctly explain a CNN model's decision on many ground objects in remote sensing (RS) images. Wang et al. [16] deemed gradient-based saliency methods suffers from "attribution vanishing". It is still questionable why Grad-CAM makes mistakes and whether advanced Grad-based techniques can work well on RS images.

As for evaluating visual explanation methods of neural networks, most existing methods such as DeConvNet, GBP, CAM, and Grad-CAM, are based on human visual assessment or localization error with respect to bounding boxes. Recently, Adebayo et al. [17] focused on measuring the sensitivity of explanations to model and input perturbations. Yang et al. [18] proposed three complementary metrics to evaluate more aspects of attribution methods with a priori knowledge of relative feature importance. Wang et al.

[16] proposed Disturbed Weakly Supervised Object Localization (D-WSOL) to evaluate the reliability of visual saliency methods. Their requirement for ground truth limits the applicability [8]. Zhang et al. [8] targeted at a unified evaluation without ground-truth, and evaluated objectiveness, completeness, robustness and commonness of the explanation methods. However, pixel-wised metrics of objectiveness and completeness cannot be applied to gradient-based methods such as Grad-CAM. Chattopadhyay et al. [12] proposed to evaluate the faithfulness of Grad-CAM++ by inferring the accuracy drop, confidence increase, and contrastive win between two compared methods upon unimportant regions occlusion. We argue these metrics cannot fully evaluate the precision of explanations, as a method producing boarder saliency regions can also achieve high faithfulness.

From the literatures, two main questions are drawn out: (1) How do Grad-CAM and its variations behave in visual explanation for CNN models on remote sensing images? (2) Can we provide an efficient inference level visual explanation method for CNN models? In this paper, while solving these questions, we make the below contributions:

- A median pooling Grad-CAM that can better localize objects than Grad-CAM in a saliency map. The median pooling Grad-CAM has much lower cost than Grad-CAM++, but almost identical performance.
- A new evaluation metric for gradient-based visual explanation method, named confidence drop %. It directly observes the model's confidence drop in occluding important regions that are determined by the visual explanation method. A higher drop means a more precise importance localization in a saliency map.
- An extensive experiment to compare different Grad-CAM based visual explanation techniques. It demonstrates the performance of these techniques in different remote sensing image datasets and for different CNN models.

2   Gradient-Based Visual Explanation on Remote Sensing Image Classification

This section aims to demonstrate how the gradient-based techniques visually explain a CNN model in the task of RS image classification, and discuss the underlying reasons.

2.1   Introduction to Grad-CAM Techniques

Grad-CAM. Given the score y^c for class c, the gradient matrix G_k^c of y^c with respect to kth feature map $A^k \in \mathbb{R}^{u \times v}$ can be computed by (1). n is the size of a feature map in one layer of CNN model. Grad-CAM computes global-average-pooled G_k^c to obtain the weight of kth feature map, α_k^c. It can be expressed by (2) and (3), where W is a weight matrix filled by the inverse of the number of matrix elements in G_k^c, and α_k^c is the sum of Hadamard multiplication ($\circ$) of G_k^c and W_k^c. The α_k^c is regarded as a key parameter to decide contributions of different parts of feature map in Grad-CAM. The class-specific saliency map L^c is then computed as a weighted combination of feature maps and remains only positive values by a ReLU. This coarse saliency map is then normalized for final visualization and interpretation as a heat map. Except that, Grad-CAM fuses the nature of GBP via point-wise multiplication to produce high-resolution pixel-space visual explanation.

$$
G_k^c = \begin{bmatrix} \frac{\partial y^c}{\partial A_{11}^k} & \cdots & \frac{\partial y^c}{\partial A_{1n}^k} \\ \vdots & \ddots & \vdots \\ \frac{\partial y^c}{\partial A_{n1}^k} & \cdots & \frac{\partial y^c}{\partial A_{nn}^k} \end{bmatrix} \tag{1}
$$

$$
W_k^c = \begin{bmatrix} \frac{1}{n^2} & \cdots & \frac{1}{n^2} \\ \vdots & \ddots & \vdots \\ \frac{1}{n^2} & \cdots & \frac{1}{n^2} \end{bmatrix} \tag{2}
$$

$$
\alpha_k^c = \sum_i \sum_j \left(W_k^c \circ G_k^c \right)_{i,j} \tag{3}
$$

$$
L^c = ReLU \left(\sum_k \alpha_k^c A^k \right) \tag{4}
$$

Grad-CAM++. In Grad-CAM++, W_k^c is a weight matrix about the contribution of each location. Its element $w_{ij}^{c,k}$, as expressed by (5), is computed with the 1st, 2nd and 3rd order partial derivatives with respect to A_{ij}^k, and simplified using S^c as the penultimate layer scores for class c, $y^c = \frac{\exp(S^c)}{\sum_k \exp(S^k)}$. Then, α_k^c is computed by (6).

$$
w_{ij}^{c,k} = \frac{\frac{\partial^2 y^c}{\left(\partial A_{ij}^k \right)^2}}{2 \frac{\partial^2 y^c}{\left(\partial A_{ij}^k \right)^2} + \sum_a \sum_b A_{ab}^k \frac{\partial^3 y^c}{\left(\partial A_{ij}^k \right)^3}} = \frac{\left(\frac{\partial S^c}{\partial A_{ij}^k} \right)^2}{2 \left(\frac{\partial S^c}{\partial A_{ij}^k} \right)^2 + \sum_a \sum_b A_{ab}^k \left(\frac{\partial S^c}{\partial A_{ij}^k} \right)^3} \tag{5}
$$

$$
\alpha_k^c = \sum_i \sum_j \left(W_k^c \circ ReLU \left(G_k^c \right) \right)_{i,j} \tag{6}
$$

Smooth Grad-CAM++ combines the methods from two techniques, Grad-CAM++ and SmoothGrad. It adds Gaussian noise to the original input to generate multiple noised sample images. The process of obtaining L^c is the same as Grad-CAM++, except the G_k^c and W_k^c are calculated by averaging all noised inputs. Benefiting from small perturbations of a given image, Smooth Grad-CAM++ performs better in both object localization and multiple occurrences of an object of same class than Grad-CAM and Grad-CAM++. However, the cost is high computing complexity.

2.2 Performance Analysis of Grad-CAM Techniques

To analyze the performance of Grad-CAM techniques on RS image classification, we follow the previous study [15] and used UC Merced Land Use (UCGS) [2] as the dataset, of which details are described in Sect. 4.1. The image classification CNN model is the VGG16 [1] pretrained with ImageNet dataset. It was retrained using UCGS. Then both

Grad-CAM and Grad-CAM++ were implemented on the VGG16. Finally, we generated saliency maps by Grad-CAM or Grad-CAM++ for analysis. Smooth Grad-CAM++ is not presented here because its basic process is the same as Grad-CAM++.

Figure 1 shows some examples of Grad-CAM and Grad-CAM++, including a heat/saliency map and a guided propagation map for each image. As shown in Fig. 1, the heatmaps generated by Grad-CAM highlight the regions that do not match human interpretation for some RS images (e.g., "airplane" and "harbor" in the first two rows of Fig. 1). This is consistent with the conclusion in [15]. Chattopadhyay et al. [12] deemed the drawback of Grad-CAM lies in it cannot localize the entire region of a single object and multiple occurrences of the same class. Based on its performance on RS images, Grad-CAM localizes wrong regions sometimes. Previous study [15] considered this was because the effect of negative gradients is filtered out by the ReLU operation. However, Grad-CAM ++ using the same operation gets right results. We are interested in the real reason behind this phenomenon.

Comparing the formulas (2) and (5) given in Sect. 2.1, we know that the weights of gradients W_k^c determine the difference between Grad-CAM and Grad-CAM++. To intuitively illustrate how they are working, we take one of k feature maps of the last convolutional layer, and compute its gradient matrix for a specific class as well as the weight matrices used by Grad-CAM and Grad-CAM++, shown in 3D charts in Fig. 2. In Fig. 2(a), the gradients of a 14×14 pixel feature map have different values, where negative values can be ignored as they will be eliminated by a later ReLU operation. Grad-CAM that multiplies the gradient matrix (Fig. 2(a)) with a constant weight matrix (Fig. 2(b)), treats all the gradients equally, against the fact that not all of gradients contribute to explain a specific class. The consequence is some of gradients that possibly belong to another class are considered equally, while the importance of useful gradients is reduced by averaging. Furthermore, when Grad-CAM accumulates the average-weighted gradients of k feature maps to produce the saliency map, it hardly represents the different importance of different feature maps. As a result, some important gradients may have smaller values than other irrelevant gradients, leading to a totally opposite saliency map or confusing conclusion in visual explanation, like Fig. 1 "airplane". The weight matrix of Grad-CAM++ shown in Fig. 2(c) enlarges the impact of some gradients while eliminates other gradients to zero. According to the formula (5), the weight matrix based on the 1st and 2nd order partial derivatives of gradients, will sharpen the sensitivity of gradients for the specific class and avoid the influence of gradients belonging to other classes. Our analysis concludes that the problem of Grad-CAM is mainly due to its global average pooling for gradients. Grad-CAM++ is effective in visual explanation but is computational complicated.

3   Approach

3.1   Median-Pooling Grad-CAM

Considering the above-mentioned issues, we propose a solution: using median-pooled gradients of feature maps to approximate the contributions of feature maps with respect to a specific class. An overview of median-pooling Grad-CAM (MP Grad-CAM for short) is shown in Fig. 3. It should be pointed out that any convolutional layer can be visualized, simply by backpropagating the classification score to that layer.

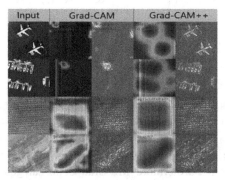

Fig. 1. Saliency maps generated by Grad-CAM and Grad-CAM++ for VGG16. (From top to bottom are airplane, harbor, agriculture and beach).

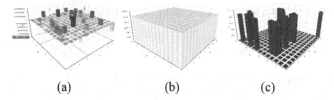

$$(a) \qquad\qquad (b) \qquad\qquad (c)$$

Fig. 2. (a) the gradients of kth feature map; (b) the global averaging weights of Grad-CAM; (c) the combined weights computed by Grad-CAM++.

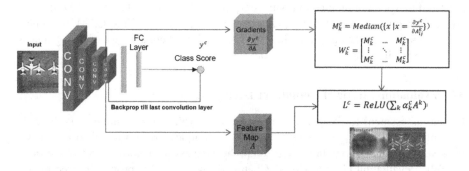

Fig. 3. An overview of Median-pooling Grad-CAM.

In details, we use the values of median-pooled gradients to fill the weight matrix W_k^c, as computed by (7) and (8). Because of the existence of zero values in backpropagation of a specific class, wiping out invalid values before median-pooling is necessary. Then, substitute (8) into (3) and (4) to compute α_k^c and L^c.

$$M_k^c = Median(\left\{ x | x = \frac{\partial y^c}{\partial A_{ij}^k} \ \& \ x \neq 0, i, j \in [1, n] \right\}) \tag{7}$$

$$W_k^c = \begin{bmatrix} M_k^c \ \dots \ M_k^c \\ \vdots \ \ddots \ \vdots \\ M_k^c \ \dots \ M_k^c \end{bmatrix} \tag{8}$$

The median of gradients is the value of the middle data point in ascending order. It can measure the central tendency of the gradients in each feature map, it is especially good for skewed distribution of data and not affected by outliers (/noises). Considering remote sensing images are vulnerable to pulse noises, median pooling is suitable to reduce the effect of noises. Furthermore, MP Grad-CAM can generate different weight matrices for different feature maps, and thus it can capture the importance of the feature maps in representing objects. Although the median-pooled matrix gives the same weight to the gradients in one feature map, when the mean value of gradients is small it can better stretch the absolute difference of the gradients than average pooling. This may help restrain relative importance of irrelevant gradients for a specific class.

3.2 Computation Complexity

MP Grad-CAM has a great advantage in computation complexity over Grad-CAM++ and its variations. Since the difference among the gradient-based methods is mainly about the weight matrix W_k^c, we only need to analyze the time complexity of computing this matrix. Let n be the number of gradients in each of k feature maps. For Grad-CAM++, it takes $O(n^4)$ to compute weight matrix W_k^c using (5). For Smooth Grad-CAM++, if using m noised images, the time complexity is $O(mn^4)$. The simplest Grad-CAM only takes $O(1)$, but it gives unacceptable explanation sometimes. In MP Grad-CAM, the time to find the median is $O(n \log n)$, and the n will be small as there are a lot of zero gradients which are not involved into computation.

3.3 Evaluation Metric of Confidence Drop

Human assessment to the performance of visual explanation techniques is direct by viewing the heatmaps generated by various techniques, being limited to the viewed images. It is required to obtain more comprehensive assessment by objective metrics.

When evaluating Grad-CAM++, Chattopadhay et al. [12] proposed three metrics to measure the faithfulness change of a CNN model after occluding parts of the input image according to their importance in model decision-making as determined by the visual explanation method. These metrics are: (i) Average Drop % measures the total percentage of decrease of model's confidence after occlusion in a given set of images; (ii) % Increase in Confidence measures the number of times in a given set of images, the model's confidence increased upon occlusion; (iii) Win % compares the contrastive effectiveness of the explanation maps of two methods. It measures the number of times in a given set of images, the fall in the model's confidence for an explanation map generated by method A is less than that by method B. (Refer to [12] for more details).

As all the three metrics are upon occluding unimportant parts of an image, they focus on whether unhighlighted regions are important to the model decision. Considering the following situation: if a method generates a slightly larger area of the unimportant regions than another method, the faithfulness of these two methods probably will not be different based on the three metrics. However, the former has a less precision of the highlighted regions.

To directly reveal the precision of visual explanation to the importance in model decision-making, we propose a new metric: confidence drop % measures the decrease of confidence for a particular class in images after occluding important regions. It is formally expressed as $\left(\sum_{i=1}^{N} \frac{\max(0, Y_i^c - H_i^c)}{Y_i^c} \right) \times 100$. Y_i^c refers to the model's output score (confidence) for class c for the ith image. H_i^c is the score to the same model's confidence in class c with important regions of ith input image occluded. The summation is over all the images in a given dataset. The max function ensures no negative values are accounted. Occluding parts of an image would mostly lower the confidence of the model in its decision. If the occluded parts are significant for the model decision, the fall of the model output score is expected to be high. Therefore, the higher the confidence drop %, the better the visual explanation technique.

4 Experiments

We carried out experiments to compare the performance of MP Grad-CAM with Grad-CAM, Grad-CAM++ and Smooth Grad-CAM for explaining CNN models. Besides, in order to elaborate why max-pooling is not considered, we also implemented a max-pooling Grad-CAM (hereafter MX Grad-CAM). It takes the same structure of MP Grad-CAM, except that it uses the maximum value of gradients in the weight matrix. Other recent methods [10, 16] are not compared because they are integrated with the original classification model, modifying the training loss of the original model.

4.1 Datasets

Two different remote sensing datasets are used. One is UC Merced Land Use dataset (UCGS) [2], which 21 kinds of land scenes cover various urban areas across the United States. A total of 2100 images in UCGS, each is 256 × 256 pixels and each pixel resolution is 1 foot. The other is SIRI-WHU dataset [20], which is sourced from Google Earth and mainly covers urban areas in China in 2 m of spatial resolution. The SIRI-WHU is composed of 2400 images organized in 12 classes. Each image is 200 × 200 pixels. Since the two remote sensing datasets are acquired by different sensors and contain scene images of different regions, the experiments with them can further demonstrate the ability of different visual explanation methods. To train a CNN network, data argument was executed by rotating 90°, 180°, 270° and mirroring the raw images. The split ratio of training and testing dataset is 8:2. The following results are on the testing dataset.

4.2 Results on Remote Sensing Images

To ensure the results are not affected by CNN models, the experiments were run for two classic CNN models: VGG-16 [1] and Resnet-34 [19]. We demonstrate visual explanation results in Fig. 4 and Fig. 5 and objective evaluation results in Table 1, 2, 3 and 4.

Visual Explanation. Figure 4 presents the visual explanation maps from the last convolution layer of VGG16 for some remote sensing images from UCGS and SIRI-WHU, respectively. We can see that MP Grad-CAM outputs similar results with Grad-CAM++ and Smooth Grad-CAM++, and it successfully fixes the problem of Grad-CAM in localizing "airplane". MX Grad-CAM is similar with Grad-CAM and cannot produce correct saliency map on the "airplane" image. For the Resnet-34 network, Fig. 5 show the results from the 27th layer, in order to testify the capability of visualizing a layer at the interference level of the model prediction process.

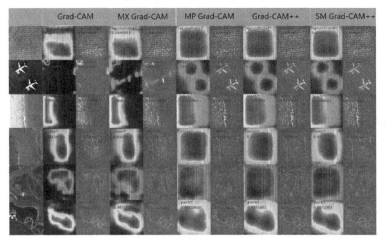

Fig. 4. Comparison of visual explanation results generated by Grad-CAM, MX Grad-CAM, MP Grad-CAM, Grad-CAM++ and Smooth Grad-CAM++ for VGG16 model. The images from top to bottom are agriculture, airplane, beach (from UCGS) and river, overpass, park (from SIRI-WHU).

We found that although MP Grad-CAM produces the heatmaps covering less areas than Grad-CAM++ and Smooth Grad CAM++, its highlights are reasonable and its guided propagation maps extract objects completely in most cases. Comparing Fig. 4 with Fig. 5, it can also be observed the heatmaps of different CNN models (VGG16 and ResNet-34) are different. In addition, MX Grad-CAM is sensitive to the change of CNN models.

Objective Evaluation. Table 1 for VGG16 and Table 2 for ResNet-34 list the scores calculated using both the three metrics suggested by Chattopadhay et al. [12] and our proposed new metric confidence drop %. Except for Average Drop %, which is the lower the better, other metrics are the higher the better. When calculating Win %, we first computed the contrastive value between Grad-CAM and each of other four methods,

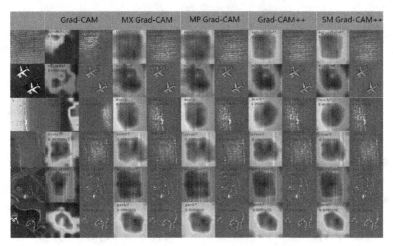

Fig. 5. Comparison of visual explanation results generated by Grad-CAM, MX Grad-CAM, MP Grad-CAM, Grad-CAM++ and Smooth Grad-CAM++ for Resnet-34 model

then rescaled each pair of wins by taking the win of Grad-CAM as 1, finally recalculated the ratio of each method to the sum of wins. This results in the accumulated win % of all the methods is 100, allowing comparison among more than two methods. These objective results confirm that MP Grad-CAM greatly improves the visual explanation with global average pooling. It is also much better than MX Grad-CAM. The objective scores are consistent with those demonstrated in Fig. 4 and Fig. 5.

Table 1. Objective evaluation for VGG16 network

Method	Dataset	Average drop %	% Incr. in confidence	Win%	Confidence drop %
Grad-CAM	USGS	52.32	1.43	6.03	39.29
	SIRI-WHU	34.44	4.55	4.67	32.17
MX Grad-CAM	USGS	48.44	2.38	10.88	47.16
	SIRI-WHU	28.00	6.57	22.55	35.70
MP Grad-CAM	USGS	32.50	5.87	27.87	56.92
	SIRI-WHU	18.68	10.10	24.17	36.63
Grad-CAM++	USGS	**29.10**	**9.21**	**37.61**	57.26
	SIRI-WHU	**18.05**	**11.62**	**24.36**	**37.58**
Smooth Grad-CAM++	USGS	32.26	7.62	28.49	**57.29**
	SIRI-WHU	20.04	11.11	24.25	37.49

For the VGG16 network in Table 1, MP Grad-CAM has achieved a close performance to Grad-CAM++ and Smooth Grad-CAM++, based on the index of Confidence

Table 2. Objective evaluation for ResNet-34 network

Method	Dataset	Average drop %	% Incr. in confidence	Win%	Confidence drop %
Grad-CAM	USGS	53.372	4.202	7.31	34.819
	SIRI-WHU	56.692	1.010	12.94	12.841
MX Grad-CAM	USGS	58.139	0.935	7.11	39.530
	SIRI-WHU	57.605	1.010	15.53	23.442
MP Grad-CAM	USGS	51.633	3.244	12.69	41.168
	SIRI-WHU	46.120	1.515	23.66	38.812
Grad-CAM++	USGS	**41.380**	**7.742**	**42.39**	**51.357**
	SIRI-WHU	**35.291**	**2.525**	19.49	**50.539**
Smooth Grad-CAM++	USGS	43.372	7.395	30.51	49.297
	SIRI-WHU	42.096	2.020	**28.38**	45.986

Drop%. This indicates that MP Grad-CAM is able to localize the important parts for explaining the model's decision. Meanwhile, it is slightly worse based on the metrics upon occlusions in unimportant regions (e.g., % Incr. In Confidence). This means it is less accurate to weight the significance of the important regions. This is expected, given the fact the weight matrix by MP Grad-CAM is constant for each feature map.

Comparing Table 1 and Table 2, Resnet-34 show worse performance than VGG16. This is because a layer visualization only explains what have been learned by a CNN model at a certain level. As the coverage of the heatmaps are different (as shown in Fig. 5), the score differences among these visual explanation techniques are large, even between Grad-CAM++ and Smooth Grad CAM++. This raises a question whether these occlusion-based metrics are appropriate to measure the performance of a visual technique in explaining layer-wise (and/or neuron-wise) decision of a deep learning model.

Overall, Grad-CAM++ achieves the best performance. This is different from our expectation—the Smooth Grad-CAM++ would be the best because it is an advanced version of Grad-CAM++. This might be because the RS image classification is more dependent on a wholistic view of the image (e.g., the class "agriculture"), which is reached more by Grad-CAM++ than Smooth Grad-CAM++. Meanwhile, MP Grad-CAM gives a good wholistic view due to the use of median-pooling, as a result it is suitable for visual explanation of a model decision in RS classification tasks.

5   Conclusion

An efficient visual explanation technique can help our understanding of internal workings of a trained deep convolutional neural network at the inference stage. In this paper, through the analysis on Grad-CAM and Grad-CAM++ methods for the task of RS image classification, we provide better insights on these visual explanation techniques and the workings of weighting gradients of feature maps. Then, we proposed MP Grad-CAM, an efficient visual mapping for deep convolutional neural network in RS image classification. We also proposed a new metric of confidence drop % to provide more comprehensive assessment of the performance of visual explanation techniques. The proposed MP Grad-CAM with the computing complexity $O(n\log n)$, is more efficient than the comparable methods Grad-CAM++ and Smooth Grad-CAM++. Comparison experiments on different RS datasets and different CNN networks show that MP Grad-CAM performs well in object localization. During the experiments, we identified the existing evaluation metrics for visual explanation methods are problematic to judge the layer-wise or neuron-wise interpretation of a model decision. This should be considered in the future work. Future works also involve further investigations to extend this technique to handle various application scenarios and network architectures.

Acknowledgement. This work was supported by NSFC (601702323) and Marine Science Research Special Project of Hebei Normal University of Sci & Tech (2018HY020).

References

1. Simonyan, K., Zisserman, A.: Very deep convolutional networks for large-scale image recognition. arXiv preprint arXiv:1409.1556 (2014)
2. Yang Y., Newsam, S.: Bag-of-visual-words and spatial extensions for land-use classification. In: ACM GIS 2010, p. 270. ACM, San Jose (2010)
3. Lecun, Y., Bottou, L., Bengio, Y., Haffner, P.: Gradient-based learning applied to document recognition. Proc. IEEE **86**(11), 2278–2324 (1998)
4. Molnar, C.: Interpretable machine learning - a guide for making black box models explainable. https://christophm.github.io/. Accessed 21 Feb 2020
5. Zeiler, M.D., Fergus, R.: Visualizing and understanding convolutional networks. In: Fleet, D., Pajdla, T., Schiele, B., Tuytelaars, T. (eds.) ECCV 2014. LNCS, vol. 8689, pp. 818–833. Springer, Cham (2014). https://doi.org/10.1007/978-3-319-10590-1_53
6. Springenberg, J.T., Dosovitskiy, A., Brox, T., Riedmiller, M.: Striving for simplicity: the all convolutional net. In: Proceedings of International Conference on Learning Representations (2015)
7. Ribeiro, M. T., Singh, S., Guestrin, C.: Why should I trust you?: explaining the predictions of any classifier. In: Proceedings of the 22nd ACM SIGKDD, pp. 1135–1144. ACM (2016)
8. Zhang, H., Chen, J., Xue, H., Zhang, Q.: Towards a unified evaluation of explanation methods without ground truth. arXiv:1911.09017 (2019)
9. Zhou, B., Khosla, A., Oliva, L. A., A., Torralba, A.: Learning deep features for discriminative localization. In: Proceedings of CVPR, pp. 2921–2929. IEEE, Las Vegas (2016)
10. Fukui H., Hirakawa T., Yamashita T., et al.: Attention branch network: learning of attention mechanism for visual explanation. In: Proceedings of 2019 IEEE/CVF Conference on Computer Vision and Pattern Recognition (CVPR), pp. 10697–10706. IEEE (2019)
11. Selvaraju, R. R., Das, A., Vedantam, R., Cogswell, M., Parikh, D., Batra, D.: Grad-CAM: visual explanations from deep networks via gradient-based localization. In: 2017 ICCV, pp. 618–626. IEEE, Venice (2017)
12. Chattopadhay, A., Sarkar, A.: Grad-CAM++: generalized gradient-based visual explanations for deep convolutional networks. In: WACV, pp. 839–847. IEEE, Lake Tahoe (2018)
13. Omeiza, D., Speakman, S., Cintas C., Weldemariam, K.: Smooth grad-CAM++: an enhanced inference level visualization technique for deep convolutional neural network models. arXiv preprint arXiv:1908.01224v1 (2019)

14. Smilkov, D., Thorat, N., Kim, B., et al.: SmoothGrad: removing noise by adding noise. arXiv preprint arXiv:1706.03825v1 (2017)

15. Song, W., Dai, S.Y., Wang, J., Huang, D., Liotta, A., Di Fatta, G.: Bi-gradient verification for grad-CAM towards accurate visual explanation for remote sensing images. In: 2019 ICDMW, pp. 473–479, Beijing, China (2019)

16. Wang, Y., Su, H., Zhang, B., Hu, X.: Learning reliable visual saliency for model explanations. IEEE Trans. Multimed. 22(7), 1796–1807 (2020)

17. Adebayo, J., Gilmer, J., Muelly, M., et al.: Sanity checks for saliency maps. In: Advances in Neural Information Processing Systems, pp. 9505–9515. Curran Associates, Inc. (2018)

18. Yang M., Kim, B.: Benchmarking attribution methods with relative feature importance. arXiv preprint arXiv:1907.09701 (2019)

19. He, K., Zhang, X., Ren, S., Sun, J.: Deep residual learning for image recognition. In: 2016 CVPR, pp. 770–778. IEEE, Las Vegas (2016)

20. Zhao, B., Zhong, Y., Xia, G.S., Zhang, L.: Dirichlet-derived multiple topic scene classification model for high spatial resolution remote sensing imagery. IEEE Trans. Geosci. Remote Sens. 54(4), 2108–2123 (2016)

Multi-granularity Recurrent Attention Graph Neural Network for Few-Shot Learning

Xu Zhang[1]([✉]), Youjia Zhang[1], and Zuyu Zhang[2]

[1] Department of Computer Science and Technology,
Chongqing University of Posts and Telecommunications, Chongqing, China
zhangx@cqupt.edu.cn, s180201080@stu.cqupt.edu.cn
[2] Department of Computer Science and Technology, Inha University,
Incheon, South Korea
changjoey56@gmail.com

Abstract. Few-shot learning aims to learn a classifier that classifies unseen classes well with limited labeled samples. Existing meta learning-based works, whether graph neural network or other baseline approaches in few-shot learning, has benefited from the meta-learning process with episodic tasks to enhance the generalization ability. However, the performance of meta-learning is greatly affected by the initial embedding network, due to the limited number of samples. In this paper, we propose a novel Multi-granularity Recurrent Attention Graph Neural Network (MRA-GNN), which employs Multi-granularity graph to achieve better generalization ability for few-shot learning. We first construct the Local Proposal Network (LPN) based on attention to generate local images from foreground images. The intra-cluster similarity and the inter-cluster dissimilarity are considered in the local images to generate discriminative features. Finally, we take the local images and original images as the input of multi-grained GNN models to perform classification. We evaluate our work by extensive comparisons with previous GNN approaches and other baseline methods on two benchmark datasets (i.e., *mini*ImageNet and CUB). The experimental study on both of the supervised and semi-supervised few-shot image classification tasks demonstrates the proposed MRA-GNN significantly improves the performances and achieves the state-of-the-art results we know.

Keywords: Few-shot learning · Multi-granularity · Graph neural network

1 Introduction

Deep learning has shown great success in various computer vision applications such as image classification and object detection. However, a large amount of labeled data and tedious manual tuning is needed for training to achieve excellent performance. Hence, various challenges such as data shortage and expensive

© Springer Nature Switzerland AG 2021
J. Lokoč et al. (Eds.): MMM 2021, LNCS 12573, pp. 147–158, 2021.
https://doi.org/10.1007/978-3-030-67835-7_13

manual annotation limit the applicability of deep learning models to new categories. Few-shot learning (FSL) is proposed to tackle this problem. Using prior knowledge, FSL can rapidly generalize to new tasks containing only a few samples with supervised information [22]. Recently, many studies have sprung up on one/few-shot learning and zero-shot learning [9,18,20,23,24]. In contrast to the conventional deep learning methods, few-shot learning aims to recognize a set of target classes by using the prior knowledge learned from sufficient labelled samples from a set of source classes and only few labelled samples from the target classes. Few-shot learning is confronted with the challenges of data scarcity and poor generalization ability [22,23]. To some extent, data enhancement and regularization can reduce the risk of overfitting, however, they still cannot solve it. In few-shot learning, the training stage typically adopts the meta-learning method, so that the model can identify new classes automatically and efficiently with few labels only through fine-tuning.

Recently, graph neural networks have become increasingly interesting to researchers in the field of deep learning. The concept of GNN was first proposed by Gori et al. [7] to process graph structure data in a recursive neural network manner. Subsequently, Scarselli et al. [17] further elaborated on GNN. Since then, GNN has been continuously improved, expanded, and advanced [2,12]. GNN is a deep learning method in non-Euclidean fields, a neural network that runs directly on graphs. In essence, each node of the graph is related to the label and depends on each other. One typical application of GNN is to classify nodes in the graph structure. The GNN [5] for few-shot Learning is a node-labeling framework, which predicts the node-labels on the graph. Obviously, it is also reasonable to classify the edges in the graph structure. The EGNN is an edge-labeling framework, which predicts the edge-labels on the graph by iteratively updating the edge-labels. Due to GNN's powerful ability to model the dependence of nodes in the graph, GNN has also made a significant contribution to the study of few-shot learning [5,9,13].

In this paper, we propose a novel Multi-granularity Recurrent Attention Graph Neural Network (MRA-GNN) for Few-shot learning. Firstly, foreground images are extracted from the sample images by the saliency network to mitigate the negative impact of complicated environment. Furthermore, we employ the Local Proposal Network based on attention mechanism to generate local images with intra-cluster similarity and the inter-cluster dissimilarity from foreground images. Then, the local images and original images are embedded as multi-granularity images and accepted as the input of GNN models. The classification results of the multi-granularity images in the query set are fused. Different from the GNN [5] or the EGNN [9], the proposed MRA-GNN employs LPN to enlarge the number of effective samples to reduce the risk of overfitting and enhance the generalization ability of model instead of just training the embedding network, despite the adoption of reasoning mechanism based on graph neural network.

Our contributions can be summarized as follows:

- We propose a novel Multi-granularity Recurrent Attention Graph Neural Network (MRA-GNN) for Few-shot learning.

- We propose a significant local proposal network based on attention mechanism, which can generate additional local images with intra-cluster similarity and the inter-cluster dissimilarity.
- We acquire the local images via LPN to expand the volume of effective samples, and then the multi-granularity images are embedded into GNN, which can reduce the risk of overfitting and enhance the generalization ability.
- We evaluate our approach on two benchmark datasets for few-shot learning, namely *mini*ImageNet and CUB. On both of the supervised and semi-supervised image classification tasks for few-shot learning, the experimental results show that the proposed MRA-GNN outperforms the state-of-the-art methods on both datasets.

2   Related Works

Few-Shot Learning. In recent years, the approach for few-shot image classification based on representation learning has gradually become the mainstream method. Siamese Network [11] train a two-way neural network in a supervised manner, and exploits simple metric learning to represent the distance of sample pairs. Compared with the Siamese Network, Match Network [20] constructs different feature encoders for the support set and query set by end-to-end nearest neighbor classification, calculates the similarity between test samples and support samples, and the final classification of test samples is a weighted sum of the predicted values of the support set. The Prototype Network [18] takes the center of the mean value of the support set of each category in the embedded space as the prototype expression of the category, which turns the problem of few-shot learning into the nearest neighbor problem in the embedded space. Relation Network [19] changes the previous single and fixed measurement method, and improves the similarity expression between samples from different tasks by training a neural network (such as CNN). Node-labeling graph neural network [5] and edge-labeling graph neural network [9] integrate the similarity measurement between samples into the graph network and serve as the basis for updating nodes.

Graph Neural Network. The Graph Neural Network is the natural extension of the Convolutional Neural Network in non-Euclidean space. GNN was first proposed by Scarselli et al. [17], and many variants have been generated since then. Recently, some GNN methods for few-shot learning have been proposed. To be specific, Garcia and Bruna [5] first proposed to build a graphical model for few-shot learning where all examples of the support set and the query set are densely connected. The embedding features of each labeled and unlabeled sample (e.g., the outputs of a convolutional neural network, etc.) and the given label (e.g., one-hot encoded label) are used as nodes of the graph network. Then, the features of nodes are updated from neighborhood aggregation. Through the messaging inference mechanism of the graph neural network, the labels of unlabeled samples are predicted. The EGNN [9] explicitly models the intra-cluster similarity and inter-cluster dissimilarity, and takes the similarity as the edge, and propagates the encoded label between the whole support set and the query set through the previous edge feature value and the updated similarities. WDAE-GNN [6] performs knowledge transfer between the base classifier and the new classifier by transmitting the weights of all classifiers to a Graph Neural Network-based Denoising Autoencoder.

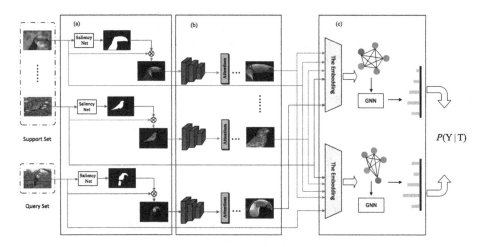

Fig. 1. Pipeline of proposed framework (a) the pre-training saliency network, (b) local proposal network, and (c) multi-granularity graph neural network.

3   The Proposed Method

In this section, we describes the Multi-granularity Recurrent Attention Graph Neural Network (MRAGNN) for Few-shot learning in detail, as it is shown in Fig. 1.

3.1   Model

The proposed MRAGNN framework shown in Fig. 1 is mainly composed of three parts :(a) the pre-training Saliency Network, which is used to generate foreground images, (b) the Local Proposal Network, which extracts local images with significant intra-cluster similarity and the inter-cluster dissimilarity, (c) the Multi-granularity Graph Neural Network, which contains an embedding layer and GNN for final.

The Pre-training Saliency Network. Human vision can actively allocate limited resources to useful information so as to prioritize the most valuable data, when processing input. Similarly, machine vision can also detect the significance region to judge the importance of the visual information. Generally, scenes with more abrupt features shall attract human attention more than other regions. Therefore, we expect to obtain the image foreground related to human attention through a Saliency Network when carrying out the few-shot classification task. As one of the main research contents in the field of object detection, Salient Object Detection has relatively advanced theories and techniques. Therefore, in this paper, we learn the saliency network S from the work of Xuebin Qin et al. [14], and obtain the foreground images in an unsupervised way.

Consider an image **I** which is passed through the saliency network **S** to extract the corresponding saliency map **S(I)**, the foreground **F** can be expressed as follows:

$$\mathbf{F} = \mathbf{S}(\mathbf{I}) \odot \mathbf{I} \tag{1}$$

Where $\odot$ is the Hadamard product.

Local Proposal Network. The Local Proposal Network is a designed to generate significant local foreground images based on attentional mechanism, which can represent intra-cluster similarity and the inter-cluster dissimilarity. The specific structure of LPN is shown in Fig. 2.

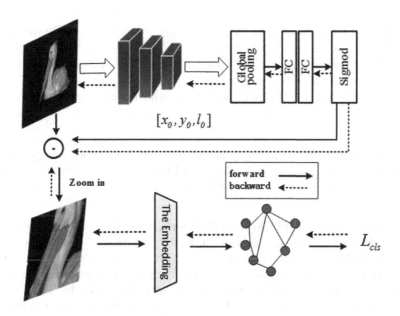

Fig. 2. Structure of the Local Proposal Network.

As it is shown in Fig. 2, we extracts features from the input image and obtains the significant local region. In detail, the foreground image of the original images was taken as the input x. After a series of convolution operations, the feature information of the input image was extracted represented as $\varphi(x)$, where $\varphi(\cdot)$ represents the feature extraction network. To obtain the coordinates of the significant region of the input image, we add a global pooling layer and two full connection layers. The last full connection layer has three channels and returns the center of the local region. It can be expressed as follows:

$$
\begin{aligned}
[x_0, y_0, l_0] &= g\left(\varphi(x)\right) \\
x_l &= x_0 - l_0, x_r = x_0 + l_0 \\
y_1 &= y_0 - l_0, y_r = y_d + l_0
\end{aligned}
\tag{2}
$$

$[x_0, y_0, l_0]$ represents the center coordinates of the selected square local region. x_l and x_r are the boundary of this region on the x-axis respectively. y_l and y_r are the boundary of this region on the y-axis respectively.

Then we enlarge the significant local region identified by the locating, including two stages: *cropping* and *zooming*. With the coordinates of local region, the coarse-grained image can be cropped and zoomed in to get a finer local image. To ensure that the LPN network can be optimized during training, a continuous attention mask function $M(\cdot)$ [4] is set up.

$$
\begin{aligned}
M(\cdot) &= [h(x - x_l) - h(x - x_r)] * [h(y - y_l) - h(y - y_r)] \\
h(x) &= \tfrac{1}{1+e^{-kx}}
\end{aligned}
\tag{3}
$$

where $h(x)$ is a logistic regression function. When k is large enough, if the pixel is in the selected square region (i.e., $x_l \leq x \leq x_r$ and $y_l \leq y \leq y_r$), the value of the function $M(\cdot)$ are one, whereas $M(\cdot) = 0$. Therefore, the cropping operation can be implemented by an element-wise multiplication be-tween the original image at coarser scales and an attention mask $M(\cdot)$, which can be computed as:

$$
x^* = x \odot M(\cdot)
\tag{4}
$$

we use bilinear interpolation to zoom in the x^*. Therefore, we can obtain the discriminatively local image x' through LPA network, which can represent intra-cluster similarity and the inter-cluster dissimilarity of the samples.

Multi-granularity Graph Neural Network. The Multi-granularity classification model is described in this section, as it is shown in Fig. 1 part (c). Formally, the Multi-granularity Graph Neural Network (MGNN) is a multi-layer network that operates on graphs $G = (V, E)$ by structuring their computations according to the graph connectivity. i.e., at each MGNN layer the feature responses of a node are computed based on the neighboring nodes defined by the adjacency graph. The MGNN input consists of unlabeled and labeled images in our methods. Let $G = (V, E; \mathcal{T})$ be the graph constructed with samples from the task $\mathcal{T}$, where $V := \{V_i\}_{i=1,\ldots,|N|}$ and $E := \{E_{ij}\}_{i,j=1,\ldots,|N|}$ denote the set of nodes and edges of the graph, respectively. $|N| = C \times K + M$ is the total number of samples in the task $\mathcal{T}$. And, for all the images $X_i = \left\{x_i, x_i'\right\} \in \mathcal{T}$, the one-hot encoding of the label y_i is concatenated with the embedding features of the image as the input of the MGNN.

$$v_i^{(0)} = (\psi(X_i), h(y_i)) \tag{5}$$

Here $i = 1, \ldots, |N|$ and $h(y_i) \in \mathbb{R}_+^k$ is the one-hot encoding of the label. The $\psi(\cdot)$ is a embedding network composed of four convolutional blocks and one fully-connected layer, shown in the Fig. 3.

Given all the nodes with different granularity, a fully-connected graph can be initially constructed, where each node represents each sample, and each edge represents the similarity between two connected nodes. So, we can get the edge feature e_{mn} as follows:

$$e_{mn}^{(0)} = f_{sim}\left(v_m^{(0)}, v_n^{(0)}\right) \tag{6}$$

Where $v_m^{(0)}, v_n^{(0)} \in \left\{v_i^{(0)}\right\}_{i=1,\ldots,|N|}$ and $f_{sim}(\cdot)$ is a symmetric metric function. In this work, we consider it as a neural network. Then, we update the node representation at layer l with the neighbor node $v_n^{(t-1)}$ of each node and the edge feature $e_{mn}^{(l-1)}$ from the layer $l-1$.

$$v_m^{(l)} = v_m^{(l-1)} + \sum_n e_{mn}^{(l-1)} \cdot v_n^{(l-1)} \tag{7}$$

After L layer of node and edge feature up-dates, the prediction of our model can be obtained from the final edge feature. In order to obtain the final classification labels, we weighted average the result vectors of the input of multi-granularity at the last layer as follows:

$$y = \delta \cdot y^{(L)} + y'^{(L)} \tag{8}$$

The δ is the weight parameter. $y'^{(L)}$ and $\tilde{y}^{(L)})$ represent the prediction label of the significant local samples through LPN network and the original testing sample in the last layer respectively.

3.2   Training

Given T training tasks $\{\mathcal{T}_t^{train}\}_{t=1}^T$ at a certain iteration during the episodic training, the parameters of the proposed MRA-GNN are trained by minimizing the following loss function:

$$L(X) = L_{cls}(\tilde{Y}, Y) \tag{9}$$

$\tilde{Y}$ denotes the final predicted label for the all samples in the union support set and query set, and Y denotes the ground truth label vector, respectively. And, the classification loss function L_{cls} is computed as:

$$P(\tilde{y}_i = j|x_i) = \frac{\exp(y_{ij}^*)}{\sum_{k=1}^c \exp(y_{ik}^*)}$$
$$L_{cls}(\tilde{Y}, Y) = \sum_{i=1}^{C \times K + M} \sum_{j=1}^C -\sigma(y_i == j)\log(P(\tilde{y}_i = j|x_i)) \tag{10}$$

Where y_i means the ground-truth label of x_i and $\sigma(\cdot)$ is an indicator function, $\sigma(a) = 1$ if a is true and 0 otherwise. And y_{ij}^* denotes the jth component of predicted label y_i^* from label propagation.

4 Experiments

For the supervised and semi-supervised experiments of few-shot learning, in this section, we evaluated and compared our proposed method with several state-of-art work on *mini*ImageNet [20] and CUB-200-2011 [21].

4.1 Datasets

miniImageNet. The miniImageNet dataset is a benchmark dataset for few-shot image classification by Vinyalset al. [20] derived from the original ILSVRC-12 dataset Krizhevsky et al. [16]. It contains 60,000 images from 100 different classes with 600 samples per class. And, all the images are RGB colored, and resized to 84 × 84 pixels, are 84 × 84 RGB images. In order to ensure the fairness of the experiment, we also use the splits proposed by Ravi & Larochelle [15], which divides the dataset into 64 classes for training, 16 classes for validation, and 20 classes for testing.

CUB. The CUB-200-2011 dataset [21] (referred to as the CUB hereafter) has widely used for FSL. The CUB dataset consists of 200 bird species and 11,788 images in total. And, the number of images per class is less than 60. Following the evaluation protocol of Hilliard et al. [8], we randomly split CUB into 100 training classes, 50 validation classes, and 50 test classes. Each image is also resized to 84 × 84 pixels.

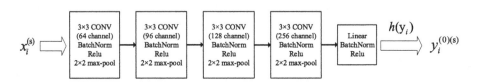

Fig. 3. Detailed architecture of the embedding.

4.2 Experimental Setups

Following Snell et al. [18], we adopt the episodic training procedure, i.e, we sample a set of C-way K-shot training tasks to mimic the C-way K-shot test problems. We conducted 5-way 5-shot and 5-way 1-shot experiments for both datasets, which are the standard few-shot learning settings. For feature embedding module, we adopt a widely-used CNN [18,19] in most few-shot learning models as the feature embedding function module described in detail in Fig. 3.

Moreover, All our models were trained with Adam [10] and an initial learning rate of 10^{-3}. For miniImageNet, we cut the learning rate in half every 15,000 episodes while for CUB, we cut the learning rate every 5,000 episodes. Although both transductive and non-transductive test strategies are followed in [9], we only take the non-transductive test strategy on board for fair comparison, since most of the state-of-the-art FSL methods are non-transductive.

4.3 Few-Shot Classification

The few-shot classification performance of the proposed MRA-GNN model for miniImageNet and CUB is compared with several state-of-the-art models in Table 1 and Table 2 (All results average accuracies with 95% confidence intervals).

Table 1. Few-shot classification accuracies on miniImageNet.

Model	5-way Acc	
	1-shot	5-shot
MAML (Finn et al. 2017)	48.70	63.11
PROTO NET (Higher Way)(Snell et al. 2017)	49.42	68.20
RELATION NET c(Sung et al. 2018)	51.38	67.07
TPN(Liu et al. 2019)	49.42	68.20
EGNN(Kim et al. 2019)	52.86	66.85
GNN(Garcia et al. 2018)	50.33	66.41
MRAGNN-LPN	**46.39**	**62.73**
MRA-GNN	**61.58**	**78.05**

As it is shown in Table 1, MRA-GNN we proposed shows the best performance in 5-way 5-shot and 5-way 1-shot seting on *mini*Imagenet. Especially, the proposed MRA-GNN method has also achieved excellent results in the more challenging 10-way experiments on miniImagenet. Moreover, comparing with node-labeling GNN [5] and edge-labeling EGNN [9], MRA-GNN surpasses those Graph Neural Network methods with a large margin [5,9] about Few-Shot Learning.

The comparative results under FSL for CUB are shown in Table 2. It can be seen that: (1) Our MRA-GNN yields 8–12% improvements over the latest GNN-based FSL methods [5,9,13], validating the effectiveness of multi-granularity network for GNN-based FSL. (2) Comparing with other methods over the state-of-the-art FSL baselines [1,3,18,19], our method ranges from 8% to 23% improvements, showing that MRA-GNN has a great potential for FSL. Furthermore, the outstanding performance on two benchmark datasets also means that the MRA-GNN achieve better generalization ability for few-shot learning.

Table 2. Few-shot classification accuracies on CUB.

Model	5-way Acc	
	1-shot	5-shot
MAML (Finn et al. 2017)	55.92	72.09
PROTO NET (Snell et al. 2017)	51.31	70.77
RELATION NET c(Sung et al. 2018)	62.45	76.11
Baseline(Wei-Yu Chen et al. 2019)	47.12	64.16
EGNN(Kim et al. 2019)	60.05	74.58
AdarGCN(Jianhong Zhang et al. 2020)	–	78.04
GNN(Garcia et al. 2018)	60.19	75.90
MRAGNN-LPN	**63.41**	**77.45**
MRA-GNN	**70.99**	**86.70**

4.4   Ablation Studies

One question that may be of concern is whether the benefits of the proposed network improvement come from the signficant local region generated by LPN. To prove the effectiveness of the proposed method, we train the LPN network with labeled data on *mini*Imagenet dataset and CUB dataset. As shown in Tables 1 and 2, the MRAGNN-LPN only include local images in the training. Surprisingly, its results are close to other baseline on *mini*Imagenet and even surpasses many advanced methods on CUB dataset. And, the experimental results are satisfactory, although the local image contain only part information of the sample. Our method combines local features with global features to improve the results significantly. As detailed analyzed in [1], current few-shot learning methods can not efficiently transfer the domain of learning, i.e., the training domain can not have huge gap with the testing set. In this paper, a transfer ability test is also conducted by pre-training the embedding network on ImageNet and applied on CUB dataset. As shown in Table 3, the proposed method with ImageNet

Table 3. 5-shot accuracy under the cross-domain scenario.

*mini*Imagenet ⟶ CUB	
MAML (Finn et al. 2017)	51.34
MatchingNet (Oriol Vinyals et al. 2016)	53.07
PROTO NET (Higher Way)(Snell et al. 2017)	62.02
RELATION NET c(Sung et al. 2018)	57.71
Baseline(Wei-Yu Chen et al. 2019)	65.57
Baseline++(Wei-Yu Chen et al. 2019)	62.04
GNN(Garcia et al. 2018)	58.77
MRA-GNN	73.22

pre-trained embedding network can be efficiently transferred to CUB dataset and gain 7.65% to 21.88% improvements over other mainstream approaches.

5   Conclusion

We proposed the Multi-granularity Recurrent Attention Graph Neural Network for Few-shot learning. This work addressed the problem of few-shot learning, especially on the few-shot classification task. In the process of MRA-GNN, we obtain significant local images through the LPN network, which better show the intra-cluster similarity and inter-cluster dissimilarity. Besides, our model takes advantage of the information transmission mechanism of GNN to propagate label information from labeled multi-granularity samples towards the unlabeled multi-granularity query image for few-shot learning. We obtained the state-of-the-art results of the supervised few-shot image classification tasks on *mini*ImageNet and CUB. In the future work, one of the directions we can consider is to sparse the graph and lightweight our model so that our model can be applied more widely.

Acknowledgment. This research is supported by National Natural Science Foundation of China (41571401), Chongqing Natural Science Foundation (cstc2019jscx-mbdxX0021).

References

1. Chen, W.Y., Liu, Y.C., Kira, Z., Wang, Y.C., Huang, J.B.: A closer look at few-shot classification. In: International Conference on Learning Representations (2019)
2. Defferrard, M., Bresson, X., Vandergheynst, P.: Convolutional neural networks on graphs with fast localized spectral filtering. In: Advances in Neural Information Processing Systems, pp. 3844–3852 (2016)
3. Finn, C., Abbeel, P., Levine, S.: Model-agnostic meta-learning for fast adaptation of deep networks. In: Proceedings of the 34th International Conference on Machine Learning-Volume 70, pp. 1126–1135. JMLR. org (2017)
4. Fu, J., Zheng, H., Mei, T.: Look closer to see better: recurrent attention convolutional neural network for fine-grained image recognition. In: Proceedings of the IEEE Conference on Computer Vision and Pattern Recognition, pp. 4438–4446 (2017)
5. Garcia, V., Estrach, J.B.: Few-shot learning with graph neural networks. In: 6th International Conference on Learning Representations, ICLR 2018 (2018)
6. Gidaris, S., Komodakis, N.: Generating classification weights with gnn denoising autoencoders for few-shot learning. In: Proceedings of the IEEE Conference on Computer Vision and Pattern Recognition, pp. 21–30 (2019)
7. Gori, M., Monfardini, G., Scarselli, F.: A new model for learning in graph domains. In: Proceedings. 2005 IEEE International Joint Conference on Neural Networks, 2005, vol. 2, pp. 729–734. IEEE (2005)
8. Hilliard, N., Phillips, L., Howland, S., Yankov, A., Corley, C.D., Hodas, N.O.: Few-shot learning with metric-agnostic conditional embeddings. IEEE Trans. Image Process. (2018)

9. Kim, J., Kim, T., Kim, S., Yoo, C.D.: Edge-labeling graph neural network for few-shot learning. In: Proceedings of the IEEE Conference on Computer Vision and Pattern Recognition, pp. 11–20 (2019)

10. Kingma, D.P., Ba, J.: Adam: a method for stochastic optimization. In: International Conferenceon Learning Representations, vol. 5 (2014)

11. Koch, G., Zemel, R., Salakhutdinov, R.: Siamese neural networks for one-shot image recognition. In: ICML Deep Learning Workshop, vol. 2. Lille (2015)

12. Li, Y., Tarlow, D., Brockschmidt, M., Zemel, R.: Gated graph sequence neural networks. In: International Conference on Learning Representations (2016)

13. Liu, Y., et al.: Learning to propagate labels: transductive propagation network for few-shot learning. In: International Conference on Learning Representations (2019)

14. Qin, X., Zhang, Z., Huang, C., Gao, C., Dehghan, M., Jagersand, M.: Basnet: boundary-aware salient object detection. In: Proceedings of the IEEE Conference on Computer Vision and Pattern Recognition, pp. 7479–7489 (2019)

15. Ravi, S., Larochelle, H.: Optimization as a model for few-shot learning. In: International Conference on Learning Representations (ICLR) (2017)

16. Russakovsky, O., et al.: Imagenet large scale visual recognition challenge. Int. J. Comput. Vision **115**(3), 211–252 (2015)

17. Scarselli, F., Gori, M., Tsoi, A.C., Hagenbuchner, M., Monfardini, G.: The graph neural network model. IEEE Trans. Neural Netw. **20**(1), 61–80 (2008)

18. Snell, J., Swersky, K., Zemel, R.: Prototypical networks for few-shot learning. In: Advances in Neural Information Processing Systems, pp. 4077–4087 (2017)

19. Sung, F., Yang, Y., Zhang, L., Xiang, T., Torr, P.H., Hospedales, T.M.: Learning to compare: relation network for few-shot learning. In: Proceedings of the IEEE Conference on Computer Vision and Pattern Recognition, pp. 1199–1208 (2018)

20. Vinyals, O., Blundell, C., Lillicrap, T., Wierstra, D., et al.: Matching networks for one shot learning. In: Advances in Neural Information Processing Systems, pp. 3630–3638 (2016)

21. Wah, C., Branson, S., Welinder, P., Perona, P., Belongie, S.: The caltech-ucsd birds-200-2011 dataset (2011)

22. Wang, Y., Yao, Q., Kwok, J.T., Ni, L.M.: Generalizing from a few examples: a survey on few-shot learning. ACM Comput. Survey **53**(3), 1–34 (2020)

23. Zhang, H., Zhang, J., Koniusz, P.: Few-shot learning via saliency-guided hallucination of samples. In: Proceedings of the IEEE Conference on Computer Vision and Pattern Recognition, pp. 2770–2779 (2019)

24. Zhang, J., Zhang, M., Lu, Z., Xiang, T., Wen, J.: Adargcn: Adaptive aggregation GCN for few-shot learning. arXiv preprint arXiv:2002.12641 (2020)

EEG Emotion Recognition Based on Channel Attention for E-Healthcare Applications

Xu Zhang[1(✉)], Tianzhi Du[1], and Zuyu Zhang[2]

[1] Department of Computer Science and Technology,
Chongqing University of Posts and Telecommunitcations, Chongqing, China
zhangx@cuqpt.edu.cn, s180231915@stu.cqupt.edu.cn
[2] Department of Computer Science and Technology, Inha University,
Incheon, South Korea
changjoey56@gmail.com

Abstract. Emotion recognition based on EEG is a critical issue in Brain-Computer Interface (BCI). It also plays an important role in the e-healthcare systems, especially in the detection and treatment of patients with depression by classifying the mental states. Unlike previous works that feature extraction using multiple frequency bands leads to a redundant use of information, where similar and noisy features extracted. In this paper, we attempt to overcome this limitation with the proposed architecture, Channel Attention-based Emotion Recognition Networks (CAERN). It can capture more critical and effective EEG emotional features based on the use of attention mechanisms. Further, we employ deep residual networks (ResNets) to capture richer information and alleviate gradient vanishing. We evaluate the proposed model on two datasets: DEAP database and SJTU emotion EEG database (SEED). Compared to other EEG emotion recognition networks, the proposed model yields better performance. This demonstrates that our approach is capable of capturing more effective features for EEG emotion recognition.

Keywords: E-Healthcare · EEG emotion recognition · Channel attention · Deep residual networks

1 Introduction

Emotions are an essential part of people's daily lives, and emotion recognition is also a key technology in the field of artificial intelligence. Positive emotions can improve human health and work efficiency, while negative emotions may cause health problems [1]. Emotion recognition has been applied in many fields, such as: safe driving [2], mental health monitoring [3], social safety [4].

Emotion recognition methods can be divided into two categories according to signal acquisition: non-physiological signal based method (NPSM) and physiological signal based method (PSM). The NPSM include facial expressions [5]

© Springer Nature Switzerland AG 2021
J. Lokoč et al. (Eds.): MMM 2021, LNCS 12573, pp. 159–169, 2021.
https://doi.org/10.1007/978-3-030-67835-7_14

and speech [6]. The PSM includes electroencephalogram (EEG) [7], electromyogram (EMG) [8], functional magnetic resonance imaging (FMRI) [9], functional near-infrared spectroscopy (fNIRS) [10]. The difference is that non-physiological signals are easy to disguise and hide, while the physiological signals are spontaneously generated by humans, and they are more objective and authentic. Among these physiological signals, the EEG is directly produced in the human brain and can directly reflect the emotional state of people. And, the advantages of EEG are high temporal resolution, non-invasiveness, and relatively low financial cost. So, EEG is widely used in research involving neural engineering, neuroscience, and biomedical engineering [11].

Many researchers have paid attention to emotion recognition based on EEG. Duan et al. discovered that the differential entropy feature can better distinguish emotional states [12]. Zheng et al. used deep neural networks to find that the important frequency ranges in emotion recognition include beta and gamma bands [13]. Tripathi et al. used deep neural networks and convolutional neural networks to perform emotion recognition tasks on the DEAP database [14]. Yang et al. retained the spatial information between the electrodes, integrated the multi-band EEG to construct the 3-dimensional features as the input of the convolutional neural network [15]. Wang et al. used the phase lock value method [16] in the brain network to construct a graph convolutional neural network model [17]. Chao et al. extracted Power Spectral Density (PSD) from different frequency bands and combined multiband feature matrix, and used a capsule network (CpasNet) to recognize human emotional states [18]. However, most of the studies extract features from multiple frequency bands with redundant information, which may lead to sub-optimal.

Recently, the attention mechanism has been proven to bring benefits in various tasks such as natural language processing, image caption and speech recognition. The goal of the attention mechanism is to select information that is relatively critical to the current task from all input [19]. Motivated by these studies, we employ deep residual networks [21,22] with efficient channel attention [20] to extract more critical and effective EEG emotional features.

This paper is organized as follows. In Sect. 2, the emotional EEG databases are given. Section 3 introduces feature extraction and model construction in detail. Section 4 gives the experimental settings and results, comparing our model with other models. Finally, Sect. 4.3 discusses and concludes our work.

2   Emotional EEG Databases

In this paper, we conduct extensive experiments on two emotional EEG databases that are commonly used in EEG emotion recognition to evaluate the effectiveness of the proposed Channel Attention-based Emotion Recognition Networks (CAERN) model. One is Public multimodal DEAP Database [23] and the other is SJTU Emotion EEG Database (SEED) [13].

The DEAP database contains EEG signals and other physiological signals from 32 subjects (16 males and 16 females). The subjects watched 40 selected

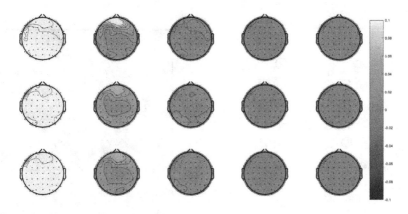

Fig. 1. The brain activity of the same subject under three different emotions.

music videos of 1 min, and collected the EEG signals of watching each video. After watching the video, the subjects will perform self-assessment for arousal and valence, and they will score the stimulus video, and the scoring range is 1–9. The EOG is also removed. The data is down-sampled 128 Hz and passed through a 4 45 Hz band-pass filter. Each experiment is divided into 60 s experimental data and 3 s preliminary experimental data.

The SEED database contains 15 subjects (7 males and 8 females). The subjects watched 3 different emotional types (negative, neutral, positive) Chinese movie clips, each movie clip is about 4 min. The EEG data collection lasts for 3 different periods corresponding to 3 sessions, and each session corresponds to 15 trials of EEG data such that there are totally 45 trials of EEG data for each subject.

3   Method

3.1   Feature Extraction

We first visualized the brain activity of the same subject under three different emotions, which use power spectral density to measure brain activity. As shown in Fig. 1, from top to bottom, they are positive, neutral, and negative emotions. From left to right. Followed by the δ, θ, α, β, γ band. It can be seen from Fig. 1 that the difference between different emotions is relatively small, and the brain activity in each frequency band is also different. So the task of emotion recognition based on EEG is very challenging.

In order to verify the effectiveness of our proposed model, our feature extraction adopted the same method with [15] in the DEAP database.

Differential entropy (DE) features have been proven to be very suitable for emotion recognition tasks [12,13]. The definition of DE is as follows.

$$h\left(X\right) = -\int_{-\infty}^{\infty} \frac{1}{\sqrt{2\pi\sigma^2}} e^{\frac{(x-\mu)^2}{2\sigma^2}} \log(\frac{1}{\sqrt{2\pi\sigma^2}} e^{-\frac{(x-\mu)^2}{2\sigma^2}}) dx = \frac{1}{2}\log\left(2\pi e\sigma^2\right) \quad (1)$$

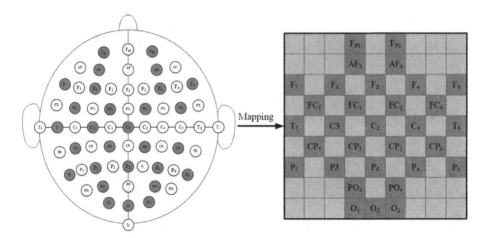

Fig. 2. EEG electrode map and 9 × 9 square matrix.

A Butterworth filter is used to decompose the original signals into 4 frequency bands (θ band, α band, β band, γ band). After decomposition, we divide the EEG data of 4 frequency bands into 1 s non-overlapping segments as a sample. So, in the DEAP database, we get 2400 samples on each participant. We directly extract the differential entropy (DE) feature of each segment. So the features of a specific frequency band of each segment can be represented by a 1D (1-dimensional) vector.

To preserve spatial information among multiple adjacent channels. Firstly, the 32 electrodes used in the DEAP database were mapped into a 9×9 square matrix [14,18]. Simultaneously, 1D DE feature vector is transformed to 2D (2-dimensional) plane (9×9). The gray points are unused electrodes and are filled with zeros. The process is depicted in Fig. 2 For each EEG segment, we obtain 4 2D plans corresponding to 4 frequency bands. The process is depicted in Fig. 3 The 4 2D planes are stacked as a 3D (3-dimensional) cubes and considered as color image input to our model. Particularly, in the DEAP database, the 3-seconds baseline signal is divided into 3 1-s segments and transformed into 4 DE feature vectors $base\_v \in R^{32}$. Then the mean DE feature value of these 3 EEG cubes saves as the DE feature of baseline signals. The DE deviation is calculated to represent the emotional state feature of the segment. This step can be expressed as:

$$ final\_v_j^i = exper\_v_j^i - \frac{\sum_{k=1}^{3} base\_v_k^i}{3} \left\{ final\_v_j^i, exper\_v_j^i, base\_v_k^i \right\} \in R^{32}, \quad (2) $$

where $exper\_v_j^i$ denotes the DE feature vector for frequency band i on segment j. $base\_v_k^i$ is the DE feature vector for frequency band i on baseline signals segment k. $final\_v_j^i$ is the final emotional state feature vector for frequency band i on segment j [15].

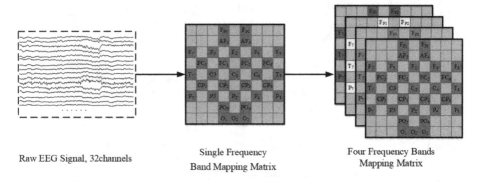

Raw EEG Signal, 32channels   Single Frequency   Four Frequency Bands
                                    Band Mapping Matrix   Mapping Matrix

Fig. 3. The mapping process of four feature matrix according to the raw electroencephalogram (EEG) signals of 32 channels.

The SEED database has provided the extracted DE features of EEG signals with five frequency bands (δ band: 1–3 Hz, θ band: 4–7 Hz, α band: 8–13 Hz, β band: 14–30 Hz, γ band: 31–50 Hz). Different from the DEAP database, the SEED database has 62 channels and no baseline signals. We also converted the DE features of the SEED database into 3D cubes as the input of our model according to the method mentioned earlier.

3.2 Attention Mechanisms and Deep Residual Networks

Deep residual networks have led to a series of breakthroughs for image classification [21,22]. The attention mechanisms have been proven to significantly improve the performance of deep convolutional neural networks [20,24]. Inspired by this, we employ deep residual network with effective channel attention [20] to perform emotion recognition tasks.

Hu et al. proposed SE (Squeeze-and-Excitation) module, which comprises a lightweight gating mechanism which focuses on enhancing the representational power of the network by modelling channel-wise relationships [24]. Figure 4 illustrates the overview of SE module. Let the output of one convolution block be $\chi \in R^{W \times H \times C}$, where W, and H and C are width, height and channel dimension. The weights of channels in SE module can be computed as

$$\omega = \sigma \left(f_{\{W_1, W_2\}} \left(g\left(\chi\right)\right)\right), \tag{3}$$

where $g\left(\chi\right) = \frac{1}{WH}\sum_{i=1,j=1}^{W,H} \chi_{ij}$ is channel-wise global average pooling (GAP) and σ is Sigmoid function. Let $y = g\left(\chi\right), f_{\{W_1, W_2\}}$ takes the from

$$f_{\{W_1, W_2\}}\left(y\right) = W_2 \text{ReLU}\left(W_1 y\right), \tag{4}$$

where ReLU indicates the Rectified Linear Unit [25], sizes of W_1 and W_1 are set to $C \times \left(\frac{C}{r}\right)$ and $\left(\frac{C}{r}\right) \times C$, respectively. For the good balance between accuracy and complexity, the parameter r is set to 16 in our experiments. Then, the weights are applied to the feature maps to generate the output of the SE module.

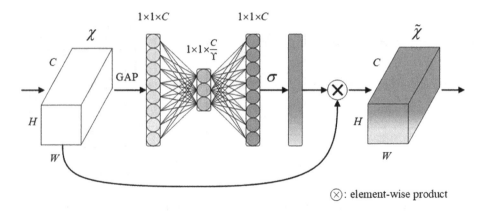

Fig. 4. Squeeze-and-Excitation (SE) module.

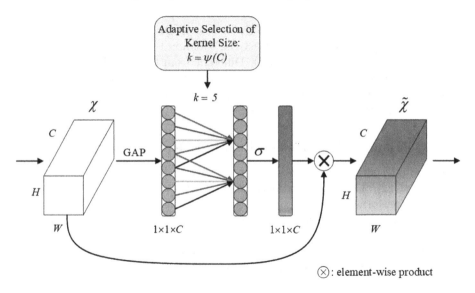

Fig. 5. Efficient Channel Attention (ECA) module.

Different from SE module, the effective channel attention via 1D convolution proved appropriate cross-channel interaction can preserve performance while significantly decreasing model complexity, and provide a method to adaptively select kernel size of 1D convolution [20]. Effective channel attention aims to appropriately capture local cross-channel interaction, so the coverage of interaction needs to be determined. Wang et al. introduced a solution to determine adaptively kernel size k according to channel dimension C [20].

$$k = \psi(C) = \left| \frac{\log_2(C)}{\gamma} + \frac{b}{\gamma} \right|_{odd}, \tag{5}$$

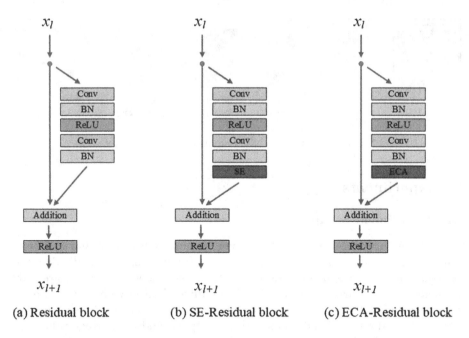

(a) Residual block   (b) SE-Residual block   (c) ECA-Residual block

Fig. 6. Comparison of Residual block and SE-Residual block, ECA-Residual block.

where $|t|_{odd}$ indicates the nearest odd number of t. γ and b are set to 2 and 1 throughout all the experiments, respectively. Obviously, through the non-linear mapping ψ, high-dimensional channels have a wider range of interactions than low-dimensional channels.

Figure 5 illustrates the overview of ECA module. Let the output of one convolution block be $\chi \in R^{W \times H \times C}$, after aggregating convolution features χ using global average pooling (GAP), ECA module first adaptively determines kernel size k, and then performs 1D convolution followed by a Sigmoid function (σ) to learn channel attention. The attention is applied to features γ to generate $\widetilde{\chi} \in R^{W \times H \times C}$ as the output of the ECA module.

In Fig. 6, we compare the Residual blocks of original ResNet [21, 22], SE-Residual blocks, and ECA-Residual blocks. In this paper, as shown in Fig. 7, and we propose Channel Attention-based Emotion Recognition Networks (CAERN). Multiple frequency bands mapping matrix as input to our model, and the first step is the standard convolution operation; a 3×3 convolution kernel with a stride size of 1 and a ReLU activation function is employed. The second step is to directly stack 3 ECA-Residual blocks of different channel dimensions, respectively named ECAResb1, ECAResb2, ECAResb3. The third step is to reduce dimension and use the global pooling layer, and the output feature maps are reshaped to be in a feature vector $f \in R^{256}$. Then the following softmax layer receives f to predict human emotional state.

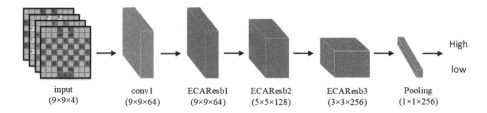

input (9×9×4) → conv1 (9×9×64) → ECAResb1 (9×9×64) → ECAResb2 (5×5×128) → ECAResb3 (3×3×256) → Pooling (1×1×256) → High / low

Fig. 7. Channel attention-based emotion recognition networks

4   Experiments

4.1   Experiment Setup

In the DEAP database, we adopt the experimental protocol as that of [15] to evaluate the proposed method. The EEG data (under stimulus) is divided into segments and they are all 1 s data length. So in our experiment, there are 2400 segments on each participant. The labels are divided into two binary classification according to the threshold set to 5 in arousal and valence dimensions. We perform 10-fold cross-validation on each participant, and the average value represents the result of this participant. The final experimental result is the average of these 32 subjects.

In the SEED database, our experiments adopt the same experimental setting with [13]. The training set includes the first 9 trials of EEG data, and the testing set include remaining 6 trials. There are totally 3 (session) ×15 (subject) = 45 experiments. Removing the worst result in 3 session of the same subject. Finally, the mean accuracy of the remaining 30 experiments is used as the final result.

We implemented the ResNets, SE-ResNets and Channel Attention-based Emotion Recognition Networks (CAERN) in the same settings, which initial learning rate was 10^{-4} . The model architecture of the ResNets, SE-ResNets, and CAERN is the same, except that ECA-Residual block is replaced by Residual block, and SE-Residual block.

4.2   Performance Comparison Among Relevant Methods

In the DEAP database, we designed two cases to verify the effect of our model and compare with other methods. Case1 represents the DE features of the baseline signals are not used while Case2 represents the DE features of the baseline signals are used. To examine the importance of different frequency bands for emotion recognition, we have designed different experiments for a single frequency band and all frequency bands. The result is shown in Tables 1 and 2. Compare with these methods [15,17], our proposed model can achieve higher accuracy in the DEAP database in the dimensions of arousal and valence. And we focused on comparing ResNets, SE-ResNets and Channel Attention-based Emotion Recognition Networks (CAERN). On the whole, the CAERN method achieves better results. It proved that the attention mechanism helps emotion recognition tasks.

Table 1. Classification result on arousal class.

Method	Case	θ	α	β	γ	all (θ, α, β, γ)
MLP [15]	1	66.38	64.22	64.27	64.91	69.51
	2	70.68	67.7	69.43	68.37	88.68
CCNN [15]	1	62.99	63.63	67.37	70.29	69.55
	2	78.09	80.81	79.97	80.98	90.24
P-GCNN [17]	1	69.02	69.21	68.56	69.19	71.32
	2	81.33	80.61	80.09	81.10	91.70
ResNets	1	61.23	62.05	67.86	72.46	70.35
	2	81.25	83.18	83.64	84.90	92.06
SE-ResNets	1	62.10	62.38	67.70	72.19	69.74
	2	81.11	83.43	83.43	84.57	92.51
Ours	1	62.23	63.97	69.17	72.86	70.63
	2	80.92	83.41	83.40	84.73	92.77

Table 2. Classification result on valence class.

Method	Case	θ	α	β	γ	all (θ, α, β, γ)
MLP [15]	1	63.61	58.00	58.52	61.51	68.11
	2	68.70	61.97	65.90	64.67	87.82
CCNN [15]	1	58.90	60.15	64.84	68.55	68.56
	2	75.66	78.73	78.13	79.83	89.45
P-GCNN [17]	1	67.42	67.41	66.79	67.09	70.10
	2	79.97	80.14	79.16	80.11	91.28
ResNets	1	58.17	59.15	65.84	70.68	68.81
	2	76.92	80.61	81.32	83.07	91.00
SE-ResNets	1	58.28	59.32	65.76	70.40	69.09
	2	78.19	81.26	82.16	83.89	91.46
Ours	1	59.19	60.19	67.05	71.70	70.34
	2	78.25	81.32	82.21	83.70	91.87

In the SEED database, The result is shown in Tables 3, we mainly compared the methods [13,17]. In all frequency bands, our proposed model achieves a relatively high effect. We believe that all frequency bands features including richer information are more suitable for our proposed model. The experimental results also prove that effective channel attention can improve the accuracy of emotion recognition.

Table 3. Classification result on SEED database.

Method	δ	θ	α	β	γ	all $(\delta, \theta, \alpha, \beta, \gamma)$
SVM [13]	60.50	60.95	66.64	80.76	79.56	83.99
DNN [13]	64.32	60.77	64.01	78.92	79.19	86.08
P-GCNN [17]	73.05	75.49	75.66	82.32	83.55	84.08
ResNets	63.87	64.09	68.90	77.01	79.10	85.82
SE-ResNets	65.45	65.74	69.91	78.43	80.22	85.01
Ours	65.09	66.16	71.40	79.50	80.67	86.43

4.3   Conclusion

In this work, we presented a novel deep model CAERN for EEG-based emotion recognition. The proposed deep architecture is based on ResNets with a channel attention module. The ResNets can extract richer features on multi-band differential entropy features. While the channel attention module focus on the more important features for tasks. Our experiment results demonstrate the ability of our model to efficiently exploit the more critical and effective features for addressing challenging EEG-based emotion recognition tasks.

Acknowledgment. This research is supported by National Natural Science Foundation of China (41571401), Chongqing Natural Science Foundation (cstc2019jscx-mbdxX0021).

References

1. Shu, L., Xie, J., Yang, M., et al.: A review of emotion recognition using physiological signals. Sensors **18**(7), 2074 (2018)
2. De Nadai, S., Dinca, M., Parodi, F., et al.: Enhancing safety of transport by road by on-line monitoring of driver emotions. In: Service Oriented Software Engineering, pp. 1–4 (2016)
3. Guo, R., Li, S., He, L., et al.: Pervasive and unobtrusive emotion sensing for human mental health. In: International Conference on Pervasive Computing, pp. 436–439 (2013)
4. Verschuere, B., Crombez, G., Koster, E.H., et al.: Psychopathy and physiological detection of concealed information: a review. Psychologica Belgica **46**, 99–116 (2006)
5. Zhang, Y., Yang, Z., Lu, H., et al.: Facial emotion recognition based on biorthogonal wavelet entropy, fuzzy support vector machine, and stratified cross validation. IEEE Access **4**, 8375–8385 (2016)
6. Mao, Q., Dong, M., Huang, Z., et al.: Learning salient features for speech emotion recognition using convolutional neural networks. IEEE Trans. Multimed. **16**(8), 2203–2213 (2014)
7. Wang, X.W., Nie, D., Lu, B.L.: Emotional state classification from EEG data using machine learning approach. Neurocomputing **129**(apr.10), 94–106 (2014)

8. Cheng, B., Liu, G.: Emotion recognition from surface EMG signal using wavelet transform and neural network. In: International Conference on Bioinformatics and Biomedical Engineering, pp. 1363–1366 (2008)

9. Juárez-Castillo, E., Acosta-Mesa, H.G., Fernandez-Ruiz, J., et al.: A feature selection method based on a neighborhood approach for contending with functional and anatomical variability in fMRI group analysis of cognitive states. Intell. Data Anal. **21**(3), 661–677 (2017)

10. Tak, S., Ye, J.C.: Statistical analysis of fNIRS data: a comprehensive review. Neuroimage **85**, 72–91 (2014)

11. Craik, A., He, Y., Contreras-Vidal, J.L.: Deep learning for Electroencephalogram (EEG) classification tasks: a review. J. Neural Eng. **16**, 031001 (2019). https://doi.org/10.1088/1741-2552/ab0ab5

12. Duan, R., Zhu, J., Lu, B., et al.: Differential entropy feature for EEG-based emotion classification. In: International IEEE/EMBS Conference on Neural Engineering, pp. 81–84 (2013)

13. Zheng, W., Lu, B.: Investigating critical frequency bands and channels for EEG-based emotion recognition with deep neural networks. IEEE Trans. Auton. Mental Dev. **7**(3), 162–175 (2015)

14. Tripathi, S., Acharya, S., Sharma, R.D., et al.: Using deep and convolutional neural networks for accurate emotion classification on DEAP dataset. In: Innovative Applications of Artificial Intelligence, pp. 4746–4752 (2017)

15. Yang, Y., Wu, Q., Fu, Y., et al.: Continuous convolutional neural network with 3D input for EEG-based emotion recognition. In: International Conference on Neural Information Processing, pp. 433–443 (2018)

16. Piqueira, J.R.C.: Network of phase-locking oscillators and a possible model for neural synchronization. Commun. Nonlinear Sci. Numer. Simul. **16**(9), 3844–3854 (2011)

17. Wang, Z., Tong, Y., Heng, X., et al.: Phase-locking value based graph convolutional neural networks for emotion recognition. IEEE Access **7**, 93711–93722 (2019)

18. Chao, H., Dong, L., Liu, Y., Lu, B.: Emotion recognition from multiband EEG signals using CapsNet. Sensors **19**, 2212 (2019)

19. Guo, S., Lin, Y., Feng, N., et al.: Attention based spatial-temporal graph convolutional networks for traffic flow forecasting. In: Proceedings of the AAAI Conference on Artificial Intelligence, vol. 33, pp. 922–929 (2019)

20. Wang, Q., Wu, B., Zhu, P., Li, P., Zuo, W., Hu, Q.: ECA-Net: Efficient Channel Attention for Deep Convolutional Neural Networks (2019)

21. He, K., Zhang, X., Ren, S., et al.: Deep residual learning for image recognition. In: Computer Vision and Pattern Recognition, pp. 770–778 (2016)

22. He, K., Zhang, X., Ren, S., Sun, J.: Identity mappings in deep residual networks. In: Leibe, B., Matas, J., Sebe, N., Welling, M. (eds.) ECCV 2016. LNCS, vol. 9908, pp. 630–645. Springer, Cham (2016). https://doi.org/10.1007/978-3-319-46493-0_38

23. Koelstra, S., Muhl, C., Soleymani, M., et al.: DEAP: a database for emotion analysis; using physiological signals. IEEE Trans. Affect. Comput. **3**(1), 18–31 (2012)

24. Hu, J., Shen, L., Albanie, S., Sun, G., Wu, E.: Squeeze-and-excitation networks. IEEE Trans. Pattern Anal. Mach. Intell. **42**(8), 2011–2023 (2020). https://doi.org/10.1109/TPAMI.2019.2913372

25. Nair, V., Hinton, G.E.: Rectified linear units improve restricted Boltzmann machines. In: ICML (2010)

The MovieWall: A New Interface for Browsing Large Video Collections

Marij Nefkens and Wolfgang Hürst[✉]

Utrecht University, Utrecht, The Netherlands
huerst@uu.nl

Abstract. Streaming services offer access to huge amounts of movie and video collections, resulting in the need for intuitive interaction designs. Yet, most current interfaces are focused on targeted search, neglecting support for interactive data exploration and prioritizing speed over experience. We present the *MovieWall*, a new interface that complements such designs by enabling users to randomly browse large movie collections. A pilot study proved the feasibility of our approach. We confirmed this observation with a detailed evaluation of an improved design, which received overwhelmingly positive subjective feedback; 80% of the subjects enjoyed using the application and even more stated that they would use it again. The study also gave insight into concrete characteristics of the implementation, such as the benefit of a clustered visualization.

Keywords: Video browsing · Interaction design · Data exploration

1 Introduction

Streaming videos and movies has become very popular. Netflix currently has over 90 million subscribers worldwide [10] and the number is still growing. The interfaces of these streaming applications are generally designed for targeted search and focused on providing users with a movie or TV show to watch without too much time or effort spent looking around. This is typically achieved via some recommendation mechanism that presents a limited number of movies that are expected to fit users' interests. This approach makes sense for a television interface controlled by a remote, where complex navigation is often undesirable. However, nowadays a multitude of devices are used to browse and watch video content, including smartphones and tablets. For such mobile touch-based devices, which offer new and rich interaction possibilities, the lack of a convenient way to explore beyond the recommendations seems limiting. Providing an alternative and engaging means to explore all movies on offer would nicely complement the targeted, efficiency-based approach of modern movie browsing interfaces.

To address this lack of support for experience-focused large data browsing, we introduce the *MovieWall* interface concept: a movie browsing interface for visual exploration of large movie collections, targeted for mobile touch-based devices. The essence of the concept is rather simple; a large grid with each cell containing

© Springer Nature Switzerland AG 2021
J. Lokoč et al. (Eds.): MMM 2021, LNCS 12573, pp. 170–182, 2021.
https://doi.org/10.1007/978-3-030-67835-7_15

a movie poster (the movie wall), that can be explored using common zoom and pan touch gestures. To establish the feasibility of our idea, we conducted an informal pilot study with a prototype implementation of the interface. After verifying this was the case, the design was improved upon and its potential was further evaluated in a detailed user study. Additionally, the influence of different arrangements of the movies was examined in this user study. Our work makes the following contributions: First, we provide a novel, engaging and exploratory movie browsing interface that could complement interfaces of existing movie streaming services as it has the following advantages: 1. The ability to show a large movie collection in its whole and to easily explore related movies based on movie metadata. 2. The ability to easily navigate between different parts of the collection, for example, between genres. Second, we prove the feasibility and potential of the interface concept, and demonstrate the influence of clustering and randomization on the browsing experience and behaviour of users.

2   Related Work

While movie browsing has become very relevant in commercial areas due to the popularity of online streaming services, there is limited scientific research on this topic. Low et al. [8] recently proposed an interface for exploratory movie browsing using map-based exploration that allows users to easily find interesting links between movies. Other approaches [3,9] allow users to browse through a cloud of words (e.g. moods, emotions) that were previously extracted from the audio and subtitles of movies. The interface was perceived as interesting and fresh, but slightly overwhelming.

If we look at two related fields, image and video browsing, applications often use a two-dimensional scrollable storyboard layout (e.g. standard mobile gallery applications). Yet, this design has limitations (more elaborately discussed in [17]), mainly that it can only show a limited number of items, which makes it hard to get a good overview of a collection and any potential structure. To overcome these issues, many alternatives have been proposed. We see interfaces that layout all images on a large canvas, based on similarity [11,18,19]. Such a visualization is often created using methods like multi-dimensional scaling (MDS), which can be computationally expensive when used globally, especially with large data sets. Moreover, project images based on the outcome of methods such as MDS, often produces overlap. Rodden et al. [13] showed that overlap made it hard for users to recognise some images as edges and other important details are regularly occluded. For touch-based devices overlap can also make it harder to select items [4]. To explore large collections, graph-based and hierarchical approaches are common, sometimes combined [2]. A hierarchical layout typically allows its user to start their search very high-level, by providing an overview of images that represent the entire collection, as seen in [7,15]. The user can then direct their search to a region of interest, revealing more similar content. This design is quite suited for displaying large collections, the exploratory similarity-based image browsing application proposed in [7] could store millions of images.

Graph-based visualizations have been proposed to allow the user to browse a collection step by step, where selecting one item typically links to similar content, such as previously seen in [8]. There have been some three-dimensional visualizations as well, arguing that these make better use of the available screen space and can hence show more at a glance [17]. Two prime examples of 3D visualizations are shown in the interfaces in [16] and [1,15], which respectively map a grid of images to a horizontal ring-shape and a globe-shape. Both performed quite similar on tablets and smartphones [5,17], and outperformed the standard two-dimensional scrollable storyboard in terms of efficiency.

In the arrangement of images or videos we often encounter some degree of sorting or clustering, which can be very useful to conveniently locate specific content [8,13]. It allows users to quickly assess the available data and judge the relevance of entire areas instead of single items. In the area of image browsing, images are often sorted on visual features (e.g. colour). However, one downside to visual similarity sorting is shown in [14], where target images were sometimes overlooked when placed alongside similar images. In a random arrangement these images stood out more, and thus were more easily found. Visual-based sorting makes less sense for movie posters [6] as visual similarity does not guarantee conceptual similarity. To capture conceptual similarity, caption-based sorting is an option. In [14] this was even slightly preferred over visual sorting. However, caption-based arrangements rely on good image annotations. Luckily, movies generally have a lot of available metadata.

Fig. 1. Left: screenshot of the initial and at the same time "most-zoomed-out" state of the prototype application. Right: "most-zoomed-in" state of the prototype application. In this case a movie is selected and thus a panel is shown containing the details of the selected movie. (*Movie images have been blurred in this publication for copyright reasons. In the experiments, a high resolution display was used, resulting in perfect visibility even at very small image sizes.*)

3   Pilot Study

As the foundation of the interface was built on assumptions, it is important to verify its feasibility with a pilot study. The main question to answer is: *Does the MovieWall interface concept have potential for exploratory movie browsing?*

First, users need to be able to handle it, i.e., *are users able to (easily) interact with the interface?* Also, *are users not too overwhelmed by it?* It's important that users are not negatively influenced by either. Second, it should provide an engaging experience. Thus, another important variable to *assess is if users enjoy interacting with it.* Finally, we wanted to analyse the interaction behaviour of participants. *How do they use the available functionalities? What zoom levels do they use?* The latter could give more insight in what size images people are comfortable browsing. A large size may suggest low potential of our idea, as it limits the amount of items that can be presented at a time.

Implementation. The prototype application is developed for Android and designed for tablet use, with reference resolution 1920 × 1200 pixels (16:10). Images and data used were downloaded from The Movie Database (TMDb, https://www.themoviedb.org/) and used under the fair usage policy.

For the pilot study, a basic version of the interface concept was implemented. It features the core element of the concept: a large wall of 8640 movie posters (144 × 60 movies), which are the 8640 most popular movies in TMDb. The user can change the position in the grid and the number of movie posters on the screen by scrolling and zooming, respectively. The initial screen is shown in Fig. 1 (left). In case of the reference resolution, all movies are shown on the screen, each poster image with an average size of 13.3 × 20 pixels. Based on earlier research on recognition of small images [20], we assumed it would be nearly impossible for users to recognise anything at this point and that this would encourage them to use zoom in and stop at a zoom level on which they could comfortably browse. The most-zoomed-in version of the interface is shown in Fig. 1 (right). In this case the movie posters are 248 × 372 pixels on screen and easy to recognize.

Selecting a movie marks its poster in the grid to indicate the current focus, and opens a panel with the main movie details, such as actors and plot (Fig. 1, right). Instead of taking up the entire screen space, the panel only occupies the right side, allowing the users to still interact with the movie grid on the left side.

The arrangement of movies in the grid is random. We did not expect the sorting of the movies to significantly influence the feasibility or potential of the concept. Furthermore, we hypothesized that a random arrangement could actually provide an interesting environment which allows for serendipitous discoveries.

Procedure. The experiments took place in a controlled environment with an observing party. Participation was entirely voluntary and not reimbursed. Prior to the experiment, subjects filled out a consent form and received a short introduction to the experiment and interface. The test device used was a Samsung Galaxy Tab A with a screen resolution of 1920 × 1200 pixels. Before the actual trial, the participants were asked to play around with the application for a short while (max. 2 min). This was both to familiarize them with the interface and to observe how they interacted with it when using it for the first time. Afterwards, they were asked to read a list of all functionalities of the interface and state if they missed anything. Next, each subject was asked to use the application with the following task (max. 5 min): *"Assume you are about to board a long flight, so you will be offline for a couple of hours. Hence, you want to download a couple* *of movies to watch when you are in the air. Now use the interface to search for, let's say 3–5 movies that you would probably download."* Afterwards, they filled out a questionnaire and a semi-structured interview was conducted about the answers given in the questionnaire and the observed behaviour while conducting the task. During both trials, the subjects were observed and each interaction was logged by the application. Ten people participated in the pilot study, nine male and one female, ages 21–26 (average of 23.2 years old).

Results and Discussion. The pilot study proved the feasibility of the concept but also highlighted several shortcomings. We discuss the results with respect the questions introduced above.

Are Users able to Handle the Interface? Generally all functionalities were quite clear to the participants. For two it took a while to realize they could zoom in. All were asked to rate four statements (S1-4) on a five-point Likert scale about their experience. Every subject thought the interface was easy to use (S1), and most also thought it was intuitive to interact with (S2). Most stated that the interface worked fluently and according to expectation. One was neutral about this statement, as this participant was unsure what to do with the application initially. Most subjects stated they were (slightly) overwhelmed by the interface (S3), mostly at the beginning. Only one participant disagreed with this statement. The main reasons for this were the vast number of movies, the lack of a starting point and the lack of structure. Three out of the nine subjects that stated they were overwhelmed, viewed this as a clearly negative experience.

Do Users Enjoy Using the Interface? Subjects were divided about enjoying the browsing experience (S4). Participants who liked it, mentioned that they appreciated casually browsing around and that they discovered new and unexpected movies. Participants who did not like it that much, stated that it was too overwhelming, that they encountered too many uninteresting movies or that they did not like the lack of options to direct their search.

Most subjects stated they would find such an interface a useful complement in existing streaming services, such as Netflix, especially with the addition of more structure or another means to direct their search to some degree. Participants could see themselves use it when not knowing or caring what to watch, or when picking a movie with multiple people, for example friends or family.

How do Users Interact with the Interface? Observations during the experiment and the log data indicate that the subjects used the zoom function mainly to get to a comfortable size and generally remained on this level. This suspicion was confirmed in the interviews.

Conclusion (pilot study). All in all, the pilot study results indicate that the concept has potential, but there are some clear issues that need to be addressed. Initially, the interface was quite overwhelming for most users, which seemed to be primarily caused by the fact that the wall was completely zoomed out at the beginning. Starting with a larger initial zoom level could diminish this feeling. Another cause for the overwhelmed feeling was the lack of structure. While

Fig. 2. Visualization of a clustered (left) and a semi-clustered arrangement (right).

the random arrangement of movies helped users to be more open-minded and was often appreciated in the beginning, after a while subjects generally wanted more structure. There was no means for users to direct their search to some degree, which was frustrating for some. Sorting the movies based on metadata is a possible improvement here. Yet, we should keep in mind that the random arrangement did allow for more serendipitous discoveries, and arranging the movies in clusters might negate this effect.

4   Detailed User Study

The pilot study identified two potential issues of the prototype interface. First, it was considered overwhelming, likely caused by (a) the number of movies, (b) the initial zoom level and (c) the lack of structure. Second, there was a lack of structure, giving users no control over what content they came across, causing some users to encounter a lot of uninteresting movies. We hypothesize that these issues can be resolved by (a) reducing the number of movies in total, (b) initially showing less movies on screen, and (c) (semi-)sorting the collection and/or adding filter functionality.

While the pilot study suggests a preference for a sorted arrangement over a random one, this needs further investigation. Arranging movies in clusters based on genre provides a means for users to direct their search and might make the interface less overwhelming. However, this is likely at the cost of serendipity. A potential middle ground is to cluster the movies, but introduce a certain degree of randomness. This approach can provide structure for the user, but prevent a narrow vision. To test these assumptions, we compared three different arrangements of movies in this study: *Random* (R): Same as in the pilot study; all movies are randomly distributed. *Clustered* (C): All movies are strictly clustered on genre, with the most popular movies in the center of the cluster. A possible arrangement is shown in Fig. 2, left. *Semi-clustered* (CR): Same as (C), but with a certain degree of randomization in the outer areas of each cluster and within the popularity sorting. A possible arrangement is shown in Fig. 2, right.

Fig. 3. Left: initial state of the improved interface. Right: The movie wall with multiple filters applied: *Animation* (genre) and *Walt Disney Pictures* (company), highlighting related movies. (*Again, images are blurred in this publication for copyright reasons.*)

Another option to direct the exploration process is to allow users to filter on metadata like genre or actors. In contrast to clustering, this is an optional tool for structure. Moreover, it can be combined with any arrangement. All in all, the following was investigated in this user study: 1) The general usefulness and acceptance of the interface concept. 2) The usefulness of filter functionality. 3) The influence of different arrangements of the movie collection, more specifically: a) The *interaction with the movie collection*; i.a. the number of movies selected while browsing the interface. b) The *satisfaction of movies encountered* while browsing the interface. We suspect this is higher for the clustered arrangements, as the users are in control of what content they encounter. However, serendipitous discoveries could increase the satisfaction for the random arrangement as well. And c) the *intrinsic motivation* of users to use the interface with each arrangement. We assume that users will like the (semi-clustered) arrangements better than the random one, and thus have a higher intrinsic motivation to use these.

4.1   Implementation and Study Design

Implementation. The application used in the second study was an improved version of the one from the pilot study with the following changes. First, the grid contains less movies, 2160 in total (72×30). At the smallest zoom level each image is approximately 26×39 pixels. This decrease in movies potentially makes the interface less overwhelming. Furthermore, by decreasing the total number of movies, the average popularity of the movies increases, likely resulting in a more interesting and relevant collection. Additionally, the initial zoom level greatly decreased. Figure 3, left, shows the new initial state. Second, filters were added so users can look for movies with similar metadata as the currently selected movie. Movie details in the panel can be selected to highlight all movies with, for example, a common actor. Filters can be combined so the user can easily search for movies based on multiple criteria, without having to enter complex queries. Figure 3, right, shows an example. The following metadata can be applied as a filter: genres (up to five), top actors (up to three), top directors (up to two), production companies (up to six). Finally, clustered and semi-clustered arrangements were added. Moreover, when zoomed out far enough, labels are shown indicating the location of each genre cluster, helping the user identify the structure of the collection.

Procedure. Experiments took place in a controlled environment with an observing party. Participation was entirely voluntary and not reimbursed. Prior to the experiment, subjects filled out a consent form, a list of general background questions and received a short introduction. The same device (Samsung Galaxy Tab A) was used as during the pilot study.

Each subject tested all three different arrangements. The order in which the arrangements were presented to each participant were rotated between subjects, resulting in a total of six different orders. To familiarize subjects with the application prior to the actual experiment, they were asked to use it for a short while (max. 2 min) with the task to figure out all functionalities. Afterwards, they were informed about anything they possibly missed.

For the upcoming three trials, the subjects received the following scenario: *"Imagine you're going on trip next week, you're going to visit South-Africa and subsequently Brazil. After your trip you're going back home, meaning you'll have three long flights ahead of you. During these flights you want to watch some movies. There is no WiFi on the plane, so you'll have to download the movies in advance. Assume that you're at the airport gate for your first/second/third flight and you have some spare time. Find some movies to download that you might like to watch during your the flight and add them to your watchlist."*

For each different arrangement the subject used the application for three minutes, starting with the same arrangement as used while familiarizing with the application. Prior to each new trial, the new arrangement of movies was globally explained to the participant. In case the subject started with the random arrangement, the cluster feature was explained prior to the second trial. During each trial, all interaction was logged by the application. After each trial, the participant rated several statements on a seven-point Likert scale, consisting of the interest/enjoyment subscale of the Intrinsic Motivation Inventory (IMI) [12] and specific questions about movies encountered.

After completing the three trials, a semi-structured interview was conducted with the subject concerning their opinion on the different arrangements and the interface in general. Thirty people participated in the study, of which twenty-two were male and eight female, ages 20 to 29 (on average 23.3 years old). Five of them had also participated in the pilot study.

4.2   Results and Discussion

General Usefulness and Acceptance: All functionalities were quite clear to the participants. Positive aspects mentioned include that it worked smoothly

(8 subjects) and was easy/intuitive to use (5). In contrast to the pilot study, not all subjects (5) realized it was possible to zoom. This could be caused by the difference in initial zoom level. Yet, none of these five commented on the small image size (initially 77 × 115 pixels per image), suggesting they were not bothered by it.

Looking at the zoom levels, the suspicion that users are comfortable browsing with relatively small images seems further confirmed. In the second study we saw that over all arrangements, roughly two-thirds of the time a zoom level between 50–150 was used. This ranges from an image width of 50 × 75 pixels to 150 × 225 pixels. With the test device used, this roughly translates to between 60 and 610 movies on screen at once with the panel closed. While there are clear similarities in zoom levels between the two studies, subjects of the pilot study used larger zoom levels on average. We hypothesize that this is mainly caused by the difference in initial zoom level.

In the interview following the experiment, subjects were mostly positive about the interface. Comments included: *"I saw a lot of movies (in a short time)"* (10); *"it works smoothly"* (8); *"I liked the filters"* (8); *"I encountered a lot of unexpected/interesting movies"* (7); *"I liked the use of movie posters"* (6). The main negative comment was that one could only judge a movie by its poster (at first sight) (5). Two subjects mentioned they felt like this caused them to only pick familiar movies. When asked whether they enjoyed using the interface, 24 subjects gave an affirmative answer, 4 were neutral about the statement and 2 said no. Both of them stated that they did not like looking for movies in general. When asked whether they thought they would use an interface like this again, 26 subjects said they probably would. When asked what they would improve, the most common answer was the addition of a search function (14). Many mentioned they appreciated having a "starting point" in the wall, for example a movie they liked and from where they could search for more related movies, which a search function could help accommodate. Another requested feature was to keep the filters active while switching or deselecting movies (9).

Filters: The filter functionality was frequently used. One or more filters were used in 66 of the 90 trials (73.3%) and only two subjects did not use them in any of their trials. As expected, genre filters were used most. Yet, less common search criteria like director or production company were also used quite often. This indicates a desire to find related movies based on diverse sorts of criteria. Participants also mentioned they would like to be able to search on more kinds of metadata like collections (e.g. Star Wars), script writers or story origin.

Comparison of Different Movie Arrangements: To compare the interaction of users with the movie collection between the different arrangements, the number of unique movies inspected during each trial and the number of movies in the watchlist at the end of each trial were recorded. We expected that both these variables could be strongly influenced for the second and third trials, since in each trial the same data set was used. Thus, only the results of the first arrangement used by each subject were considered for this comparison. Table 1 contains the difference in inspected movies (left) and the difference in movies added to the

Table 1. Descriptives for the number of movies inspected during the first trial (left) and numbers of movies in the watchlist at the end of the first trial (right).

	N	Mean	SD	Min	Max
R	10	10.9	3.51	4	15
C	10	9.3	2.83	5	14
CR	10	7.1	2.079	3	10
Total	30	9.1	3.188	3	15

	N	Mean	SD	Min	Max
R	10	6.3	3.917	1	11
C	10	5.9	2.331	3	10
CR	10	4.8	1.476	3	8
Total	30	5.67	2.746	1	11

watchlist between the three arrangements (right). A One-Way ANOVA shows a significant difference in the number of movies inspected, $F(2,27) = 4.429$, $p = 0.022$. This number is significantly higher with the random arrangement than with the semi-clustered one. No significant results were found when inspecting R vs. C or CR vs. C ($p = 0.223$, and $p = 0.098$ respectively). The numbers of movies added to the watchlist between the different arrangements are not significantly different; $F(2,27) = 0.788$, $p = 0.465$.

After each trial subjects were asked to rate seven statements on a 7-point Likert scale. None of these ratings were significantly different between the arrangements. Besides specific questions about movies encountered and picked during the trials, the subjects were asked to rate the seven statements of the interest/enjoyment subscale of the Intrinsic Motivation Inventory after each trial. Yet, a significant difference could not be identified.

In contrast to the previous results, the interview showed a clear favourite arrangement though. One-third of the participants preferred either the clustered or semi-clustered version; of those most mentioned they did not really see the difference between the two arrangements (7 subjects). Only one person specifically preferred the semi-clustered version and ten the strictly clustered version. The random version was least liked, with twenty participants stating it was their least favourite. However, five subjects liked it best. Three had no preference and stated they would use the different arrangements in different situations ("without specific intent I'd use the random version, otherwise one of the clustered ones").

The interviews with the subjects indicated that the randomization in the semi-clustered version mostly stayed unnoticed. Only five subjects explicitly mentioned they were surprised with the placement of certain movies, of which three preferred the strictly clustered version. While mostly unnoticed, subjects did inspect and pick randomized movies. Fourteen subjects had at least one randomized movie in their watchlist. Interestingly enough, two out of the three subjects that preferred the strictly clustered version over the semi-clustered version because of the randomization, had randomized movies in their watchlist.

Discussion. The results of this study suggest that the two main issues identified in the pilot study (overwhelming and lacking structure) have been resolved. Consequently, that seems to have lead to a more enjoyable, engaging browsing experience. When asked whether they enjoyed using the application, 80% of the participants gave an affirmative answer, as opposed to only half of the subjects in the pilot study. While the study further confirmed the usefulness of the interface concept, the influence of the different arrangements on the tested variables was less apparent than expected. We hypothesized that the semi-clustered version could provide a nice balance between structure and allowing the user to discover unexpected content. However, there is little evidence this was the case. The issue with the semi-clustered arrangement is most likely that it opposes the expectations of users, which can be undesirable. When participants noticed the randomization in the clustered version, it was generally perceived as strange or faulty, both rather negative characteristics.

5   Conclusion

In this work, we presented the *MovieWall*, an interface concept for engaging, exploratory movie browsing. By using a rather simple, yet detailed overview of a huge amount of movies, it aims at complementing existing search approaches for situations where exploration and discovery of new data is more important than speed. One concern of this concept was the number of movies on screen could be too large and the sizes of the movie posters too small. However, participants were quite able to handle this. The interaction of users with it suggests that they were comfortable browsing with relatively small images and thus many movies on their screen at once. Previous research already showed that humans are quite proficient at recognising small images, but the fact that they are also willing to browse with such small images is new. Furthermore, the large majority stated they enjoyed using the interface and would probably use it again.

The filters added in the second version of the interface were used frequently and explicitly mentioned by users as a very useful feature. While users mostly filtered on genre, we saw a moderate use of the cast, company and director filters as well. This indicates a desire of at least some users to navigate by diverse kinds of metadata, which is often not encouraged or focused on in state-of-the-art movie browsing applications. Some subjects stated they would like even more sorts of metadata to navigate by. For future work, it would be interesting to further analyse what types of metadata people would use while browsing movies, and how to best facilitate this. As for the arrangement of movies, the results were inconclusive. In contrary to our prediction however, the randomized arrangement did not perform significantly different than the other arrangements in terms of movies found or satisfaction about movies encountered. However, when asked what arrangement subjects preferred, the strictly clustered arrangement was the clear favourite, and the random arrangement was liked least.

References

1. Ahlström, D., Hudelist, M.A., Schoeffmann, K., Schaefer, G.: A user study on image browsing on touchscreens. In: Proceedings of the 20th ACM International Conference on Multimedia, pp. 925–928. ACM (2012)

2. Barthel, K.U., Hezel, N., Mackowiak, R.: Graph-based browsing for large video collections. In: He, X., Luo, S., Tao, D., Xu, C., Yang, J., Hasan, M.A. (eds.) MMM 2015. LNCS, vol. 8936, pp. 237–242. Springer, Cham (2015). https://doi.org/10.1007/978-3-319-14442-9_21
3. Gil, N., Silva, N., Duarte, E., Martins, P., Langlois, T., Chambel, T.: Going through the clouds: search overviews and browsing of movies. In: Proceeding of the 16th International Academic MindTrek Conference, pp. 158–165. ACM (2012)
4. Gomi, A., Itoh, T.: Mini: a 3D mobile image browser with multi-dimensional datasets. In: Proceedings of of the 27th Annual ACM Symposium on Applied Computing, pp. 989–996. ACM (2012)
5. Hudelist, M.A., Schoeffmann, K., Ahlstrom, D.: Evaluation of image browsing interfaces for smartphones and tablets. In: 2013 IEEE International Symposium on Multimedia (ISM), pp. 1–8. IEEE (2013)
6. Kleiman, Y., Goldberg, G., Amsterdamer, Y., Cohen-Or, D.: Toward semantic image similarity from crowdsourced clustering. Visual Comput. 32(6–8), 1045–1055 (2016)
7. Kleiman, Y., Lanir, J., Danon, D., Felberbaum, Y., Cohen-Or, D.: Dynamicmaps: similarity-based browsing through a massive set of images. In: Proceedings of the 33rd ACM Conference on Human Factors in Computing Systems, pp. 995–1004. ACM (2015)
8. Low, T., Hentschel, C., Stober, S., Sack, H., Nürnberger, A.: Exploring large movie collections: comparing visual berrypicking and traditional browsing. In: Amsaleg, L., Guðmundsson, G.Þ., Gurrin, C., Jónsson, B.Þ., Satoh, S. (eds.) MMM 2017. LNCS, vol. 10133, pp. 198–208. Springer, Cham (2017). https://doi.org/10.1007/978-3-319-51814-5_17
9. Martins, P., Langlois, T., Chambel, T.: Movieclouds: content-based overviews and exploratory browsing of movies. In: Proceedings of the 15th International Academic MindTrek Conference: Envisioning Future Media Environments, pp. 133–140. ACM (2011)
10. Netflix: Netflix website targeted for investors. https://ir.netflix.com/index.cfm
11. Pečenović, Z., Do, M.N., Vetterli, M., Pu, P.: Integrated browsing and searching of large image collections. In: Laurini, R. (ed.) VISUAL 2000. LNCS, vol. 1929, pp. 279–289. Springer, Heidelberg (2000). https://doi.org/10.1007/3-540-40053-2_25
12. Plant, R.W., Ryan, R.M.: Intrinsic motivation and the effects of self-consciousness, self-awareness, and ego-involvement: an investigation of internally controlling styles. J. Pers. 53(3), 435–449 (1985)
13. Rodden, K., Basalaj, W., Sinclair, D., Wood, K.: Evaluating a visualisation of image similarity as a tool for image browsing. In: 1999 IEEE Symposium on Information Visualization, 1999. (Info Vis 1999) Proceedings, pp. 36–43. IEEE (1999)
14. Rodden, K., Basalaj, W., Sinclair, D., Wood, K.: Does organisation by similarity assist image browsing? In: Proceedings of the SIGCHI Conference on Human Factors in Computing Systems, pp. 190–197. ACM (2001)
15. Schaefer, G.: A next generation browsing environment for large image repositories. Multimed. Tools Appl. 47(1), 105–120 (2010)
16. Schoeffmann, K., Ahlström, D.: Using a 3D cylindrical interface for image browsing to improve visual search performance. In: 2012 13th International Workshop on Image Analysis for Multimedia Interactive Services (WIAMIS), pp. 1–4. IEEE (2012)
17. Schoeffmann, K., Ahlström, D., Hudelist, M.A.: 3-D interfaces to improve the performance of visual known-item search. IEEE Trans. Multimed. 16(7), 1942–1951 (2014)

18. Strong, G., Hoeber, O., Gong, M.: Visual image browsing and exploration (Vibe): user evaluations of image search tasks. In: An, A., Lingras, P., Petty, S., Huang, R. (eds.) AMT 2010. LNCS, vol. 6335, pp. 424–435. Springer, Heidelberg (2010). https://doi.org/10.1007/978-3-642-15470-6_44

19. Tardieu, D., et al.: Browsing a dance video collection: dance analysis and interface design. J. Multimod. User Interfaces 4(1), 37–46 (2010)

20. Torralba, A., Fergus, R., Freeman, W.T.: 80 million tiny images: a large data set for nonparametric object and scene recognition. IEEE Trans. Pattern Anal. Mach. Intell. 30(11), 1958–1970 (2008)

Keystroke Dynamics as Part
of Lifelogging

Alan F. Smeaton[1,2](✉) (iD), Naveen Garaga Krishnamurthy[2],
and Amruth Hebbasuru Suryanarayana[2]

[1] Insight Centre for Data Analytics, Dublin City University,
Glasnevin, Dublin 9, Ireland
Alan.Smeaton@DCU.ie
[2] School of Computing, Dublin City University, Glasnevin, Dublin 9, Ireland

Abstract. In this paper we present the case for including keystroke
dynamics in lifelogging. We describe how we have used a simple keystroke
logging application called Loggerman, to create a dataset of longitudinal
keystroke timing data spanning a period of up to seven months for four
participants. We perform a detailed analysis of this data by examining
the timing information associated with bigrams or pairs of adjacently-
typed alphabetic characters. We show how the amount of day-on-day
variation of the keystroke timing among the top-200 bigrams for participants varies with the amount of typing each would do on a daily basis.
We explore how daily variations could correlate with sleep score from
the previous night but find no significant relationship between the two.
Finally we describe the public release of a portion of this data and we
include a series of pointers for future work including correlating keystroke
dynamics with mood and fatigue during the day.

Keywords: Keystroke dynamics · Sleep logging · Lifelogging

1 Introduction

Lifelogging is the automatic gathering of digital records or logs about the activities, whereabouts and interactions of an ordinary person doing ordinary things
as part of her/his ordinary day. Those records are gathered by the person, for
the exclusive use of the person and not generally shared. Lifelogs are a personal
multimedia record which can be analysed either directly by the person collecting
the data, or by others [17]. This is done in order to gain insights into long term
behaviour and trends for wellness or behaviour change. Lifelogs also support
searching or browsing for specific information from the past.

Lifelogging as a practical activity has been around for many years and has
matured as the technology to ambiently and passively capture daily activities

Alan Smeaton is partially supported by Science Foundation Ireland under Grant Number SFI/12/RC/2289_P2.

J. Lokoč et al. (Eds.): MMM 2021, LNCS 12573, pp. 183–195, 2021.
https://doi.org/10.1007/978-3-030-67835-7_16

has evolved [3]. Technologies for capturing multimedia lifelog data are wide ranging and well-documented and can be broadly classified into wearable or on-body devices, and off-body logging. The first class includes wearable cameras, location trackers or physiological trackers for heart rate, respiration, etc. while the second includes sensors which form part of our environment such as passive IR sensors for presence detection and contact sensors on doors and windows in the home. Off body logging would also include using software such as measures for cumulative screentime viewing, productivity at work or online media consumption. Whichever lifelog technologies are used, it is when these are combined and fused together into a multimedia archive that we get the best insights into the person as it is well accepted that so many aspects of our lives interact with, and depend on each other.

In a recent article by Meyer *et al.* [13] the authors highlighted several current issues for longer term self-tracking. Some of these are technical including incompleteness of data leading to data gaps, implicit tracking with secondary sources such as social networks and online services, and multiple interpretations of our data, beyond behaviour support. There are also issues of self-tracking for secondary users such as children or people with special needs with consequent ethical, legal, and social implications [6].

If we regard a lifelog collection as a multimedia or a multi-modal artifact then it can take advantage of progress made in other areas of multimedia analysis, such as computer vision to analyse images from wearable cameras [19]. Progress in areas such as computer vision have depended upon the easy availability of large datasets on which new ideas can be evaluated and compared to previous work. Initiatives such as ImageNet have helped to catalyse these developments. Yet when it comes to the general availability of lifelog collections, these are much rarer precisely because the data is personal.

In related research areas which use the same technologies such as wearable sensors for sleep or gait analysis [7] or wearable cameras for measuring exposure to different food types [15] then lifelog data collections exist but in these cases the wearers are anonymised. In lifelogging it is the accumulation of data drawn together from across different sources and then fused together, that makes the lifelog and that does not reconcile well with the idea of anonymisation.

In this paper we provide a brief review of past work on using keystroke information for user authentication, for identifying different stages of writing strategies, for measuring stress and emotion. We advocate for greater use of keystroke dynamics in lifelogging and we describe a dataset of longitudinal keystroke and sleep data gathered from four participants over a period of more than six months. We describe an analysis of this dataset examining its daily consistency over time both within and across participants and we address the anonymisation of this data by releasing part of it. The paper is organised as follows: in the next section we provide an overview of keystroke dynamics. We then describe the dataset we collected and provide an analysis of this showing its consistency over time and its comparison to sleep score data. Finally, we summarise the case for greater use of keystroke dynamics in multimedia lifelogging and point to future work.

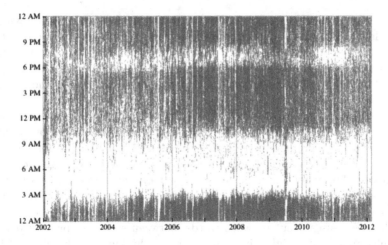

Fig. 1. Insights into timing for Stephen Wolfram's keystrokes over 10 years

2   Keystroke Dynamics

In 2009 Stephen Wolfram reported that he had been using a keystroke logger collecting his every keystroke for the previous 22 years[1]. This was in the form of the key pressed and the date and time of pressing. By 2012 this had grown to be a record of 100 million keystrokes[2] and from all this he was able to generate interesting visualisations on his life, such as the one shown in Fig. 1. This shows his interesting work patterns – he basically works all day, and evening, stopping at about 3AM before resuming at about 10AM the following day with a break of a couple of hours, sometimes, in the evening for dinner. We can also see his various trips where he switched to local timezones such as Summer of 2008 spent in Europe. There are other interesting facts such as the average fraction of keys he types that are backspaces has consistently been about 7%.

While this kind of raw visualisation and analysis may be interesting, when we add detailed timing information we can get further insights. For example, recording keystroke times to the nearest millisecond allows us to look at inter-keystroke times, i.e. the time needed to type two or more adjacent characters.

The original application for keystroke dynamics with accurate timing information was for user authentication. Work in this area goes back over four decades, from 1980 onwards, with regular re-visits to the topic [1,2,8]. The security application for keystroke dynamics is based on the premise that each of us have unique timing information as we interact with GUIs and that includes the timings of our keystrokes, mouse movements, and mouse clicks [4].

An advantage of using keystroke dynamics for security and authentication would be that we would never need to remember passwords, and passwords could never be hacked because they would be replaced by our keystroke dynamics.

---

[1] https://quantifiedself.com/blog/stephen-wolfram-keystroke-logg/.
[2] https://writings.stephenwolfram.com/2012/03/the-personal-analytics-of-my-life/.

However the way authentication for access to computer systems has developed over the last half-century is that they present as tests to be overcome at the point of entry, similar to the way a passport is used at an airport. Keystroke dynamics take some time for baseline timing patterns to emerge. Thus they are not useful for authentication at point of entry, which is why we do not see it in common use today.

Keystroke logging has had other more successful applications including identifying different kinds of author writing strategies and understanding cognitive processes [12]. The premise here is that we establish a baseline for keystroke timing gathered over a long period and at any given period during the day we can compare the current dynamics with the baseline to see if we are typing faster, or slower, perhaps indicating that we are in full creative flow or that we are pondering our thoughts as we write. This also exploits pause location as we type, whether pauses occur between words, between sentences or even between paragraphs and what insights into the author's thinking can be gleaned from such pauses [11].

Keystroke timing information has been used for measuring stress [18] where the authors found it is possible to classify cognitive and physical stress conditions relative to non-stress conditions, based on keystroke and text features. It has also been used for emotion detection where [9] provides a review of almost a dozen published papers addressing this specific topic, and that review was from 2013.

What previous work shows is that keystroke dynamics can provide insights into our behaviour in a way which is non-intrusive, requires no investment in hardware and uses a minuscule amount of computer resources. Yet this is a data source that we have largely ignored to date. Most of the reported work focuses on keystrokes from a keyboard rather than across devices though we are now seeing work appearing on keystroke dynamics on mobile devices. In this paper we argue for keystroke logging as a data source for lifelogging and we illustrate our case using keystroke information collected from four participants over more than six months.

3   Collecting Keystroke Data

For collecting keystroke dynamics we used Loggerman [5] a comprehensive logging tool which can capture many aspects of our computer usage including keyboard, mouse and interface actions. This information is gathered ambiently and stored on the local computer. For keystrokes, Loggerman can record complete words typed by the participant, though when the participant is typing a password, recording is automatically disabled. Once installed, Loggerman records information to log files. A sample of the logfile is shown below with the first number being the typed character using the Mac Virtual Button Standard and the second being the Unix timestamp, in milliseconds.

Table 1. Keystroke information for four participants gathered between 3 and 7 months

Participant	Total keystrokes	No. logged days	Avg. keystrokes/day
1	2, 174, 539	219	9,929
2	1, 147, 285	189	6,070
3	802, 515	227	3,535
4	92, 967	90	1,033

```
14,1586015629644
15,1586015629700
27,1586015630139
2,1586015630421
34,1586015631219
...
```

From an examination of Loggerman files across participants we see that participants regularly make use of autocomplete, they make typing errors and then use backspace or they re-position their cursor to change a previously mis-typed word or fix a spelling error. Thus the number of fully-typed and correctly-typed words in Loggerman's word file is lower than we anticipated. The keystroke dynamics associated with such instances of cursor navigation and re-positioning will not be reflective of the ideal creative flow that we would like when we type and thus keystroke timing information for the overall logging period will have been "polluted" by this necessity of correcting typing errors. That is unfortunate, and some participants may have more of this than others and even a given participant may have periods of more or less of the "flow" experience.

To illustrate the potential of keyboard dynamics in lifelogging we gathered information using Loggerman from four participants covering 1 January 2020 to 24 July 2020 (206 days) and we present an analysis of data from those subjects. Table 1 shows the amount of data generated. The number of days logged varies per participant because participants might disable logging and forget to resume it. We see an almost tenfold variation in the number of daily keystrokes typed between participants 1 and 4, with 2 and 3 in between. Raw keystroke usage data with over 2.5M timed keystrokes for one of our participants for the period covered here plus some additional logging, is available at [16].

When we analysed the participants' log files we found that many of the most frequently used characters are special characters such as punctuation marks and numbers as well as keys for cursor navigation. For the purpose of our timing analysis we will not consider these special characters since they are not part of normal typing flow, thus we consider only alphabetic characters A to Z. This reduces the number of keystrokes by almost half, so for participant 1 the total of 2,174,539 keystrokes reduces to 1,220,850 typed characters. For timing purposes we treat uppercase and lowercase as equal. A rationale for doing this is because it reduces the number of possible 2-character strings (bigrams) we work with to $26 \times 26 = 676$ possible combinations.

Table 2. 10 most frequently used bigrams for participants

Participant	Bigram									
1	TH	IN	HE	AN	RE	ER	ON	AT	ES	ND
2	SS	IN	TH	RE	AT	CV	ES	ER	HE	ON
3	IN	RE	AN	AT	ES	ER	SS	CV	TI	ON
4	IN	TH	AN	RE	ER	HE	AT	ON	ES	TE
Norvig's analysis	TH	HE	IN	ER	AN	RE	ON	AT	EN	ND

As mentioned earlier, a lifelog's usefulness increases when there are multiple sources of multimedia logged data gathered by the participant. Logging data on mood, emotion, stress or writing style at a given time were beyond the scope of this work which focuses on keystroke dynamics only. However, in addition to keystroke logging we also gathered information on participants' sleep.

There are a range of sleep tracking devices available off-the-shelf [14] and we used the Ōura ring [10]. This is a smart ring with in-built infrared LEDs, NTC temperature sensors, an accelerometer, and a gyroscope all wrapped into a ring form factor which gathers data for up to seven days between charges. During sleep it measures heart rate including heart rate variability, body temperature, and movement, from which it can calculate respiration rate. From its raw data it computes a daily activity score, average METs, walking equivalent, a readiness score and for sleep it records bedtime, awake time, sleep efficiency, total sleep and several other metrics, including an overall sleep score. From among all these options we use the overall score, a measure in the range 0 to 100 calculated using a proprietary algorithm which is a function of total sleep, sleep efficiency, restfulness, REM and deep sleep, latency and timing. Ōura's interpretation of the sleep score is that if it is 85 or higher that corresponds to an excellent night of sleep, 70–84 is a good night of sleep while under 70 means the participant should pay attention to their sleep.

Our participants used a sleep logger for most of the 206 days of logging and for nights when the logger was not used we used simple data imputation to fill the gaps.

4   Data Analysis

In 2013 Peter Norvig published the results of his computation of letter, word and n-gram frequencies drawn from the Google Books collection of 743,842,922,321 word occurrences in the English language[3]. In this he found the top five most frequently occurring bigrams in English are TH, HE, IN, ER and AN, though some of the possible 676 bigrams will never or almost never appear, such as JT, QW or ZB. In our first analysis we focus on participant 1 as s/he gathered the largest volume of log data. From among the 369,467 individual words typed over

---

[3] http://norvig.com/mayzner.html.

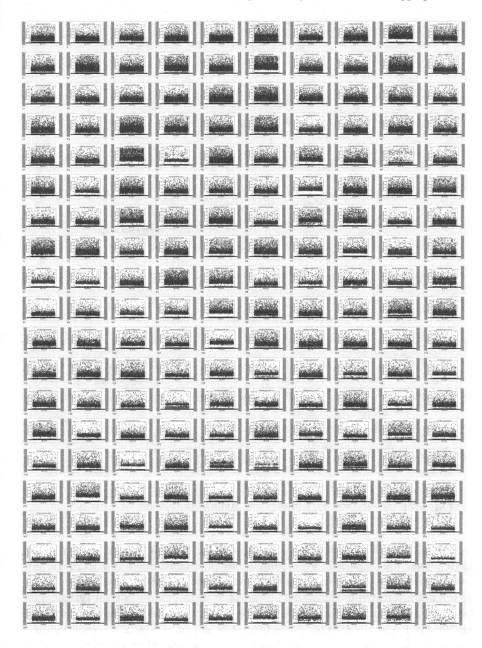

Fig. 2. Timing intervals over 206 days for participant 1 for each of the top-200 bigrams ranked by mean overall speed

206 days, the top 10 most frequently occurring bigrams for all four participants are shown in Table 2, along with the top 10 as found from Norvig's analysis.

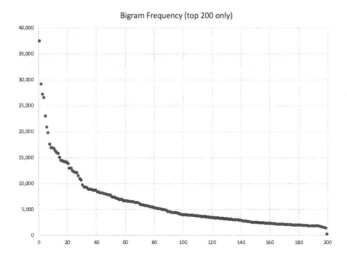

Fig. 3. Frequency of occurrence for most frequent bigrams (top 200 only) for participant 1

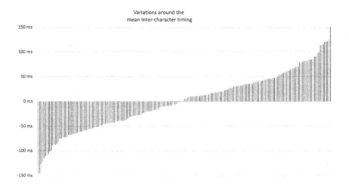

Fig. 4. Spread or variability of meaning timing information for participant 1

This shows there is little overlap among the top 10 bigrams as typed by participants. Participant 1, for example, has nine of Norvig's top 10 in Table 2 with a correlation of 0.94 on the ordering while participant 2 has only seven of 10. We are not interested in the actual bigrams but in the timing of their typing. The distributions of timing information for each of the 200 most frequent bigrams over the 206 day logging period for participant 1 is shown as Fig. 2. These individual graphs are too small to see any detail, but it is clear that the actual timing patterns for bigrams vary quite a lot among these top 200. For these graphs and the subsequent ones in this paper, we do not include inter-character timing gaps greater than 1,000 ms and the graphs show the time taken for instances of each bigram plotted left to right from 1 January to 24 July.

Figure 3 shows the frequencies of occurrence for the top-200 most frequently used bigrams from participant 1. This highlights that there are a small number of

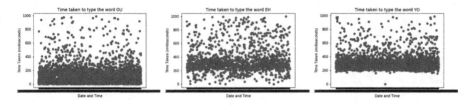

Fig. 5. Timing information for fastest mean OU (ranked 1st), slowest mean from top 200 EH (ranked 200th), and slowest baseline (never fast) YO (ranked 57th for participant 1)

very frequently occurring bigrams and then it tails off, in a Zipfian-like manner. This pattern is repeated for our other participants. When we look at how mean typing speeds for these top-200 bigrams from across the 206 days vary compared to the overall mean for participant 1, which is 204 ms, there are a very small number of bigrams up to 150 ms faster than the average and a small number up to 150 ms slower than the average. Most of the rest of these, approx 80%, are between 75 ms faster and 75 ms slower than the average. Thus a clustering of approximately 80% of bigram mean timings are within an overall range of only 150 ms as shown in Fig. 4.

The mean and standard deviations for some bigram timings for participant 1 are shown in Fig. 5. The fastest average of the 676 bigrams is OU with a mean time of 58 ms but with very large standard deviation of 83.3 while the slowest from among the top 200 bigrams is EH with a mean time of 358 ms and standard deviation of 191. The bigram YO (ranked 179th most frequently occurring) with a mean of 283 ms and standard deviation of 81 has an interesting characteristic of never, ever, being faster than about 200 ms. This can only be explained as a quirky characteristic of the keyboard typing of participant 1.

For participant 1 we found that some bigrams (XV and VV) have an average timing which is over 500 ms slower than the overall average, indicating that this participant has trouble finding the XV and VV character combinations. Other bigrams such as EI and IN are 162 ms and 151 ms faster than the average. This might be due to the fingers usually used by her/him to type these particular character combinations. We would expect that when using the middle and index fingers consecutively on adjacent keys on the same row of the keyboard would be faster to type than, say, using the little and index fingers on keys on a lower and then a higher row of the keyboard. On checking with the participant as to which fingers s/he uses to type the fastest of the bigrams we find that it is indeed the middle and index fingers for keys which are on the same row of the keyboard. Thus some of the timing characteristics is explained by keyboard layout and the particular fingering that a subject will use for pressing different keys.

We also discovered a banding effect for this participant's timing information, shown in Fig. 6. For bigrams AS (ranked 31th), IM (90th), EW (99th), PL (124th), GH (146th) and DC (188th) there is a lower (faster) band of rapidly typed characters spanning right across the 207 day logging period with a gap in timing

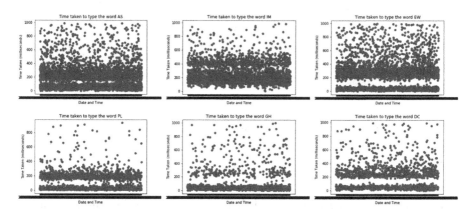

Fig. 6. Timing for AS, IM, EW, PL, GH and DC for participant 1 showing banding effect

Table 3. Correlation among top bigrams ranked by frequency for logging period

Number of bigrams	Participant 1	Participant 2	Participant 3	Participant 4
Top 5	0.793	0.515	0.371	0.251
Top 10	0.921	0.456	0.437	0.311
Top 25	0.867	0.233	0.239	0.357
Top 50	0.829	0.16	0.234	0.318
Top 200	0.626	0.214	0.238	0.302
Avg. keystrokes per day	9,929	6,070	3,535	1,033

before the more regular characteristic pattern of dense occurrences leading to more scattered occurrences as we approach 1,000 ms. Our only explanation for this is that it is to do with the rapid typing of a regularly used word among this participant's wordlist but this needs to be investigated further.

We now look at consistency of bigram timing characteristics for participants across each day of their logging period. If we take the top-200 most frequent bigrams and rank order them by their mean speed for each of their logging days, then correlate the bigram rankings for each day, the average pairwise correlation for participant 1 is 0.6262. The reader is reminded that these correlations are the averaged across 206×206 individual daily correlations of the top 200 bigrams so these are day-on-day variations in rank ordering of mean bigram typing speeds. If we reduce the number of bigrams to the top 50, 25 and then 10 we see this correlation across the days increases, as shown in the first column of Table 3. When we get to the top five bigrams, the correlation drops to 0.7937. This is explained by the fact that the top-5 mean fastest bigrams may vary from day to day. This is highlighted in our comparison to Peter Norvig's analysis earlier. Table 3 also shows the same analysis applied to the other three participants and from this we see the others have a much lower correlation among the bigrams they have typed fastest.

Also included in Table 3 and taken from Table 1 earlier, is the average number of keystrokes typed by each participant per day. We can see that the more a participant types, the greater the consistency of their typing speeds. Putting this in other words, no matter what day it is, participant 1 has almost the same ordering of her/his ranked bigram timings whereas for the others, that ordering will vary more, probably because they type less than participant 1.

As mentioned previously, a lifelog becomes most useful when there are multiple sources of data which are then cross-referenced to gain insights into the participant's life. We saw in Sect. 2 how keystroke dynamics has been used for measuring stress [9], emotion [9] and even our level of creative flow when writing [11]. Using the sleep score data gathered by the Ōura ring, we explored whether sleep score correlates with bigram timing for any or all of our top-200 bigrams. Mean daily timing data for the bigram TH had a +0.209 correlation with sleep score from the previous night while mean daily timing data for CV had a correlation of −0.18. The average of these bigram correlations with sleep score was +0.014 which leads us to conclude that there is no correlation between daily typing speed of any bigrams and sleep score. This means that participant 1 is consistent in typing speed, even when tired from poor sleep the previous night. We applied this to other participants and found the same result. Perhaps if we explored windowing sleep score as a moving average over a number of days, or used other metrics to measure fatigue then that might correlate with timing information.

Another possibility is that bigram timing information might vary during the day, differing from morning to evening as fatigue effects might alternate with bursts of energy or enthusiasm or as the participant's level of stress, emotion or creativity might vary. However that would require a more comprehensive lifelog and goes back to the point we made earlier about the best kind of lifelog being a multimodal artifact, a fusion across multiple, diverse information sources. The exercise reported in this paper has served to illustrate the possibilities that keystroke dynamics have as one of those information sources.

5   Conclusions

In this paper we present the case for greater use of keystroke dynamics in lifelogging. We recap on several previous applications for keystroke dynamics and we describe how we used a tool called Loggerman to gather keystroke data for four participants over more then six months. We are particularly interested in the timing information associated with keystrokes and we showed how timing information between bigram keystrokes can vary for the same participant across different days. We also showed how the relative speeds with which bigrams are typed varies hugely for the same participant and across different participants. This shows how useful keystroke dynamics can be for security and authentication applications.

Keystroke dynamics has been shown to correlate with stress, fatigue and writing style. In this preliminary analysis we explored whether keystroke timing was correlated with fatigue, as measured by sleep score from the previous night. Unfortunately we found no correlation between these suggesting that a simple sleep score is insufficient to measure participant fatigue. We need more fine-grained measures which would allow levels of fatigue which vary throughout the day, to be measured.

For future work there are a range of ways in which data from keystroke dynamics could be used as part of a lifelog, especially to gain insights into the more complex cognitive processes in which we engage each day. Keystroke timing information has been shown to reveal writing strategies on the conventional keyboard/screen/mouse setup and it would be interesting to explore keystroke dynamics on mobile devices and see how that correlates with stress, cognitive load from multi-tasking, fatigue and distraction.

References

1. Bergadano, F., Gunetti, D., Picardi, C.: User authentication through keystroke dynamics. ACM Trans. Inf. Syst. Secur. (TISSEC) **5**(4), 367–397 (2002)
2. Gaines, R.S., Lisowski, W., Press, S.J., Shapiro, N.: Authentication by keystroke timing: some preliminary results. Technical report, Rand Corp Santa Monica CA (1980)
3. Gurrin, C., Smeaton, A.F., Doherty, A.R.: Lifelogging: personal big data. Found. Trends Inf. Retrieval **8**(1), 1–125 (2014). https://doi.org/10.1561/1500000033
4. Hinbarji, Z., Albatal, R., Gurrin, C.: User identification by observing interactions with GUIs. In: Amsaleg, L., Guðmundsson, G.Þ., Gurrin, C., Jónsson, B.Þ., Satoh, S. (eds.) MMM 2017. LNCS, vol. 10132, pp. 540–549. Springer, Cham (2017). https://doi.org/10.1007/978-3-319-51811-4_44
5. Hinbarji, Z., Albatal, R., O'Connor, N., Gurrin, C.: LoggerMan, a comprehensive logging and visualization tool to capture computer usage. In: Tian, Q., Sebe, N., Qi, G.-J., Huet, B., Hong, R., Liu, X. (eds.) MMM 2016. LNCS, vol. 9517, pp. 342–347. Springer, Cham (2016). https://doi.org/10.1007/978-3-319-27674-8_31
6. Jacquemard, T., Novitzky, P., O'Brolcháin, F., Smeaton, A.F., Gordijn, B.: Challenges and opportunities of lifelog technologies: a literature review and critical analysis. Sci. Eng. Ethics **20**(2), 379–409 (2014). https://doi.org/10.1007/s11948-013-9456-1
7. Johansson, D., Malmgren, K., Murphy, M.A.: Wearable sensors for clinical applications in epilepsy, Parkinson's disease, and stroke: a mixed-methods systematic review. J. Neurol. **265**(8), 1740–1752 (2018). https://doi.org/10.1007/s00415-018-8786-y
8. Joyce, R., Gupta, G.: Identity authentication based on keystroke latencies. Commun. ACM **33**(2), 168–176 (1990)
9. Kołakowska, A.: A review of emotion recognition methods based on keystroke dynamics and mouse movements. In: 2013 6th International Conference on Human System Interactions (HSI), pp. 548–555. IEEE (2013)
10. Koskimäki, H., Kinnunen, H., Kurppa, T., Röning, J.: How do we sleep: a case study of sleep duration and quality using data from Ōura ring. In: Proceedings of the 2018 ACM International Joint Conference and 2018 International Symposium on Pervasive and Ubiquitous Computing and Wearable Computers, pp. 714–717 (2018)

11. Leijten, M., Van Horenbeeck, E., Van Waes, L.: Analysing keystroke logging data from a linguistic perspective. In: Observing Writing, pp. 71–95. Brill (2019)

12. Leijten, M., Van Waes, L.: Keystroke logging in writing research: using Inputlog to analyze and visualize writing processes. Written Commun. **30**(3), 358–392 (2013)

13. Meyer, J., Kay, J., Epstein, D.A., Eslambolchilar, P., Tang, L.M.: A life of data: characteristics and challenges of very long term self-tracking for health and wellness. ACM Trans. Comput. Healthcare **1**(2) (2020). https://doi.org/10.1145/3373719

14. Shelgikar, A.V., Anderson, P.F., Stephens, M.R.: Sleep tracking, wearable technology, and opportunities for research and clinical care. Chest **150**(3), 732–743 (2016)

15. Signal, L., et al.: Children's everyday exposure to food marketing: an objective analysis using wearable cameras. Int. J. Behav. Nutr. Phys. Act. **14**(1), 137 (2017). https://doi.org/10.1186/s12966-017-0570-3

16. Smeaton, A.F.: Keystroke timing information for 2,522,186 characters typed over several months. https://doi.org/10.6084/m9.figshare.13157510.v1. Accessed 29 Oct 2020

17. Tuovinen, L., Smeaton, A.F.: Remote collaborative knowledge discovery for better understanding of self-tracking data. In: 2019 25th Conference of Open Innovations Association (FRUCT), pp. 324–332. IEEE (2019)

18. Vizer, L.M.: Detecting cognitive and physical stress through typing behavior. In: CHI 2009 Extended Abstracts on Human Factors in Computing Systems, CHI EA 2009, pp. 3113–3116. Association for Computing Machinery, New York (2009)

19. Wang, P., Sun, L., Smeaton, A.F., Gurrin, C., Yang, S.: Computer vision for lifelogging: characterizing everyday activities based on visual semantics (chap. 9). In: Leo, M., Farinella, G.M. (eds.) Computer Vision for Assistive Healthcare, pp. 249–282. Computer Vision and Pattern Recognition, Academic Press (2018)

HTAD: A Home-Tasks Activities Dataset with Wrist-Accelerometer and Audio Features

Enrique Garcia-Ceja[1]([✉]), Vajira Thambawita[2,3], Steven A. Hicks[2,3], Debesh Jha[2,4], Petter Jakobsen[5], Hugo L. Hammer[3], Pål Halvorsen[2], and Michael A. Riegler[2]

[1] SINTEF Digital, Oslo, Norway
enrique.garcia-ceja@sintef.no
[2] SimulaMet, Oslo, Norway
michael@simula.no
[3] Oslo Metropolitan University, Oslo, Norway
[4] UIT The Arctic University of Norway, Tromsø, Norway
[5] Haukeland University Hospital, Bergen, Norway

Abstract. In this paper, we present HTAD: A Home Tasks Activities Dataset. The dataset contains wrist-accelerometer and audio data from people performing at-home tasks such as sweeping, brushing teeth, washing hands, or watching TV. These activities represent a subset of activities that are needed to be able to live independently. Being able to detect activities with wearable devices in real-time is important for the realization of assistive technologies with applications in different domains such as elderly care and mental health monitoring. Preliminary results show that using machine learning with the presented dataset leads to promising results, but also there is still improvement potential. By making this dataset public, researchers can test different machine learning algorithms for activity recognition, especially, sensor data fusion methods.

Keywords: Activity recognition · Dataset · Accelerometer · Audio · Sensor fusion

1 Introduction

Automatic monitoring of human physical activities has become of great interest in the last years since it provides contextual and behavioral information about a user without explicit user feedback. Being able to automatically detect human activities in a continuous unobtrusive manner is of special interest for applications in sports [16], recommendation systems, and elderly care, to name a few. For example, appropriate music playlists can be recommended based on the user's current activity (exercising, working, studying, etc.) [21]. Elderly people at an early stage of dementia could also benefit from these systems, like by monitoring their hygiene-related activities (showering, washing hands, or brush

© Springer Nature Switzerland AG 2021
J. Lokoč et al. (Eds.): MMM 2021, LNCS 12573, pp. 196–205, 2021.
https://doi.org/10.1007/978-3-030-67835-7_17

teeth) and sending reminder messages when appropriate [19]. Human activity recognition (HAR) also has the potential for mental health care applications [11] since it can be used to detect sedentary behaviors [4], and it has been shown that there is an important association between depression and sedentarism [5]. Recently, the use of wearable sensors has become the most common approach to recognizing physical activities because of its unobtrusiveness and ubiquity, specifically, the use of accelerometers [9,15,17], because they are already embedded in several commonly used devices like smartphones, smart-watches, fitness bracelets, etc.

In this paper, we present HTAD: a Home Tasks Activities Dataset. The dataset was collected using a wrist accelerometer and audio recordings. The dataset contains data for common home tasks activities like *sweeping, brushing teeth, watching TV, washing hands,* etc. To protect users' privacy, we only include audio data after feature extraction. For accelerometer data, we include the raw data and the extracted features.

There are already several related datasets in the literature. For example, the epic-kitchens dataset includes several hours of first-person videos of activities performed in kitchens [6]. Another dataset, presented by Bruno et al., has 14 activities of daily living collected with a wrist-worn accelerometer [3]. Despite the fact that there are many activity datasets, it is still difficult to find one with both: wrist-acceleration and audio. The authors in [20] developed an application capable of collecting and labeling data from smartphones and wrist-watches. Their app can collect data from several sensors, including inertial and audio. The authors released a dataset[1] that includes 2 participants and point to another website (http://extrasensory.ucsd.edu) that contains data from 60 participants. However, the link to the website was not working at the present date (August-10-2020). Even though the present dataset was collected by 3 volunteers, and thus, is a small one compared to others, we think that it is useful for the activity recognition community and other researchers interested in wearable sensor data processing. The dataset can be used for machine learning classification problems, especially those that involve the fusion of different modalities such as sensor and audio data. This dataset can be used to test data fusion methods [13] and used as a starting point towards detecting more types of activities in home settings. Furthermore, the dataset can potentially be combined with other public datasets to test the effect of using heterogeneous types of devices and sensors.

This paper is organized as following: In Sect. 2, we describe the data collection process. Section 3 details the feature extraction process, both, for accelerometer and audio data. In Sect. 4, the structure of the dataset is explained. Section 5 presents baseline experiments with the dataset, and finally in Sect. 6, we present the conclusions.

2   Dataset Details

The dataset can be downloaded via: https://osf.io/4dnh8/.

---

[1] https://www.kaggle.com/yvaizman/the-extrasensory-dataset.

The home-tasks data were collected by 3 individuals. They were 1 female and 2 males with ages ranging from 25 to 30. The subjects were asked to perform 7 scripted home-task activities including: *mop floor, sweep floor, type on computer keyboard, brush teeth, wash hands, eat chips* and *watch TV*. The *eat chips* activity was conducted with a bag of chips. Each individual performed each activity for approximately 3 min. If the activity lasted less than 3 min, an additional trial was conducted until the 3 min were completed. The volunteers used a wrist-band (Microsoft Band 2) and a smartphone (Sony XPERIA) to collect the data.

The subjects wore the wrist-band in their dominant hand. The accelerometer data was collected using the wrist-band internal accelerometer. Figure 1 shows the actual device used. The inertial sensor captures motion from the x, y, and z axes, and the sampling rate was set to 31 Hz. Moreover, the environmental sound was captured using the microphone of a smartphone. The audio sampling rate was set at 8000 Hz. The smartphone was placed on a table in the same room where the activity was taking place.

An in-house developed app was programmed to collect the data. The app runs on the Android operating system. The user interface consists of a dropdown list from which the subject can select the home-task. The wrist-band transfers the captured sensor data and timestamps over Bluetooth to the smartphone. All the inertial data is stored in a plain text format.

Fig. 1. Wrist-band watch.

3   Feature Extraction

In order to extract the accelerometer and audio features, the original raw signals were divided into non-overlapping 3 s segments. The segments are not overlapped. A three second window was chosen because, according to Banos *et al.* [2], this is a typical value for activity recognition systems. They did comprehensive tests by trying different segments sizes and they concluded that small segments produce better results compared to longer ones. From each segment, a set of features were calculated which are known as *feature vectors* or *instances*. Each *instance* is characterized by the audio and accelerometer features. In the following section, we provide details about how the features were extracted.

3.1 Accelerometer Features

From the inertial sensor readings, 16 measurements were computed including: The *mean, standard deviation, max* value for all the x, y and z axes, *pearson correlation* among pairs of axes (xy, xz, and yz), *mean magnitude, standard deviation of the magnitude,* the *magnitude area under the curve* (AUC, Eq. 1) , and *magnitude mean differences* between consecutive readings (Eq. 2). The *magnitude* of the signal characterizes the overall contribution of acceleration of x, y and z. (Eq. 3). Those features were selected based on previous related works [7, 10, 23].

$$AUC = \sum_{t=1}^{T} magnitude(t) \tag{1}$$

$$meandif = \frac{1}{T-1} \sum_{t=2}^{T} magnitude(t) - magnitude(t-1) \tag{2}$$

$$Magnitude(x, y, z, t) = \sqrt{a_x(t)^2 + a_y(t)^2 + a_z(t)^2} \tag{3}$$

where $a_x(t)^2$, $a_y(t)^2$ and $a_z(t)^2$ are the squared accelerations at time t.

Figure 2 shows violin plots for three of the accelerometer features: mean of the x-axis, mean of the y-axis, and mean of the z-axis. Here, we can see that overall, the mean acceleration in x was higher for the *brush teeth* and *eat chips* activities. On the other hand, the mean acceleration in the y-axis was higher for the *mop floor* and *sweep* activities.

3.2 Audio Features

The features extracted from the sound source were the Mel Frequency Cepstral Coefficients (MFCCs). These features have been shown to be suitable for activity classification tasks [1, 8, 12, 18]. The 3 s sound signals were further split into 1 s windows. Then, 12 MFCCs were extracted from each of the 1 s windows. In total, each instance has 36 MFCCs. In total, this process resulted in the generation of 1, 386 instances. The tuneR R package [14] was used to extract the audio features. Table 1 shows the percentage of instances per class. More or less, all classes are balanced in number.

4 Dataset Structure

The main folder contains directories for each user and a *features.csv* file. Within each users' directory, the accelerometer files can be found (*.txt* files). The file names are comprised of three parts with the following format: *timestamp-acc-label.txt. timestamp* is the timestamp in Unix format. *acc* stands for accelerometer and *label* is the activity's label. Each *.txt* file has four columns: timestamp and the acceleration for each of the x, y, and z axes. Figure 3 shows an example

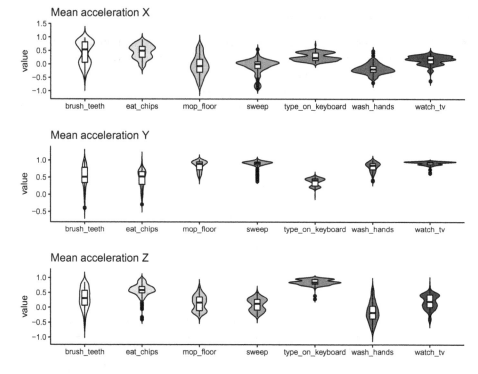

Fig. 2. Violin plots of mean acceleration of the x, y, and z axes.

Table 1. Distribution of activities by class.

Class	Proportion
Brush teeth	12.98%
Eat chips	20.34%
Mop floor	13.05%
Sweep	12.84%
Type on keyboard	12.91%
Wash hands	12.98%
Watch TV	14.90%

of the first rows of one of the files. The *features.csv* file contains the extracted features as described in Sect. 3. It contains 54 columns. *userid* is the user id. *label* represents the activity label and the remaining columns are the features. Columns with a prefix of *v1_* correspond to audio features whereas columns with a prefix of *v2_* correspond to accelerometer features. In total, there are 36 audio features that correspond to the 12 MFCCs for each second, with a total of 3 s and 16 accelerometer features.

```
1468360517664,-0.12915039,0.9797363,-0.21191406
1468360517693,-0.13500977,0.98168945,-0.21118164
1468360517743,-0.1496582,0.9819336,-0.20336914
1468360517763,-0.16894531,0.9892578,-0.21606445
1468360517788,-0.18847656,0.99658203,-0.20581055
1468360517818,-0.1850586,0.97998047,-0.21362305
1468360517857,-0.19140625,0.97216797,-0.21533203
1468360517904,-0.18066406,0.9692383,-0.21411133
1468360517921,-0.1730957,0.9560547,-0.21435547
1468360517962,-0.17871094,0.9626465,-0.2163086
```

Fig. 3. First rows of one of the accelerometer files.

5 Baseline Experiments

In this section, we present a series of baseline experiments that can serve as a starting point to develop more advanced methods and sensor fusion techniques. In total, 3 classification experiments were conducted with the HTAD dataset. For each experiment, different classifiers were employed, including ZeroR (baseline), a J48 tree, Naive Bayes, Support Vector Machine (SVM), a K-nearest neighbors (KNN) classifier with $k = 3$, logistic regression, and a multilayer perceptron. We used the WEKA software [22] version 3.8 to train the classifiers. In each experiment, we used different sets of features. For experiment 1, we trained the models using only *audio features*, that is, the MFCCs. The second experiment consisted of training the models with only the 16 *accelerometer features* described earlier. Finally, in experiment 3, we combined the *audio and accelerometer* features by aggregating them. 10-fold cross-validation was used to train and assess the classifier's performance. The reported performance is the weighted average of different metrics using a one-vs-all approach since this is a multi-class problem.

Table 2. Classification performance (weighted average) with audio features. The best performing classifier was KNN.

Classifier	False-Positive Rate	Precision	Recall	F1-Score	MCC
ZeroR	0.203	0.041	0.203	0.069	0.000
J48	0.065	0.625	0.623	0.624	0.559
Naive Bayes	0.049	0.720	0.714	0.713	0.667
SVM	0.054	0.699	0.686	0.686	0.637
KNN	0.037	0.812	0.788	0.793	0.761
Logistic regression	0.062	0.654	0.652	0.649	0.591
Multilayer perceptron	0.041	0.776	0.769	0.767	0.731

Tables 2, 3 and 4 show the final results. When using only audio features (Table 2), the best performing model was the KNN in terms of all performance

Table 3. Classification performance (weighted average) with accelerometer features. The best performing classifier was KNN.

Classifier	False-Positive Rate	Precision	Recall	F1-Score	MCC
ZeroR	0.203	0.041	0.203	0.069	0.000
J48	0.036	0.778	0.780	0.779	0.743
Naive Bayes	0.080	0.452	0.442	0.447	0.365
SVM	0.042	0.743	0.740	0.740	0.698
KNN	0.030	0.820	0.820	0.818	0.790
Logistic regression	0.031	0.800	0.802	0.800	0.769
Multilayer perceptron	0.031	0.815	0.812	0.812	0.782

Table 4. Classification performance (weighted average) when combining all features. The best performing classifier was Multilayer perceptron.

Classifier	False-Positive Rate	Precision	Recall	F1-Score	MCC
ZeroR	0.203	0.041	0.203	0.069	0.000
J48	0.035	0.785	0.785	0.785	0.750
Naive Bayes	0.028	0.826	0.823	0.823	0.796
SVM	0.020	0.876	0.874	0.875	0.855
KNN	0.014	0.917	0.911	0.912	0.899
Logistic regression	0.022	0.859	0.859	0.859	0.837
Multilayer perceptron	0.014	0.915	0.914	0.914	0.901

metrics with a Mathews correlation coefficient (MCC) of 0.761. We report MCC instead of accuracy because MCC is more robust against class distributions. In the case when using only accelerometer features (Table 3), the best model was again KNN in terms of all performance metrics with an MCC of 0.790. From these tables, we observe that most classifiers performed better when using accelerometer features with the exception of Naive Bayes. Next, we trained the models using all features (accelerometer and audio). Table 4 shows the final results. In this case, the best model was the multilayer perceptron followed by KNN. Overall, all models benefited from the combination of features, of which some increased their performance by up to ≈0.15, like the SVM which went from an MCC of 0.698 to 0.855.

All in all, combining data sources provided enhanced performance. Here, we just aggregated the features from both data sources. However, other techniques can be used such as late fusion which consists of training independent models using each data source and then combining the results. Thus, the experiments show that machine learning systems can perform this type of automatic activity detection, but also that there is a large potential for improvements - where the

HTAD dataset can play an important role, not only as an enabling factor, but also for reproducibility.

6   Conclusions

Reproducibility and comparability of results is an important factor of high-quality research. In this paper, we presented a dataset in the field of activity recognition supporting reproducibility in the field. The dataset was collected using a wrist accelerometer and captured audio from a smartphone. We provided baseline experiments and showed that combining the two sources of information produced better results. Nowadays, there exist several datasets, however, most of them focus on a single data source and on the traditional *walking, jogging, standing, etc.* activities. Here, we employed two different sources (accelerometer and audio) for home task activities. Our vision is that this dataset will allow researchers to test different sensor data fusion methods to improve activity recognition performance in home-task settings.

References

1. Al Masum Shaikh, M., Molla, M., Hirose, K.: Automatic life-logging: a novel approach to sense real-world activities by environmental sound cues and common sense. In: 11th International Conference on Computer and Information Technology, ICCIT 2008, pp. 294–299, December 2008. https://doi.org/10.1109/ICCITECHN.2008.4803018
2. Banos, O., Galvez, J.M., Damas, M., Pomares, H., Rojas, I.: Window size impact in human activity recognition. Sensors **14**(4), 6474–6499 (2014). https://doi.org/10.3390/s140406474. http://www.mdpi.com/1424-8220/14/4/6474
3. Bruno, B., Mastrogiovanni, F., Sgorbissa, A., Vernazza, T., Zaccaria, R.: Analysis of human behavior recognition algorithms based on acceleration data. In: 2013 IEEE International Conference on Robotics and Automation, pp. 1602–1607. IEEE (2013)
4. Ceron, J.D., Lopez, D.M., Ramirez, G.A.: A mobile system for sedentary behaviors classification based on accelerometer and location data. Comput. Ind. **92**, 25–31 (2017)
5. Ciucurel, C., Iconaru, E.I.: The importance of sedentarism in the development of depression in elderly people. Proc. - Soc. Behav. Sci. **33** (Supplement C), 722–726 (2012). https://doi.org/10.1016/j.sbspro.2012.01.216. http://www.sciencedirect.com/science/article/pii/S1877042812002248. pSIWORLD 2011
6. Damen, D., et al.: Scaling egocentric vision: the dataset. In: Ferrari, V., Hebert, M., Sminchisescu, C., Weiss, Y. (eds.) ECCV 2018. LNCS, vol. 11208, pp. 753–771. Springer, Cham (2018). https://doi.org/10.1007/978-3-030-01225-0_44
7. Dernbach, S., Das, B., Krishnan, N.C., Thomas, B.L., Cook, D.J.: Simple and complex activity recognition through smart phones. In: 2012 8th International Conference on Intelligent Environments (IE), pp. 214–221, June 2012. https://doi.org/10.1109/IE.2012.39
8. Galván-Tejada, C.E., et al.: An analysis of audio features to develop a human activity recognition model using genetic algorithms, random forests, and neural networks. Mob. Inf. Syst. **2016**, 1–10 (2016)

9. Garcia, E.A., Brena, R.F.: Real time activity recognition using a cell phone's accelerometer and Wi-Fi. In: Workshop Proceedings of the 8th International Conference on Intelligent Environments. Ambient Intelligence and Smart Environments, vol. 13, pp. 94–103. IOS Press (2012). https://doi.org/10.3233/978-1-61499-080-2-94

10. Garcia-Ceja, E., Brena, R.: Building personalized activity recognition models with scarce labeled data based on class similarities. In: García-Chamizo, J.M., Fortino, G., Ochoa, S.F. (eds.) UCAmI 2015. LNCS, vol. 9454, pp. 265–276. Springer, Cham (2015). https://doi.org/10.1007/978-3-319-26401-1_25

11. Garcia-Ceja, E., Riegler, M., Nordgreen, T., Jakobsen, P., Oedegaard, K.J., Tørresen, J.: Mental health monitoring with multimodal sensing and machine learning: a survey. Pervasive Mob. Comput. **51**, 1–26 (2018). https://doi.org/10.1016/j.pmcj.2018.09.003.   http://www.sciencedirect.com/science/article/pii/S1574119217305692

12. Hayashi, T., Nishida, M., Kitaoka, N., Takeda, K.: Daily activity recognition based on DNN using environmental sound and acceleration signals. In: 2015 23rd European Signal Processing Conference (EUSIPCO), pp. 2306–2310, August 2015. https://doi.org/10.1109/EUSIPCO.2015.7362796

13. Khaleghi, B., Khamis, A., Karray, F.O., Razavi, S.N.: Multisensor data fusion: a review of the state-of-the-art. Inf. Fusion **14**(1), 28–44 (2013). https://doi.org/10.1016/j.inffus.2011.08.001.   http://www.sciencedirect.com/science/article/pii/S1566253511000558

14. Ligges, U., Krey, S., Mersmann, O., Schnackenberg, S.: tuneR: Analysis of music (2014). http://r-forge.r-project.org/projects/tuner/

15. Mannini, A., Sabatini, A.M.: Machine learning methods for classifying human physical activity from on-body accelerometers. Sensors **10**(2), 1154–1175 (2010). https://doi.org/10.3390/s100201154. http://www.mdpi.com/1424-8220/10/2/1154

16. Margarito, J., Helaoui, R., Bianchi, A.M., Sartor, F., Bonomi, A.G.: User-independent recognition of sports activities from a single wrist-worn accelerometer: a template-matching-based approach. IEEE Trans. Biomed. Eng. **63**(4), 788–796 (2016)

17. Mitchell, E., Monaghan, D., O'Connor, N.E.: Classification of sporting activities using smartphone accelerometers. Sensors **13**(4), 5317–5337 (2013)

18. Nishida, M., Kitaoka, N., Takeda, K.: Development and preliminary analysis of sensor signal database of continuous daily living activity over the long term. In: 2014 Asia-Pacific Signal and Information Processing Association Annual Summit and Conference (APSIPA), pp. 1–6. IEEE (2014)

19. Richter, J., Wiede, C., Dayangac, E., Shahenshah, A., Hirtz, G.: Activity recognition for elderly care by evaluating proximity to objects and human skeleton data. In: Fred, A., De Marsico, M., Sanniti di Baja, G. (eds.) ICPRAM 2016. LNCS, vol. 10163, pp. 139–155. Springer, Cham (2017). https://doi.org/10.1007/978-3-319-53375-9_8

20. Vaizman, Y., Ellis, K., Lanckriet, G., Weibel, N.: Extrasensory app: data collection in-the-wild with rich user interface to self-report behavior. In: Proceedings of the 2018 CHI Conference on Human Factors in Computing Systems, pp. 1–12 (2018)

21. Wang, X., Rosenblum, D., Wang, Y.: Context-aware mobile music recommendation for daily activities. In: Proceedings of the 20th ACM International Conference on Multimedia, pp. 99–108. ACM (2012)

22. Witten, I.H., Frank, E., Hall, M.A.: Data Mining: Practical Machine Learning Tools and Techniques. Morgan Kaufmann Series in Data Management Systems, 3rd edn. Morgan Kaufmann, Burlington (2011)
23. Zhang, M., Sawchuk, A.A.: Motion primitive-based human activity recognition using a bag-of-features approach. In: ACM SIGHIT International Health Informatics Symposium (IHI), Miami, Florida, USA, pp. 631–640, January 2012

MNR-Air: An Economic and Dynamic Crowdsourcing Mechanism to Collect Personal Lifelog and Surrounding Environment Dataset. A Case Study in Ho Chi Minh City, Vietnam

Dang-Hieu Nguyen[1,3], Tan-Loc Nguyen-Tai[1,3], Minh-Tam Nguyen[1,3], Thanh-Binh Nguyen[2,3], and Minh-Son Dao[4(✉)]

[1] University of Information Technology, Ho Chi Minh City, Vietnam
{hieund.12,locntt.12,tamnm.12}@grad.uit.edu.vn
[2] University of Science, Ho Chi Minh City, Vietnam
ngtbinh@hcmus.edu.vn
[3] Vietnam National University in Ho Chi Minh City,
Ho Chi Minh City, Vietnam
[4] National Institute of Information and Communications Technology,
Koganei, Japan
dao@nict.go.jp

Abstract. This paper introduces an economical and dynamic crowdsourcing mechanism to collect personal lifelog associated environment datasets, namely MNR-Air. This mechanism's significant advantage is to use personal sensor boxes that can be carried on citizens (and their vehicles) to collect data. The MNR-HCM dataset is also introduced in this paper as the output of MNR-Air and collected in Ho Chi Minh City, Vietnam. The MNR-HCM dataset contains weather data, air pollution data, GPS data, lifelog images, and citizens' cognition of urban nature on a personal scale. We also introduce AQI-T-RM, an application that can help people plan their travel to avoid as much air pollution as possible while still saving time on travel. Besides, we discuss how useful MNR-Air is when contributing to the open data science community and other communities that benefit citizens living in urban areas.

Keywords: Particulate matter · $PM_{2.5}$ · AQI · Sensors · Sensor data · Smart navigation · Crowdsourcing

1 Introduction

Air pollution has been a rising concern in Vietnam during the last few years. It has become one of the critical reasons that impact many aspects of people's life from society, healthcare, to the economy, especially in urban areas and industrial zones [1].

© Springer Nature Switzerland AG 2021
J. Lokoč et al. (Eds.): MMM 2021, LNCS 12573, pp. 206–217, 2021.
https://doi.org/10.1007/978-3-030-67835-7_18

A better understanding of air pollution's influence on human health to find a suitable policy to cope with quickly has been the emerging demand from both government and scientists [2]. In [3], the authors brought significant evidence of the strong correlation between air pollution and the risk of respiratory and cardiovascular hospitalizations, especially in the most populous city in Vietnam. Among different subjects, children are the vulnerable subjects easily harmed by air pollution, especially on their respiratory system and human mentality [4]. Air pollution also impacts tourism, which is an essential component of Vietnam's GDP. In [5], the authors claimed that short-term exposure to air pollutants while visiting regions with high air pollution levels can lead to acute health effects. Many reasons lead to a high level of air pollution in urban areas. In [1], the authors pointed out that in Vietnam, particularly in urban areas, the top reasons for air pollution came from transportation and construction, while, in industrial zones, the main reasons originated from factories. In [5], the authors mentioned that air pollution is mostly environmental tobacco smoke, commuting, and street food stands.

Although air pollution is the emerging requirement, not much effort has been made to gather air pollution with smaller granularity and high density. It is difficult for researchers to access a good source of air pollution data, except for some public websites and apps aimed at serving ordinary citizens such as AirVisual [6], EnviSoft [7], IQAir [8], AQIcn [9], and Breezometer [10], to name a few. Unfortunately, not all are free, and researchers must pay lots of money to have the necessary data set for their research. Besides, these data sources do not provide a smaller granularity and high density. Most mentioned sources only give one value for a whole city regardless of its size. Another option for researchers is to look for datasets created by other countries. Unfortunately, these datasets mostly come from meteorological or satellite sources that may probably not reflect precisely the ground situation [11,12].

Vietnamese researchers and their foreign colleagues have tried to create their data set to avoid significantly depending on open data sources. To et al. [13] setup air pollution stations to collect $PM_{2.5}$ and PM_{10} hourly at five places in the city, including urban, residential, and sidewalk areas. In [5], the authors used portable instrumentation to collect $PM_{2.5}$ and black carbon conducted during the wet season in 2014 in Ho Chi Minh City, Da Lat, and Nha Trang. In [12], the authors have integrated ground-based measurements, meteorological, and satellite data to map temporal PM concentrations at a 10×10 km grid over the entire of Vietnam.

Inspired by the emerging problems mentioned above, we propose an economical and dynamic method to build a data crowdsourcing mechanism to collect air pollution with smaller granularity and high density. We design a small sensor box assembled various sensors of weather, air pollution, GPS, and digital camera to collect data at a small granularity level (i.e., less than 10 m). This sensor box can be carried by pedestrians or by motorbikers who travel around a city. When these sensor boxes are spread out a whole city, they become mobile stations leading to have a grid map where each vertex is a sensor box, and each edge is the connection between two nearest vertices. We also develop an algorithm that can interpolate the value of AQI based on this grid map to have AQI value even if there are no sensors. We also develop an app to get feedback from users to see the relationship between their cognition of the surrounding environment and the real air pollution.

The contributions of our solution are:

- Provide an economical and dynamic mechanism for crowdsourcing using cheap materials and popular transportation services in Ho Chi Minh City, Vietnam (i.e., motorbike). Nevertheless, these sensor boxes can be embedded/mounted in other transportation types such as cars, buses, or bicycles.
- Collect air pollution, weather, GPS, and image data at the small granularity level with the first-perspective view. The grid map created by these sensor boxes' positions can increase the density of AQI value throughout the city map.
- Reflect the correlation between urban areas and their urban nature, air pollution, and people's behaviors.
- Attract researchers in multimedia information retrieval, machine learning, AI, data science, event-based processing and analysis, multimodal multimedia content analysis, lifelog data analysis, urban computing, environmental science, and atmospheric science.
- Provide the air pollution risk map to support people to plan their travel with less air pollution.

This paper is organized as follows. In Sect. 2, we introduce the system and hardware. Section 3 expresses how the dataset is collected, processed, and stored. Section 4 denotes the risk map application. Finally, in Sect. 5, we briefly give our conclusion and further works.

2   System Architecture and Hardware

In this section, we introduce our system architecture and hardware for the data collection. Figure 1 illustrates our system architecture, including sensors, networks, storage, data format, and data flow information. All data collected from our sensors and devices are uploaded to the cloud (e.g., Thingspeak) periodically (e.g., 5 min) and backed up in memory cards whose content can be later updated to our storage at the end of the day. This policy helps us to avoid missing data due to internet connection and cloud problems. We also apply the privacy preservation policy to protect personal data (e.g., face blurring, data encryption).

We design three components to collect data from people and surrounding areas: (1) lifelog camera, (2) smartphone app, and (3) sensor box. The first one is a lifelog camera (Camera SJCAM SJ4000 AIR 4K Wifi) that can be embedded on a helmet or a shoulder strap to make sure it can capture images under the first-perspective view. We set 1pic/3s mode to record the surrounding areas during traveling. The second one is an app that displays questionnaires to capture the surrounding environment's cognition of people. Table 1 denotes the content of the questionnaires. The volunteers answer questionnaires whenever they stop by

Table 1. The list of questionnaires

Group	Questionnaires
Common: Reflect the personal feeling of the surrounding environment	**Greenness Degree**
	1 (building) → 5 (greenness)
	Cleanliness Degree
	1 (filthy) → 5 (cleanliness)
	Crowdedness Degree
	1 (very light) → 5 (high dense)
	Noisy Degree
	1 (very quiet) → 5 (very noisy)
	Obstruction: (multi-choice)
	Flood/Rain/Congestion/Traffic light/Pothole/Construction site
A: For predicting personal feeling of the surrounding atmosphere	**Skin feeling degree**
	1 (worst) → 5 (best)
	Stress Degree
	Stressed/Depressed/Calm/Relaxed/Excited
	Personal AQI degree
	1 (fresh air) → 5 (absolute pollution)
	How do you feel of the air pollution
	(smoky, smelly, pollen, eye itchy)
B: For predicting the reputation of surrounding atmosphere for a specific activity	**Safety degree degree**
	1 (not want at all) → 5 (absolute want)
	Do you want to use this route so that you can protect your health
	and safety (i.e., avoid air pollution, congestion,and obstruction)
	What is the most suitable vehicle for using the route?
	[multichoice]: Motorbike/Bus/Car/Bicycling

red lights, traffic jams, or break time (i.e., they recognize something relating to the bad surrounding environment) during the patrol.

The last one is the set of sensors that collect environmental data. We use Arduino (Arduino Uno R3 SMD (CH340 + ATMEGA328P-AU) and version 1.8.10 of Arduino Software (IDE)) as the mainboard to connect and handle different sensors, as depicted in Fig. 2, 3 and described in Table 2.

One of our primary targets is to collect the essential factor of air pollution, $PM_{2.5}$ with low-cost sensors. Hence, we have to balance the qualification of low-cost sensors and the accuracy of these sensors' data. To meet our desire, we leverage the research presented in [16] to select and to integrate suitable sensors (e.g., humidity, particle matter) and calculate the value of $PM_{2.5}$ accuracy. We utilize the formula mentioned in [16] to estimate $PM_{2.5}$ using data collected by our sensors, as follows:

$$PM_{2.5,c} = \frac{PM_{2.5}}{1.0 + 0.48756 \left(\frac{H}{100}\right)^{8.60068}} \qquad (1)$$

$$PM_{10,c} = \frac{PM_{10}}{1.0 + 0.81559 \left(\frac{H}{100}\right)^{5.83411}} \qquad (2)$$

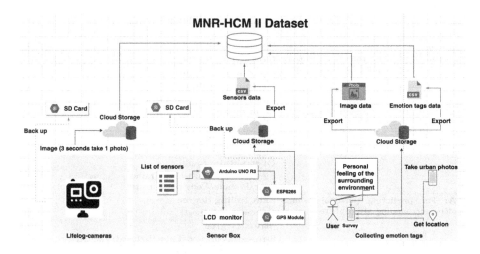

Fig. 1. MNR-Air System Architecture

Device 1	Device 2	Detail	Device 2 Volt (Using)	Range Convert	Pin type	Pin Device 1	Pin Device 2 (Use)
Arduino UNO	UVM-30A	Get UV value	3-5V (5V)	200nm – 370nm => 1 – 11index	Analog	A0	OUT
Arduino UNO	MICS-6814	Get CO value	5V (5V)	1 – 1000ppm	Analog	A1	CO
		Get NO2 value		0.05 – 10ppm	Analog	A4	NO2
		Get NH3 value		1 – 500ppm	Analog	A5	NH3
Arduino UNO	MQ-136 (SO2)	Get SO2 value	5V (5V)	1 – 200ppm	Analog	A2	AO
Arduino UNO	MQ-131 (O3)	Get O3 value	5V (5V)	10 – 1000ppm	Analog	A3	AOUT
Arduino UNO	SDS011 (PM2.5, PM10)	Get PM2.5 and PM10 values	5V (5V)	0.0-999.9µg/m3	Digital	2,3	TXD, RXD
Arduino UNO	DHT21 AM2301 (Humidity,	Get Humidity, Temperature,	3.3 - 5V (5V)	-	Digital	4	OUT
Arduino UNO	LCD	Show values sensor [PM2.5, Tem, Hum, UV]	5V (5V)	String	Digital	SCL, SDA	SCL, SDA
ESP8266 NodeMCU	GPS U-Blox NEO-6M and Anten GPS 1575.42Mhz SMA	Get longitude and latitude	3.3 - 5V (5V)	-	Digital	D3, D4	TXD, RXD
Arduino UNO	ESP8266 (Tx-Rx)	Transmit data	5V (5V)	String	Digital	D12, D13	D2, D1
ESP8266 NodeMCU	SD	Store data	5V (5V)	String	Digital	D8, D7, D6, D5	CS, MOSI, MISO, SCK
ESP8266 NodeMCU	LED	Led Write SD success	3.3V	-	Digital	D0	-

Fig. 2. Sensor Box Architecture

Table 2. Sensors

No	Name	Specification
1	SDS011	$PM_{2.5}$, PM_{10}, Range: 0.0–999.9 µg/m³
2	MQ-131	O_3, Range: 10–1000 ppm
3	MQ-136	SO_2, Range: 1–200 ppm
4	GPS U-Blox NEO-6M, Anten GPS 1575.42Mhz SMA	Longitude and Latitude
5	MICS-6814	NO_2, Range: 0.05–10 ppm CO, Range: 1–1000 ppm
6	UVM-30A	UV, Range: 200–370 nm
7	DHT21 AM2301	Humidity and Temperature
8	Grove - 16x2 LCD	Display information
9	ESP8266 NodeMCU Lua CP2102	Connect WIFI and push data to cloud
10	2 x Arduino Uno R3	Control data transmission

where:

- $PM_{2.5,c}$ is the corrected $PM_{2.5}$ value.
- $PM_{2.5}$ is the measured quantity.
- $PM_{10,c}$ is the corrected PM_{10} value.
- PM_{10} is the measured quantity.
- H is the measured humidity.

3 The MNR-HCM Dataset

We utilize the MNR-Air to collect both the personal-sized surrounding environment (e.g., air pollution, weather) and the lifelog (i.e., images, cognitive, urban nature, congestion) data. We choose Ho Chi Minh City, Vietnam, as the testing field because it is the most populous and polluted in Vietnam [1,3,4,11,13]. We have carried on the data collection campaign for a while on the predefined routes patrolled three times per day (e.g., morning, noon, evening) by motorbikers, the most popular transportation method in Vietnam. The sensing data are transferred to the cloud in real-time mode while the images data are uploaded from memory cards to the cloud at the end of the day after privacy protection. In this small-and-mobile scale, each person wearing our sensor box and interacting with our apps becomes a dynamic and intelligent sensor that, if connected, can generate a network of sensors to collect data of a whole city continuously and endlessly.

3.1 Data Collection

The set of personal-sized data collectors (i.e., lifelog camera, a smartphone with apps, sensor box) is carried on by a volunteer who rides a motorbike patrolling along the predefined route. The route, approximately 17 km, is a closed-loop where two volunteers patrol in the opposite direction. The average speed is within [20–30] km/h. The volunteers have to patrol three times per day: morning, noon, and late afternoon. The campaign had conducted from July 12, 2020, to July 31, 2020.

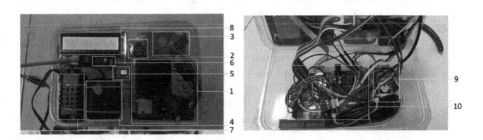

Fig. 3. Sensor Box Layout. Please refer Table 2 for mapping sensor's ID and the corresponding description.

3.2 Data Collection Route

To capture the diversity of urban nature that might correlate to air pollution, we create the route to contain parks, traffic density, small roads, and riverside scene. We also select the rush-hour period for collecting data to prove our hypothesis of the association between transportation and between air pollution, stress, and weather and congestion. Figure 4 illustrates and Table 3 denotes the route map and route's urban nature.

Table 3. Selected routes and their urban nature

No	Route	Urban Nature
1	Nguyen Tat Thanh - Khanh Hoi bridge - Ton Duc Thang -Nguyen Huu Canh - Nguyen Binh Khiem - Thi Nghe bridge	The main route with Riverside path and parks having a lot of trees
2	Xo Viet Nghe Tinh - Dien Bien Phu - Dinh Tien Hoang -Vo Thi Sau - Cong Truong Dan Chu Roundabout	The main route with heavy traffic, having less trees and many buildings
3	Cach Mang Thang Tam - Nguyen Thi Minh Khai -Hai Ba Trung - Le Thanh Ton - Nguyen Thi Nghia	There are many large parks with lots of trees and crowded traffic
4	Nguyen Thi Nghia - Ong Lanh bridge - Hoang Dieu	Nearby the river, traffic, less trees, lots of high-rise buildings

3.3 Data Description

The following are the data and folder structure description.

- **Image data:** stored in the following folders:
    - **image_tag:** Contain images taken by smartphones during four routes. Most of the images are taken at the predefined checkpoints. The information of each image is recorded in userX-emotion-tags-dd-mm-yyyy.csv.
    - **photo:** Contain images were taken by lifelog cameras (i.e., each sub-folder contains images were taken in survey day "yyyymmdd").
        * The images were recorded as 1pic/3s from July 12, 2020, to July 31, 2020.
- **Emotion tags:** stored in
    - **emotion_tags:** Contain information on all checkpoints of four routes in a day userX-emotion-tags-dd-mm-yyyy.csv.
- **Sensor data:** stored in:
    - **Sensor:** The parameters' sensor values are measured during the day in four routes, including urban nature, weather variables, the concentration of pollutants, and psycho-physiological. (i.e., each sub-folder contains sensor value taken in measured day "yyyymmdd").
        * Contain the value of Time, Longitude, Latitude, Humidity, Temperature, Fahrenheit, $PM_{2.5}$, PM_{10}, UV, NO_2, CO, SO_2, O_3, Heartbeat, Image.
        * All data stored in "userX/yyyy-mm-dd.csv" files were recorded every 2 or 3 s.

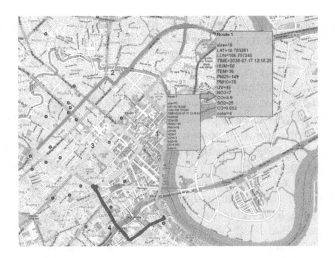

Fig. 4. Data Collection Route in Ho Chi Minh City, Vietnam

4   Air Pollution and Traffic Risk Map

In this section, we introduce the application that utilized the MNR-HCM dataset to serve society.

4.1   Motivation and Purposes

Among many applications that can be built by utilizing the MNR-HCM dataset, we decide to build the "Air pollution and Traffic Risk Map" (AQI-T-RM) application to support people to avoid as much as possible the impact of air pollution and congestion in Ho Chi Minh City, Vietnam.

The AQI-T-RM aims to provide people the ability to plan their traveling to protect their health and save their time. The core of the AQI-T-RM is the algorithm that can predict air qualification index (AQI) quickly and the algorithm that can find the optimal route between two points on the map (in terms of less AQI than other routes). We assume that AQI and congestion have a tight association: the more vehicle traveling on the route, the more AQI. This hypothesis based on research published in [1,5,13].

Currently, there is no such application as AQI-T-RM running on the air pollution dataset in Ho Chi Minh City. People mostly rely on AQI information provided freely by some open websites/sources [6–10]. Unfortunately, these sources only provide the value of AQI for the large area (e.g., the whole city), and have no function to plan their travels based on air pollution data.

4.2   Methodology

We develop a simple but effective algorithm to support people in finding the shortest path between two places A and B with less impact of bad AQI: (1)

utilize Google Map API to locate three shortest routes between A and B, (2) calculate the total AQI for each route Using the MNR-HCM dataset with the suitable interpolation and prediction algorithms, and (3) select the path with the smallest sum of AQI as a recommendation plan for people.

We utilize the algorithm presented in MediaEval 2019 [18] to predict AQI along each route. In this case, instead of using data from air pollution stations, we use the MNR-HCM dataset provided by mobile stations. After having three shortest routes with the approximate time needed to finish these routes, we apply the mentioned algorithm to predict AQI from T to $T+m^i$ (with m^i is the time needed to complete the $i^t h$ path). We can then quickly estimate the total AQI for each route and visualize it to people, including routes, total distance, travel times, and an air quality map as the color of the line. The line's color represents the level of air quality (Blue: Very Good, Green: Good, Yellow: Moderate, Brown: Bad, Red: Very Bad), as depicted in Fig. 6. In terms of the line's color representing the AQI level, the thresholds applied to identify the Very Good/Good/Moderate/Bad/Very Bad categories are calculated using the Taiwan AQI [15].

4.3   AQI-T-RM Architecture

We design AQI-T-RM as a client-server application (Fig. 5) that connects to the MNR-HCM database and allows people to interact with the system to get their optimal traveling plan.

- Client (ReactJS/React Native):
    - is a high performance, consistent language for both web and mobile to speed development.
    - can be easily developed into Progressive Web App later.
- Load balancer:
    - Use multiple server instances and load balancers for concurrent requests.
    - Use Apache for stability.
- Container (Docker):
    - Use Docker for a fast deploying server and ensure uniformity between environments.
- Server:
    - Use NestJS to ensure the stabilization and extendability of applications.
- Database:
    - Use MongoDB to access and manage large amounts of data and low latency collected from sensors. Figure 6 illustrates the current interface of the application.

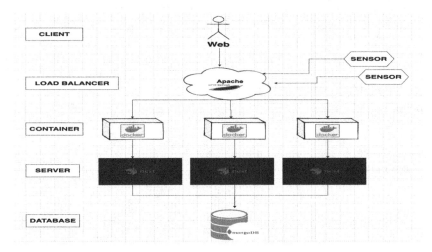

Fig. 5. The architecture of AQI-T-RM

Table 4. The comparison with another method

Model	Dataset	Number of sensors	Simulated Trips	Pollutants	Exposure reduction	Distance Increase
AQI-T-RM	MNR-HCM dataset	2	100	AQI	30.25%	<0.5 km (average)*
CAR [17]	Big Taipei	2963**	4,364	PM2.5	17.1%	2.5%

*Distance increase percentage not available for this work.
**Based on a homepage [14], the number of $PM_{2.5}$ sensors deployed around Taiwan.

4.4 Discussion

We compare our dataset and model with the dataset and model introduced in [17], namely CAR. The CAR utilizes the dataset collected by air pollution stations, namely Big Taipei, spread out the whole city (Fig. 7). It provides the spatiotemporal linear interpolation algorithm that can support people to have the optimal route to avoid as much bad AQI as possible while still saving traveling time. Theoretically, the MNR-HCM dataset contains data collected at every corner of a city while the Big Taipei dataset cannot. Besides, the number of (mobile) stations of MNR-Air can be easily increased comparing to Big Taipei stations. The AQI-T-RM proved useful even when using very few sensors. Hence, we can say that AQI-T-RM can work better than CAR. Table 4 denotes the comparison between AQI-T-RM with the MNR-HCM dataset and CAR with the Big Taipei dataset. In the three routes of Google Maps suggest, we compare the total AQI level of the three routes, get the best route, compare it with the worst route, and obtain the exposure reduction. The results show better benefits for citizens when using our dataset and our application.

Fig. 6. A screen-shot of the user interface of AQI-T-RM

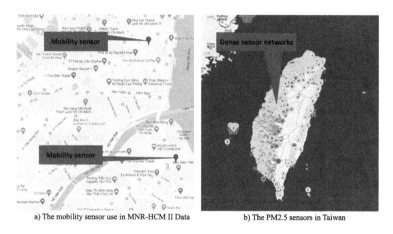

a) The mobility sensor use in MNR-HCM II Data   b) The PM2.5 sensors in Taiwan

Fig. 7. A distributed network of sensors.

5   Conclusion and Future Work

We have introduced the MNR-Air mechanism to carry on data crowdsourcing. The fundamental background of MNR-Air is to provide an economical and dynamic solution for collecting air pollution data and citizens' cognition of the surrounding environment and contributing to the open data science community. The significant advantage of the MNR-HCM dataset, the output of MNR-Air, is that data can be collected at every corner of a city as long as people can reach there. Besides, with the low price of building a sensor box and the significant amount of motorbikers traveling every day in the populous city, the dataset can capture most of the city's life. Hence, scientists can have a rich dataset to discover various interesting and useful things for society.

We keep collecting data and finding financial support to increase the number of sensor boxes and enlarge routes' scope. We believe that when our sensor boxes can cover the whole city, we can build the AQI risk map precisely, economically, and usefully.

References

1. Morisson, A., Gilabert, P.J.: Air pollution in Vietnam. United Nations Industrial Development Organization (UNIDO) (2015). https://open.unido.org/api/documents/5105345/download/AIR%20POLLUTION%20IN%20VIET%20NAM
2. Sato, T., Dao, M.-S., Kuribayashi, K., Zettsu, K.: SEPHLA: challenges and opportunities within environment - personal health archives. In: Kompatsiaris, I., Huet, B., Mezaris, V., Gurrin, C., Cheng, W.-H., Vrochidis, S. (eds.) MMM 2019. LNCS, vol. 11295, pp. 325–337. Springer, Cham (2019). https://doi.org/10.1007/978-3-030-05710-7_27
3. Phung, D., et al.: Air pollution and risk of respiratory and cardiovascular hospitalizations in the most populous city in Vietnam. Sci. Total Environ. **557**, 322–330 (2016)
4. Luong, L.T.M., Dang, T.N., Huong, N.T.T., Phung, D., Tran, L.K., Thai, P.K.: Particulate air pollution in Ho Chi Minh city and risk of hospital admission for acute lower respiratory infection (ALRI) among young children. Environ. Pollut. **257**, 113424 (2020)
5. Pant, P., Huynh, W., Peltier, R.E.: Exposure to air pollutants in Vietnam: assessing potential risk for tourists. J. Environ. Sci. **73**, 147–154 (2018)
6. Airvisual. https://www.airvisual.com/
7. Envisoft. http://cem.gov.vn/
8. IQAir. https://www.iqair.com/
9. AQIcn. https://aqicn.org/
10. Breezometer. https://breezometer.com/air-quality-map/
11. Ho, B.Q., Vu, K.H.N., Nguyen, T.T., et al.: Study loading capacities of air pollutant emissions for developing countries: a case of Ho Chi Minh City, Vietnam. Sci. Rep. **10**, 5827 (2020)
12. Nguyen, T.T., et al.: Particulate matter concentration mapping from MODIS satellite data: a Vietnamese case study. Environ. Res. Lett. **10**(9), 095016 (2015)
13. Hien, T.T., Chi, N.D.T., Nguyen, N.T., Takenaka, N., Huy, D.H.: Current status of fine particulate matter (PM2. 5) in Vietnam's most populous city, Ho Chi Minh City. Aerosol. Air Qual. Res. **19**(10), 2239–2251 (2019)
14. AirMap-Taiwan. https://v5.airmap.g0v.tw/#/map
15. Air Quality Index-Taiwan. https://airtw.epa.gov.tw/ENG/Information/Standard/AirQualityIndicator.aspx
16. Rothkugel, S., Strum, P.: Spontaneous networking with CORBA, Jini and JavaCards in RoamX, a mobile X-desktop. In: IEEE International Symposium on Distributed Objects and Applications, DOA 2000, pp. 325–333 (2000)
17. Mahajan, S., Tang, Y., Wu, D., Tsai, T., Chen, L.: CAR: the clean air routing algorithm for path navigation with minimal PM2.5 exposure on the move. IEEE Access **7**, 147373–147382 (2019)
18. Nguyen, T.T.L., Nguyen, M.T., Nguyen, D.H.: Predicting missing data by using multimodal data analytics. In: MediaEval 2019 (2019)

Kvasir-Instrument: Diagnostic and Therapeutic Tool Segmentation Dataset in Gastrointestinal Endoscopy

Debesh Jha[1,2]($\boxtimes$), Sharib Ali[9], Krister Emanuelsen[3], Steven A. Hicks[1,5], Vajira Thambawita[1,5], Enrique Garcia-Ceja[10], Michael A. Riegler[1], Thomas de Lange[4,6,7], Peter T. Schmidt[8], Håvard D. Johansen[2], Dag Johansen[2], and Pål Halvorsen[1,5]

[1] SimulaMet, Oslo, Norway
debesh@simula.no
[2] UIT The Arctic University of Norway, Tromsø, Norway
[3] Simula Research Laboratory, Oslo, Norway
[4] Augere Medical AS, Oslo, Norway
[5] Oslo Metropolitan University, Oslo, Norway
[6] Medical Department, Sahlgrenska University Hospital-Mölndal, Gothenburg, Sweden
[7] Department of Medical Research, Bærum Hospital, Gjettum, Norway
[8] Karolinska University Hospital, Solna, Sweden
[9] Department of Engineering Science, University of Oxford, Oxford, UK
[10] Sintef Digital, Oslo, Norway

Abstract. Gastrointestinal (GI) pathologies are periodically screened, biopsied, and resected using surgical tools. Usually, the procedures and the treated or resected areas are not specifically tracked or analysed during or after colonoscopies. Information regarding disease borders, development, amount, and size of the resected area get lost. This can lead to poor follow-up and bothersome reassessment difficulties post-treatment. To improve the current standard and also to foster more research on the topic, we have released the "Kvasir-Instrument" dataset, which consists of 590 annotated frames containing GI procedure tools such as snares, balloons, and biopsy forceps, etc. Besides the images, the dataset includes ground truth masks and bounding boxes and has been verified by two expert GI endoscopists. Additionally, we provide a baseline for the segmentation of the GI tools to promote research and algorithm development. We obtained a dice coefficient score of 0.9158 and a Jaccard index of 0.8578 using a classical U-Net architecture. A similar dice coefficient score was observed for DoubleUNet. The qualitative results showed that the model did not work for the images with specularity and the frames with multiple tools, while the best result for both methods was observed on all other types of images. Both qualitative and quantitative results show that the model performs reasonably good, but there is potential for further improvements. Benchmarking using the dataset provides an opportunity for researchers to contribute to the field of automatic endoscopic diagnostic and therapeutic tool segmentation for GI endoscopy.

© Springer Nature Switzerland AG 2021
J. Lokoč et al. (Eds.): MMM 2021, LNCS 12573, pp. 218–229, 2021.
https://doi.org/10.1007/978-3-030-67835-7_19

Keywords: Gastrointestinal endoscopy · Tool segmentation ·
Endoscopic tools · Convolutional neural network · Benchmarking

1   Introduction

Minimally Invasive Surgery (MIS) is a commonly used technique in surgical procedures. The advantage of MIS is that small surgical incisions are made in the patient for endoscopy that causes less pain, reduced time of the hospital stay, fast recovery, reduced blood loss, and less scaring process as compared to the traditional open surgery. The nature of the operation is complex, and the surgeons have to precisely tackle hand-eye coordination, which may lead to restricted mobility and a narrow field of view [5].

However, unlike the treatment of accessory organs such as liver and pancreas, no incision is required forGastrointestinal (GI) tract organs (*oesophagus, stomach, duodenum, colon, and rectum*). GI procedures also include both minimally invasive surveillance and treatment (*including surgery*) procedures. A varied number of tools are used as per the requirement of these procedures. For example, balloon dilatation to help open the GI surface, biopsy forceps for tissue sample collection, polyp removal with snares, and submucosal injections.

A computer and robotic-assisted surgical system can enhance the capability of the surgeons [9]. It can provide the opportunity to gain additional information about the patient, which can be useful for decision making during surgery [6]. However, it is difficult to understand the spatial relationship between surgical instruments, cameras, and anatomy for the patient [11]. In GI endoscopy, it is vital to track and guide surgeons during tumor resection or biopsy collection from a defined site and help to correlate the biopsied samples and treatment locations post-diagnostic and therapeutic or surgical procedures. While most datasets and automated-algorithm developments for instrument segmentation are mostly focused on laparoscopy-based surgical removal, automatic guidance of tools for GI surgery has not been addressed before.

New developments in the area of robot-assisted systems show that there is potential for developing a fully automated robotic surgeon [14]. The da Vinci robot is a surgical system that is considered the de-facto standard-of-care for certain urological, gynecological, and general procedures [4]. Thus, it is critical to have information regarding intra-operative guidance, which plays an essential role in decision making. However, there are specific challenges, such as limited field of view and difficulties with the surgeons handling the instruments during surgery [13]. Therefore, image-based instrument segmentation and tracking are gaining more and more attention in both robotic and non-robotic minimally invasive surgery. Previous work targeting instrument segmentation, detection, and tracking on endoscopic video images failed on challenging images such as images with blood, smoke, and motion artifacts [13]. Other reasons that make semantic segmentation of surgical instruments a challenging task are the presence of images containing shadows, specular reflections, blood, camera lens fogging, and the complex background tissue [14]. The segmentation masks of these images can be useful for instrument detection and tracking.

Similarly, in the GI tract procedures, from tissue sample collection to surgical removal of pathologies is performed in low field-of-view areas. Visual clutter such as artifacts, moving objects, and fluid, hinders the localisation of the target site during surgical procedures. Additionally, currently, there is no way of correlating the tissue sample collection with biopsied location and assessing surgical procedure effectiveness or even post-treatment recovery analysis. Automated localisation and tracking of tools can help guide the endoscopists and surgeons to perform their tasks more effectively. Also, post-procedure video analysis can be done using these automated methods to track such tools, thus enabling improved surgical procedures or surveillance and their post-assessment. Currently, this is an open problem in the research community, where most procedures are not automated in GI tract endoscopy.

While there is an open research question for automated tool detection and guidance in GI procedures, there is a lack of available public datasets. We aim to initiate the development of automated systems for the segmentation of GI tract diagnostic and therapeutic endoscopy tools. This research direction will enable tracking and localisation of essential tools used in endoscopy and help to improve targeted biopsies and surgeries in complex GI tract organs. To accomplish this, and to address the lack of publicly available labeled datasets, we have publicly released 590 pixel-level annotated frames that comprise of tools such as balloon dilation for facilitating the opening of GI organs, biopsy forceps for tissue sample collection, polyp removal with snares, submucosal injections, radiofrequency ablation of dysplastic mucosa using probes and some other related surgical/diagnostic procedures. The released video frames will allow for building automated Machine Learning (ML) algorithms that can be applied during clinical procedures or post-analyses. To commence this effort, we provide a baseline benchmark on this dataset. U-Net [12] is a common semantic segmentation based architecture for medical image segmentation tasks. In this paper, we therefore present results utilising two U-Net based architectures. The provided dataset is open and can be used for research and development, and we invite medical imaging, computer vision, ML and multimedia researchers to develop novel algorithms on the provided dataset. The main contributions of this paper are:

- The release of 590 annotated images with bounding boxes and segmentation masks of GI diagnostic and surgical tool dataset. To the best of our knowledge, this is the first dataset of segmented tools used in the GI tract.
- A benchmark of the provided dataset using the U-Net [12] and Double-UNet [10] architectures for semantic segmentation is provided.

2   Related Work

Surgical vision is evolving as a promising technique to segment and track instruments using endoscopic images [6]. To gather researchers on a single platform, the *Endoscopic vision (EndoVis) challenge* has been organized since 2015 at Medical Image Computing and Computer Assisted Intervention Society (MICCAI)

Table 1. Similar available datasets

Dataset	Content	Task type	Procedure
Instrument segmentation and tracking (2015) [6]	Rigid and robotic instruments	Segmentation and tracking	Laparoscopy
Robotic Instrument Segmentation (2017) [4]	Robotic surgical instruments	Binary segmentation, part based segmentation, instrument segmentation	Abdominal porcine
Robotic Scene Segmentation (2018) [3]	Surgical instruments and other	Multi-instance segmentation	Robotic nephrectomy
Robust Medical instrument segmentation (2019) [13]	Laparoscopic instrument	Binary segmentation, multiple instance detection, multiple instance segmentation	Laparoscopy
Kvasir-Instrument (Ours)	Diagnostic and therapeutic tools in endoscopic images	Binary segmentation, detection and localization	Gastroscopy & colonoscopy

with an exception in 2016. The EndoVis challenge hosts different sub-challenges. The year-wise information about the hosted sub-challenge can be found on the challenge website[1].

Bodenstedt et al. [6] organized "EndoVis 2015 Instrument sub-challenge" for developing new techniques and benchmarking ML algorithms for segmentation and tracking of the instruments on a common dataset. The organizers challenged on two different tasks, i.e., (1) Segmentation and (2) Tracking. The goal of the challenge was to address the problem related to segmentation and tracking of articulated instruments in both laparoscopic and robotic surgery[2]. A comprehensive evaluation of the methods used in instrument segmentation and tracking task for minimally invasive surgery is summarized in this work [6]. The extensive evaluation showed that deep learning works well for instrument segmentation and tracking tasks.

In 2017, a follow up to the previous 2015 challenge was organized called "Robotic Instrument Segmentation Sub-Challenge"[3]. The challenge was part of the Endoscopic vision challenge that was organized at MICCAI 2017. This challenge offered three tasks: (1) Binary segmentation, (2) Parts based segmentation, and (3) Instrument type segmentation. The goal of the binary segmentation

---

[1] https://endovis.grand-challenge.org/.

[2] https://endovissub-instrument.grand-challenge.org/EndoVisSub-Instrument/.

[3] https://endovissub2017-roboticinstrumentsegmentation.grand-challenge.org/.

task was to separate the image into an instrument and background. Parts segmentation challenged the participants to divide the binary instrument into a shaft, wrist, and jaws. Type segmentation challenged the participants to identify different instrument types. A detailed description of the challenge tasks, dataset, methodologies used by ten participating teams in different tasks, challenge design, and limitation of the challenge can be found in the challenge summary paper [4].

In 2019, a similar challenge called "Robust Medical Instrument Segmentation Challenge 2019"[4] was organized by Roß et al. [13]. This challenge offered three tasks (1) Binary segmentation, (2) Multiple instance detection, and (3) Multiple instance segmentation. The challenge was focused on addressing two key issues in surgical instruments, *Robustness* and *Generalization*, and benchmark medical instrument segmentation and detection on the provided surgical instrument dataset. Endoscopic artefact detection challenge (EAD2019) challenge focused on endoscopic artifact detection primarily but also included instrument class in their detection, segmentation, and "out-of-sample" generalisation tasks. The challenge outcome revealed that most methods performed well for instrument detection and segmentation class [2]. However, this dataset mostly consisted of large biopsy forceps.

In Table 1, we present available instrument datasets in the field of tool segmentation. All of the datasets were designed for hosting challenges. The training dataset is released for all the datasets (except ROBUST-MIS); however, the test dataset is not provided by the challenge organizers. Thus, it makes it difficult to calculate and compare the results on the test dataset. However, experiments are still possible by splitting the training dataset into train, validation, and testing sets. The Robust Medical instrument segmentation dataset is yet not public. However, the participants who have participated in the challenge have the opportunity to download the training dataset. Usually, there are certain practicalities to download the dataset, such as signing the agreement and getting permission from the owner, which takes time, and it is inconvenient. Moreover, to participate in the challenge, the participants have to signup in a particular year, and usually, it often takes a very longtime before they publish the dataset. Thus, the significance of the datasets becomes less as the technology is changing rapidly. More information on available instrument datasets, contents, and offered tasks by the organizers and about the availability can be found from Table 1.

The literature review shows that there are only a few open-access datasets for MIS instrument segmentation. Moreover, to the best of our knowledge, GI tract tools have never been explored. This is the first attempt to provide the community with a curated and annotated public dataset that comprises diagnostic and therapeutic tools in the GI tract. We believe that the presented dataset and the widely used U-Net based algorithm benchmark will encourage the researchers to develop robust and efficient algorithms using the provided dataset that can help clinical procedures in endoscopy.

---

[4] https://robustmis2019.grand-challenge.org/.

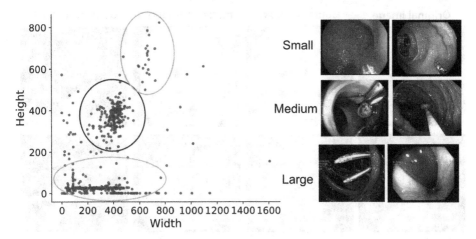

Fig. 1. Distribution of Kvasir-Instrument dataset. On left: Small (green), medium (blue) and large (pink) sized tool clusters. On right: sample images with variable tool size in images. (Color figure online)

3 Kvasir-Instrument Dataset

In this section, we introduce the Kvasir-Instrument dataset with details on how the data was collected, the annotation protocol, and the dataset's structure. The dataset was collected from endoscopic examinations performed at Bærum Hospital in Norway. The unlabelled images' frames are selected from the HyperKvasir dataset [7].

HyperKvasir provides frame-level annotations for 10,662 frames for 23 different classes. However, the majority of the images (99,417 frames) are not labeled. We trained a model using the labeled samples of this dataset and tried to predict the classes of the unlabeled samples. Although our algorithm [15,16] could not classify all the images correctly; however, we were able to classify the presence of instrument or tool out of thousands of provided image frames. However, in order to perform segmentation, pixel-wise masks and bounding boxes were missing. This is what is provided in the proposed dataset, and below, we present the acquisition and annotation protocols used in the data preparation:

3.1 Data Acquisition

The images and videos were collected using standard endoscopy equipment from Olympus (Olympus Europe, Germany) and Pentax (Pentax Medical Europe, Germany) at Bærum Hospital, Vestre Viken Hospital Trust, Norway. All the data used in this study were obtained from videos for procedures that had followed the patient consenting protocol of Bærum Hospital. Additionally, no patient information was available. We have performed a random naming for each publicly released image for further effective annonymisation.

Original image	Bounding box	Image annotation	Generated mask

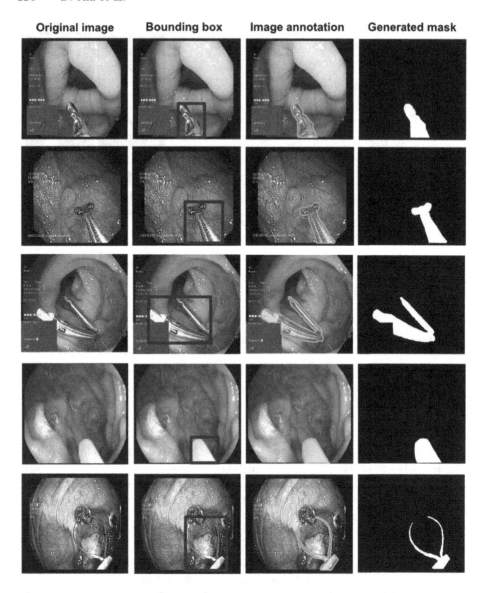

Fig. 2. Kvasir-Instrument dataset: first two rows represent frames with biopsy forceps, the middle row consist of metallic clip, the fourth row is a radio-frequency ablation probe and the last row depicts the crescent and hexagonal shaped snares for polyp removal.

3.2   Annotation Strategy

We have uploaded the Kvasir-Instrument dataset to labelbox[5] and labeled the Region of Interest (ROI) in the image frames, i.e., the ROI of diagnostic and

---
[5] https://www.labelbox.com/.

therapeutic tools in our cases, and generated all the ground truth masks. Figure 2 shows the example images, bounding box, image annotation, and generated masks for the Kvasir-Instrument dataset. All annotations were then exported in a JSON format, which was used to generate masks for each of the annotations. Related source codes and more information about the dataset can be found at https://github.com/DebeshJha/Kvasir-Instrument.

The exported file contained the information of the images along with the coordinate points that were used for mask and bounding box generation. All annotations were performed using a three-step strategy:

1. The selected samples were labeled by two experienced research assistants.
2. The annotated samples were cross-validated for their delineation quality by two experienced GI experts (more than 10 years of work experience in colonoscopy).
3. The suggested changes were incorporated using the comments from the experts.

The Kvasir-Instrument dataset includes 590 frames consisting of various GI endoscopy tools used during both endoscopic surveillance and therapeutic or surgical procedures. A thorough annotation strategy (detailed above) was used to create bounding boxes and segmentation masks. The dataset consists of variable tool size with respect to image height and width, as presented in Fig. 1. The majority of the tools are small and medium-sized. The sample bounding box annotation, precise area delineation, and extracted masks are shown in Fig. 2.

Our dataset is publicly available and can be accessed at https://datasets.simula.no/kvasir-instrument/. It consists of original image samples (in JPEG format), their corresponding masks (in PNG format), and bounding box information (in JSON format).

4   Benchmarking, Results and Discussion

In this section, we explore encoder-decoder based classical models for baseline algorithm benchmarking, their implementation details for reproducibility, details on evaluation metric used for quantitative analysis, and results and discussion.

4.1   Baseline Methods

U-Net [12] has been explored in the past through many biomedical segmentation challenges and has shown strength towards an effective supervised segmentation model. In this paper, we, therefore, use U-Net based architectures on our Kvasir-Instrument dataset to provide a baseline result for future comparisons. U-Net uses an encoder-decoder architecture, that is, a contractive feature extraction path and expansive path with a classifier to perform binary classification of each image pixel in an upsampled feature map. In our previous work, we have shown that the strength of supervised classification can be amplified by using the output mask from one U-Net [12] architecture to the other by proposing DoubleUNet [10]. In addition, the DoubleUNet architecture uses VGG-19 pretrained on ImageNet as one of the encoder blocks, squeeze and excite block, and Atrous spatial pyramid pooling (ASPP) block. All other components in the network remain the same as the U-Net. For both networks, dice loss gives a $1 - DSC$, where DSC is the dice similarity coefficient (see Eq. 1 below).

4.2 Implementation Details

We have implemented the U-Net-based and DoubleUNet based architectures using the Keras framework [8] with TensorFlow [1] as backend running on the Experimental Infrastructure for Exploration of Exascale Computing (eX3), NVIDIA DGX-2 machine. We have resized the training dataset into 512×512. We set the batch size of 8 for training. Both architectures are optimized by using Adam optimizer. We have made use of dice loss as the loss function. We split the dataset using 80% of the dataset for training and the remaining 20% for the testing (evaluation). The same split is also provided in the dataset for the further research. We performed basic augmentation, such as horizontal flip, vertical flip, and random rotation. Moreover, we have also provided the train-test split so that others can improve the methods on the same dataset.

4.3 Evaluation Metrics

In this medical image segmentation approach, each pixel of the diagnostic and therapeutic tool either belongs to a tool or non-tool region. The Dice similarity coefficient (DSC) is the mainly used for result evaluation in medical image segmentation. Additionally, we calculate other standard metrics such as Jaccard similarity coefficient (JC) (also known as the intersection over union (IoU)), precision, recall, overall accuracy, F2, and frames per second (FPS). Using tp, fp, tn, and fn to represent the true positives, false positives, true negatives, and false negatives, respectively, the mathematical formulas for them are as follows:

$$DSC = \frac{2 \cdot tp}{2 \cdot tp + fp + fn} \tag{1}$$

$$JC \text{ or } IoU = \frac{tp}{tp + fp + fn} \tag{2}$$

$$Recall \ (r) = \frac{tp}{tp + fn} \tag{3}$$

$$Precision \ (p) = \frac{tp}{tp + fp} \tag{4}$$

$$F2 = \frac{5p \times r}{4p + r} \tag{5}$$

$$Overall \ accuracy \ (Acc.) = \frac{tp + tn}{tp + tn + fp + fn} \tag{6}$$

$$Frame \ Per \ Second \ (FPS) = \frac{\#frames}{sec} \tag{7}$$

Table 2. Baseline results for tool segmentation

Method	JC	DSC	F2-score	Precision	Recall	Acc.	FPS
U-Net [12]	**0.8578**	**0.9158**	**0.9320**	**0.8998**	**0.9487**	**0.9864**	**20.4636**
DoubleUNet [10]	0.8430	0.9038	0.9147	0.8966	0.9275	0.9838	10.0000

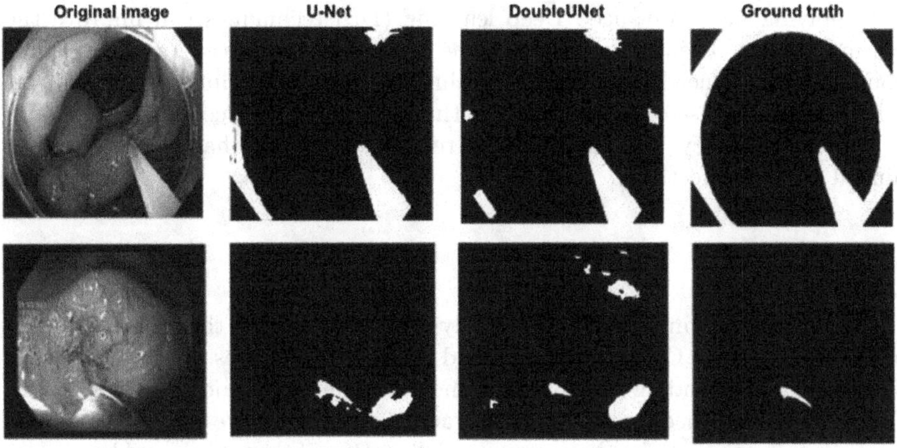

Fig. 3. Failed cases: cap region (top) is under-segmented and small clip area is over-segmented and consist of large number of false positives (bottom).

4.4 Quantitative and Qualitative Results

Table 2 shows the results of the baseline methods for the tool segmentation on the proposed Kvasir-Instrument dataset. From the table, we can observe that the UNet achieved a high JC of 0.8578 and DSC of 0.9158, which is slightly above than the DoubleUNet that yielded JC of 0.8430 and DSC of 0.9038. Also, UNet achieved a speed of 20.4636 FPS, whereas computational time is double for DoubleUNet with only 10 FPS. Similarly, both the recall and precision scores are very comparable for both U-Net ($p = 0.8998, r = 0.9487$) and DoubleUNet ($p = 0.8966, r = 0.9275$).

Figure 3 shows the qualitative result on two challenging sample images. It can be observed that both UNet and DoubleUNet are under-segmenting the cap region (top) and over-segmenting the small clip area (bottom). Some parts of these images are confused because of the presence of saturation areas. However, both models were able to segment well with most endoscopic tool samples in the dataset. This is also evident from the quantitative results. However, even better models are still needed to motivate further research.

4.5 Discussion

From the experimental results in Table 2, we can validate that the classical U-Net architecture outperforms DoubleUNet model. Additionally, U-Net is 2×

faster than the DoubleUNet. This is because U-Net uses basic convolution blocks, whereas DoubleUNet uses pre-trained encoders, ASPP, squeeze, and excite blocks, all of which increase the inference latency. Here, the UNet is optimized by dice loss instead of binary cross-entropy loss, which showed improved performance during our experiments.

Further, fine-tuning on other similar datasets, rigorous data augmentation, and applying more advanced Deep learning (DL) techniques can improve the baseline results - eventually achieving the detection, localisation, and segmentation performance needed to make the technology useful in a clinical environment. Additionally, the use of DL networks with fewer parameters could increase computational efficiency, thereby enabling real-time systems that can be used in clinical settings effectively.

5   Conclusion

We have curated, annotated, and publicly released a dataset that contains *endoscopic tools* used in GI examinations and surgical procedures. The dataset consists of images, bounding boxes, and segmentation masks of endoscopy tools used during different procedures in the GI tract. Additionally, we provided baseline segmentation methods for the automatic delineation of these tools and have compared them using standard computer vision metrics. In the future, we plan to continuously increase the amount of data and also call for multimedia challenges using the presented dataset.

Acknowledgements. This work is funded in part by the Research Council of Norway, project number 263248 (Privaton) and project number 282315 (AutoCap). We performed all computations in this paper on equipment provided by the Experimental Infrastructure for Exploration of Exascale Computing (eX^3), which is financially supported by the Research Council of Norway under contract 270053.

References

1. Abadi, M., et al.: TensorFlow: a system for large-scale machine learning. In: Proceedings of USENIX Symposium on Operating Systems Design and Implementation, pp. 265–283 (2016)
2. Ali, S., et al.: An objective comparison of detection and segmentation algorithms for artefacts in clinical endoscopy. Sci. Rep. **10**(1), 1–15 (2020)
3. Allan, M., Azizian, M.: Robotic scene segmentation sub-challenge. arXiv preprint arXiv:1902.06426 (2019)
4. Allan, M., et al.: 2017 robotic instrument segmentation challenge. arXiv preprint arXiv:1902.06426 (2019)
5. Bernhardt, S., Nicolau, S.A., Soler, L., Doignon, C.: The status of augmented reality in laparoscopic surgery as of 2016. Med. Image Anal. **37**, 66–90 (2017)
6. Bodenstedt, S., et al.: Comparative evaluation of instrument segmentation and tracking methods in minimally invasive surgery. arXiv preprint arXiv:1805.02475 (2018)

7. Borgli, H., et al.: Hyperkvasir, a comprehensive multi-class image and video dataset for gastrointestinal endoscopy. Sci. Data **7**(1), 1–14 (2020)
8. Chollet, F., et al.: Keras (2015)
9. Cleary, K., Peters, T.M.: Image-guided interventions: technology review and clinical applications. Annu. Rev. Biomed. Eng. **12**, 119–142 (2010)
10. Jha, D., Riegler, M., Johansen, D., Halvorsen, P., Håvard, J.: DoubleU-net: a deep convolutional neural network for medical image segmentation. In: Proceedings of 33rd International Symposium on Computer-Based Medical Systems, pp. 558–564 (2020)
11. Pakhomov, D., Premachandran, V., Allan, M., Azizian, M., Navab, N.: Deep residual learning for instrument segmentation in robotic surgery. In: Suk, H.-I., Liu, M., Yan, P., Lian, C. (eds.) MLMI 2019. LNCS, vol. 11861, pp. 566–573. Springer, Cham (2019). https://doi.org/10.1007/978-3-030-32692-0_65
12. Ronneberger, O., Fischer, P., Brox, T.: U-net: convolutional networks for biomedical image segmentation. In: Navab, N., Hornegger, J., Wells, W.M., Frangi, A.F. (eds.) MICCAI 2015. LNCS, vol. 9351, pp. 234–241. Springer, Cham (2015). https://doi.org/10.1007/978-3-319-24574-4_28
13. Ross, T., et al.: Robust medical instrument segmentation challenge 2019. arXiv preprint arXiv:2003.10299 (2020)
14. Shvets, A.A., Rakhlin, A., Kalinin, A.A., Iglovikov, V.I.: Automatic instrument segmentation in robot-assisted surgery using deep learning. In: Proceedings of International Conference on Machine Learning and Applications, pp. 624–628 (2018)
15. Thambawita, V., et al.: The medico-task 2018: disease detection in the gastrointestinal tract using global features and deep learning. arXiv preprint arXiv:1810.13278 (2018)
16. Thambawita, V., et al.: An extensive study on cross-dataset bias and evaluation metrics interpretation for machine learning applied to gastrointestinal tract abnormality classification. arXiv preprint arXiv:2005.03912 (2020)

CatMeows: A Publicly-Available Dataset of Cat Vocalizations

Luca A. Ludovico[1]([✉]) [iD], Stavros Ntalampiras[1] [iD], Giorgio Presti[1] [iD],
Simona Cannas[2] [iD], Monica Battini[3] [iD], and Silvana Mattiello[3] [iD]

[1] Department of Computer Science, University of Milan, Milan, Italy
{luca.ludovico,stavros.ntalampiras,giorgio.presti}@unimi.it
[2] Department of Veterinary Medicine, University of Milan, Milan, Italy
simona.cannas@unimi.it
[3] Department of Agricultural and Environmental Science, University of Milan,
Milan, Italy
{monica.battini,silvana.mattiello}@unimi.it

Abstract. This work presents a dataset of cat vocalizations focusing on the meows emitted in three different contexts: brushing, isolation in an unfamiliar environment, and waiting for food. The dataset contains vocalizations produced by 21 cats belonging to two breeds, namely Maine Coon and European Shorthair. Sounds have been recorded using low-cost devices easily available on the marketplace, and the data acquired are representative of real-world cases both in terms of audio quality and acoustic conditions. The dataset is open-access, released under Creative Commons Attribution 4.0 International licence, and it can be retrieved from the Zenodo web repository.

Keywords: Audio dataset · Audio signal processing · Bioacoustics · Cat vocalizations · Meows

1 Introduction

Domestic cats (*Felis silvestris catus*) produce different vocalizations, with closed mouth (e.g., purrs and trills), with the mouth held open (e.g., spitting or hissing), and while the mouth is gradually closing or opening (i.e. meows) [2]. Meows are considered as the most relevant type of vocalizations used by domestic cats to communicate with humans [2], whereas undomesticated felids rarely meow to humans [5], despite being common as an intra-specific vocalization [3]. Consequently, the analysis of meows can be useful to investigate cat-human communication.

Concerning widespread companion animals, who are often perceived as social partners [11,13], few studies have been carried out to understand the mechanisms of their vocal communication with humans. For example, Pongrácz analyzes the effects of domestication on the vocal communication of dogs [23], McComb *et al.* focus on cat purrs [15], and Owens *et al.* propose a visual classification of feral cat vocalizations [18].

© Springer Nature Switzerland AG 2021
J. Lokoč et al. (Eds.): MMM 2021, LNCS 12573, pp. 230–243, 2021.
https://doi.org/10.1007/978-3-030-67835-7_20

In 2019, in the framework of a multi-disciplinary project including veterinarians, psychologists, and experts in sound computing, a study was conducted on the automatic analysis of meows emitted in three different contexts: waiting for food, isolation in an unfamiliar environment, and brushing. The idea was to extract the audio characteristics of the sounds produced in response to heterogeneous stimuli, where the request for food was expected to induce excitement, isolation to cause discomfort, and brushing to provoke either positive or negative responses, depending on the specimen. The results of the project have been published in [17]. After the design phase, the first step of the project was the production of a suitable dataset, called *CatMeows*, now available to the scientific community for research purposes.

The goal of this paper is to describe *CatMeows* from different perspectives, focusing on the protocol used to gather the data (Sect. 2), the internal composition of the dataset (Sect. 3), and some possible application scenarios (Sect. 4). Our conclusions will be summarized in Sect. 5.

As detailed below, *CatMeows* is open access, and it can be retrieved from Zenodo,[1] a general-purpose open-access repository developed under the European OpenAIRE program and operated by CERN. The dataset has been assigned the following DOI: 10.5281/zenodo.4007940.

2   Building the Dataset

The goal of this section is to provide details about the whole protocol adopted to obtain the *CatMeows* dataset. Section 2.1 will discuss some design choices, Sect. 2.2 will focus on technical issues related to meow capturing, and Sect. 2.3 will describe post-processing operations.

2.1   Design Choices

For our experiments, one of the critical problems to solve was recording vocalizations while minimizing the potential influences caused by environmental sounds and other noises. Moreover, the characteristics of the room (e.g., its topological properties, reverberation effects, the presence of furniture, etc.) should not affect the captured audio signals. A possible solution would have been to record sounds in a controlled anechoic environment, but, unfortunately, an unfamiliar space was expected to influence the behavior of the cat in the waiting for food and brushing scenarios. In fact, the meowing when placed in an unknown environment, without other stimuli, was one of the reactions our experiment tried to capture and analyse.

Technology in use consists in reliable, commercially available, mass-production sensors, so as to be representative of the sound quality which can be realistically implemented at the time of writing. The use of such an equipment should also ease technological transfer.

---

[1] https://zenodo.org/.

Another constraint regarded the relative position of the microphone with respect to the sound source, that had to be fixed in terms of distance and angle, as many characteristics to be measured in vocalizations could depend on such parameters. As a matter of fact, this requirement prevented the adoption of easy solutions, such as a single omnidirectional microphone or an array of fixed microphones suitably mounted in the room. The input device had to move together with the sound source, keeping the distance and angle as constant as possible.

Fig. 1. Three of the cats participating in the experiment. A Bluetooth microphone is mounted on the collar so as to point upwards and keep an almost fixed distance from their vocal apparatus.

The idea was to capture sounds very close to the source. A number of solutions were explored, including wearable devices somehow attached to the cat's back. Being a whole computing system embedding the recording, processing, and storing chain within a single miniaturized board, such a solution would have been quite cheap, technologically sound, and extremely easy to adopt. Unfortunately, this approach could not be applied to cats, since, according to veterinarians, the perception of a device in close contact with the fur could have produced biased behavioral results. Moreover, the problem of capturing sound in proximity of the vocal apparatus would have persisted.

Thus, a refinement of the original idea was to adopt a very small and lightweight microphone to be placed under the cat's throat through a collar, as shown in Fig. 1. The collar is an object the animal is generally already familiar with, so it was not expected to alter its behavior; in other case, the cat was trained to wear the collar for some days before data recording, until no sign of discomfort (e.g., scratching the collar, shaking, lip licking, yawning, etc.) was reported [7,8,16,19,20].

Since the placement of the microphone was the same for all cats, sound modifications due to angle, distance and relative position of the mouth were consistent across all recordings, thus marginalizing the effects of these parameters. Some types of information, such as the pitch, did not change, whereas other audio features resulted biased by a constant factor across recordings.

Finally, it could be argued that different situations could alter cats' postures, thus influencing sound production. We did not consider this as a bias, since we were interested in the final vocalization outcome, regardless of how the cat produced it. A change in posture was considered a legitimate and realistic consequence of the situation. It should be mentioned that if the recording device was able to catch such an aspect, it would have been considered as a useful additional information.

2.2 Capturing Audio Signals

Concerning the input device for audio signals, we turned to Bluetooth connectivity-based products, due to their small dimensions and weight, low power consumption, good transmission range, low price, and acceptable recording quality.

Bluetooth microphones are usually low-budget devices packed with mono earphones and optimized for human voice detection. For example, the frequency range correctly acquired is typically very limited when compared to high-quality microphones. Nevertheless, in this category of devices, there are products presenting all the features required by our experimentation, specifically for the development of automatic classification systems.

Fig. 2. Detailed view of the *QCY Q26 Pro* Bluetooth headset. The microphone is visible in the form of a small hole in the right part of the image.

For the goals of the project, we selected the *QCY Q26 Pro* Bluetooth headset (see Fig. 2), presenting a dynamic range of 98 ± 3 dB. It transmits audio signals via Bluetooth with HFP [25] or HSP [6] profiles, introducing a logarithmic A-law quantization[2] at a sampling frequency of 8 kHz. As a consequence, the actual range of frequencies we could rely on was 0 to 4 kHz. The fundamental frequency emitted by cats when meowing fall within that range, but some higher-frequency harmonics could be cut.

The size ($15 \times 10 \times 23$ mm) and weight (50 g) of the device were sufficiently small to be carried by a cat without significant behavioral implications. Interestingly, this microphone represented a very low-budget solution: the whole headset, including an earphone (obviously not employed in the experimentation), at the moment of writing can be found in the marketplace for 12 to 25 €.

---

[2] The A-law algorithm is a standard used in European 8-bit PCM digital communications systems to optimize the dynamic range of an analog signal for digitizing.

The adoption of the Bluetooth communication protocol greatly simplified the recording chain. In the experimentation, the headset was paired to a smartphone equipped with *Easy Voice Recorder PRO*, a high-quality audio recording application supporting the LPCM[3] encoding. This aspect was fundamental to get audio files in a non-supervised environment, e.g. at home, also from people not expert in the IT field, like most cat owners.

2.3 Post-processing

Non-meowing sounds produced by the cats were discarded (13 instances, still available in a separate folder of the dataset), and the remaining sounds were cut so that each file contained a single vocalization, with 0.5 s of leading and trailing silence segments, or *handles*.

Table 1. File naming policy: explanation of the pattern C_NNNNN_BB_SS_OOOOO_RXX.

Placeholder	Description	Values
C	Emission context	B: brushing
		F: waiting for food
		I: isolation
NNNNN	Cat name	Cat's unique ID
BB	Breed	MC: Maine Coon
		EU: European Shorthair
SS	Sex	FI female, intact
		FN: female, neutered
		MI: male, intact
		MN: male, neutered
OOOOO	Cat owner	Cat owner's unique ID
R	Recording session	1, 2 or 3
XX	Vocalization counter	01...99

Files are named after the pattern C_NNNNN_BB_SS_OOOOO_RXX, where each placeholder corresponds to a part of the information carried by the file (see Table 1). Included files follow the wav format type.

3 Composition of the Dataset

The cats recruited for the original project described in [17] were 10 adult *Maine Coon* cats (1 intact male, 3 neutered male, 3 intact females and 3 neutered

---

[3] LPCM stands for Linear Pulse-Code Modulation, a standard method to digitally represent sampled analog signals. In a LPCM stream, the amplitude of the analog signal is sampled regularly at uniform intervals, and each sample is quantized to the nearest value within a range of digital linearly spaced steps.

females) belonging to a single private owner and housed under the same conditions, and 11 adult *European Shorthair* cats (1 intact male, 1 neutered male, no intact females and 9 neutered females) belonging to different owners and housed under heterogeneous conditions.

In presence of at least one veterinarian, cats were repeatedly exposed to three different contexts that were expected to stimulate the emission of meows:

1. *Brushing*—Cats were brushed by their owners in their home environment for a maximum of 5 min;
2. *Isolation in an unfamiliar environment*—Cats were transferred by their owners into an unfamiliar environment (e.g., a room in a different apartment or an office). Distance was minimized and the usual transportation routine was adopted so as to avoid discomfort to animals. The journey lasted less than 30 min and cats were allowed 30 min with their owners to recover from transportation, before being isolated in the unfamiliar environment, where they stayed alone for maximum 5 min;
3. *Waiting for food*—The owner started the routine operations that preceded food delivery in the usual environment the cat was familiar with. Food was given at most 5 min after the beginning of the experiment.

The typical vocalization in response to a single stimulus was composed by a number of separated vocalizations. In this case, sound files have been cut according to the strategy explained in Sect. 2.3.

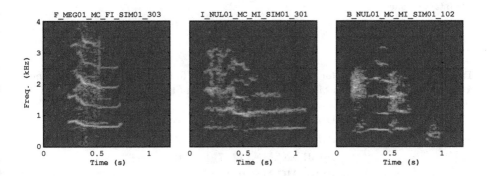

Fig. 3. Time-frequency spectrograms of meows originated by the three considered classes.

The final corpus contains 440 files, organized as shown in Tables 2 and 3. The total number of files referring to *Maine Coon* specimens is 188, and 252 for *European Shorthair* cats. The average length of each file is 1.83 s, with a variance of 0.36 s. About 1s of each file contains only background noise, due to 0.5s handles before and after the vocalization.

From the analysis of sounds emitted in response to the same stimulus, some common characteristics emerge, as detailed in Sect. 4.2. For example, Fig. 3 shows the spectrograms of samples that resulted to be the medioids of the three emitting context classes. Note that the dataset is imbalanced across contexts, due to the tendency of cats to not vocalize uniformly. Thus, such imbalances are indicative of their behavior during cat-human interactions.

Additional files can be found in the repository in a separate folder, with the following naming convention: C_NNNNN_BB_SS_00000_F*SEQ*X.wav. These were the cases where the recorded cat meowed repeatedly, in sequences of many meows separated only by few seconds of silence.

4   Application Scenarios

CatMeows is open and available to the scientific community for research purposes. In this section, we present possible research directions starting from either our dataset or a similar collection of sounds, which can be built by following the protocol described in this paper.

Table 2. Number of files in the *CatMeows* dataset, grouped by situation, breed (MC = Maine Coon; EU = European Shorthair), and sex.

	Food (92)		Isolation (221)		Brushing (127)	
	MC (39)	EU (53)	MC (92)	EU (129)	MC (57)	EU (70)
Intact males (19)	–	5	10	–	4	–
Neutered males (76)	14	8	18	15	17	4
Intact females (68)	21	–	28	–	19	–
Neutered females (277)	4	40	36	114	17	66

Table 3. Number of files for each cat (F: waiting for food, I: isolation, B: brushing). A horizontal line separates Maine Coon and European Shorthair cats.

ID	F	I	B	Total
ANI01_MC_FN	–	4	6	10
BAC01_MC_MN	12	7	–	19
BRA01_MC_MN	2	10	7	19
BRI01_MC_FI	8	–	7	15
DAK01_MC_FN	4	32	4	40
JJX01_MC_FN	–	–	7	7
MEG01_MC_FI	4	10	–	14
NIG01_MC_MN	–	1	4	5
NUL01_MC_MI	–	10	4	14
WHO01_MC_FI	9	18	12	39

(*continued*)

Table 3. (*continued*)

ID	F	I	B	Total
BLE01_EU_FN	3	45	–	48
CAN01_EU_FN	2	26	26	54
CLE01_EU_FN	–	20	–	20
IND01_EU_FN	10	–	11	21
LEO01_EU_MI	5	–	–	5
MAG01_EU_FN	14	3	10	27
MAT01_EU_FN	6	10	10	26
MIN01_EU_FN	4	6	4	14
REG01_EU_FN	1	1	3	5
SPI01_EU_MN	8	15	4	27
TIG01_EU_FN	–	3	2	5

4.1 Proposed Scenarios

Among many application scenarios, the most obvious one is probably to support comparative and repeatable experimentation concerning our original project, that aimed at the automatic classification of cat vocalizations [17]. In detail, the idea is to have a fully automatic framework with the ability to process vocalizations and reveal the context in which they were produced, by using suitable audio signal processing techniques and pattern recognition algorithms.

Another usage of the *CatMeows* dataset may concern feature analysis in general. In this scenario, key features are detected in order to better understand what is meaningful in cat-human communication. Under this perspective, the original situations in which sounds have been recorded are only examples of contexts that cause cats to emit vocalizations. Moreover, a statistical analysis of the dataset's audio features will be presented and discussed in Sect. 4.2.

Besides, since both cats and situations are univocally identified in the dataset, it is possible to plan listening sessions aiming to test the ability of humans in: i) recognizing specimens across different recordings, ii) clustering vocalizations emitted by different cats in the same context, iii) distinguishing between male/female cats or specimens belonging to different breeds, etc. In this sense, some early experiments have already been carried out [24].

Finally, a different scenario implies the investigation of meowing in relation with cat behavior. While our research took a given stimulus as the input and the corresponding vocalization as the output, deliberately ignoring behaviors and emotional states of cats, such aspects could be addressed in other research projects. To this end, even if in the current version of the dataset sounds are not tagged from the behavioral or emotional point of view, we are planning to make the corresponding video recordings available, thus facilitating in-depth affective analysis as well.

4.2 Example

To provide some insights on *CatMeows* and exemplify one of its possible uses, we carried out a statistical analysis of the dataset's audio features, and specifically: Vocalization duration D, Level Envelope E, Pitch F_0, Tristimulus T_{1-3}, Roughness R, Brightness B, and Spectral Centroid C (for further information about audio features, please refer to [1]).

- D is just the duration of the vocalisation;
- E is the envelope profile of audio level, carrying information about dynamics (*i.e.* slow amplitude modulations);
- F_0 is the fundamental frequency of the sound. As we are dealing with meows, that are harmonic sounds, such a frequency is supposed to be always present;
- T_{1-3} is a feature borrowed from color perception and adapted to the audio domain. T_1 is the ratio between the energy of F_0 and the total energy, T_2 is the ratio between the second, third, and fourth harmonics taken together and the total energy, and T_3 is the ratio between remaining harmonics and the total energy [22]. Since the sampled signal is strongly band-limited, T_3 would have considered only few harmonics – if any – thus it was discarded. T_{1-2} was chosen to represent information regarding formants, since with a such band-limited signal, containing only a reduced number of harmonics, the actual computation of formants, either with filterbanks or linear prediction techniques, resulted to be reliable only in a reduced number of cases. The downside of using T_{1-3} is that this feature strongly depends on the quality of F_0 estimation;
- R is defined by [9] as the sensation for rapid amplitude variations, which reduces pleasantness, and whose effect is perceived as dissonant [21]. This is a timbral perceptual descriptor which may convey information about unpleasantness;
- C is the *center of mass* of the energy distribution across the spectrum. This feature provides relevant information about the spectral content, without considering harmonic aspects [12].

The pre-processing consisted in trimming the first and last 0.4 s of the audio file, and to perform a linear fade-in and fade-out of 0.05 s each. This was made in order to discard possible background sound events occurred during the 0.5 s handles before and after the cat vocalisation.

Except for D, all the above-mentioned features are time-varying signals, extracted either from the short-term Fourier Transform of the sound (for the frequency domain features), or an equal frame subdivision of the signal (for time-domain features), so as to have the same temporal resolution for every feature. In particular we used a window size of 512 samples, with an overlap of 448 samples (that is, an hop-size of 64 samples), and a Hanning windowing function.

- D is expressed in seconds, and computed as the length of the recording minus one (to remove the duration of silence handles);

- E is computed as the Root Mean Square (RMS) of each audio frame, and it is expressed in decibels;
- F_0 is computed with the SWIPE$'$ method [4], limiting the results in the range 100 Hz–1250 Hz, and ignoring values where the confidence reported by SWIPE$'$ is below 0.15. A median filtering with a span of 5 frames helped in removing artifacts. A visual inspection of the results showed that these settings works well, with only limited and sporadic octave-errors. F_0 is expressed in a logarithmic frequency scale;
- T_{1-2} are computed via retrieving the energy around F_0 and the corresponding harmonics using a triangular window of 9 FFT bins to mitigate potential F_0 errors and the FFT finite resolution. T_{1-2} are expressed in decibels;
- R is computed using the corresponding function of the MIRToolbox [14] and more specifically using the strategy proposed in [26].
- C is the weighted mean of the frequencies, using the energy over each frequency as the weight. C is expressed in a logarithmic frequency scale;

To avoid measuring features over silent portions of the signal which typically burden the modelling process, features that do not rely on F_0 (which implicitly ignores silence and non-periodic portions) were ignored in the instants where $E < 0.0005$ (-66 dB). This value has been empirically found to provide almost perfect silence detection over the entire dataset[4].

In order to reduce the time-varying features to a set of global values, and to mitigate the effects of local errors (of any magnitude), median values and inter-quartile ranges are calculated for every feature. To grasp information about temporal aspects, this has been carrier out also to the delta-features (i.e. the differential of the feature in the temporal dimension). This led to a final feature space of a dimensionality equal to 25 variables.

Finally, a One-way ANOVA test has been carried out to assess which features are meaningful in distinguishing three different partitions of the dataset: Male v.s. Female; European Shorthair v.s. Maine Coon; Brushing v.s. Isolation v.s. Waiting for food. Results are reported in Table 4, where columns represent the comparison of two classes, rows represent the feature space, and a "•" mark points out where ANOVA revealed significant differences ($p < 0.05$).

From Table 2 it can be seen how isolation is the situation that elicits the majority of meows, and that European Shorthair cats tend to vocalize more than Maine Coon. Nevertheless, Table 3 shows a great variability among cats' reactions to given stimuli. Finally, Table 4 shows that every class can be distinguished: the emitting context affects all features; the breed mainly affects D, P, E, and C; the sex mainly affects E, and C.

---

[4] Please note that different envelope functions and frame sizes may have different optimal thresholds.

Table 4. ANOVA results, where "•" denotes cases with significant differences between classes. m subscript indicates the median; i subscript indicates interquartile range; Δ prefix indicates features' delta. FI = Food-Isolation; FB = Food-Brushing; IB = Isolation-Brushing; EM = European Shorthair-Maine Coon; FM = Female-Male.

	FI	FB	IB	EM	FM
D	•	•		•	
P_m	•	•		•	
P_i			•	•	
ΔP_m			•	•	
ΔP_i	•	•	•	•	
E_m	•		•	•	
E_i				•	•
ΔE_m	•		•	•	•
ΔE_i	•		•	•	•
C_m				•	
C_i	•		•	•	•
ΔC_m	•		•	•	•
ΔC_i			•		
$T1_m$	•				
$T1_i$	•	•		•	
$\Delta T1_m$					
$\Delta T1_i$		•	•		
$T2_m$	•				
$T2_i$					
$\Delta T2_m$					
$\Delta T2_i$	•	•	•		
R_m	•		•		
R_i	•		•	•	•
ΔR_m					
ΔR_i	•		•		

5   Conclusions

This paper described the motivation, recording protocol, and contents of a dataset encompassing cat vocalizations in response to diverse stimuli. It comprises a fundamental step in cataloging and modeling cat-human communication, while, to the best of our knowledge, *CatMeows* is a unique corpus of audio documents.

After extensive analysis of the dataset's contents, we provided potential application scenarios as well as an initial statistical analysis of physically/perceptually meaningful features which could address them. This does not limit the usage of unsupervised feature-extraction mechanisms based on deep learning technologies [10].

Given the wide range of applications, *CatMeows* has not been divided in train-test sets beforehand; however, cross-validation techniques are recommended in order to evaluate the generalization capabilities of potential models. Focusing on reproducibility of the obtained results, which comprises a strong requirement in modern machine learning-based systems, the dataset is available to the scientific community. We hope that the availability of *CatMeows* will encourage further research in the scientific field of cat-human interaction and animal welfare in general.

Ethical Statement

The present project was approved by the *Organism for Animals Welfare* of the University of Milan (approval n. OPBA_25_2017). The challenging situations cats were exposed to were required in order to stimulate the emission of meows related to specific contexts. They were conceived considering potentially stressful situations that may occur in cats' life and to which cats can usually easily adapt. In order to minimize possible stress reactions, preliminary information on the normal husbandry practices (e.g., brushing or transportation) to which the experimental cats were submitted and on their normal reactions to these practices were collected by interviews to the owners. Additionally, cats were video recorded using a camera connected to a monitor for 5 min before the stimulus, during the stimulus and for 5 min after the stimulus, in order to monitor their behavior during the isolation challenge, with the idea of stopping the experiment if they showed signs of excessive stress; however, such a situation never occurred.

Intellectual Property of the Dataset

The dataset is one of the outcomes of an interdepartmental project at [OMITTED]. The dataset is open access, released under Creative Commons Attribution 4.0 International licence.[5] It can be retrieved from the following DOI: 10.5281/zenodo.4007940.

The dataset can be freely used for research and non-commercial purposes, provided that the authors are acknowledged by citing the present paper.

References

1. Alías, F., Socoró, J.C., Sevillano, X.: A review of physical and perceptual feature extraction techniques for speech, music and environmental sounds. Appl. Sci. **6**(5), 143 (2016)

---

[5] https://creativecommons.org/licenses/by/4.0/legalcode.

2. Bradshaw, J., Cameron-Beaumont, C.: The signalling repertoire of the domestic cat and its undomesticated relatives. In: The Domestic Cat: The Biology of Its Behaviour, pp. 67–94 (2000)

3. Brown, S.L.: The social behaviour of neutered domestic cats (Felis catus). Ph.D. thesis, University of Southampton (1993)

4. Camacho, A., Harris, J.G.: A sawtooth waveform inspired pitch estimator for speech and music. J. Acoust. Soc. Am. **124**(3), 1638–1652 (2008)

5. Cameron-Beaumont, C.: Visual and tactile communication in the domestic cat (Felis silvestris catus) and undomesticated small-felids. Ph.D. thesis, University of Southampton (1997)

6. Car Working Group: Headset Profile (HSP) 1.2. Bluetooth SIG (2008)

7. Carney, H., Gourkow, N.: Impact of stress and distress on cat behaviour and body language. In: Ellis, S., Sparkes, A. (eds.) The ISFM Guide to Feline Stress and Health. Tisbury (Wiltshire): International Society of Feline Medicine (ISFM) (2016)

8. Casey, R.: Fear and stress in companion animals. In: Horwitz, D., Mills, D., Heath, S. (eds.) BSAVA Manual of Canine and Feline Behavioural Medicine, pp. 144–153. British Small Animal Veterinary Association, Guarantee (2002)

9. Daniel, P., Weber, R.: Psychoacoustical roughness: implementation of an optimized model. Acta Acust. United Acust. **83**(1), 113–123 (1997)

10. Dara, S., Tumma, P.: Feature extraction by using deep learning: a survey. In: 2018 Second International Conference on Electronics, Communication and Aerospace Technology (ICECA), pp. 1795–1801 (2018)

11. Eriksson, M., Keeling, L.J., Rehn, T.: Cats and owners interact more with each other after a longer duration of separation. PLOS One **12**(10), e0185599 (2017). https://doi.org/10.1371/journal.pone.0185599

12. Grey, J.M., Gordon, J.W.: Perceptual effects of spectral modifications on musical timbres. J. Acoust. Soc. Am. **63**(5), 1493–1500 (1978)

13. Karsh, E.B., Turner, D.C.: The human-cat relationship. Domestic Cat: Biol. Behav. 159–177 (1988)

14. Lartillot, O., Toiviainen, P.: A matlab toolbox for musical feature extraction from audio. In: International Conference on Digital Audio Effects, pp. 237–244. Bordeaux (2007)

15. McComb, K., Taylor, A.M., Wilson, C., Charlton, B.D.: The cry embedded within the purr. Curr. Biol. **19**(13), R507–R508 (2009)

16. Notari, L.: Stress in veterinary behavioural medicine. In: BSAVA Manual of Canine and Feline Behavioural Medicine, pp. 136–145. BSAVA Library (2009)

17. Ntalampiras, S., et al.: Automatic classification of cat vocalizations emitted in different contexts. Animals **9**(8), 543.1–543.14 (2019). https://doi.org/10.3390/ani9080543

18. Owens, J.L., Olsen, M., Fontein, A., Kloth, C., Kershenbaum, A., Weller, S.: Visual classification of feral cat Felis silvestris catus vocalizations. Curr. Zool. **63**(3), 331–339 (2017)

19. Palestrini, C.: Situational sensitivities. In: Horwitz, D., Mills, D., Heath, S. (eds.) BSAVA Manual of Canine and Feline Behavioural Medicine, pp. 169–181. British Small Animal Veterinary Association, Guarantee (2009)

20. Palestrini, C., et al.: Stress level evaluation in a dog during animal-assisted therapy in pediatric surgery. J. Veterinary Behav. **17**, 44–49 (2017)

21. Plomp, R., Levelt, W.J.M.: Tonal consonance and critical bandwidth. J. Acoust. Soc. Am. **38**(4), 548–560 (1965)

22. Pollard, H.F., Jansson, E.V.: A tristimulus method for the specification of musical timbre. Acta Acust. United Acust. **51**(3), 162–171 (1982)
23. Pongrácz, P.: Modeling evolutionary changes in information transfer: effects of domestication on the vocal communication of dogs (Canis familiaris). Eur. Psychol. **22**(4), 219–232 (2017)
24. Prato Previde, E., et al.: What's in a meow? A study on human classification and interpretation of domestic cat vocalizations. Animals **10**, 1–17 (2020). in press
25. Telephony Working Group: Hands-Free Profile (HFP) 1.7.1, Bluetooth Profile Specification. Bluetooth SIG (2015)
26. Vassilakis, P.: Auditory roughness estimation of complex spectra-roughness degrees and dissonance ratings of harmonic intervals revisited. J. Acoust. Soc. Am. **110**(5), 2755 (2001)

Search and Explore Strategies for Interactive Analysis of Real-Life Image Collections with Unknown and Unique Categories

Floris Gisolf[1,2](✉) ⓘ, Zeno Geradts[1,3] ⓘ, and Marcel Worring[1] ⓘ

[1] University of Amsterdam, Amsterdam, The Netherlands
[2] Dutch Safety Board, The Hague, The Netherlands
f.gisolf@safetyboard.nl
[3] Netherlands Forensic Institute, The Hague, The Netherlands

Abstract. Many real-life image collections contain image categories that are unique to that specific image collection and have not been seen before by any human expert analyst nor by a machine. This prevents supervised machine learning to be effective and makes evaluation of such an image collection inefficient. Real-life collections ask for a multimedia analytics solution where the expert performs search and explores the image collection, supported by machine learning algorithms. We propose a method that covers both exploration and search strategies for such complex image collections. Several strategies are evaluated through an artificial user model. Two user studies were performed with experts and students respectively to validate the proposed method. As evaluation of such a method can only be done properly in a real-life application, the proposed method is applied on the MH17 airplane crash photo database on which we have expert knowledge. To show that the proposed method also helps with other image collections an image collection created with the Open Image Database is used. We show that by combining image features extracted with a convolutional neural network pretrained on ImageNet 1k, intelligent use of clustering, a well chosen strategy and expert knowledge, an image collection such as the MH17 airplane crash photo database can be interactively structured into relevant dynamically generated categories, allowing the user to analyse an image collection efficiently.

Keywords: Image collections · Exploration · Search · Strategy · Interactive

1 Introduction

Human analysts can quickly grasp the meaning of a small set of complex images, but it is difficult for them to analyze a large, unorganized image collection. Images in image collections often are related to each other in many possible

ⓒ Springer Nature Switzerland AG 2021
J. Lokoč et al. (Eds.): MMM 2021, LNCS 12573, pp. 244–255, 2021.
https://doi.org/10.1007/978-3-030-67835-7_21

ways, such as in time, location, objects and persons, making the collections very complex with $O(n^2)$ relations, even for a relatively small number of images. There are highly successful automatic categorization methods, e.g. [14,21]. However, many real-life image collections contain image categories that are unique and have not been seen before by analyst nor by the machine. With no training data available, this prevents supervised machine learning to be effective in classification. The analyst can learn a new category with only a few examples, however, has a limited working memory and suffers from fatigue during repetitive tasks. Therefore, to categorize the collection and gain insight in an efficient way, interplay between the human expert and a machine is essential and for that a good understanding of the analytical process is important.

In [27], the process is modeled by the exploration-search axis. **Search** is a sequence of query-response pairs where both the analyst and the system have a fixed model of the data. **Exploration** is the process of the analyst uncovering some structure and points of interest within the image collection, where the analyst and the system work with a dynamic model of the data, which can change over time based on what is deemed relevant and what is not.

Generally, some exploration needs to take place before the analyst can start searching for specific items of relevance. The individual components for exploring [12,13,18] and searching [17] exist and have been studied extensively. However, it remains unclear how a user can go from a complex, unstructured image collection to exploring the data, bringing structure to it, searching relevant items and ultimately gain insight. To model the analytical process, we should not only consider the individual components, but also a comprehensive strategy of exploration and search.

Categorizing relevant images is the umbrella task for the exploration-search axis [28] and allows the analyst to perform all other exploration and search tasks more efficiently. Browsing the data becomes more meaningful; the data can be summarized by its categories; and search tasks can be performed using the categories as a rough decomposition.

We develop and evaluate several explore and search strategies that allow the analyst to efficiently perform this umbrella task. A strategy is more successful when the analyst has to assess fewer images in order to find what she is looking for. A demonstration video, the code and the application[1] are available.

The unique opportunity of having access to and expert knowledge of a real life accident investigation photo database of the MH17 airplane crash on 17 July 2014 [4] is used to design and evaluate the proposed methods. To evaluate robustness of the methods an additional image collection is constructed using the Open Image Database [11] (OID).

The main contributions of this paper consist of (1) a method to explore and search through an image collection containing images of unknown and unique categories in an easy to use and intuitive way where the user can control the process using only a single parameter; (2) the evaluation of several explore and

---

[1] Demonstration video on https://youtu.be/73-ExDd2lco, code and application on https://tinyurl.com/imexMMM.

search strategies using an artificial user model to show that the right strategy can make a large difference in how many images the user has to inspect to find the relevant items; (3) user studies performed with investigators of the Dutch Safety Board (DSB) and Forensic Science students to verify the results obtained through the experiments with the simulated users.

2   Related Work

As there is no comparable method that covers both exploration and search and allows for structuring an image collection, this section mainly looks at the different components of the proposed method as discussed in the previous section, and into the expertise of humans versus machines.

Neural networks now perform with almost human accuracy on certain image classification tasks [17], but need many training examples. While one-shot and few-shot learning for machines has been studied, it is not yet on par with human capabilities [5,22,24]. It is thus necessary to combine human expertise with that of the computer in order to achieve insight in large image collections [27].

Zero-shot learning (ZSL) tries to tackle the problem of not having training data for new categories [25,29] through attributes, where a new category may consist of a combination of attributes already present in the training set. However, such extensive attributes need to be available for the categories needed by the analyst, which is a costly process. This makes ZSL unsuitable for the problem discussed in this paper.

Methods such as relevance feedback [30] and active learning [19] can help the analyst with finding new instances of a certain category through interaction. VITRIVR [7] is a complete search method for both images and video, but is less suitable for exploration. (Meta-)transfer learning [20,26] takes an existing neural network and uses additional training for new categories in order to classify images. These search methods do require that the analyst already knows what she is looking for, thus they need to be preceded by an exploration phase.

Worring et al. [23] proposed a framework to explore and visualize data, using content-based image features. However, they concentrate on the use of meta data and forensics, making it less suitable for other domains. If meta data is present at all, it is fairly unreliable and easily changed (on purpose or by accident). Furthermore, its main focus is on browsing of image collections, which is only part of the exploration-search axis. MediaTable [16] allows analysts to browse and categorize the data based on a variety of features, but it leaves the clustering and exploring mostly up to the user, which means it does not scale as well with more data. ImageX [10] uses a hierarchical graph, but offers limited search capabilities and no way to store any user progress or structure.

To evaluate algorithms, most papers use one of the in general very good publicly available databases. Life events such as VBS and LSC are becoming more commonplace too. However, in the case of real-life image collections that require expert knowledge to gain insight, such publicly available databases and events cannot completely capture the difficulties an analyst may have to deal with.

3   Proposed Method

The foundation of our proposed method consists of features extracted using a convolutional neural network (CNN); followed by clustering; exploration by the analyst, assigning clusters and images to categories as an interplay between expert and machine; and using the insight gained through exploration for searching additional relevant items Fig. 1. Other work [1,3] has shown that a pretrained CNN has a sufficient number of useful features to differentiate between images of unknown and unique categories. Resnet152 [9] is the neural network we used to extract 2048 features per image.

Clustering. The features are used to generate clusters. This initial clustering helps the analyst in identifying structure and relations within the image collection. The clustering algorithm needs to be fast, so that the user can quickly adjust the clustering results; any user set parameter should be intuitive; and larger, high quality clusters are preferred for faster analysis and reducing user fatigue.

For initial experimentation we used K-means and DBSCAN as a baseline, as they are established clustering algorithms. We also used DESOM [6] (self-organizing map) [2] and DCEC (convolutional autoencoder) [8] as recent methods. Despite recommended hyperparameters as well as others, DCEC performed not much above chance and needs long training time, making it unsuitable. DESOM performed similar to k-means but slower, despite k-mean's already significant computation time. Due to limited space, DCEC and DESOM results have not been included. DBSCAN was by far the fastest. It requires the user to set a distance (threshold) between 0 and 1, which we find more intuitive than the number of clusters. Unfortunately, DBSCAN was too sensitive to the distance function used and would either mark most images as outliers, or put most images in a single cluster for any threshold between 0 and 1.

As none of the clustering algorithms meets all requirements, we developed Correlation-based Clustering (CC, Algorithm 1), loosely based on DBSCAN. A distance matrix is calculated for all image feature vector pairs using correlation (to us the most intuitive distance metric). Next, the image that correlates above threshold t with most other images is used as the start of the first cluster. The highest correlating image is first added to the cluster. Then, the cluster center is recalculated and correlations for all images are recalculated with respect to this new center. The highest correlating image is added to the cluster. This is repeated until no image correlates with the cluster center above t. Of the remaining images not in a cluster, the image with most other images correlating above t is the start of the second cluster. This repeats until all images are clustered or until no image has a correlation above t. In the latter case, t is lowered by 0.1 and the process repeats until no images remain. This process means clusters can have arbitrary shapes and that early clusters will be of high quality, while later clusters will be of lower quality. We did not look into optimizing memory usage of the distance matrix, but CC can be batched with an additional step of merging clusters of different batches. For the remainder of the paper, k-means and CC were used as the clustering algorithms. In our method the user is presented with the images of one cluster at a time on a basic scrollable grid. For CC, clusters are shown in order of creation. This means the analyst sees the highest quality clusters first. Images within the cluster are sorted from most to least similar to the first image. For k-means cluster order is random, and images are shown in order of distance to the cluster center.

ALGORITHM 1: Correlation-based clustering

Calculate correlation matrix
while *there are correlations between image pairs in Correlation Matrix* **do**
    if *max(Correlation Matrix) > threshold* **then**
        initialize new *cluster* with I_{most} // I_{most} is the image that is correlated with most other images above *threshold*
        center = feature vector of I_{most}
        while *max(Correlation(center, other images) > threshold* **do**
            add I_{max} to *cluster* // I_{max} is the image that has the highest correlation with *center*
            center = average feature vector of images in *cluster*
        end
    else
        decrease *threshold*
    end
end

Buckets. An important part of the method is the set of buckets, in which the analyst can place images belonging to a self defined category. These buckets can be created on the fly and may change over time, as the insight of the analyst changes. Creation of buckets is part of exploration, where the data model is not fixed.

Strategies. The user needs a strategy to go from knowing nothing about an image collection given some basic structure through the clustering algorithm, to a structured, categorized image collection. The fewer images a user has to see to attain a high recall, the better the strategy. Evaluating the results of a strategy in such a way implicitly takes precision into account as well. The most basic strategy is browsing clusters and assigning images or clusters to buckets. This can take a long time, as relevant images may be located in clusters shown last. We therefore propose several explore and search methods. Methods were chosen based on common explore and search tasks [28]: structuring, summarization, sorting and querying. These methods are combined into strategies.

An artificial (simulated) analyst (AA) is used to evaluate all strategies, as this is infeasible with human analysts. The AA has access to the ground truth (the annotations of the images in the image collection). The goal is to categorize all images in the image collection (m) into predetermined buckets (one bucket for each class in the annotations). Each strategy (S) is repeated for each bucket (B), resulting in a total of $S * B$ simulations per image collection. For each simulation, the image collection is divided into relevant images $RI_{0,1,...n}$ (the images belonging to the current B), and irrelevant images $II_{0,1...m-n}$ (all images not belonging in B). For each simulation, the AA will see each image once. The recall of S is calculated after each image the AA has seen and then averaged for all B per S. Success of a strategy is determined by how many images need to be visually inspected to reach 80% recall of all RI over all B. 80% was chosen as it is the majority of images in a category, and should give the analyst a clear idea of the relevant images.

Each strategy consists of five choices. The **_first choice_** is between k-means or CC. The **_second choice_** is to set the threshold t or k for k-means. The AA uses 0.3 and 0.7 for t when the first choice is CC. If the first choice is k-means, k is based on the number of clusters generated by CC using these thresholds. All t between 0 and 1 with steps of 0.1 have been tested, 0.3 and 0.7 were chosen to show the difference between high and low t due to limited space in the paper.

Exploration. We propose two exploration methods. The AA chooses whether to use them or not. Using the overview is the **_third choice._** The overview is a grid showing one representative image for each cluster to quickly get an idea of the contents of the image collection, and allows the analyst to select relevant clusters to inspect further. The representative image is the image whose feature vector is closest to the average feature vector of the cluster. The **_fourth choice_** is to use sorting. Once at least one image is added to a bucket, the analyst can sort all clusters based on similarity to that bucket, making it more likely that clusters with relevant images show up first. _Search_ There are several ways to search through an unannotated image collection. Most are based on similarity queries. The following is a list of search methods that we consider for our strategies and is the **_fifth choice._** The AA picks one or no search method. Search methods that rank images make use of the correlation matrix.

Expand Cluster/Bucket: A cluster with relevant images can be recalculated with a lower threshold, expanding the cluster with more images that may be relevant. Similarly, more images can be found using the images in a bucket. This can be repeated until no more relevant images are found. The AA reduces t by 0.2 for each expansion. If additional images found through the expansion contained at least 30% relevant images, expansion is repeated with a reduced threshold. _Query (external) image/bucket/part of image:_ query (part of) an image or the average feature vector of multiple images to rank all images from most similar to least similar to the queried image (the analyst selects which part of an image she is interested in if required). Each query is processed until 8 II in a row are encountered. For Query Image the AA selects 10 random images, for Query Bucket, the query is repeated if at least 10 RI have been added. Query Part of Image is not used by the AA, as it would require bounding box annotations that are not available. Query External Image (image not in the image collection) is also not used as it would require human judgment to select a useful external image. _Select from projection:_ the feature vectors are used to calculate a 2D representation of the features using UMAP [15]. The analyst can then select with a bounding box which images to view. For the AA, the UMAP 2D representation was normalized. The AA will calculate the modal image from the images in bucket within the 2D representation and draw a bounding box of 0.2 by 0.2, centered on the modal image. Images within the bounding box will then be assessed. Combining the baseline with these choices gives us our strategies. Figure 1 shows the complete method.

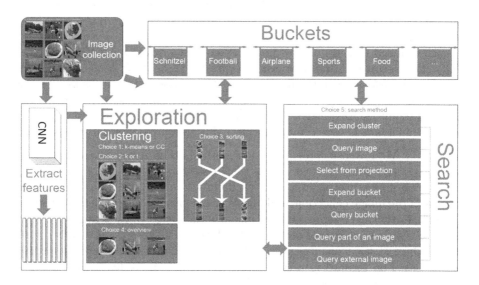

Fig. 1. Schematic overview of the method

Image Collections. To evaluate the strategies, this paper makes use of two image collections. The *MH17 image collection (MIC)*: the real-life accident investigation database [4] of the MH17 airplane crash. The MIC contains 14,579 images and contains classified information, thus only the first author, part of the DSB, had full access. Co-authors have seen examples of content. Images shown are from public sources. For evaluation, annotation used by the DSB was used to reflect the actual situation. Each image can be annotated with one or more of 27 categories; some examples are given in Fig. 2. *OID image collection (OIC)*: a selection of images from the Open Image Database (OID). OIC consists of 37 image categories with objects, locations and activities that were not present in the training set for the CNN. Only images verified by human annotators were used.

4   Evaluation and Discussion

Table 1 shows strategy 1 requires the analyst to assess 70–80% of all images to reach a recall of 80% for all categories. Adding a search method (strategies 2–6) can reduce this to around 50%. Adding sorting of clusters (strategies 7–12)

Table 1. Strategies and fraction of images the analyst has to assess to reach 80% recall on all categories. CC0.3 is CC with a threshold of 0.3, km0.3 is k-means with k equal to the number of clusters CC0.3 generated.

Strategy	Sorting	Overview	Search	Fraction of images seen at recall of 0.8							
				MH17				OID			
				CC0.3	CC0.7	km0.3	km0.7	CC0.3	CC0.7	km0.3	km0.7
1	No	No	None	0.68	0.74	0.80	0.82	0.85	0.85	0.81	0.79
2			Expand cluster	0.69	0.74	0.61	0.68	0.82	0.84	0.43	0.55
3			Query image	0.69	0.74	0.78	0.80	0.71	0.62	0.57	0.57
4			Projection	0.56	0.72	0.61	0.59	0.76	0.53	0.28	0.51
5			Expand bucket	0.67	0.74	0.70	0.82	0.78	0.84	0.56	0.77
6			Query bucket	0.68	0.74	0.78	0.80	0.81	0.67	0.68	0.45
7	Yes	No	None	0.58	0.66	0.57	0.55	0.62	0.51	0.24	0.29
8			Expand cluster	0.58	0.62	0.50	0.51	0.59	0.51	0.26	0.23
9			Query image	0.57	0.63	0.52	0.49	0.59	0.51	0.23	0.25
10			Projection	0.57	0.69	0.60	0.60	0.67	0.53	0.29	0.52
11			Expand bucket	0.56	0.66	0.57	0.55	0.59	0.51	0.23	0.28
12			Query bucket	0.54	0.66	0.71	0.55	0.59	0.50	0.23	0.17
13	No	Yes	None	0.60	0.54	0.49	0.60	0.62	0.36	0.21	0.13
14			Expand cluster	0.55	0.44	0.65	0.50	0.28	0.18	0.10	0.12
15			Query image	0.47	0.39	0.65	0.50	0.25	0.10	0.04	0.10
16			Projection	0.48	0.54	0.52	0.34	0.23	0.11	0.06	0.10
17			Expand bucket	0.62	0.42	0.63	0.60	0.56	0.35	0.11	0.12
18			Query bucket	0.40	0.28	0.55	0.47	0.52	0.10	0.06	0.09
19	Yes	Yes	None	0.46	0.28	0.26	0.25	0.15	0.10	0.05	0.09
20			Expand cluster	0.30	0.30	0.43	0.36	0.19	0.10	0.9	0.10
21			Query image	0.42	0.26	0.30	0.24	0.08	0.10	0.04	0.09
22			Projection	0.46	0.33	0.33	0.30	0.14	0.11	0.06	0.10
23			Expand bucket	0.52	0.28	0.40	0.25	0.15	0.10	0.07	0.09
24			Query bucket	0.33	0.24	0.26	**0.23**	0.09	0.10	**0.04**	0.09

Fuselage

Main landing gear

Vertical Tail

Fig. 2. Some examples of the categories in the MH17 image collection

reduces the number of images that have to be assessed and likely results in a better browsing experience. Using the overview (strategies 13–18) has a stronger impact. Combining all choices (strategies 19–24) means the user only has to assess about a quarter of the MIC to find 80% of all images of a specific category, with querying the bucket being the best search method. For OIC the user only

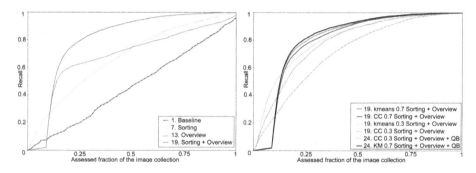

Fig. 3. Recall vs. # of images assessed of strategies employed on the MIC. (left) Exploration strategies are shown, using k-means with $k = 1221$, which has the best performance on the MIC as shown in Table 1. (right) The difference between k-means and CC, the effects of high and low thresholds, and the effect of the best performing search method. The graph of strategies using the overview have a distinct shape at the start; this is where the user assesses the representative images of the clusters.

has to assess 4% of the image collection using strategy 24. The table shows the MIC is more difficult than the OIC, with users having to assess a fraction at least 5 times as large to reach the same results. While k-means performs better than CC on OIC, the difference on the MIC is minimal and depends on the applied strategy. Figure 3 shows the complete curve of recall versus images assessed.

Table 1 shows that increasing the number of clusters achieves better results for MIC, as smaller clusters are generally more precise. However, when using strategy 13 with $k \geq 2000$ for k-means recall will increase less fast than with fewer than 2000 clusters, because the overview requires the user to view a representative image, which counts towards the images seen. And while the AA does not get tired of assessing many small clusters, it is our experience that for a human analyst this increases fatigue compared to fewer, larger clusters.

Clustering 10,000 images on an Intel Xeon E3-1505M v5 using CC ($t = 0.5$) takes 12.7 s (2.6 s for the matrix and 10.1 s for clustering), resulting in 449 clusters. The Scikit implementation of k-means took approximately 349.2 s with $k = 449$.

4.1   User Experiments

To validate whether the results obtained with our method through the AA are actually useful for human analysts, we performed two user experiments. One with a group of domain experts, another with a group of Forensic Science students (to-be domain experts).

Domain Expert User Experiment. The main goal of the domain expert user experiment is to find out whether the experts think the clusters provide value for their work. The strategy used in this user experiment resembles mostly strategy 1. An application was built around the method (see footnotes). Four accident investigators from the DSB were given 2 hours to organize the MIC as they would in an investigation. Two users were blindly assigned clustering results for the MIC from CC and two users k-means. t for CC was set to 0.5 resulting in 1150 clusters, which was also used as k for k-means. Consensus among expert users was that the method as implemented in the application worked, that clusters were of usable and of high quality, and that it would significantly increase their efficiency when working with large image collections. In most cases it was clear why certain images were in a cluster, giving confidence in the method.

Students User Experiment. The second user experiment involved 16 Forensic Science students who had access all explore and search functions in Sect. 3. In pairs they were asked to perform 3 tasks in 1 hour with a written report about their approaches.

Find Image of Old Blue Car. Strategies: Query an external image they found on the internet of a blue car; use the overview to find a cluster with cars, then query images with cars to find the blue car; find an image of a wheel and query that part of the image to find the blue car. A group using the external image query was the fastest to find the image.

Find 40 Images That Best Summarizes the Image Collection. Evaluated by measuring the minimum distance of the features of all images to the features of the 40 selected images. Strategies: change t such that only 40 clusters were generated, then choose 1 image of each cluster; look through the overview from the default t and pick 40 images; browse through clusters and select 40 images. Adjusting t gave the best results.

Find the Food Item Most Common in the Image Collection and Find the Most Images with This Item. 8 different food items were present. 4 of 6 groups found the right food group. Strategies: all used the overview to find food clusters and divided them over several buckets. Finding more items was done through querying individual items or buckets. Best group recalled 83% with a false positive rate of 0.1. Groups not identifying the correct food item both chose the second most common food item.

The experiment showed that the proposed method was fairly intuitive to use, since all groups managed to use several explore and search functions and combine them to complete the tasks.

5 Conclusion

In this paper, a method for analyzing image collections is proposed to help expert analysts. The method covers the exploration-search axis, allowing the analyst to quickly find relevant images in an image collections containing unique, never seen before image categories. For this, the analyst only needs to set a single intuitive parameter. A clustering algorithm (CC) was designed to meet the criteria of serving the expert user at interactive speed, with priority on showing large and high quality clusters. Quantitative results are similar to k-means, with much lower computation time. We show how effective explore and search strategies can greatly reduce the number of images the analyst needs to see in order to find relevant images. Through the user experiment it is demonstrated that with combining expert knowledge, computer vision and the right strategy, the analyst has the tools needed to face the challenges posed by the ever increasing amount of data in a complex environment, such as a real life accident investigation database.

References

1. Babenko, A., Slesarev, A., Chigorin, A., Lempitsky, V.: Neural codes for image retrieval. In: Fleet, D., Pajdla, T., Schiele, B., Tuytelaars, T. (eds.) ECCV 2014. LNCS, vol. 8689, pp. 584–599. Springer, Cham (2014). https://doi.org/10.1007/978-3-319-10590-1_38
2. Barthel, K.U., Hezel, N.: Visually exploring millions of images using image maps and graphs, pp. 251–275. John Wiley and Sons Inc. (2019)
3. Caron, M., Bojanowski, P., Joulin, A., Douze, M.: Deep clustering for unsupervised learning of visual features. In: European Conference on Computer Vision (2018)
4. Dutch Safety Board: Investigation crash mh17, 17 July 2014, October 2015
5. Fei-Fei, L., Fergus, R., Perona, P.: One-shot learning of object categories. IEEE Trans. Pattern Anal. Mach. Intell. **28**, 594–611 (2006)
6. Forest, F., Lebbah, M., Azzag, H., Lacaille, J.: Deep embedded SOM: joint representation learning and self-organization. In: ESANN 2019 - Proceedings, April 2019
7. Gasser, R., Rossetto, L., Schuldt, H.: Multimodal multimedia retrieval with Vitrivr. In: Proceedings of the 2019 on International Conference on Multimedia Retrieval, ICMR 2019, pp. 391–394. Association for Computing Machinery, New York (2019)
8. Guo, X., Liu, X., Zhu, E., Yin, J.: Deep clustering with convolutional autoencoders. In: Liu, D., Xie, S., Li, Y., Zhao, D., El-Alfy, E.S. (eds.) ICONIP 2017. LNCS, vol. 10635, pp. 373–382. Springer, Cham (2017). https://doi.org/10.1007/978-3-319-70096-0_39
9. He, K., Zhang, X., Ren, S., Sun, J.: Deep residual learning for image recognition. In: 2016 IEEE Conference on Computer Vision and Pattern Recognition (CVPR), pp. 770–778, June 2016
10. Hezel, N., Barthel, K.U., Jung, K.: ImageX - explore and search local/private images. In: Schoeffmann, K., et al. (eds.) MMM 2018. LNCS, vol. 10705, pp. 372–376. Springer, Cham (2018). https://doi.org/10.1007/978-3-319-73600-6_35
11. Krasin, I., et al.: OpenImages: a public dataset for large-scale multi-label and multi-class image classification (2017). https://github.com/openimages
12. Kratochvíl, M., Veselý, P., Mejzlík, F., Lokoč, J.: SOM-hunter: video browsing with relevance-to-SOM feedback loop. In: Ro, Y.M., et al. (eds.) MMM 2020. LNCS, vol. 11962, pp. 790–795. Springer, Cham (2020). https://doi.org/10.1007/978-3-030-37734-2_71
13. Leibetseder, A., et al.: LifeXplore at the lifelog search challenge 2019. In: Proceedings of the ACM Workshop on Lifelog Search Challenge, pp. 13–17. Association for Computing Machinery, New York (2019)
14. Liu, C., et al.: Progressive neural architecture search. In: Ferrari, V., Hebert, M., Sminchisescu, C., Weiss, Y. (eds.) ECCV 2018. LNCS, vol. 11205, pp. 19–35. Springer, Cham (2018). https://doi.org/10.1007/978-3-030-01246-5_2

15. McInnes, L., Healy, J., Melville, J.: UMAP: uniform manifold approximation and projection for dimension reduction (2018)

16. de Rooij, O., van Wijk, J.J., Worring, M.: MediaTable: interactive categorization of multimedia collections. IEEE Comput. Graph. Appl. **30**(5), 42–51 (2010)

17. Russakovsky, O., Deng, J., Su, H., Krause, J., Satheesh, S., Ma, S., et al.: Imagenet large scale visual recognition challenge. Int. J. Comput. Vis. **115**(3), 211–252 (2015)

18. Schoeffmann, K.: Video browser showdown 2012–2019: a review. In: 2019 International Conference on Content-Based Multimedia Indexing (CBMI), pp. 1–4 (2019)

19. Settles, B.: Active learning literature survey. Computer Sciences Technical report 1648, University of Wisconsin-Madison (2009)

20. Sun, Q., Liu, Y., Chua, T.S., Schiele, B.: Meta-transfer learning for few-shot learning. In: The IEEE Conference on Computer Vision and Pattern Recognition (CVPR), June 2019

21. Touvron, H., Vedaldi, A., Douze, M., Jégou, H.: Fixing the train-test resolution discrepancy. In: Advances in Neural Information Processing Systems (NeurIPS) (2019)

22. Wang, Y., Chao, W.L., Weinberger, K.Q., van der Maaten, L.: SimpleShot: revisiting nearest-neighbor classification for few-shot learning (2019)

23. Worring, M., Engl, A., Smeria, C.: A multimedia analytics framework for browsing image collections in digital forensics. In: Proceedings of the 20th ACM International Conference on Multimedia, MM 2012, pp. 289–298. ACM, New York (2012)

24. Yan, M.: Adaptive learning knowledge networks for few-shot learning. IEEE Access **7**, 119041–119051 (2019)

25. Yang, G., Liu, J., Xu, J., Li, X.: Dissimilarity representation learning for generalized zero-shot recognition. In: Proceedings of the 26th ACM International Conference on Multimedia, MM 2018, pp. 2032–2039. Association for Computing Machinery, New York (2018)

26. Yosinski, J., Clune, J., Bengio, Y., Lipson, H.: How transferable are features in deep neural networks? In: Proceedings of the 27th International Conference on Neural Information Processing Systems, NIPS 2014, vol. 2. pp. 3320–3328. MIT Press, Cambridge (2014)

27. Zahálka, J., Worring, M.: Towards interactive, intelligent, and integrated multimedia analytics. In: 2014 IEEE Conference on Visual Analytics Science and Technology (VAST), pp. 3–12, October 2014

28. Zahálka, J., Rudinac, S., Worring, M.: Analytic quality: evaluation of performance and insight in multimedia collection analysis. In: Proceedings of the 23rd ACM International Conference on Multimedia, MM 2015, pp. 231–240. ACM, New York (2015)

29. Zhang, Z., Saligrama, V.: Zero-shot learning via joint latent similarity embedding. In: 2016 IEEE Conference on Computer Vision and Pattern Recognition (CVPR), pp. 6034–6042, June 2016

30. Zhou, X.S., Huang, T.S.: Relevance feedback in image retrieval: a comprehensive review. Multimedia Syst. **8**(6), 536–544 (2003)

Graph-Based Indexing and Retrieval
of Lifelog Data

Manh-Duy Nguyen[1]([✉]), Binh T. Nguyen[2,3,4], and Cathal Gurrin[1]

[1] Dublin City University, Dublin, Ireland
manh.nguyen5@mail.dce.ie
[2] AISIA Research Lab, Ho Chi Minh City, Vietnam
[3] University of Science, Ho Chi Minh City, Vietnam
[4] Vietnam National University, Ho Chi Minh City, Vietnam

Abstract. Understanding the relationship between objects in an image is an important challenge because it can help to describe actions in the image. In this paper, a graphical data structure, named "Scene Graph", is utilized to represent an encoded informative visual relationship graph for an image, which we suggest has a wide range of potential applications. This scene graph is applied and tested in the popular domain of lifelogs, and specifically in the challenge of known-item retrieval from lifelogs. In this work, every lifelog image is represented by a scene graph, and at retrieval time, this scene graph is compared with the semantic graph, parsed from a textual query. The result is combined with location or date information to determine the matching items. The experiment shows that this technique can outperform a conventional method.

Keywords: Lifelog · Scene graph · Information retrieval

1 Introduction

As explained in [15], a lifelog is a digital archive gathered by an individual reflecting their real-world life experiences. Lifelogs are typically media-rich, comprising digital images, documents, activities, biometrics, and many other data sources. Such lifelogs have been deployed for many use-cases, such as dietary monitoring [4], memory assistance [10], epidemiological studies [27] and marketing analytics [17]. Regardless of the application, a basic underlying technology is a retrieval mechanism to facilitate content-based access to lifelog items. Research into lifelogging has been gaining in popularity with many collaborative benchmarking workshops taking place recently - the NTCIR Lifelog task [12], the Lifelog Search Challenge (LSC) [13] and the ImageCLEFlifelog [23]. In all of these activities, the query process is similar; a textual query is provided, which acts as the information need for a new generation of retrieval engines that operate over multimodal lifelogs.

In this paper, we introduce a new approach to lifelog retrieval by utilizing a scene graph data structure [18] as the primary indexing mechanism, which

© Springer Nature Switzerland AG 2021
J. Lokoč et al. (Eds.): MMM 2021, LNCS 12573, pp. 256–267, 2021.
https://doi.org/10.1007/978-3-030-67835-7_22

could represent both the objects visible in lifelogging images and the interactions between the objects. Textual user queries are mapped into the same graph space to be compared with the scene graph generated in the previous step to produce the ranked results. Non-visual lifelog data is integrated to support faceted filtering over the generated ranked list. In our experiments, the proposed system and a baseline were evaluated by eight volunteers in an interactive retrieval experiment. We highlight that this paper's contribution is a first lifelog retrieval system to index the lifelog data in a graph-space and map a textual query to the graph space to facilitate similarity calculations. The original query dataset and the experiment design for the lifelog retrieval are also introduced. To facilitate repeatable science, we release our code for community use[1] and we evaluate using accessible datasets.

2   Related Work

Many interactive lifelog retrieval systems have been proposed in recent years, with MyLifeBits [11] being one of the pioneers, which considered the lifelog retrieval problem as a application of database inquiry [15]. Many novel retrieval approaches followed MyLifeBits, such as Doherty et al. [7], who build a linkage graph for lifelog events and presented a basic interactive browsing system. Lemore [24] was an early interactive retrieval system which enriched lifelog images by incorporating object labels and facilitated retrieval via textual descriptive queries. Recently, many systems also followed this idea by annotating images with detected items and their semantic concepts in the corresponding metadata. Some of them used this textual information as a filter mechanism to enhance retrieval results produced with a visual-based input sketched by users [16,21]. The number of matching concepts between a query and the annotation of images can be used as a ranking criterion in retrieval systems [8]. With different considerations, Myscéal [28] viewed this as a document retrieval problem by indexing textual annotations and matching with textual queries. Embedding techniques are also commonly based on the idea of encoding concepts from both queries and images tags into the same vector space to calculate the similarity between them [19,20]. Regarding using graphs, LifeGraph [25] applied knowledge graph structure with the nodes representing detected things or scenes recognized in images. These entities can be linked with corresponding images and external sources to expand the information with frequent activities and relevant objects.

Generally, all the above systems do not focus on the interaction between objects in the lifelog data or the query. Some systems did make progress by encoding the entire textual input or generating captions for lifelogging images to describe activities appearing in them [30,31]. However, these ideas did not focus on the association between objects in lifelog images. Some approaches have been proposed to describe visual relations within an image, such as [2,5]. It was not until the scene graph structure [18] was introduced that there was

---

[1] To ensure repeatability, our code is publicly available on GitHub for references: https://github.com/m2man/MMM2021-LifelogRetrieval.

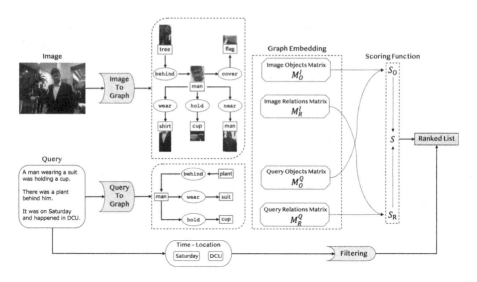

Fig. 1. The overview of our indexing approach for lifelog data and lifelog queries for retrieval purposes. Each lifelog image was converted to a graph, followed by an embedding stage to be stored as an object and relation matrices. A query was then parsed to a graph and encoded to matrices, which would be compared to those of each image by the scoring function. The result, combined with the filtering mechanism of time and location, was returned as a ranked list.

a clear and comprehensive solution to express object relationships in an image and initiate an interesting field for the research community [3]. The proposed graph structure can represent an image as a directed graph comprising nodes and edges where nodes describe objects appearing in the image, and edges indicate the relationship between objects. Many studies have tried applying scene graphs in image retrieval and achieved better results compared to using objects features only [18,26].

In this paper, we address the lifelog retrieval challenge by indexing both images and textual queries as graphs, as depicted in Fig. 1. We then ranked the matching results based on the similarity between these graphs. Given that we work with multimodal lifelogs, the graph matching process's outcome could be filtered by other information, such as geolocation or time, which are automatically extracted from the query. This approach facilitated the capture of the interactions between objects in images and the comparison of them with those described in the textual input. Eight users evaluated the proposed method in an experiment comparing the proposed graph-based approach with a recent baseline method using visual concept indexing. For this experiment, we used the LSC'18/19 dataset [14], and we created a new set of twenty semantic queries, including ten randomly chosen topics from the LSC'19 dataset (representing conventional lifelog queries) and ten manually created topics that focus on visually describing a known-item from a lifelog. It is noticeable that [6] also followed the concept of using a scene graph for lifelogging visual data. However, this system used such a graph as a supplement to the retrieval process and did not consider a query as a graph like our proposed method.

3   Dataset

The lifelog data we used is the official data provided by the recent LSC'18 and 19 [1] comparative benchmarking challenges, which incorporated multimodal lifelog data from a single lifelogger who wore a small camera that passively captured images at the resolution of 1024 × 768 every 30 s for 27 days, leading to the collection of more than 40, 000 images. All identifiable information in the dataset was removed by blurring faces and readable textual content. The data also came with the biometric data (heart rate, galvanic skin response, etc.), physical activities (standing, walking, etc.), and GPS location along with its timestamps. We currently used the visual data with its location and date for this work, though future research will incorporate more aspects of the dataset.

Each of the twenty queries represents a lifelogger's textual description to recall a specific moment that happened during one particular time covered by the test collection. The result of a topic could be a single image or a sequence of images. An example of a topic, noted as LSC31, is *"[LSC31] Eating fishcakes, bread, and salad after preparing my presentation in PowerPoint. It must have been lunchtime. There was a guy in a blue sweater. I think there were phones on the table. After lunch, I made a coffee."*. Additionally, we also built ten new topics that better describe the information need in terms of visual relationships, which we call Descriptive Interaction Topics (DITs). A sample DIT query, named DIT02, is *"[DIT02] I was eating a pizza. My hand was holding a pizza. There was a guy wearing a pink shirt talking to me. There was a black box on a table. It was on Friday morning. It happened at my workplace"*. The answers to those topics can be illustrated in Fig. 2. In our experimental analysis, we report separately on the results using both types of queries.

4   Graph Generation

As our system aimed to solve the interaction between objects within an image and a semantic query by using a scene graph, it raised a challenge of how to represent these two distinct types of data into a standard graph structure.

4.1   Image to Graph

Although there are many methods for generating a scene graph for an given image, Neural Motifs [33] was chosen due to its accurate performance [3]. A predicted scene graph, noted as G, contains a set O indicating detected objects in the image, with its corresponding set of bounding box B, and the set of visual relations R where:

(a) Sample result for a LSC31   (b) Sample result for DIT02

Fig. 2. Sample results for the example queries for the two mentioned queries. A result for a single topic could contain more than one images.

- $O = \{o_1, ..., o_m\}$: m labels of recognised objects in the image. Each object o_i was a single node in the graph G.
- $B = \{b_1, ..., b_m\}$: m bounding boxes of O respectively in which

$$b_i = \{x_i, y_i, width, height\} \in \mathbb{R}^4$$

where (x_i, y_i) is the top-left coordinates of the object $o_i \in O$.
- $R = \{r1, ..., r_n\}$: n detected relationships between objects. Each r_k is a triplet association of a start object $o_i \in O$, a end object $o_j \in O$, and a predicate $p_{i \rightarrow j} \in \mathcal{P}$ where $\mathcal{P}$ is a set including all labels of predicates in the Visual Genome. These relations could be considered as edges in G.

All elements in each set are assigned with their confidence score after running the Neural Motif model. We firstly remove inaccurate prediction by setting a threshold for object and relation. To expand the graph to obtain more interaction information in the image that could be not entirely captured by the model, we then create a fully connected graph of G, called G_{fc}, in which there was an edge connecting any two nodes. G_{fc} can be obtained by building missing edges in G with the procedure of visual dependency representations (VDR) [9] based on the bounding box set B. The starting node and the ending node of the constructed edge can be decided based on their predicted scores, whose higher score would be the subject and lower was the object. One drawback of this G_{fc} is that there are many noisy and unimportant relations since not all objects correlated to others. We apply Maximum Spanning Tree on the graph G_{fc} to remove the least meaningful edges with low weights to get G_{mst}. The weight of an edge is the sum of both nodes' scores and that of the predicate connecting them. The score of a predicate could be the score of r_j if this edge was in G or 0 if it was created by the VDR. It is worth noting that we only filter out edges from VDR and retain the original relations in G. In general, the expansion stages of getting the G_{mst} was to enlarge the set R and left two sets O and B intact. The entire process can be illustrated in Fig. 3.

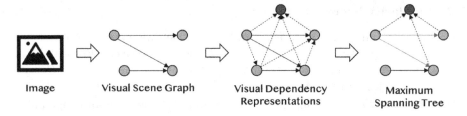

Fig. 3. Image To Graph Pipeline. A visual scene graph was firstly generated from an input image by Neural Motif [33] in which a blue node and an arrow represent an object and a relation respectively. The VDR [9] was applied to generate a fully connected graph with new relations illustrated by dotted arrows. A green node depicted a detected object in the image but not included in the scene graph. Finally, the maximum spanning tree process removed undesirable predicates. Red arrows show relations discarded by the tree but still kept as they were originally in the scene graph. (Color figure online)

4.2 Query to Graph

Besides images, a textual query may also contain several objects and the relations between them. For our proposed process, it is vital to have a graph describing the context of the topic. We applied the rule-based approach proposed in [26] to generate a semantic graph representing query items and their interactions, as described in the query text. Before this, any location and time query information can be extracted and archived for the later filtering mechanism by analyzing part-of-speech tagging from the topic, as utilized in Myscéal [28]. All words from the query are pre-processed to exclude stopwords and then lemmatized.

5   Image Retrieval

After building the graph structure for both images and a query, the retrieval problem became how to calculate the similarity between the semantic graph of the topic with the set of scene graphs of lifelogging photos. This section will describe how a graph is embedded and, based on that, find the similarity score.

5.1   Graph Embedding

Recall that a scene graph of an image G_{mst} has a set of detected objects $O = \{o_1, ..., o_m\}$ and the expanded relation set $R = \{r1, ..., r_k\}$ gained after the spanning process. We now represent the graph in two embedded matrices: M_O^I and M_R^I describing the objects and relations information accordingly. $M_O^I \in \mathbb{R}^{m \times d}$ created with each row is the d-dimension feature vector of a label name of the corresponding object encoded by the Word2Vec [22] model ($d = 300$). Similarly, $M_R^I \in \mathbb{R}^{k \times 3d}$ can be obtained with each row, which is a concatenated embedded vector of a subject, predicate, and an object in the relation. With the same method, a description query can also be encoded into M_O^Q and M_R^Q.

5.2   Similarity Score

Our similarity function is adopted from [32] in which we match the object and relation matrices of a text to those of an image, respectively. Regarding the object matrices, we take the score of the most relevant object in the M_O^I with each of the objects in the M_O^Q. After that, we get the average for the object's similarity score S_O. Assuming there are n_O^I and n_O^Q objects in an image and a query, the S_O can be calculated as:

$$S_O = \frac{1}{n_O^Q} \sum_{t=1}^{n_O^Q} MaxRow[M_O^Q * Transpose(M_O^I)], \tag{1}$$

where $*$ is the normal matrix multiplication, $MaxRow(X)$ is the function to calculate the highest value of each row of a matrix X, and $Transpose(X)$ is the matrix operation to find X^T. Likewise, suppose that there are n_R^I and n_R^Q relations detected in an image and a query, the relation similarity scores S_R can be measured as follows:

$$S_R = \frac{1}{n_R^Q} \sum_{t=1}^{n_R^Q} MaxRow[M_R^Q * Transpose(M_R^I)] \tag{2}$$

Finally, the similarity score between two graphs, S, can be defined as $S = \alpha * S_O + \beta * S_R$ where α, β are obtained from empirical experimentation.

6   Experiments

Since this graph version only focused on indexing data as graphs and applying graph operations, we currently do not take the temporal retrieval issue into account. The baseline for the comparison was the modified version of the Myscéal system [29] that had been used in the ImageCLEFlifelog2020 benchmarking workshop and achieved third place out of six participants [23]. We chose this system as the baseline because the top two teams have not released their code at the time we were doing this research. The baseline utilised a standard design of a typical lifelog retrieval system that facilitates textual queries and generates a ranked list by utilizing a scoring function inspired by the TF-IDF. Both baseline and proposed methods were configured to return the top 100 images matching a given query. The users could revise their inquiries until they thought the answers were on the list. There was a time limit of five minutes for a volunteer to solve a single query in each system. The users were trained on each system using several sample queries before the official experiment.

As mentioned in Sect. 3, there were a total of twenty queries used in the experiment, which were divided into four smaller runs, namely A, B, C, and D, with each run containing five topics from either the LSC or DIT types. Eight volunteers were asked to perform two runs, one for each lifelog retrieval system. It means that a single volunteer would use a system to do five queries and then use the other system to find the answers to another five queries. To avoid any potential learning bias between the first and second runs, we designed the experiment according to Table 1. With this configuration, we could ensure that each setting's pair would be performed twice with different orders of the systems used to do the retrieval. For example, the couple of A-B experiment was done two times with User 1 and User 7 in a distinct context. While User 1 did run A with the baseline first followed by run B with the proposed system, User 7 used the proposed system for run A before doing run B with the baseline. This configuration allowed the entire query set to be executed twice on each system.

Table 1. The assignment of query subsets and systems for each user in our experiment in which A, B, C, D were our 4 runs (groups of five topics).

	Baseline	Proposed		Proposed	Baseline
User 1	A	B	User 5	C	D
User 2	B	C	User 6	D	A
User 3	C	D	User 7	A	B
User 4	D	A	User 8	B	C

7   Results and Discussion

To evaluate the retrieval system's accuracy, we used the Mean Reciprocal Rank (MRR) on the top 100 images found by the systems' users within the experimental timeframes. We chose this metric because it is sensitive to the ranking position, which was also the main criterion in our assessment. We illustrate the scores in Table 2. The graph-based method achieved a higher result (MRR of 0.28) compared to 0.15 for the concept-based system by considering all queries. By examining specific query types, the graph technique also obtained better scores. Due to a competitive MMR on DIT queries from both systems, the proposed method surpassed the baseline with the MMR of 0.41 and 0.2, respectively. The proposed system also got a higher score of 0.15 compared to that of the baseline with 0.1. It can be seen that both methods performed better on DIT topics than LSC topics. This might be because the DIT described the lifelog events in more detail than those in LSC as they had more objects and interactions in the queries. However, there was only a minuscule change in the baseline with an increase of 0.1 in the metric. In contrast, the graph-based retrieval engine witnessed an increase in the scores between two types of topics since this technique could capture the relationships between objects in the query and images, which was the critical point in the DIT set. The MMR of this system on DIT was nearly three times higher than LSC, which were 0.41 and 0.15 accordingly. Figure 4 illustrated the result of both systems for the DIT02 topic.

Fig. 4. Top 5 images of the baseline (above) and the proposed system (bottom) for the DIT02 query. The correct answer is marked with the red boundary. (Color figure online)

Table 2. Mean Reciprocal Rank scores of 2 systems on each type of queries and entire dataset.

	LSC	DIT	Entire
Baseline	0.1087	0.2027	0.1557
Proposed	**0.1548**	**0.4166**	**0.2857**

Figure 5 depicts the distribution of reciprocal rank on every query. It was interesting that the baseline system showed less variance than the proposed approach. The variance in the latter system became stronger for DIT topics. It might indicate that the new system was not easy to use as the baseline, making its scores fluctuate between users and queries. The parsing of a query into a graph stage could be the reason. As this step required users to input description in a certain format to fully catch the relations in a query, the volunteers needed to have more time to get familiar with using the graph-based system most efficiently.

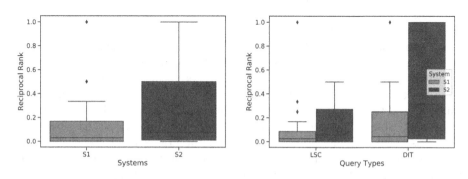

Fig. 5. The distribution of reciprocal rank of each query in overall (left) and on each query type (right) of two systems. S1 and S2 were the baseline and the proposed method accordingly.

8   Conclusion

In this paper, we employ a new perspective on the challenge of lifelog retrieval, where we transform into the graphs similarity matter. We applied graph indexing techniques in which lifelog images and queries are transformed into graphs, which were encoded into matrices in later stages, to capture visual relations between objects, hence improving the retrieved result's accuracy. We designed the experiments to evaluate our approach and compared it with an object-based baseline system with specific settings to reduce the bias of users' behaviors. The experimental results show that the graph-based retrieval approach outperformed the conventional method on both queries focusing on relationships and the ordinary topics used in the official lifelog search competition. Although there are some drawbacks, we believe that using relation graphs in the lifelog challenge is promising in this compelling field, especially when visual data of image contents are integrated into the graph structure. It poses an interesting avenue for future research on the topic of lifelog retrieval and related fields.

Acknowledgments. This publication has emanated from research supported in party by research grants from Science Foundation Ireland under grant numbers SFI/12/RC/2289, SFI/13/RC/2106, and 18/CRT/6223.

References

1. LSC 2019: Proceedings of the ACM Workshop on Lifelog Search Challenge. Association for Computing Machinery, New York (2019)
2. Aditya, S., Yang, Y., Baral, C., Fermuller, C., Aloimonos, Y.: From images to sentences through scene description graphs using commonsense reasoning and knowledge. arXiv preprint arXiv:1511.03292 (2015)
3. Agarwal, A., Mangal, A., et al.: Visual relationship detection using scene graphs: a survey. arXiv preprint arXiv:2005.08045 (2020)
4. Aizawa, K., Maruyama, Y., Li, H., Morikawa, C.: Food balance estimation by using personal dietary tendencies in a multimedia food log. IEEE Trans. Multimed. **15**(8), 2176–2185 (2013)
5. Aksoy, E.E., Abramov, A., Wörgötter, F., Dellen, B.: Categorizing object-action relations from semantic scene graphs. In: 2010 IEEE International Conference on Robotics and Automation, pp. 398–405. IEEE (2010)
6. Chu, T.T., Chang, C.C., Yen, A.Z., Huang, H.H., Chen, H.H.: Multimodal retrieval through relations between subjects and objects in lifelog images. In: Proceedings of the Third Annual Workshop on Lifelog Search Challenge, LSC 2020, pp. 51–55. Association for Computing Machinery, New York (2020)
7. Doherty, A.R., et al.: Experiences of aiding autobiographical memory using the SenseCam. Hum.-Comput. Interact. **27**(1–2), 151–174 (2012)
8. Duane, A., Gurrin, C., Huerst, W.: Virtual reality lifelog explorer: lifelog search challenge at ACM ICMR 2018. In: Proceedings of the 2018 ACM Workshop on the Lifelog Search Challenge, LSC 2018, pp. 20–23. Association for Computing Machinery, New York (2018)

9. Elliott, D., Keller, F.: Image description using visual dependency representations. In: Proceedings of the 2013 Conference on Empirical Methods in Natural Language Processing, pp. 1292–1302 (2013)

10. Gelonch, O., et al.: Acceptability of a lifelogging wearable camera in older adults with mild cognitive impairment: a mixed-method study. BMC Geriatr. 19(1), 110 (2019)

11. Gemmell, J., Bell, G., Lueder, R.: MyLifeBits: fulfilling the Memex vision. In: Proceedings of the Tenth ACM International Conference on Multimedia, pp. 235–238 (2002)

12. Gurrin, C., Joho, H., Hopfgartner, F., Zhou, L., Albatal, R.: NTCIR lifelog: the first test collection for lifelog research. In: Proceedings of the 39th International ACM SIGIR Conference on Research and Development in Information Retrieval, pp. 705–708 (2016)

13. Gurrin, C., et al.: Comparing approaches to interactive lifelog search at the lifelog search challenge (LSC2018). ITE Trans. Media Technol. Appl. 7(2), 46–59 (2019)

14. Gurrin, C., et al.: A test collection for interactive lifelog retrieval. In: Kompatsiaris, I., Huet, B., Mezaris, V., Gurrin, C., Cheng, W.-H., Vrochidis, S. (eds.) MMM 2019. LNCS, vol. 11295, pp. 312–324. Springer, Cham (2019). https://doi.org/10.1007/978-3-030-05710-7_26

15. Gurrin, C., Smeaton, A.F., Doherty, A.R., et al.: Lifelogging: personal big data. Found. Trends® Inf. Retr. 8(1), 1–125 (2014)

16. Heller, S., Amiri Parian, M., Gasser, R., Sauter, L., Schuldt, H.: Interactive lifelog retrieval with vitrivr. In: Proceedings of the Third Annual Workshop on Lifelog Search Challenge, LSC 2020, pp. 1–6. Association for Computing Machinery, New York (2020)

17. Hughes, M., Newman, E., Smeaton, A.F., O'Connor, N.E.: A lifelogging approach to automated market research (2012)

18. Johnson, J., et al.: Image retrieval using scene graphs. In: Proceedings of the IEEE Conference on Computer Vision and Pattern Recognition, pp. 3668–3678 (2015)

19. Kovalčík, G., Škrhak, V., Souček, T., Lokoč, J.: Viret tool with advanced visual browsing and feedback. In: Proceedings of the Third Annual Workshop on Lifelog Search Challenge, LSC 2020, pp. 63–66. Association for Computing Machinery, New York (2020)

20. Le, T.K., et al.: Lifeseeker 2.0: interactive lifelog search engine at LSC 2020. In: Proceedings of the Third Annual Workshop on Lifelog Search Challenge, LSC 2020, pp. 57–62. Association for Computing Machinery, New York (2020)

21. Leibetseder, A., Schoeffmann, K.: Lifexplore at the lifelog search challenge 2020. In: Proceedings of the Third Annual Workshop on Lifelog Search Challenge, LSC 2020, pp. 37–42. Association for Computing Machinery, New York (2020)

22. Mikolov, T., Sutskever, I., Chen, K., Corrado, G.S., Dean, J.: Distributed representations of words and phrases and their compositionality. In: Advances in Neural Information Processing Systems, pp. 3111–3119 (2013)

23. Ninh, V.T., et al.: Overview of ImageCLEF lifelog 2020: lifelog moment retrieval and sport performance lifelog. In: CLEF2020 Working Notes. CEUR Workshop Proceedings, CEUR-WS.org, Thessaloniki, 22–25 September 2020. http://ceur-ws.org

24. de Oliveira Barra, G., Cartas Ayala, A., Bolaños, M., Dimiccoli, M., Giró Nieto, X., Radeva, P.: LEMoRe: a lifelog engine for moments retrieval at the NTCIR-lifelog LSAT task. In: Proceedings of the 12th NTCIR Conference on Evaluation of Information Access Technologies (2016)

25. Rossetto, L., Baumgartner, M., Ashena, N., Ruosch, F., Pernischová, R., Bernstein, A.: Lifegraph: a knowledge graph for lifelogs. In: Proceedings of the Third Annual Workshop on Lifelog Search Challenge, LSC 2020, pp. 13–17. Association for Computing Machinery, New York (2020)
26. Schuster, S., Krishna, R., Chang, A., Fei-Fei, L., Manning, C.D.: Generating semantically precise scene graphs from textual descriptions for improved image retrieval. In: Workshop on Vision and Language (VL15). Association for Computational Linguistics, Lisbon (September 2015)
27. Signal, L., et al.: Kids'Cam: an objective methodology to study the world in which children live. Am. J. Prev. Med. **53**(3), e89–e95 (2017)
28. Tran, L.D., Nguyen, M.D., Binh, N.T., Lee, H., Gurrin, C.: Myscéal: an experimental interactive lifelog retrieval system for LSC'20. In: Proceedings of the Third Annual Workshop on Lifelog Search Challenge, pp. 23–28 (2020)
29. Tran, L.D., Nguyen, M.D., Nguyen, B.T., Gurrin, C.: An experiment in interactive retrieval for the lifelog moment retrieval task at imagecleflifelog2020. In: CLEF2020 Working Notes. CEUR Workshop Proceedings, CEUR-WS.org, Thessaloniki (2020). http://ceur-ws.org/Vol-2696/paper_103.pdf, http://ceur-ws.org
30. Tran, M.T., et al.: First - flexible interactive retrieval system for visual lifelog exploration at LSC 2020. In: Proceedings of the Third Annual Workshop on Lifelog Search Challenge, LSC 2020, pp. 67–72. Association for Computing Machinery, New York (2020)
31. Truong, T.D., Dinh-Duy, T., Nguyen, V.T., Tran, M.T.: Lifelogging retrieval based on semantic concepts fusion. In: Proceedings of the 2018 ACM Workshop on The Lifelog Search Challenge, LSC 2018, pp. 24–29. Association for Computing Machinery, New York (2018)
32. Wang, S., Wang, R., Yao, Z., Shan, S., Chen, X.: Cross-modal scene graph matching for relationship-aware image-text retrieval. In: The IEEE Winter Conference on Applications of Computer Vision, pp. 1508–1517 (2020)
33. Zellers, R., Yatskar, M., Thomson, S., Choi, Y.: Neural motifs: scene graph parsing with global context. In: Proceedings of the IEEE Conference on Computer Vision and Pattern Recognition, pp. 5831–5840 (2018)

On Fusion of Learned and Designed Features for Video Data Analytics

Marek Dobranský and Tomáš Skopal[(✉)] [ID]

SIRET Research Group, Faculty of Mathematics and Physics,
Charles University, Prague, Czech Republic
{dobransky,skopal}@ksi.mff.cuni.cz

Abstract. Video cameras have become widely used for indoor and out-
door surveillance. Covering more and more public space in cities, the
cameras serve various purposes ranging from security to traffic monitoring, urban life, and marketing. However, with the increasing quantity
of utilized cameras and recorded streams, manual video monitoring and
analysis becomes too laborious. The goal is to obtain effective and efficient artificial intelligence models to process the video data automatically
and produce the desired features for data analytics. To this end, we propose a framework for real-time video feature extraction that fuses both
learned and hand-designed analytical models and is applicable in real-
life situations. Nowadays, state-of-the-art models for various computer
vision tasks are implemented by deep learning. However, the exhaustive
gathering of labeled training data and the computational complexity of
resulting models can often render them impractical. We need to consider the benefits and limitations of each technique and find the synergy
between both deep learning and analytical models. Deep learning methods are more suited for simpler tasks on large volumes of dense data
while analytical modeling can be sufficient for processing of sparse data
with complex structures. Our framework follows those principles by taking advantage of multiple levels of abstraction. In a use case, we show
how the framework can be set for an advanced video analysis of urban
life.

1 Introduction

We have reached a point where the automatic video analysis is achievable, and
it is a highly demanded functionality due to the enormous volume of video data.
The recent development in artificial intelligence (specifically deep learning) combined with the availability of high-performance hardware opened the possibility
to produce complex analytics with a high level of abstraction. However, there
are two prominent scalability difficulties with currently available methods. The
first one is the requirement for large quantities of complex and manually labeled
training data. The other one is the practical infeasibility of the state-of-the-art
methods. Many benchmark competitions are scored based on achieved precision
and do not consider time performance, which is not a suitable approach for many
practical applications.

© Springer Nature Switzerland AG 2021
J. Lokoč et al. (Eds.): MMM 2021, LNCS 12573, pp. 268–280, 2021.
https://doi.org/10.1007/978-3-030-67835-7_23

To address the above-mentioned problems, we sketch a framework that uses both deep learning and analytical approaches. This framework accomplishes the required tasks with minimal cost in terms of the need for training data and processing time. The fact is, that there exist deep learning models for every task related to video processing (detection, tracking, re-identification), however, the availability of training data is scarce or non-existent.

Although many researchers seek to create powerful models that directly produce the data on a high level of abstraction, we want to create a modular pipeline that uses the most efficient approach in each of its steps. The deep learning approaches are excellent in processing the large volumes of unstructured data (images/video) into the low-level features, while the analytical approaches can produce more abstract features based on those for a much lower cost.

The ability to gradually produce more complex and abstract features is also a key step in the creation of a modular and easily extensible framework. For example, in the pipeline of detection, trajectory, and re-identification, we may use the intermediate features (trajectories) for additional analyses (e.g., action detection). If we use a single and complex deep learning model that inputs the video and outputs identities, we lose this ability and would have to either employ additional model or redefine and retrain the existing one.

1.1   Running Use-Case in Urban Settings

As a running use-case we have chosen urban settings, where the framework is assembled by components to allow to monitor/improve the safety and quality of life in modern cities. We aim to increase the efficiency of the utilization of already available video data (provided by street cameras). For example, traffic flow analysis (road or pedestrian) can help to identify the bottlenecks and improve the overall throughput and prevent the formation of large crowds of people. Statistics on the number and wait time of passengers can help to optimize the public transport network. The detection of presence or suspicious behavior (loitering, vandalism, harassment, etc.) can benefit security.

We designed this framework with those applications in mind. The aim is to provide a wide range of analytics from the basic low-level detections to high-level aggregated statistical data, including the user (analyst) in the loop. With the consideration of data protection regulations, this framework can process real-time video stream without saving any sensitive data. Our goal is not to identify individuals and create a "Big Brother"-like system, but to obtain a fully anonymized statistical data. However, the framework provides both modes—the online (real-time) and offline video analytics.

2   Related Work

In this section, we briefly overview the relevant domains that contribute to the proposed framework.

2.1 Multi-modal Retrieval and Feature Fusion

Since we address combinations of various feature extractions/object detections from video data, the background domains for us are the multi-modal (multimedia) retrieval and feature fusion [3]. In multi-modal retrieval, multiple feature descriptors are extracted from each data object (e.g., an image or video keyframe), and later combined into a complex multi-feature descriptor. This process can take place either at the moment of basic data modeling and indexing (early fusion) or just at the moment a query is being processed (late fusion). In a general environment, the query is usually processed as the similarity search task [32], where feature descriptors are compared by a similarity function. The relevancy of the individual objects for the particular retrieval task (querying, joining, clustering, etc.) is determined by their similarity score. For the specific case of object detection in the image, the extracted feature could also be in the form of simple geometry (bounding box) describing the location of the detected object (see next subsections for the details). For retrieval based on spatial objects, numerous techniques have been developed in the domain of spatial databases [20].

In the case of video keyframes, a series of timestamped features and detected spatial objects is obtained. Higher-level video features could be thus acquired by fusion of lower-level features by their similarity, spatial relationships, and time. The central part of the proposed framework is a database of video features produced at different levels of abstraction, and the focus is on the fusion process.

2.2 Object Detection

In recent years, there has been rapid development in the field of computer vision and object detection. However, for the task of object detection, the best performing approaches have been based on deep learning. Although, there are many object detection models developed every year, an important step towards real-time video processing and analysis arrived when single-stage object detectors replaced older multi-stage models such as *Faster R-CNN* [25] by outperforming them both in terms of precision and processed frames per second. *SSD* [18] and multiple versions of *YOLO* [22–24] are well-known examples of this approach. At the time of writing, the state-of-the-art object detector, *EfficientDet* [28], adopts the same principles.

With the use-case of object detection transitioning from image processing to video, the available information in the processed media has also increased. The most interesting feature of the video is the continuity in time; objects generally do not suddenly change position between consecutive frames and follow a trajectory. We can find some models that build on this additional temporal information and combine object detection and object tracking tasks into a single neural network. There is, however, a penalty for such an approach in the form of time performance. We can use *T-CNN for action detection in video* [10] as an example of such a network. It achieved state-of-the-art performance by directly classifying temporal detections by action instead of generating a series of detections and further analyzing those.

A step towards increased efficiency could be decomposition of monolithic single-model detector into a detector designed as an ensemble of partial deep learning models, such as the *NoScope* system [14], that can achieve state-of-the-art detection performance for a fraction of time. From another point of view, the detection efficiency could be increased by employing database indexes into the inference process. The *Focus* system [11] provides approximate indexing of object classes, outperforming the NoScope by two orders of magnitude. In this paper, we follow both directions and resort to even more general index-supported framework where the deep learning models are just one class of modules plugged into the extraction pipeline.

2.3 Identification and Tracking

The identification and tracking are closely related tasks that complement and build on each other. To better understand the previous statement, we present an example of tracking people in surveillance video. The previously mentioned object detectors produce a set of detections (bounding boxes over detected objects) for each video frame in each camera. The tracking algorithm then needs to cluster the time series of detections with the same identity (yet anonymous) into a singular trajectory. Given the trajectories and their respective identities, we are able to match those identities across multiple cameras.

At this point, there are no universally established approaches to this task. The tracking algorithms are often based on handcrafted features extracted from the detected objects, e.g., color or texture histograms, in combination with analytical data like motion, proximity and probability (Kalman filters [12]). However, with the rise of deep learning approaches, there are attempts to merge the object detection and tracking into the singular deep learning model [21,27]. Those methods can undoubtedly achieve the state-of-the-art results in benchmarks, however, they suffer in the real-time performance and are associated with the cost of a complex manually labeled dataset. We see the similar impact of learned [15,30] and designed [26,33] features in re-identification tasks[1]. The analytical models are usually based on similarity search where the deep learning ones are trained to represent each object in the high-dimensional space with distance metrics.

2.4 General Video Analytics and Retrieval

While the previously discussed object detection, reidentification and tracking might be interpreted as specific video analytic tasks, there is also a need for general and extensible video analytics support. Traditionally, this is a domain of SQL

---

[1] A re-identified object is a previously recognized object that is identified again in different conditions (different scene/camera, lighting, color balance, image resolution, object pose, etc.).

database systems, where aggregations within SQL queries could be used to produce analytics on the relational attributes (in our case video features/detections). However, the basic SQL could be insufficient for advanced video analytics including proximity/similarity joins or clustering (instead of just exact-match GROUP BY functionality of SQL). To address this problem, the *BlazeIt* was recently presented, a video query optimizer based on the *FrameQL* (an extension of SQL for video analytics) [13]. General video retrieval is fostered by TRECVID [1] and Video Browser Showdown [19] competitions, where Ad-hoc search (AVS) and VBS known-item search (KIS) tasks are annually evaluated. The current trends involve machine learning of joint feature spaces [17], where state-of-the-art visual features are combined with approaches to represent text descriptions. However, these models are limited with the availability of training datasets. In other words, although the models currently win the competitions, their effectiveness is still not satisfactory due to the low recall.

3   Framework

We present a sketch of feature extraction and analytics framework from video data, where lower-level features are fused/aggregated into higher-level features, see Fig. 1a. The central component of the framework is the feature database, which serves as target storage for extracted features, as well as the source for extraction of higher-level features from the lower-level ones. The basic extraction module of per-frame feature extraction is denoted as the "L0 model". This model couples individual feature extractors from the raw (video) data. We assume two types of sub-modules at level zero. First, the extraction of learned features, such as bounding boxes (geometries, in general) of detected objects, using deep convolutional neural networks, and second, other features, such as color histograms, using traditional analytical feature extraction methods.

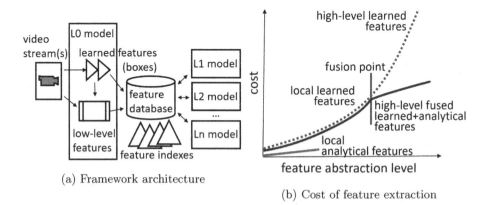

(a) Framework architecture

(b) Cost of feature extraction

Fig. 1. Framework architecture and feature extraction cost

At higher levels (see "L1 model", "L2 model" in the figure), the features are produced using the fusion of lower-level features stored in the feature database during the previous extraction steps. Hence, the feature extraction is defined as a hierarchical and recursive fusion process over the feature database. A particular pipeline within the process for our running use-case can consist of the following fusion steps:

Feature level	Features fused from lower-level features
Level 0	Timestamped learned features (e.g., bounding boxes of detected objects) and designed features (e.g., SIFTs, color/texture histograms)
Level 1	Trajectories of objects in time (bounding boxes close in time and space), crowds (bounding boxes close in space at the identical time)
Level 2	Trajectories of crowds, trajectory clusters, re-identified objects
Level 3	Action detection based on trajectory/box patterns ...
Last level	Final analytics over features

(a)   (b)

Fig. 2. (a) L0 learned features (boxes), (b) L1 fused features (red trajectories) (Color figure online)

(a)   (b)

Fig. 3. (a) L0 learned features (boxes), (b) L2, L3 fused features (green crowd, yellow crowd trajectory) (Color figure online)

In Fig. 1b, we outline the expected cost of higher-level feature extraction in terms of training dataset size, annotation efforts, and computational complexity. Although theoretically, purely machine learning models could be trained for any high-level feature extraction, the cost could reach impractical values with the increasing level of feature abstraction and greater sparseness of the training data [2,14,16].

As an alternative, the framework we present could efficiently fuse learned features and other features within the pipeline, where each step builds on the result of the previous one, no matter how the features were acquired. Since no step in this pipeline is restricted by the need for a single feature fusion model – such as one huge deep neural network or an ensemble of networks – the overall cost could be dramatically lower (see the fusion point in the figure). As a result, the proposed framework represents a multi-model database-oriented approach to efficient and effective feature extraction from video data.

3.1   Database Indexing

The efficiency of the proposed framework could be further improved by employing database indexes maintained on top of the feature database. The indexing is a key performance factor the monolithic machine learning models cannot benefit from. Based on the different feature models, various indexing methods could be utilized, such as metric indexes for general similarity search [6,32], spatial access methods for proximity search of geometries [20], and specialized indexes such as trajectory indexing [7] or indexing of clouds of points [31]. Last but not least, as feature fusion incorporates the multi-modal retrieval, compound multi-descriptor indexes for early fusion, such as the multi-metric indexing, could be added [4].

3.2   Analytics over High-Level Features

The proposed framework could as well support the analytics functionality. The simple aggregations over feature types could be easily processed using the SQL subsystem of the feature database (such as SELECT FROM WHERE GROUP BY). In order to efficiently support advanced analytics, such as the FrameQL extension of SQL [13], the system must employ higher-level database operators, e.g., trajectory search [29], spatial joins [20], similarity joins [5,9], or activity recognition [34]. Moreover, the operators' implementation could reuse the individual feature indexes.

4   Urban Use Case

To illustrate the potential application of the proposed framework, we present a possible pipeline for producing the analytics data from the video and highlight the areas of interest for our research (people tracking, in this case). We use a set of surveillance videos from multiple cameras in close proximity (e.g., in the mall or on a city square) to represent the example input data. The goal is to gather as much analytical data as possible with a universal and modular framework.

In the beginning, we process the video keyframes with a trained object detector. The viability of deep learning object detection for real-time video processing was already proven in our previous work [8]. The received features (bounding boxes) represent the "L0" learned features. The other types of the "L0" features we can gather from the video stream represent the areas inside the obtained bounding boxes (histograms, learned descriptors). Those features are then saved into a central database and made available to higher-level models. We present a visualization of "L0" features in Fig. 2(a) and Fig. 3(a). Although there are multiple types of objects detected, the people detections (blue boxes) are of particular interest for us, as they are suitable for feature fusion into higher-level real-world features (trajectories, crowds).

Using a multi-modal feature fusion, the "L0" features are aggregated into trajectories representing movements of individual objects or crowds. A crowd is a cluster of bounding boxes (see Fig. 2(b)). We have implemented a simple greedy algorithm for merging bounding boxes of detected objects (persons) into trajectories. The algorithm iteratively merges centroids of bounding boxes based on their proximity in time and space. In the case of draws (multiple bounding boxes in the neighborhood), histograms of colors extracted from the respective image patches (positioned under the boxes in the keyframe) were used to measure the object similarity. The bounding box centroid of the most similar patch was chosen to extend the trajectory. Finally, an attempt was made to connect (thus merge) the short trajectories that exhibited similar direction and had a sufficiently close endpoint. The trajectory connecting proved to be effective in situations where objects (persons) were missed by the detector in some keyframes (hence missing also their bounding boxes), or objects could not be detected because of occlusions in the scene (e.g., the person behind a car or column).

The "L1" features (trajectories) are then also written into the database. In the case of people surveillance, the generated trajectories are the base for the creation of a new "L2" object, the crowd. The crowd is a nice example of the feature level that can be commonly found as a class learned by the deep learning approaches but can also be easily obtained by our pipeline without the need to extend the object detector. The benefit of our approach is the ability to decide for each person in the perceived crowd whether they belong to it or are just passing by in a different direction. We preview the crowd detection and crowd trajectory ("L2" and "L3" features) in Fig. 3(b).

Perhaps the most interesting use-case of our layered model of features is the re-identification task. We combine the information from "L0" and "L1" features in a multi-modal retrieval based on similarity, spatial and time relationship to match the objects from multiple camera views or time distance into a single identity. The last level features can, in this case, represent behavioral patters. Hand-crafted patterns can be used to match the trajectories to determine the actions executed by the observed people (waiting, walking, running).

5   System Architecture

In this section, we preview the earlier discussed architecture from the system's point of view. Multiple modules and pipelines are illustrated on Fig. 4. The deep learning object detector is the provider of all the analytical data from the input video. For the detector itself, we are currently experimenting with the modification of the SSD object detector [18]. The product of this module is a set of bounding boxes and corresponding class labels. However, we also desire a feature descriptor for each detected object for future use in tracking and re-identification tasks. We use the bounding boxes to create cropped images from the input frame and extract the feature descriptors from those. Although our system is built around a database as a means of inter-process communication, we want to avoid transferring the cropped images via the database. Therefore we put the feature extraction module to the same pipeline as the object detector and only push the extracted features and bounding boxes to the database.

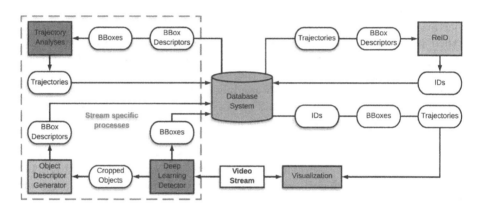

Fig. 4. System architecture. Colored boxes represent system modules, the rounded boxes describe the type of data passed between modules. (Color figure online)

The next branch that works with the data from a single video stream encompasses the trajectory module. Whereas the detector branch looks at each frame on its own, this module uses a combination of bounding box information from a sliding window of consecutive frames. We combine position, time, and extracted feature descriptors as factors for trajectory creation.

The last analytical branch is the re-identification module. Unlike the detection and trajectory modules, re-identification utilizes data from multiple streams. Therefore it is not a stream specific process, rather a global one. It also has multiple purposes, to match multiple trajectories of a single object (e.g. divided by occlusion) in one stream, and to match trajectories from multiple video streams.

The modularity of this system allows for expansion by other modules that can use the data available in the database and produce further analytics.

5.1 Architecture for Real-Time Analytics

The framework is designed to operate in both real-time and offline modes. For the real-time mode, the system architecture should be tightly coupled as a single physical machine (GPU+CPU+storage) hosting deep learning modules, database system, and other analytics modules. An SQL-based system is supposed to manage the central database, such as PostgreSQL or a commercial one. As a performance of the database is critical for real-time applications, it is of great importance to configure the database to operate in the main memory or at least on a solid-state drive. Figure 5 shows the action dependencies across modules when input video frames are being processed. For example, to obtain a visualization of a trajectory of a detected object (person) at frame T, a pipeline of actions must be processed. This includes bounding box detections at neighboring frames $(T - P_1 .. T + P_2)$. Therefore, the visualization has to work with a constant delay in the order of seconds. Note that the processing pipeline includes heavy communications with the central database, hence, the throughput of thousands of transactions (SELECT, INSERT, UPDATE) per second is necessary for (near) real-time settings.

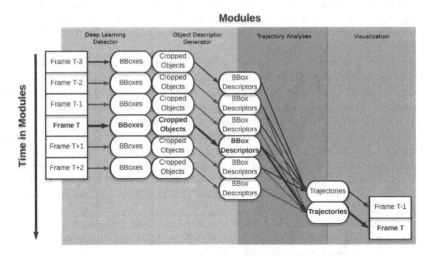

Fig. 5. Data flow in time. Simplified view of data processing, excluding database interaction and re-identification module as seen on Fig. 4.

6 Conclusions

In this paper, we have proposed a feature extraction framework from video data. We based our approach on the combination of deep learning and analytical techniques and built a multi-model database-oriented approach to extract the desired features with the highest efficiency. Our considerations include not only the processing time but also the resources and costs associated with obtaining the training data for deep learning approaches. We believe that the extensibility and ability to produce the fused features with a high level of abstraction exposes this framework to a wide variety of applications.

References

1. Awad, G., et al.: TRECVID 2017: evaluating ad-hoc and instance video search, events detection, video captioning and hyperlinking. In: Proceedings of TRECVID 2017. NIST, USA (2017)
2. Bissmark, J., Wärnling, O.: The sparse data problem within classification algorithms: the effect of sparse data on the Naïve Bayes algorithm (2017)
3. Budikova, P., Batko, M., Zezula, P.: Fusion strategies for large-scale multi-modal image retrieval. In: Hameurlain, A., Küng, J., Wagner, R., Akbarinia, R., Pacitti, E. (eds.) Transactions on Large-Scale Data- and Knowledge-Centered Systems XXXIII. LNCS, vol. 10430, pp. 146–184. Springer, Heidelberg (2017). https://doi.org/10.1007/978-3-662-55696-2_5
4. Bustos, B., Kreft, S., Skopal, T.: Adapting metric indexes for searching in multi-metric spaces. Multimed. Tools Appl. **58**(3), 467–496 (2012)
5. Čech, P., Maroušek, J., Lokoč, J., Silva, Y.N., Starks, J.: Comparing MapReduce-based k-NN similarity joins on Hadoop for high-dimensional data. In: Cong, G., Peng, W.-C., Zhang, W.E., Li, C., Sun, A. (eds.) ADMA 2017. LNCS (LNAI), vol. 10604, pp. 63–75. Springer, Cham (2017). https://doi.org/10.1007/978-3-319-69179-4_5
6. Chávez, E., Navarro, G., Baeza-Yates, R., Marroquín, J.L.: Searching in metric spaces. ACM Comput. Surv. **33**(3), 273–321 (2001)
7. Deng, K., Xie, K., Zheng, K., Zhou, X.: Trajectory indexing and retrieval. In: Zheng, Y., Zhou, X. (eds.) Computing with Spatial Trajectories. Springer, New York (2011). https://doi.org/10.1007/978-1-4614-1629-6_2
8. Dobranský, M.: Object detection for video surveillance using SSD approach (2019). http://hdl.handle.net/20.500.11956/107024
9. Dohnal, V., Gennaro, C., Zezula, P.: Similarity join in metric spaces using eD-index. In: Mařík, V., Retschitzegger, W., Štěpánková, O. (eds.) DEXA 2003. LNCS, vol. 2736, pp. 484–493. Springer, Heidelberg (2003). https://doi.org/10.1007/978-3-540-45227-0_48
10. Hou, R., Chen, C., Shah, M.: Tube convolutional neural network (T-CNN) for action detection in videos. In: Proceedings of the IEEE International Conference on Computer Vision, pp. 5822–5831 (2017)
11. Hsieh, K., et al.: Focus: querying large video datasets with low latency and low cost. In: 13th USENIX Symposium on Operating Systems Design and Implementation (OSDI 2018), pp. 269–286. USENIX Association (October 2018)
12. Kalman, R.E.: A new approach to linear filtering and prediction problems. J. Basic Eng. **82**(1), 35–45 (1960)
13. Kang, D., Bailis, P., Zaharia, M.: BlazeIt: optimizing declarative aggregation and limit queries for neural network-based video analytics. Proc. VLDB Endow. **13**(4), 533–546 (2019)
14. Kang, D., Emmons, J., Abuzaid, F., Bailis, P., Zaharia, M.: NoScope: optimizing neural network queries over video at scale. Proc. VLDB Endow. **10**(11), 1586–1597 (2017)

15. Li, W., Zhao, R., Xiao, T., Wang, X.: DeepReID: deep filter pairing neural network for person re-identification. In: Proceedings of the IEEE Conference on Computer Vision and Pattern Recognition, pp. 152–159 (2014)

16. Li, X., Ling, C.X., Wang, H.: The convergence behavior of Naive Bayes on large sparse datasets. In: 2015 IEEE International Conference on Data Mining, pp. 853–858 (November 2015). https://doi.org/10.1109/ICDM.2015.53

17. Li, X., Xu, C., Yang, G., Chen, Z., Dong, J.: W2VV++: fully deep learning for ad-hoc video search. In: Proceedings of the 27th ACM International Conference on Multimedia, MM 2019, Nice, France, October 21–25, 2019, pp. 1786–1794 (2019). https://doi.org/10.1145/3343031.3350906

18. Liu, W., et al.: SSD: single shot multibox detector. In: Leibe, B., Matas, J., Sebe, N., Welling, M. (eds.) ECCV 2016. LNCS, vol. 9905, pp. 21–37. Springer, Cham (2016). https://doi.org/10.1007/978-3-319-46448-0_2

19. Lokoč, J., Bailer, W., Schoeffmann, K., Münzer, B., Awad, G.: On influential trends in interactive video retrieval: video browser showdown 2015–2017. IEEE Trans. Multimed. **20**(12), 3361–3376 (2018). https://doi.org/10.1109/TMM.2018.2830110

20. Manolopoulos, Y.: Spatial Databases: Technologies, Techniques and Trends. IGI Global, Hershey (2005)

21. Qi, Y., et al.: Hedged deep tracking. In: Proceedings of the IEEE Conference on Computer Vision and Pattern Recognition, pp. 4303–4311 (2016)

22. Redmon, J., Divvala, S., Girshick, R., Farhadi, A.: You only look once: unified, real-time object detection. In: Proceedings of the IEEE Conference on Computer Vision and Pattern Recognition, pp. 779–788 (2016)

23. Redmon, J., Farhadi, A.: Yolo9000: better, faster, stronger. In: Proceedings of the IEEE Conference on Computer Vision and Pattern Recognition, pp. 7263–7271 (2017)

24. Redmon, J., Farhadi, A.: Yolov3: an incremental improvement. arXiv preprint arXiv:1804.02767 (2018)

25. Ren, S., He, K., Girshick, R., Sun, J.: Faster R-CNN: towards real-time object detection with region proposal networks. In: Advances in Neural Information Processing Systems, pp. 91–99 (2015)

26. Shi, Z., Hospedales, T.M., Xiang, T.: Transferring a semantic representation for person re-identification and search. In: Proceedings of the IEEE Conference on Computer Vision and Pattern Recognition, pp. 4184–4193 (2015)

27. Sun, S., Akhtar, N., Song, H., Mian, A.S., Shah, M.: Deep affinity network for multiple object tracking. IEEE Trans. Pattern Anal. Mach. Intell. **43**, 104–119 (2019)

28. Tan, M., Pang, R., Le, Q.V.: EfficientDet: scalable and efficient object detection. arXiv preprint arXiv:1911.09070 (2019)

29. Wang, H., Belhassena, A.: Parallel trajectory search based on distributed index. Inf. Sci. **388–389**, 62–83 (2017)

30. Xu, J., Zhao, R., Zhu, F., Wang, H., Ouyang, W.: Attention-aware compositional network for person re-identification. In: Proceedings of the IEEE Conference on Computer Vision and Pattern Recognition, pp. 2119–2128 (2018)

31. Yang, J., Huang, X.: A hybrid spatial index for massive point cloud data management and visualization. Trans. GIS **18**, 97–108 (2014)

32. Zezula, P., Amato, G., Dohnal, V., Batko, M.: Similarity Search: The Metric Space Approach (Advances in Database Systems). Springer, Heidelberg (2005). https://doi.org/10.1007/0-387-29151-2

33. Zhao, R., Ouyang, W., Wang, X.: Unsupervised salience learning for person re-identification. In: Proceedings of the IEEE Conference on Computer Vision and Pattern Recognition, pp. 3586–3593 (2013)

34. Zhu, Y., Zheng, V.W., Yang, Q.: Activity recognition from trajectory data. In: Zheng, Y., Zhou, X. (eds.) Computing with Spatial Trajectories. Springer, New York (2011). https://doi.org/10.1007/978-1-4614-1629-6_6

XQM: Interactive Learning on Mobile Phones

Alexandra M. Bagi[1], Kim I. Schild[1], Omar Shahbaz Khan[1], Jan Zahálka[2],
and Björn Þór Jónsson[1]($\boxtimes$)

[1] IT University of Copenhagen, Copenhagen, Denmark
bjth@itu.dk
[2] Czech Technical University, Prague, Czech Republic
jan.zahalka@cvut.cz

Abstract. There is an increasing need for intelligent interaction with
media collections, and mobile phones are gaining significant traction as
the device of choice for many users. In this paper, we present XQM,
a mobile approach for intelligent interaction with the user's media on
the phone, tackling the inherent challenges of the highly dynamic nature
of mobile media collections and limited computational resources of the
mobile device. We employ interactive learning, a method that conducts
interaction rounds with the user, each consisting of the system suggesting
relevant images based on its current model, the user providing relevance
labels, the system's model retraining itself based on these labels, and
the system obtaining a new set of suggestions for the next round. This
method is suitable for the dynamic nature of mobile media collections
and the limited computational resources. We show that XQM, a full-
fledged app implemented for Android, operates on 10K image collections
in interactive time (less than 1.4 s per interaction round), and evaluate
user experience in a user study that confirms XQM's effectiveness.

Keywords: Interactive learning · Relevance feedback · Mobile devices

1 Introduction

As media collections have become an increasingly large part of our everyday
lives, both professionally and socially, the traditional interaction paradigms of
searching and browsing have become less effective. Search requires users to have a
clear idea of the outcome of the interaction, while browsing large collections is too
inefficient. Many users have significant media collections on their *mobile devices*,
often thousands of photos, that they wish to interact with. When considering
intelligent interaction with these collections, many questions arise: How should
the system be designed to best utilize the limited computing resources of the
mobile device? How should users interact with the system to best make use of
the limited screen space? How will regular mobile device users react to this novel
interaction paradigm? In this paper, we explore these challenges.

© Springer Nature Switzerland AG 2021
J. Lokoč et al. (Eds.): MMM 2021, LNCS 12573, pp. 281–293, 2021.
https://doi.org/10.1007/978-3-030-67835-7_24

The *interactive learning* paradigm has had a revival as a viable approach for intelligent analysis of media collections. In interactive learning, the system incrementally and interactively builds a model of users' information needs, by continually using the model to suggest media items to users for feedback and then using that feedback to refine the model. In recent years, significant progress has been made towards scalable interactive learning systems, where the thrust of the work has been on handling larger and larger collections with moderate hardware whilst providing relevant results [10,17]. Interactive learning is a good fit for mobile systems, as it inherently does not rely on heavy data preprocessing, bears limited computational resources in mind, and can work with dynamic datasets.

We present and evaluate XQM, a full-fledged mobile application for interactive learning.[1] Figure 1 illustrates the overall architecture and processing of XQM. In the *interactive exploration* phase, XQM gradually builds an *interactive classifier* capable of fulfilling the user's information need. This is done by presenting the highest ranked images from the current model in the *user interface* and asking the user to provide feedback on those, labelling them as relevant or not relevant. The labels are then used to retrain the interactive classifier, and the classifier is in turn used to query the *feature database* for a new set of *suggested images* to judge. This interactive process continues until the user is satisfied or decides to start from scratch. To support the model construction, the app also has a *data processing* phase, where state-of-the-art *semantic feature extraction* is deployed on a dedicated server. When installing the app, existing images are analysed in this manner to build XQM's feature database, and subsequently the user can analyse new images to add to the feature database. In a performance study, we show that XQM can interactively explore collections of up to 10K images with response time of less than 1.4 s per interaction. Furthermore, we report on user experience in a user study with participants ranging from novice users to experienced interactive learning users.

2   Related Work

Processing large image collections makes the collaboration between humans and computers inevitable. Computers have a large memory and can process large amounts of data fast, but they lack the human's ability to extract a great amount of semantic information from visual content. This phenomenon is known as the semantic gap [14]. A large body of research has been devoted to closing the semantic gap and indeed, we are able to automatically extract more semantic information from the data than before. To that end, however, a computer still needs feature representations of the data which are meaningful to the machine, but might not be to a human. Currently, mostly convolutional neural networks (CNNs) are used [11,15,16].

---

[1] XQM is an acronym of Exquisitor Mobile, as the design of XQM relies heavily on Exquisitor, the state-of-the-art interactive learning system [10]. The XQM app is available to the research community at www.github.com/ITU-DASYALab/XQM.

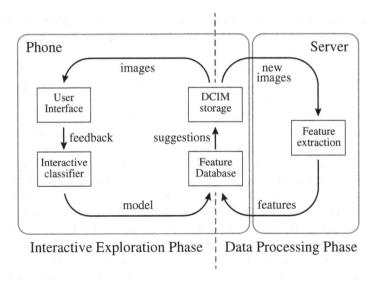

Fig. 1. Architectural overview of the XQM interactive learning mobile app.

The need for feature representations presents two key challenges for a mobile approach. Firstly, the resource-limited mobile phone needs fast access to the feature representation in addition to the raw data. Secondly, CNNs are computationally very intensive, and their performance hinges on access to state-of-the-art GPUs that either mobile phones do not possess or are unsuitable for high long-time usage due to overheating. There are approaches for deep learning on mobile phones, but their performance lags behind the desktop-based state of the art.

Most recent approaches for analyzing multimedia collections are based on retrieval: an index is built on top of the feature representation(s) and users pose queries that retrieve results based on this index. There is a large number of approaches to build the index: to name a few, product quantization [9], clustering-based approaches [8], or hashing-based approaches [2,5]. Overall, this approach works well, as the search engine is a familiar interface to the users, and the semantic quality of state-of-the-art representations is high enough to surpass human-level performance on some tasks [7]. However, an index-based approach might not be so suited for mobiles. Again, there is the limited resources challenge: an index is yet another data structure the phone needs access to. Also, multimedia collections on mobile phones tend to be highly dynamic, whereas index-based approaches favour static collections.

Interactive learning approaches, in particular user relevance feedback (URF), directly work with the user to obtain the items she finds relevant: in each interaction round, the user selects relevant and not relevant items, the interactive learning model retrains itself based on the judgment, and finally the user is presented with new and potentially relevant items for the next round. The bulk of the algorithmic research on relevance feedback and the closely related active learning techniques has been done in the 2000s [1,13,19]. Recent work has improved interactive learning to perform well on modern large-scale datasets: scaling up to interactive performance on 100 million images [17], and improving the performance to 0.29 s per interaction round whilst further reducing computational requirements [10]. A good choice of an interactive classifier is the Linear SVM [4], which combined with modern feature representations yields good performance at a modest computational cost. In terms of applications, interactive learning approaches fall under the umbrella of multimedia analytics, which strives to iteratively bring the user towards insight through an interface tightly coupled with an interactive learning machine model [18]. These approaches do not rely on a static index and computational requirements are a built-in core consideration, namely interactivity (making sure the model is able to retrain itself and produce suggestions in sub-second time). However, it is still not trivial to satisfy those requirements, especially on a mobile device.

This work aims to implement a URF system on mobile phones, therefore both the technical restrictions of these devices and the general user behaviour towards mobile applications need to be considered. There are a number of factors that prevent simple deployment of the state of the art on mobile devices. Efficient interactive learning approaches, such as Exquisitor [10], hinge on a C/C++ implementation and storage solutions not supported by mobile OS (at least Android), and the machine learning (computer vision, information retrieval, etc.) codebase for mobiles is limited. Whilst this is an implementational challenge rather than a scientific one, it is a barrier nonetheless. Moreover, user interface interactions and their convenience are somewhat different than on a computer. For example, mobile UIs make heavy use of swipes and finger gestures which do not necessarily map directly to mouse interactions and typing is more cumbersome on a screen keyboard than on a computer keyboard. Lastly, the core framework components of a native mobile application largely stipulate the way how connection can be established between the user and the system.

3   XQM Architecture

In this section, we describe the architecture of the XQM mobile app, depicted in Fig. 1. We start by presenting the user interface which supports the interactive exploration process outlined in the introduction, and then consider the underlying components in a bottom-up fashion: the semantic feature extraction, the feature database, and the interactive classifier. Note that the actual photos are stored by other applications in DCIM storage folders. When starting XQM for the first time, the user must thus grant the app access to DCIM image storage.

User Interface: Figure 2 shows the main screens of the user interface of XQM. Each exploration session starts with the *home screen* (Fig. 2(a)), which displays a non-scrollable list of 6 random images from the image collection (the largest

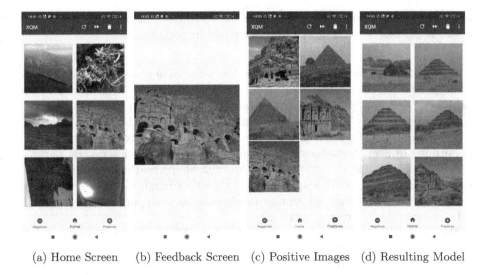

(a) Home Screen   (b) Feedback Screen   (c) Positive Images   (d) Resulting Model

Fig. 2. The interactive learning process as captured in the XQM user interface.

number of images that can be displayed on the mobile screen in adequate resolution). The user can tap on any of the six suggestions to open an enlarged image on a new *feedback screen* (Fig. 2(b)) in order to inspect it in more detail, and then potentially judge the image as a positive or negative example by swiping the image right or left, respectively. Once an image has been judged as a positive or negative example, the user returns to the home screen where the image is replaced by the image currently considered the most relevant, according to the interactive classifier.[2]

The user can, at any time, revisit the positive and negative examples by tapping on the corresponding buttons at the bottom of the home screen, to open the *positives screen* (Fig. 2(c)) or *negatives screen* (not shown). From those screens, the user can remove images from the positive or negative lists, which in turn impacts the model in the next suggestions round.

Once the interactive classifier seems good enough, the user can tap on the "fast-forward" icon at the top of the screen to fill the home screen with the most relevant images (Fig. 2(d)). In addition to the fast-forward button, the XQM app has three buttons located on the top navigation bar: the "random" icon is used to get a new set of random images;[3] the "trash" icon is used to start a new exploration from scratch; and finally the "overflow menu" icon (three dots) can

---

[2] In the current implementation, at least one positive and one negative example are needed; until these have been identified, random images replace the judged images.

[3] Loading random images is useful when the model is missing positive examples with concepts that have not yet been seen; in a future version we plan to implement search functionality to further help find positive examples.

be used to add new images to the feature database through the data processing phase, or get help as outlined in the user study in Sect. 5.

Semantic Feature Extraction: In the *data processing phase* of Fig. 1, semantic deep learning features are extracted from images and stored in a feature database. There are two options to facilitate feature extraction on mobile devices: on-device, using the CPU of the mobile phone (or GPU, in the case of high-end devices); or off-device, sending the images to be processed to a remote server. As outlined in Sect. 2, on-device processing suffers from lower semantic performance, low availability of tools/libraries, and risk of overheating causing damage to the device. The off-device approach, on the other hand, requires access to a mobile or wireless network, which may result in latency and/or usage charges [6], and can also raise security questions.

We have chosen the off-device approach for the XQM app, to (a) make use of state-of-the-art semantic features, and (b) avoid complex resource management issues on the device. We have wrapped a pretrained ResNext101 model with 12,988 classes with a web-API, which allows submitting a ZIP file with multiple images and returns a JSON file with information on the semantic features. Following [17], however, the server only returns information about the top s semantic features associated with each image (by default, $s = 6$).

During the data processing phase, the collection of images to analyse is split into batches of 100 images, which are compressed and sent synchronously to the server for processing. Upon receiving the JSON file from the server, it is parsed and the information is stored in the feature database described in the next section. When the app is first run, this process is applied to the entire collection of images on the device; subsequently, only newly added images are analysed when the user chooses to update the database.

Processing each batch of 100 images takes little over a minute with a Xiaomi Redmi Note 8 Pro smartphone and a laptop for running the extraction. Initialising the app on a mobile device with thousands of images would thus take significant time. For a production app more care must be taken to implement the data processing phase, including asynchronous and secure communication. The current process, however, is sufficient for the purposes of understanding the performance of the interactive exploration phase, which is the main emphasis of the paper.

Feature Database: The semantic feature data resulting from the analysis described above must be stored persistently on the mobile device in a format that allows efficient access in the interactive phase. However, neither the traditional multimedia approach of storing a sparse NumPy matrix in RAM nor the advanced compression mechanism of [17] are applicable in the limited environment of the Android OS. Instead, the standard approach to data storage is using the SQLite relational database, which requires careful normalisation and choice of data types to work well.

ImageName	FeatureID	Probability
storage/emulated/0/DCIM/Camera/103501.jpg	[5344, 4309, 6746, 5060, 2874, 2705]	[0.102834791, 0.063476876, 0.05930854, 0.05230371, 0.03741937, 0.0310730178]
storage/emulated/0/DCIM/Camera/101502.jpg	[6248, 5326, 2851, 5324, 5325, 5326]	[0.122722, 0.116210177, 0.059408964, 0.045039124, 0.044916617, 0.0203401245]

(a) Original Table

ImageID	ImageName	FolderPathID
0	103501.jpg	0
1	101502.jpg	0
2	103200.jpg	0
3	102200.jpg	0
4	101501.jpg	0
5	103301.jpg	0
6	103511.jpg	0

FolderPathID	FolderPath
0	storage/emulated/0/DCIM/Camera
1	storage/emulated/0/DCIM/Screenshots

ImageID	FeatureID	Probability
0	5344	102
0	4309	63
0	6746	59
0	5060	52
0	2874	37
0	2705	31
1	6248	123

(b) Path Table   (c) Image Table   (d) Feature Table

Fig. 3. Comparison of storage alternatives: (a) the original unnormalised table; and (b)–(d) final normalised tables.

Figure 3(a) shows the data that would typically be stored in (compressed) binary format in RAM. In SQLite, the IDs of the features and corresponding probabilities can be stored as a TEXT string that is parsed when reading the data to rank the images. While this approach has modest space requirements, requiring 179 kB for a database of 1,000 feature vectors, parsing the TEXT string resulted in computational overhead.

Instead, Figs. 3(b)–(d) show the final normalised SQLite database, where three tables represent the feature database, one for storing all relevant folder paths and their identifiers, one for storing image identifiers, image names and path identifiers, and the third for storing the feature identifiers and probabilities, one per row. For additional space savings, the probabilities have been converted to integers using multiplication, since the INTEGER data type requires half the storage of a FLOAT data type. With this implementation, a collection of 1,000 feature vectors requires only 116 kB.

When applying the interactive classifier to the feature vectors to rank images, only the feature table is required, as it contains the image identifiers that can be used to identify the most relevant suggestions. Once the 6 most relevant suggestions have been identified, the image table and the path table must be accessed to build the path of the image for presentation in the user interface. Since the access is based on the primary keys of both tables, reading the required data is very efficient.

Note that deletion of images is handled by detecting missing images as they are suggested to the user, and subsequently removing their information from the feature database. If an image is moved, it will be handled as a deletion and an insertion; the order will depend on when the user updates the database and when the moved image first appears as a relevant image.

Interactive Classifier: XQM uses the LIBSVM [3] implementation of linear SVM, which is considered the state of the art classifier in interactive learning

[10,17]. As the relevance judgments generally constitute a very small training set, we used generic parameter settings to avoid over-fitting. Once the model has been trained, the feature vectors are retrieved from the SQLite database, as outlined above, and fed to the model to produce a score for each image. The 6 images with the highest score, or farthest from the decision boundary on the positive side, are then returned as suggestions.

4   System Performance Evaluation

This section evaluates the efficiency of the mobile application. Due to space concerns, we focus primarily on the interactive phase, measuring the time required to retrieve new suggestions in each interaction round. We have also measured the one-time process of analysing image contents, only about one-third of the time is spent on the phone.

Experimental Setup: For the evaluation, we used a Xiaomi Redmi Note 8 Pro smartphone with 128 GB memory and 6 GB RAM, MediaTek Helio G90T 2.05 GHz (8-core) processor and Android 9.0 (Pie) OS. To obtain results that are not influenced by other design decisions, we constructed two stand-alone apps specifically for the experiments.

The first app randomly chooses positive and negative examples, retrieves their feature vectors from the SQLite database, computes the SVM model, retrieves all the features from the SQLite database, and computes the next six suggested images. The app takes three input parameters: number of feature vectors, ranging from 100 to 10,000; number of rated examples, ranging from 3 to 48; and number of suggestions to retrieve, set to 6. For each parameter combination we ran 300 iterations and report the average.

The second app repeatedly selects random images from the database and presents them on screen, as would be done in the XQM app. It takes as input a database of images, ranging in size from 100 to 10,000. To avoid warm-up effects, we first loaded 200 images, and then measured 1,000 images for each database.

We separated the two processes as the latter app is independent of many of the parameters of the first app. To simplify the presentation of results, however, we (a) only report results with 48 rated images, as computing the SVM model is efficient and is only more efficient with fewer suggestions, and (b) incorporate the time for presenting 6 images into the results from the first app.

Experiment Results: Figure 4 presents the details of the time required for each iteration of the relevance feedback process. The x-axis displays the number of images and feature vectors in the database, while the y-axis shows the time that was used for completing these tasks in seconds. As the figure shows, the time required to build the model and show the final suggestions is negligible (only visible for the smallest collection) and the time to rank images is also very small, while the majority of the time is used to retrieve feature vectors, which

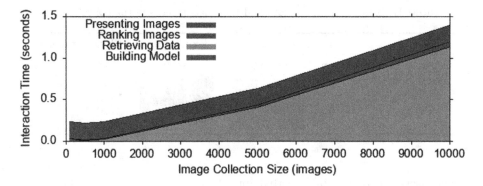

Fig. 4. The average time for subtasks in the interactive exploration phase.

is linearly dependent on database size, and display the images on screen which is independent of database size. As Fig. 4 furthermore shows, the process is very efficient for small databases, requiring less than 250 ms for 1,000 or fewer images, and that even for 10,000 images the total time per iteration is only about 1.4 s. Overall, we can conclude that for the vast majority of mobile phone users, the app will perform interactively.

5   User Interface Evaluation

XQM's design is intended to make the app easy to use for novice users who are not familiar with interactive learning. To evaluate the UI, we conducted a user study focusing on the app's usability, learnability and functionality.

Evaluation Setup: We recruited 8 users for the user study, 3 female and 5 male, all university students between 23 and 27 years of age. Three testers have limited technical knowledge, while the remaining 5 are CS students; 2 of the latter have worked with interactive learning in their thesis. All 8 users use mobile phone applications on a daily basis, but only 3 use an Android phone.

The users were provided with a Xiaomi Redmi Note 8 Pro with 2,883 images from the INRIA holiday data set[4] and the Lifelog Search Challenge.[5] The users were asked to conduct 3 sessions to find each target presented in Table 1: TV screen; exotic bay; and skyscrapers. Note that this setup is different from the intended use case scenario, as users have not created the collection themselves and thus had no information about its contents. However, to establish a consistent environment, it was necessary to use the same image collection for all users. To further guarantee comparability, an identical random seed was used to begin each interaction with the same set of random images.

---

[4] http://lear.inrialpes.fr/people/jegou/data.php.
[5] http://lsc.dcu.ie/.

Table 1. Example target images and their occurrences in the collection.

Target	TV screen	Exotic bay	Skyscrapers
Occurrences in collection	~ 100	~ 30	~ 15
Example image			

The users received a brief verbal introduction to XQM. An on-boarding modal within the app was then used to describe its functionality and usage; users could return to this information at any time using the help menu. The sessions were audio-recorded. Users were able to complete the task in 22 out of the 24 sessions; the two unsuccessful sessions are excluded from the analysis.

Quantitative Analysis: Table 2 shows how many times users took each of the main actions to accomplish the task. Due to the high variation in results, both the mean and median are provided in the table. On average, the users gave between 7 and 20 images ratings (positive or negative) for the three tasks. Showing random images was often used to accelerate the sessions, although the value is heavily skewed from one session where a user hit random 84 times; essentially simulating scrolling through the collection. Table 2 also shows that fast-forward and correcting previous ratings were rarely used, while the trash icon for starting a new session was entirely disregarded.

An analysis of the session logs indicates that novices encountered some difficulties understanding the principle of interactive learning and thus XQM's purpose. Overall, though, the results indicate that the users were largely successful in solving the tasks; in particular the fact that users never started from scratch indicates that they generally felt they were making progress in the sessions. However, the fewer instances of the target class there are the longer the sessions take, and the (sometimes extensive) use of the random button indicates that adding search functionality to find positive examples would be useful.

Qualitative Analysis: We analysed the session logs to understand and classify the concerns and suggestions of users, and report on this analysis below.

Understandability and Learnability: Most users refrained from using the buttons, mostly because they did not remember or understand their functionality. Three users had difficulties understanding which images were replaced and said they would like to understand how the model evaluates images in order to make better rating decisions. The two users with prior interactive learning experience, however, did not report such problems. One of them said: "I think XQM's strong suit is its simplistic UI. That makes it a lot easier for novices to learn." And: "The modal gave very good information of how the app works." Both experienced

Table 2. Mean/median of actions per user per motive

Target	Feedback	Random	Fast-forward	Start over	Rating change
TV	7,0/6,0	1,0/0,6	0,0/0,0	0,0/0,0	0,6/0,0
Bay	13,0/11,5	1,0/1,0	1,7/1,0	0.0/0,0	0,7/0,0
Skyscrapers	19,5/11,0	19,8/8,0	3,8/0,5	0,0/0.0	1,3/0,0

users said that they would like to have a separate screen containing a history of their search.

App Functionality: A number of suggestions were raised regarding the functionality of the app. Five users desired an "undo" function to be able to reverse their actions; such user control is desirable to allow the user to explore the app without fear of making a mistake [12]. To accelerate the sessions, two users expressed a desire to be able to scroll through the collection, since 6 images are only a small fraction of most mobile collections, and three users suggested to incorporate a search function for finding relatively rare image motives. Furthermore, two users said that they would like to swipe directly in the home screen: "It would be faster than opening the image every time you want to rate it. This would be very useful for eliminating irrelevant results." Nevertheless, after the introduction, users quickly learned how to rate: "I like the swiping feature to rate the image. That is very intuitive."

Image Features: Users noted that some images score constantly higher in the suggestion list than others without being perceived as relevant to the task. Traditional techniques, such as TF-IDF, have been reported to improve the relevance of the returned items [17] and could be relevant here. Nevertheless, overall users could find relevant results despite this issue: "It is exciting to see that the model caught up on what I am looking for."

Discussion: Overall, the results of the user study indicate that XQM succeeds in its goal of implementing interactive learning on a mobile device. Despite variation in the approach and performance of users, they were overwhelmingly able to complete the assigned tasks and find examples of the desired target items. The results also point out a number of improvements to make, including search and browsing to find positive examples, an undo button to easily take back an action, rating directly in the home screen, and improving the representative features to avoid some images occurring in every interaction session.

6 Conclusion

In this paper we presented XQM, a full-fledged mobile app for user interaction with media collections on the user's mobile device. Our interactive learning based approach is demonstrated to operate well on the dynamic mobile collections and use modest computational resources, which are at a premium on mobile devices, whilst providing relevant results in interactive time of less than 1.4 s on a 10K collection. The user study confirms that XQM is a useful tool not only to the experienced users, but also for the novice or casual users unfamiliar with interactive learning, with clear potential for further improvement. With XQM, we hope to have opened new avenues for research on advanced, intelligent approaches for media collection analytics on mobile devices.

Acknowledgments. This work was supported by a PhD grant from the IT University of Copenhagen and by the European Regional Development Fund (project Robotics for Industry 4.0, CZ.02.1.01/0.0/0.0/15 003/0000470). Thanks to Dennis C. Koelma for his help with adopting the ResNext101 model.

References

1. Aggarwal, C., Kong, X., Gu, Q., Han, J., Yu, P.: Active learning: a survey. In: Data Classification, pp. 571–605. CRC Press (2014)
2. Andoni, A., Indyk, P.: Near-optimal hashing algorithms for approximate nearest neighbor in high dimensions. Commun. ACM **51**(1), 117–122 (2008)
3. Chang, C., Lin, C.: LIBSVM: a library for support vector machines. ACM Trans. Intell. Syst. Technol. **2**, 27:1–27:27 (2011)
4. Cortes, C., Vapnik, V.: Support-vector networks. Mach. Learn. **20**(3), 273–297 (1995)
5. Datar, M., Immorlica, N., Indyk, P., Mirrokni, V.S.: Locality-sensitive hashing scheme based on p-stable distributions. In: Proceedings of SCG, pp. 253–262 (2004)
6. Ensor, A., Hall, S.: GPU-based image analysis on mobile devices. CoRR abs/1112.3110 http://arxiv.org/abs/1112.3110 (2011)
7. Geirhos, R., Temme, C.R.M., Rauber, J., Schütt, H.H., Bethge, M., Wichmann, F.A.: Generalisation in humans and deep neural networks. In: Proceedings of NIPS, pp. 7538–7550 (2018)
8. Guðmundsson, G.Þ., Amsaleg, L., Jónsson, B.Þ.: Impact of storage technology on the efficiency of cluster-based high-dimensional index creation. In: Proceedings of DASFAA, pp. 53–64 (2012)
9. Jégou, H., Douze, M., Schmid, C.: Product quantization for nearest neighbor search. IEEE PAMI **33**(1), 117–128 (2010)
10. Khan, O.S., et al.: Interactive learning for multimedia at large. In: Jose, J.M., et al. (eds.) ECIR 2020. LNCS, vol. 12035, pp. 495–510. Springer, Cham (2020). https://doi.org/10.1007/978-3-030-45439-5_33
11. Krizhevsky, A., Sutskever, I., Hinton, G.E.: ImageNet classification with deep convolutional neural networks. In: Proceedings of NIPS, pp. 1097–1105 (2012)
12. Nielsen, J.: 10 usability heuristics for user interface design (1995). https://www.nngroup.com/articles/ten-usability-heuristics/. Accessed 25 Mar 2020
13. Settles, B.: Active learning literature survey. Computer Sciences Technical Report 1648, University of Wisconsin-Madison (2009)
14. Smeulders, A.W.M., Worring, M., Santini, S., Gupta, A., Jain, R.: Content-based image retrieval at the end of the early years. IEEE PAMI **22**(12), 1349–1380 (2000)
15. Szegedy, C., et al.: Going deeper with convolutions. In: Proceedings of CVPR, pp. 1–9 (2015)

16. Xie, S., Girshick, R., Dollar, P., Tu, Z., He, K.: Aggregated residual transformations for deep neural networks. In: Proceedings of CVPR (2017)
17. Zahálka, J., Rudinac, S., Jónsson, B., Koelma, D., Worring, M.: Blackthorn: large-scale interactive multimodal learning. IEEE TMM **20**, 687–698 (2018)
18. Zahálka, J., Worring, M.: Towards interactive, intelligent, and integrated multimedia analytics. In: Proceedings of IEEE VAST, pp. 3–12 (2014)
19. Zhou, X., Huang, T.: Relevance feedback in image retrieval: a comprehensive review. Multimed. Syst. **8**, 536–544 (2003)

A Multimodal Tensor-Based Late Fusion Approach for Satellite Image Search in Sentinel 2 Images

Ilias Gialampoukidis$^{(\boxtimes)}$ ⓘ, Anastasia Moumtzidou ⓘ, Marios Bakratsas ⓘ,
Stefanos Vrochidis ⓘ, and Ioannis Kompatsiaris ⓘ

Centre for Research and Technology Hellas,
Information Technologies Institute, Thessaloniki, Greece
{heliasgj,moumtzid,mbakratsas,stefanos,ikom}@iti.gr

Abstract. Earth Observation (EO) Big Data Collections are acquired at large volumes and variety, due to their high heterogeneous nature. The multimodal character of EO Big Data requires effective combination of multiple modalities for similarity search. We propose a late fusion mechanism of multiple rankings to combine the results from several uni-modal searches in Sentinel 2 image collections. We fist create a K-order tensor from the results of separate searches by visual features, concepts, spatial and temporal information. Visual concepts and features are based on a vector representation from Deep Convolutional Neural Networks. 2D-surfaces of the K-order tensor initially provide candidate retrieved results per ranking position and are merged to obtain the final list of retrieved results. Satellite image patches are used as queries in order to retrieve the most relevant image patches in Sentinel 2 images. Quantitative and qualitative results show that the proposed method outperforms search by a single modality and other late fusion methods.

Keywords: Late fusion · Multimodal search · Sentinel 2 images

1 Introduction

The amount of Earth observation (EO) data that is obtained increases day by day due to the multitude of sources orbiting around the globe. Each satellite image has a collection of channels/bands that provide a variety of measurements for each place on Earth. This advance of the satellite remote sensing technology has led to quick and precise generation of land cover maps with concepts (snow, rock, urban area, coast, lake, river, road, etc.) that distinguish the characteristics of the underlying areas, and provide beneficial information to global monitoring, resource management, and future planning.

Searching in large amounts of EO data with respect to a multimodal query is a challenging problem, due to the diversity and size of multimodal EO data, combined with the difficulty of expressing desired queries. The multimodal character of satellite images results from the various number of channels (e.g. Red,

© Springer Nature Switzerland AG 2021
J. Lokoč et al. (Eds.): MMM 2021, LNCS 12573, pp. 294–306, 2021.
https://doi.org/10.1007/978-3-030-67835-7_25

Green, Blue, NIR, SWIR, etc.) and associated metadata (date, time, geographical location, mission, etc.). Each satellite image can be considered as a collection of satellite image patches with semantic information about each one of them, as concepts correspond to each patch (e.g. urban area, rock, water, snow, etc.). The main challenge in searching for similar satellite image patches seems to be the combination of multiple heterogeneous features (modalities) that can be extracted from collections of satellite images (e.g. low-level visual descriptors, high-level textual or visual features, etc.). The aforementioned combination process is known as multimodal fusion. The effective combination of all available information (visual patterns and concepts, spatial and temporal) results to more effective similarity search, in the case of multimodal items such as Sentinel 2 images, or patches of them. The representation of each modality is also challenging, due to the availability of several Neural Network architectures that provide feature vectors and are trained to extract concepts.

Our contribution is summarized as follows. First, we propose a novel late fusion mechanism that combines K modalities through a K-order tensor. This tensor is generated by the results of multiple single-modality searches, which then provides the final merged and unified list of retrieved results through its 2D tensor surfaces. In addition, we propose a custom neural network for concept extraction in satellite image patches which outperforms similar and standard Neural Network architectures.

The paper is organised as follows. Section 2 presents relevant works in multimodal fusion for similarity search in information retrieval. Section 3 presents our proposed methodology, where each single modality provides a list of retrieved results and the K lists of rankings are fused. In Sect. 4 we describe the dataset we have used, the settings, as well as quantitative and qualitative results. Finally, Sect. 5 concludes our work.

2   Related Work

There are two main strategies for multimodal fusion with respect to the level, at which fusion is accomplished. The first strategy is called early fusion and performs fusion at the feature level [5,12], where features from the considered modalities are combined into a common feature vector. Deep learning [7] makes use of deep auto-encoders to learn features from different modalities in the task of cross-modal retrieval. Similarly, [18] proposed a mapping mechanism for multimodal retrieval based on stacked auto-encoders. This mechanism learns one stacked auto-encoder for each modality in order to map the high-dimensional features into a common low-dimensional latent space. Modality-specific feature learning has also been introduced in [17], based on a Convolutional Neural Network architecture for early fusion. The second strategy is the late fusion that fuses information at the decision level. This means that each modality is first learned separately and the individual results are aggregated into a final common decision [19]. An advantage of early fusion inspired approaches [4] is the fact that it utilises the correlation between multiple features from different modalities at an early stage. However when the number of modalities increases, there is a decrease in their performance due to the fact that this makes it difficult to learn the cross-correlation among the heterogeneous features. On the other hand, late fusion is much more scalable and flexible (as it enables the use of the most suitable methods for analysing each single modality) than early fusion. With respect to graph-based methods and random-walk approaches [1] present a unifying multimedia retrieval framework that incorporates two graph-based methods, namely cross-modal similarities and random-walk based scores. However, the fusion is performed at the similarity level, before the retrieval of multiple ranked lists.

In remote sensing image retrieval task both traditionally extracted features and Convolutional Neural Networks (CNN) have been investigated with the latter ones presenting performance advantage. CNN models that aim for both classification prediction and similarity estimation, called classification-similarity networks (CSNs), outputs class probability predictions and similarity scores at the same time [10]. In order to further enhance performance, the authors combined information from two CSNs. "Double fusion" is used to indicate feature fusion and score fusion. Moreover, [11] proposed a feature-level fusion method for adaptively combining the information from lower layers and Fully Connected (FC) layers, in which the fusion coefficients are automatically learned from data, and not designed beforehand. The fusion is performed via a linear combination of feature vectors instead of feature concatenation. Another work is that of [16], who performed multiple SAR-oriented visual features extraction and estimated the initial relevance scores. For the feature extraction, they constructed two bag-of-visual-words (BOVWs) features for the SAR images and another SAR-oriented feature, the local gradient ratio pattern histogram. The authors calculated a set of initial relevance scores and constructed the modal-image matrix, then they estimated the fusion similarity and eventually re-ranked the results returned based on this similarity. The work of [9] uses multiple type of features to represent high-resolution remote sensing images. One fully connected graph and one corresponding locally connected graph were constructed for each type of feature. Furthermore, a fused graph was produced by implementing a cross-diffusion operation on all of the constructed graphs. Then, from the fused graph, the authors obtained an affinity value between two nodes that directly reflects the affinity between two corresponding images. Eventually, in order to retrieve the similar images retrieval, the affinity values between the query image and the other images in the image dataset are calculated. K-order tensors appear also in graph-based fusion mechanisms, as in [2], mainly for early fusion of multiple modalities for the creation and learning of a joint representation learning.

Contrary to these approaches, we perform an unsupervised late fusion of multiple rankings, without the construction of a joint representation learning at an early stage. Our late fusion approach first aims to optimize each single-modality search, either with existing features or with customized Deep Neural Network architectures. Our late fusion approach is agnostic to the representation of each modality as a vector an is easily adaptable to any meta-search engine.

3   Methodology

For the retrieval of similar-to-a-query q content in satellite image collections $\mathcal{S}$, different modalities are combined, each one representing a different aspect of the satellite images. The considered modalities are: i) visual features, ii) visual concepts, and iii) spatiotemporal information (geographical location and time).

The overall framework (Fig. 1) involves initially a multimodal indexing scheme of each satellite image as an item with multiple modalities, such as visual features from several channels, visual concepts, spatial and temporal information. Each modality provides a similarity score and a ranked list of retrieved items that need to be combined so as to obtain a unique ranked list of satellite image patches. The query in the image collection per modality then provides a ranked list of items which are relevant to the query q, and a tensor is created (Fig. 2). Thirdly, a bi-modal fusion of the retrieved results follows for each 2D surface of the created tensor and the rankings are merged in a late fusion approach, as shown in Fig. 3.

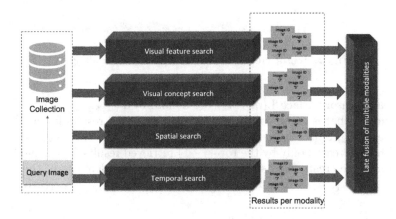

Fig. 1. Overall framework of our proposed retrieval of multiple modalities.

3.1   Late Fusion of Multiple Modalities

The proposed approach fuses the output of K modalities, where $K > 2$. For each modality, we have N retrieved results and thus we have K such lists. We set as $\mathbf{L}$ the K-order tensor of the retrieved lists, $l_1, l_2, \ldots, l_K$. A single element $\mathbf{L}_{r_1, r_2, \ldots, r_K}$ of $\mathbf{L}$ is obtained by providing its exact position through a series of indices $r_1, r_2, \ldots, r_K$, defined as follows:

$$\mathbf{L}_{r_1, r_2, \ldots, r_K} = \begin{cases} 1, & \text{if the same element is ordered as } r_1 \text{ in list } l_1, \\ & \quad \text{as } r_2 \text{ in list } l_2, \ldots, \text{ and as } r_K \text{ in list } l_K \\ 0, & \text{otherwise} \end{cases} \tag{1}$$

Given the created tensor $\mathbf{L}$ (2), one tensor 2D surface is defined for each pair of modalities $(m_1, m_2), m_1 \leq m_2, 1 \leq m_1, m_2, \leq K$.

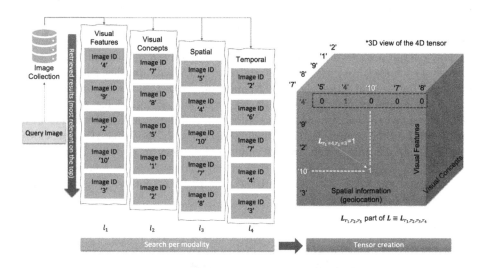

Fig. 2. Tensor creation from multiple lists of rankings.

For each tensor surface $\mathbf{L}(m_1, m_2)$, we denote by $\mathbf{L}_{r_{m_1} \leq j, r_{m_2} \leq j}(m_1, m_2)$ the tensor surface that is created only by the top-j retrieved results for the modalities m_1 and m_2. Similarly, we denote by $\mathbf{L}_{r_{m_1} \leq j-1, r_{m_2} \leq j-1}(m_1, m_2)$ the tensor surface that is created only by the top-$(j-1)$ retrieved results for the modalities m_1 and m_2. We create the list P_j, by keeping only the elements of the matrix $\mathbf{L}_{r_{m_1} \leq j, r_{m_2} \leq j}(m_1, m_2)$ which are not elements of $\mathbf{L}_{r_{m_1} \leq j-1, r_{m_2} \leq j-1}(m_1, m_2)$:

$$P_j = \mathbf{L}_{r_{m_1} \leq j, r_{m_2} \leq j}(m_1, m_2) \ominus \mathbf{L}_{r_{m_1} \leq j-1, r_{m_2} \leq j-1}(m_1, m_2) \qquad (2)$$

If $\max\{P_j\} = 1$ then there is an element that appears for the first time in more than one modalities, with rank j. In Fig. 3 we illustrate the passage from bi-modal fusion through tensor 2D surfaces to the final multimodal ranking of the retrieved results. For the tensor 2D surface that is extracted from visual concepts and visual features on the top-left we get the sequence of lists $P_1 = \{0\}$, $P_2 = \{0, 0, 0\}$, $P_3 = \{0, 0, 0, 0, 0\}$, $P_4 = \{0, 0, 0, 0, 0, 0, 0\}$, and $P_5 = \{0, 0, 1, 0, 0, 0, 0, 0, 0\}$. For the multimodal ranking of the visual concepts and visual features we get $\max\{P_j\} = 0$ for $j = 1, 2, 3, 4$, and $\max\{P_5\} = 1$, so the image ID '2' is temporarily ranked as 5th in the"Features/Concepts" list. The same procedure is followed for all pairs of modalities. Afterwards, the image IDs in each position provide altogether a merged rankings ordered list as shown in Fig. 3, where duplicates are removed and the final list is obtained.

In the following we present the uni-modal search per modality, before the creation of the unifying tensor $\mathbf{L}_{r_1, r_2, \dots, r_K}$.

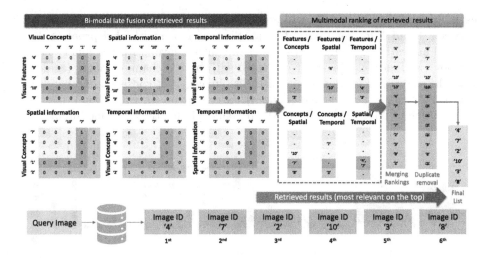

Fig. 3. Merging multiple lists from tensor surfaces.

3.2 Visual Similarity Search

For an effective search with respect to visual information, a suitable vector representation of the multi-channel Sentinel 2 image is required. After the transformation of each satellite image patch into embedding vectors, Euclidean distance calculations between the query q and the collection S represent visual similarity with respect to the content. Several visual feature representations are explored and the results are presented in Sect. 4. The visual similarity search is based on feature vectors that are extracted from pretrained networks and then Euclidean distance calculation follows on the top-N results. Feature vectors from Sentinel 2 images are extracted from specific intermediate layers of pretrained VGG19, ResNet-50 and Inception-ResNet-v2 networks [14]. Since ImageNet is a dataset of RGB images we created an input dataset of same type of images, i.e. Red (band 4), Green (band 3) and Blue (band 2) Sentinel-2 bands so as to form 3-channel patches. In VGG-19 convolutional neural network features are extracted from fc1 (dense) and fc2 (dense) layers, with feature size of 1×4096 float numbers per patch. In ResNet-50 features are extracted from avg_pool (GlobalAveraging-Pooling2) layer, with feature size of 1×2048 float numbers per patch. Finally, in Inception-ResNet-v2, which is a convolutional neural network with 164 layers the network has an image input size of 299-by-299.

3.3 Visual Concept Search

A Custom Deep Neural Network is created for the extraction of visual concept vectors and therefore to support the visual concept search by Euclidean distance. The network has a structure that resembles VGG architecture (Fig. 4). It contains blocks of convolutional layers with 3×3 filters followed by a max pooling layer. This pattern is repeating with a doubling in the number of filters with

each block added. The model will produce a 7-element vector with a prediction
between 0 and 1 for each output class. Since it is a multi-label problem, the
sigmoid activation function was used in the output layer with the binary cross
entropy loss function. For input we tested with both 3 channel images (as done
with the pretrained networks) and also with images that consisted of 5 bands of
Sentinel 2 images, i.e. the Red (band 4), Green (band 3), Blue (band 2), with
the addition of NIR (band 8) and SWIR (band 11) for the 5-channel input.

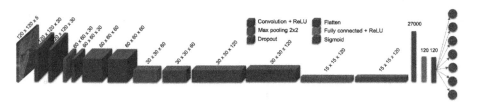

Fig. 4. The custom DCNN for visual concept search in 5-channel Sentinel 2 images.
(Color figure online)

3.4   Spatial and Temporal Search

Given a query image q, we exploit the "datetime" and "geolocation" metadata
for spatiotemporal search. Our purpose is to maximise the proximity with respect
to time and location between the query satellite image patch and the retrieved
items. Our images are indexed in a MongoDB[1], which allows spatial search
through the *geoNear* function. This function returns items ordered from the
nearest to farthest from a specified point, i.e. the centroid of the query satellite
image patch. Regarding the temporal information which is indexed in *IsoDate*
form, sorting by timestamp provides a list of items which are temporaly close
to the query image. Each modality allows unimodal search to retrieve a single
list o returns a ranked similarity list. Performing late fusion on the formed lists
returns the final sorted list with the closest images to the given query.

4   Experiments

4.1   Dataset Description

The BigEarthNet [15] dataset was selected for our experiments. The dataset
contains ground-truth annotation about Sentinel 2 level-2A satellite images and
consisted of 590,326 patches. Each image patch was annotated by the multiple land-cover classes (i.e., multi-labels) that were extracted from the CORINE
Land Cover (CLC) inventory of the year 2018. Based on the available Corine
land cover classes we group the closely related sub-classes of the CLC, forming

---

seven major classes. We selected around 130,000 patches, of resolution 120×120 pixels in order to preserve a balance among the number of items of the different classes/concepts: 1) class "rice" 2) class "urban" merges "Continuous urban fabric" and "Discontinuous urban fabric"; 3) class "bare rock"; 4) class "vineyards"; 5) class "forest" merges "Broad-leaved forest", "Mixed forest", and "Coniferous forest", 6) class "water" merges "Water courses", "Water bodies" and "Sea and ocean" classes; 7) class "snow".

4.2   Settings

For training of the custom DCNN and all pretrained deep neural networks we used Keras. Satellite image metadata and the extracted feature and concept vector are stored in MongoDB, that also allows spatial and temporal search. We select 10 test images for each of the seven classes parsed from the Corine Land Cover inventory, and thus we ended up with 70 test image patches. The procedure followed for obtaining the similarity according to the visual content involves the following steps: a) we extract feature vectors for each patch of the dataset, including the test images, b) we calculate the distance between the query image and the rest images of the dataset, c) we retrieve the images with the lowest distance from the query-test patch, and d) we calculate the mAP for the top 30 results. The learning rate for the 5-channel custom DCNN is 0.0005, the batch size is 256, and we used Adam optimizer with 200 epochs. To obtain the best possible results we enabled dropout regulation. We used the models with the best validation scores at a 5-Fold Cross-Validation.

4.3   Results

For the fusion of the results we tested our algorithm against three seminal rank fusion algorithms, namely Borda count [3], Reciprocal [13] and Condorcet [6] fusion, since they are suitable for Big Multimedia Data search [8].

Table 1. Comparison of fusion methods with mean average precision (mAP)

Classes / Method	Ours	Borda	Reciprocal	Condorcet
forest	89.56%	88.38%	60.11%	52.85%
rice	97.05%	98.92%	39.51%	66.61%
rock	62.90%	64.69%	26.53%	20.88%
snow	91.46%	89.44%	67.04%	15.22%
urban	79.96%	74.90%	53.72%	29.03%
vine	88.40%	88.25%	28.22%	19.54%
water	97.35%	97.97%	78.42%	76.05%
mAP:	**86.67%**	86.08%	50.51%	40.03%

For the evaluation of the various fusion methods we used the mean Average Precision (mAP) metric on the top-30 results that were retrieved. The comparison among our fusion method and the seminal methods of Borda, Condorcet and Reciprocal rank fusion, are shown in Table 1.

For obtaining these late fusion results, we performed experiments to identify the best performing unimodal search in visual features and concepts. The results for the pretrained and the custom neural networks are shown at Table 2 using Mean Average Precision as metric and are computed against the Corine Land Cover (CLC) annotation. The VGG19 architecture provides the optimal features for the multimodal retrieval problem. ResNet50 comes second and the Inception_v2 underperforms. Regarding the Visual concept similarity, the mAP results for the pretrained and the custom neural networks are shown at Table 3. The concepts are extracted in this case directly by the last prediction layer. The 5-channel custom DCNN obtains the best mAP results. Although it was expected that adding more channels in the DCNN architecture would lead to better performance, we observe in Table 2 that the pretrained network performs better than the custom.

Table 2. Mean average precision comparison on feature extraction of seven classes among pretrained and custom neural networks.

	Pretrained Deep Neural Networks				Custom Deep Neural Network			
classes	VGG19 fc2	VGG19 flatten	ResNet50 avg_pool	Inception ResNetV2 avg_pool	5 bands flatten	5 bands dense	3 bands flatten	3 bands dense
				top #10				
forest	83.02%	81.17%	81.66%	63.70%	76.52%	80.38%	49.22%	50.89%
rice	86.79%	75.28%	57.68%	29.21%	25.89%	17.41%	30.40%	11.57%
rock	62.21%	76.38%	59.04%	58.09%	58.37%	52.96%	86.56%	60.44%
snow	86.37%	43.96%	91.85%	88.46%	74.93%	87.79%	48.03%	79.57%
urban	68.22%	45.46%	68.25%	73.43%	73.71%	66.53%	34.60%	42.77%
vine	74.74%	76.07%	67.85%	42.75%	45.44%	47.78%	59.67%	39.51%
water	98.78%	100.00%	100.00%	96.20%	100.00%	97.11%	95.22%	92.68%
mAP	**80.02%**	71.19%	75.19%	64.55%	64.98%	64.28%	57.67%	53.92%
				top #20				
forest	78.72%	80.07%	77.63%	62.08%	76.12%	70.72%	45.98%	51.55%
rice	82.09%	72.58%	49.74%	31.58%	21.01%	15.58%	30.40%	12.80%
rock	50.41%	62.01%	51.59%	50.85%	46.30%	44.57%	83.94%	54.99%
snow	81.07%	44.04%	90.92%	88.09%	74.52%	81.62%	49.20%	66.76%
urban	61.27%	40.80%	64.92%	70.20%	69.26%	60.82%	30.85%	38.54%
vine	65.77%	70.44%	61.53%	41.98%	41.55%	43.53%	44.75%	34.45%
water	98.83%	100.00%	99.66%	97.00%	99.89%	96.58%	96.01%	92.27%
mAP	**74.02%**	67.13%	70.86%	63.11%	61.24%	59.06%	54.45%	50.19%

Table 3. Mean average precision comparison on concept extraction of seven classes among pretrained and custom neural networks.

	Pretrained Deep Neural Networks			Custom Deep Neural Network		
classes	VGG19 predictions	ResNet50 fc1000	Inception ResNetV2 fc1000	classes	5 bands dense (last)	3 bands dense (last)
			top #10			
forest	63.59%	71.28%	66.67%	forest	80.38%	49.22%
rice	19.60%	3.25%	34.16%	rice	17.41%	30.40%
rock	22.70%	14.13%	30.68%	rock	52.96%	86.56%
snow	63.48%	84.47%	91.14%	snow	87.79%	48.03%
urban	55.66%	69.17%	58.85%	urban	66.53%	34.60%
vine	44.78%	29.43%	48.55%	vine	47.78%	59.67%
water	93.44%	97.47%	99.77%	water	97.11%	95.22%
mAP	51.89%	52.74%	61.40%	mAP	**64.28%**	57.67%
			top #20			
forest	61.38%	66.58%	60.16%	forest	70.72%	45.98%
rice	18.49%	3.54%	27.07%	rice	15.58%	30.40%
rock	22.98%	15.85%	28.73%	rock	44.57%	83.94%
snow	62.33%	83.91%	87.47%	snow	81.62%	49.20%
urban	54.03%	65.62%	54.58%	urban	60.82%	30.85%
vine	39.65%	28.15%	43.32%	vine	43.53%	44.75%
water	93.49%	96.73%	98.52%	water	96.58%	96.01%
mAP	50.33%	51.48%	57.12%	mAP	**59.06%**	54.45%

We demonstrate the top-10 retrieved results for our proposed approach in Fig. 5. The satellite image patch on the left is the query and the retrieved results follow. The results show that, for the urban query, most of the misclassified results are rice and these two classes resemble to each other making difficult for the DCNNs to discriminate among them. Moreover, some of the retrieved images are of the forest class, because in some cases they depict sparse country-side areas mixed with snow. Furthermore, in the "vineyards" query, almost all the misclassified images were actually urban patches, and there is great similarity between these two classes, and even for a human it is difficult to classify. Finally, rock queries are mostly rocky areas near water, resulting to fetc.hing many water patches.

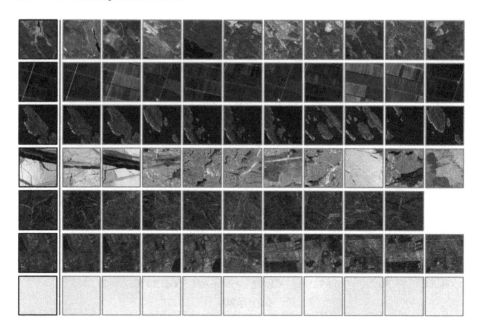

Fig. 5. Query and top-10 retrieved results with the proposed late fusion approach.

5   Conclusions

In this work we proposed a novel late fusion method that combines the outputs of K ranking lists using a K-order tensor approach. The method is agnostic to the representation of each modality in each unimodal search. However, we examined the performance of several DCNN architectures and we proposed one for concept search in 5-channel Sentinel 2 image patches. Satellite images contain more than the three optical RGB channels that can also be exploited in unimodal similarity search. Our overall framework uses 5 channels from Sentinel 2 image patches to extract concepts, and combines them with visual features and spatiotemporal information to allow multimodal similarity search scenarios. Finally, the results show the importance of combining multiple modalities of an image in similarity search and we illustrate the top-10 results per late fusion method and per query. Future work includes the use of more band-channels, and combination of more types of satellite images of different resolutions.

Acknowledgements. This work was supported by the EC-funded projects H2020-832876-aqua3S and H2020-776019-EOPEN.

References

1. Ah-Pine, J., Csurka, G., Clinchant, S.: Unsupervised visual and textual information fusion in CBMIR using graph-based methods. ACM Trans. Inf. Syst. (TOIS) **33**(2), 1–31 (2015)
2. Arya, D., Rudinac, S., Worring, M.: Hyperlearn: a distributed approach for representation learning in datasets with many modalities. In: Proceedings of the 27th ACM International Conference on Multimedia, pp. 2245–2253 (2019)
3. Aslam, J.A., Montague, M.: Models for metasearch. In: Proceedings of the 24th Annual International ACM SIGIR Conference on Research and Development in Information Retrieval, pp. 276–284 (2001)
4. Atrey, P.K., Hossain, M.A., El Saddik, A., Kankanhalli, M.S.: Multimodal fusion for multimedia analysis: a survey. Multimedia Syst. **16**(6), 345–379 (2010)
5. Caicedo, J.C., BenAbdallah, J., González, F.A., Nasraoui, O.: Multimodal representation, indexing, automated annotation and retrieval of image collections via non-negative matrix factorization. Neurocomputing **76**(1), 50–60 (2012)
6. Cormack, G.V., Clarke, C.L., Buettcher, S.: Reciprocal rank fusion outperforms condorcet and individual rank learning methods. In: Proceedings of the 32nd International ACM SIGIR Conference on Research and Development in Information Retrieval, pp. 758–759 (2009)
7. Feng, F., Wang, X., Li, R.: Cross-modal retrieval with correspondence autoencoder. In: Proceedings of the 22nd ACM International Conference on Multimedia, pp. 7–16 (2014)
8. Gialampoukidis, I., Chatzilari, E., Nikolopoulos, S., Vrochidis, S., Kompatsiaris, I.: Multimodal fusion of big multimedia data. In: Big Data Analytics for Large-Scale Multimedia Search, pp. 121–156 (2019)
9. Li, Y., Zhang, Y., Tao, C., Zhu, H.: Content-based high-resolution remote sensing image retrieval via unsupervised feature learning and collaborative affinity metric fusion. Remote Sens. **8**(9), 709 (2016)
10. Liu, Y., Chen, C., Han, Z., Ding, L., Liu, Y.: High-resolution remote sensing image retrieval based on classification-similarity networks and double fusion. IEEE J. Sel. Top. Appl. Earth Obs. Remote Sens. **13**, 1119–1133 (2020)
11. Liu, Y., Liu, Y., Ding, L.: Scene classification based on two-stage deep feature fusion. IEEE Geosci. Remote Sens. Lett. **15**(2), 183–186 (2017)
12. Magalhães, J., Rüger, S.: An information-theoretic framework for semantic-multimedia retrieval. ACM Trans. Inf. Syst. (TOIS) **28**(4), 1–32 (2010)
13. Montague, M., Aslam, J.A.: Condorcet fusion for improved retrieval. In: Proceedings of the Eleventh International Conference on Information and Knowledge Management, pp. 538–548 (2002)
14. Moumtzidou, A., et al.: Flood detection with sentinel-2 satellite images in crisis management systems. In: ISCRAM 2020 Conference Proceedings - 17th International Conference on Information Systems for Crisis Response and Management, pp. 1049–1059 (2020)
15. Sumbul, G., Charfuelan, M., Demir, B., Markl, V.: Bigearthnet: a large-scale benchmark archive for remote sensing image understanding. In: IGARSS 2019–2019 IEEE International Geoscience and Remote Sensing Symposium, pp. 5901–5904. IEEE (2019)
16. Tang, X., Jiao, L.: Fusion similarity-based reranking for SAR image retrieval. IEEE Geosci. Remote Sens. Lett. **14**(2), 242–246 (2016)

17. Wang, J., He, Y., Kang, C., Xiang, S., Pan, C.: Image-text cross-modal retrieval via modality-specific feature learning. In: Proceedings of the 5th ACM on International Conference on Multimedia Retrieval, pp. 347–354 (2015)
18. Wang, W., Ooi, B.C., Yang, X., Zhang, D., Zhuang, Y.: Effective multi-modal retrieval based on stacked auto-encoders. Proc. VLDB Endow. **7**(8), 649–660 (2014)
19. Younessian, E., Mitamura, T., Hauptmann, A.: Multimodal knowledge-based analysis in multimedia event detection. In: Proceedings of the 2nd ACM International Conference on Multimedia Retrieval, pp. 1–8 (2012)

Canopy Height Estimation from Spaceborne Imagery Using Convolutional Encoder-Decoder

Leonidas Alagialoglou[1]([⊠]), Ioannis Manakos[2]([ID]), Marco Heurich[3],
Jaroslav Červenka[4], and Anastasios Delopoulos[1]([ID])

[1] Multimedia Understanding Group, Electrical and Computer Engineering
Department, Aristotle University of Thessaloniki, 54124 Thessaloniki, Greece
`lalagial@mug.ee.auth.gr, adelo@eng.auth.gr`
[2] Information Technologies Institute, Centre for Research and Technology Hellas
(CERTH), Charilaou-Thermi Rd. 6th km, 57001 Thessaloniki, Greece
`imanakos@iti.gr`
[3] Department of Nature Protection and Research, Bavarian Forest National Park,
Freyunger Str. 2, 94481 Grafenau, Germany
`marco.heurich@npv-bw.bayern.de`
[4] Šumava National Park, Sušická 339, 34192 Kašperské Hory, Czech Republic
`jaroslav.cervenka@npsumava.cz`
`https://mug.ee.auth.gr/`

Abstract. The recent advances in multimedia modeling with deep learning methods have significantly affected remote sensing applications, such as canopy height mapping. Estimating canopy height maps in large-scale is an important step towards sustainable ecosystem management. Apart from the standard height estimation method using LiDAR data, other airborne measurement techniques, such as very high-resolution passive airborne imaging, have also shown to provide accurate estimations. However, those methods suffer from high cost and cannot be used at large-scale nor frequently. In our study, we adopt a neural network architecture to estimate pixel-wise canopy height from cost-effective spaceborne imagery. A deep convolutional encoder-decoder network, based on the SegNet architecture together with skip connections, is trained to embed the multi-spectral pixels of a Sentinel-2 input image to height values via end-to-end learned texture features. Experimental results in a study area of 942 km^2 yield similar or better estimation accuracy resolution in comparison with a method based on costly airborne images as well as with another state-of-the-art deep learning approach based on spaceborne images.

Keywords: Canopy height estimation · Deep learning · Convolutional encoder-decoder · Sentinel-2

© Springer Nature Switzerland AG 2021
J. Lokoč et al. (Eds.): MMM 2021, LNCS 12573, pp. 307–317, 2021.
https://doi.org/10.1007/978-3-030-67835-7_26

1   Introduction

The recent advances in multimedia modeling with deep learning methods have significantly affected remote sensing applications. Based on the large amounts of daily earth observations, it is now possible to timely and accurately monitor vital parameters of the ecosystem to assess its status and to support decision making towards sustainable land management. Among such variables, canopy height of forests is a fundamental structural and biophysical parameter, useful to a number of environmental studies and applications, such as biodiversity studies [5], conservation planning, biomass and carbon sources estimation [14], as well as monitoring forest degradation in large scales, such as illegal logging tracking [8].

While traditional methods for canopy height measurement involve terrestrial manual inspection, remotely sensed light detection and ranging (LiDAR) data have become the main tool for creating height maps in larger scale. Airborne LiDAR instruments can measure height maps with ground sampling distance (GSD) <1 m and high accuracy. However, those methods costly and cannot be used at large-scale and/or frequently. In the direction of acquiring LiDAR data in a global-scale, Global Ecosystem Dynamics Investigation (GEDI) mission provides reliable global-scale CHM estimates in a regular basis from a spaceborne LiDAR sensor, but its GSD is limited to 25 m. [4]

Less costly ultra high-resolution passive airborne imagery has also been used to infer CHM using machine learning techniques. In specific, Boutsoukis et al. [2] extract a number of texture features based on local variance, entropy, and binary patterns is used to discriminate 6 or 4 vegetation height classes as defined in the general habitat category (GHC) taxonomy. Airborne RGB and NIR images with GSD of 40 cm has been used as input. Classification on an object-wise manner (i.e. averaging of pixels belonging in objects delineated from a land cover map) resulted in 91.39% area-based accuracy. Similar texture features have been extracted from spaceborne WorldView-2 imagery of GSD 2 m, and tested over different classifiers for object-wise height classes estimation in the Netherlands with a variety of vegetation types, ranging from deciduous and coniferous trees to shrubs and heathlands [9,10].

Based on the success of deep learning methods in many applications, end-to-end learning solutions are being widely deployed in remote sensing. The work of Lang et al. [7] firstly introduced end-to-end learning of rich contextual feature hierarchies in large-scale CHM estimation from multi-spectral Sentinel-2 images. Based on the Xception model architecture [6], they adapt a convolutional neural network (CNN) for the regression of vegetation height map at 10m GSD. Their dataset includes LiDAR ground-truth and few Sentinel-2 snapshots from two sufficiently large areas of different geographic coordinates (Switzerland and Gabon). Correction of atmospheric effects was applied as a preprocessing step using ESA's sen2cor toolbox. The accuracy of the estimated CHM (in forest areas, after filtering out outliers and high slope areas) was assessed and achieved a root mean squared error (RMSE) between 3.4 m and 5.6 m (mean absolute error (MAE) of 1.7 m–4.3 m).

In this paper, we adopt two variants of a convolutional encoder-decoder architecture to estimate pixel-wise values of canopy height based on a single multispectral Sentinel-2 image. In order to avoid cloud covered areas, three dates with minimum cloud coverage have been used to drive the model separately and yield a median single pixel value. The model is trained and tested in a total forest area of approximately 942 km^2 and comparison results with state-of-the-art implementations, involving the same or different datasets, are given. Firstly, we aim at pixel-wise comparison with the study of Lang et al. [7], based on two bigger datasets in Switzerland and Gabon and secondly. Furthermore, we perform object-wise comparison in an aggregated level with the previous study of Boutsoukis et al. [2] in the same dataset as ours.

2 Materials

The experimental study area of the Bohemian Forest (BF) includes two national parks. Bavarian Forest National Park and Šumava National Park are located at the border between Germany (Bavaria) and Czech Republic. The forest area is characterized as mountainous with an altitude ranging between 600 and 1453 m, while the temperate climate is cold and humid. Locations of the highest altitude are covered with heavy and long-lasting snow for more than half of a year (6 to 7 months). The dominant tree species include Norway spruce (*Picea abies*), European beech (*Fagus sylvatica*), and silver fir (*Abies alba*) [3].

LiDAR measurements in the study area that yielded Digital Elevation Model (DEM) and Digital Surface Model (DSM) were used to calculate canopy height model as the ground truth for training and evaluating the model. The dataset spans in a total measured forest area of approximately 942 km^2 and has been collected in June 2017. The Riegl 680i sensor was used for the measurements (350 KHz pulse repetition rate, nominal point density of 30–40 pts/m^2, average altitude 650 m above ground) at a 0.32 m pulse footprint diameter and 300–400 m swath width (depending on flight altitude) by Milan Flug GmbH. The ground sampling distance of the measured DEM and DSM is 1m. Subtraction of the two raster layers and bilinear downsampling yields the CHM model in GSD of 10m, with the use of GDAL library.

Furthermore, Sentinel-2 Level-2A products representing bottom-of-atmosphere reflectances in cartographic geometry have been downloaded from European Space Agency's (ESA) Copernicus Hub. The whole study area is covered by tile number fields T33UUP, T33UUQ, T33UVQ, T33UQV. A cloud coverage filter <4% was applied in the datatake sensing time from April to July 2017 to select three available products for all tiles of the study area. The selected dates without clouds are 24/04, 13/06 and 13/07 of 2017.

For evaluation purposes, a land cover map, generated by local experts for a smaller part of the study area in 2012, was used to delineate the landscape patches(objects) within the image. [13] Aggregated pixel values of height for each landscape patch is calculated to allow results comparison in an object-wise manner with the past work of Boutsoukis et al. [2].

3   Methodology

In our method, we adopt two variations of a deep convolutional encoder-decoder architecture, based on SegNet as proposed by Badrinarayanan et al. [1] for semantic labeling purposes. In specific, four convolutional blocks correspond to encoding the multi-spectral input image into a multi-dimensional representation, that can be though as extracted features. Following the feature extraction layers, four symmetrical decoding convolutional blocks are applied in order to generate the height map via a final fully connected layer for each pixel.

Each convolutional block of the encoder includes a 2D convolution layer with 3×3 kernels, followed by a ReLU non-linearity. Downsampling in each encoding block is performed with max-pooling, while keeping track of the pooling indices to be used in the unpooling of the corresponding decoding blocks, as introduced by SegNet [1]. A 0.5 dropout layer during training is applied after the last two encoding blocks to avoid overfitting. We call the first variation of the network architecture (as already described) 'ConvEnc-Dec', while the second variation is named 'ConvEnc-DecSkip'. Based on the U-Net architecture [11], the additional feature in the second variation is the reuse of each encoding block's output in the corresponding decoding block via skip connections and concatenation. The complete model architecture of 'ConvEnc-DecSkip' is shown in Fig. 1. A similar schematic diagramm, without the skip connections, shall be considered for the 'ConvEnc-Dec' model.

Mean squared error $MSE = \frac{1}{n} \sum_{i=1}^{n} (\mathbf{y}_i - \hat{\mathbf{y}}_i)$, between the estimated height values $(\hat{\mathbf{y}})$ and target height values $(\mathbf{y})$ for all the training tiles is used as cost function to be optimized with Adam optimizer in batches of 16 tiles. No other regularization is used since the large training set combined with the dropout layers is sufficient to avoid overfitting.

The whole dataset is split into non overlapping tiles of size 48×48 pixels. Experiments on land cover classification, which is a similar learning task, has shown minimum impact of the size of non-overlapping tiles [12]. The tiles in the edge of the measured area have been discarded to avoid influence of no-data values in training as well as testing. The 80% (6.5 Mpixels $\rightarrow$ 650 km^2) of the tiles in the dataset is used for training, while 10% is used for validation during training to quide hyper-parameter selection and training stop time. The remaining 10% (0.8 Mpixels $\rightarrow$ 80 km^2) of the tiles (0.79 Mpixels $\rightarrow$ 79 km^2) is used for testing purposes.

The model was trained in a NVIDIA GeForce GTX 1080Ti graphics card and the whole process lasted approximately 20 h. Inference time is relatively quick as long as the dataset is downloaded and split into the appropriate tiles; thus, allowing a realistic implementation of large-scale use in web applications[1].

---

[1] For example, Samaria's Data Cube is a tool for satellite data users to monitor and evaluate land resources and land change. http://datacube.iti.gr/.

Convolutional encoder-decoder with skip connections

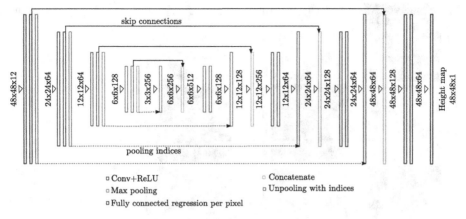

Fig. 1. Schematic diagram of the convolutional encoder-decoder architecture with skip connections. The multi-dimensional feature map of the final decoding layer is fed in a fully connected layer for the regression of the height value per pixel

4 Experimental Results

4.1 Results

A number of randomly selected 48×48 CHM tiles with GSD of 10m from the test set are shown in Fig. 2. The CHM estimations are inferred from the 'ConvEnc-DecSkip' model using the median value of each pixels from the three Sentinel-2 input images in different dates. The LiDAR measured CHM for the same tiles that is considered as ground-truth, as well as the absolute values of estimation error per pixel ($|\mathbf{y} - \hat{\mathbf{y}}|$) are given in the same figure for visual inspection. A scatterplot of the ground-truth values versus the estimated ones for all pixels of the testing set is given in Fig. 3.

Qualitative evaluation of the models' performance is performed in two manners, pixel-wise and aggregated object-wise. Based on the ground truth CHM from LiDAR measurements, RMSE and MAE is computed from the estimated CHM of all pixels in the test set ($79\,\mathrm{km}^2$). In Table 1 the calculated pixel-wise RMSE and MAE are compared with the state-of-the-art work of Lang et al. [7], that utilizes 4–12 Sentinel-2 images with <70% cloud coverage for CHM estimation in Table 1. The two studies involve different datasets, but the main results may be set side by side.

The object-wise testing method is an aggregation of pixel-wise CHM estimation based on the delineation of landscape patches (objects) of the available land cover map of 2012. Only a part of the total study area (0.21 Mpixels $\rightarrow$ $21\,\mathrm{km}^2$) is included in the object-wise evaluation that cover 2604 objects which is slightly smaller than the work of Boutsoukis et al. [2]. In [2], all testing objects have been separated into 3 size classes (large, medium, small) and distinct results

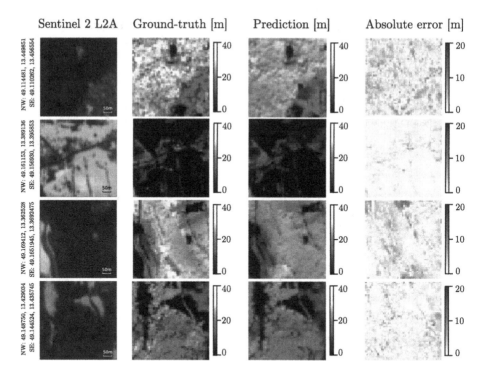

Fig. 2. Estimated canopy height tiles (Predictions) of 48 × 48 pixels with GSD of 10m. Ground truth CHM using LiDAR measurements (Target) and absolute values of estimation error per pixel. Color infrared band combination of the Sentinel 2 L2A input image is also provided with tile coordinates in WGS84. (Color figure online)

are given for each class, whereas in our work all sizes are included in a single result. A mean value of height was calculated for each delineated object from all pixels in it and the area-based accuracy was calculated based on the size of each object. For comparison reasons, the estimated value of each object is quantized in 4 classes similarly to the method of [2], with limits in height meters $[0, 0.6)$, $[0.6, 2)$, $[2, 5)$, $[5, 40)$. Object-based accuracy (i.e. percentage of correctly classified objects) and area-based results (i.e. percentage of correctly classified area) are given in Table 2.

4.2 Estimation Error Analysis

By visual inspection of the scatterplot in Fig. 3, a larger estimation error appears in higher height values (in the right part of the plot). This fact of larger error in estimating higher vegetation, has also been observed in the work of Lang et al. [7] and can be identified more clearly in the histogram of Fig. 4. The mean absolute error for groups of the ground-truth height values with interval 10 m is given in Fig. 4, demonstrating the positive correlation of estimation error and height value.

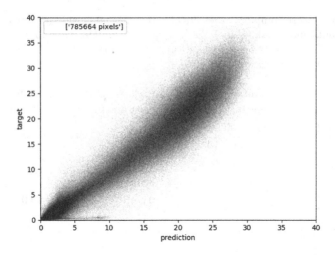

Fig. 3. Scatterplot of ground-truth versus estimated values for each pixel using 'ConvEnc-DecSkip'.

Table 1. Pixel-wise comparison results of convolutional encoder-decoder neural network with Lang et al. [7].

Method	Location	Dataset size	MAE	RMSE
Lang et al. [7]	Switzerland	91Mpx	1.7	3.4
Lang et al. [7]	Gabon	25Mpx	4.3	5.6
ConvEnc-Dec	BF	9.4Mpx	3.32	4.42
ConvEnc-DecSkip	BF	9.4Mpx	**2.29**	**3.15**

An error analysis based on the slope and aspect of each pixel location is performed for the 'ConvEnc-DecSkip' model. In Fig. 5, the mean absolute error for different groups (i.e bins) of test pixels is plotted in respect to slope and aspect. The edges of each bin have been selected in order to include the same number of training pixels (e.g. the same amount of training pixels is used with slope 17°–22° as well as 22°–60°). The rationale behind this choice is to eliminate the factor of training set size in the comparison of the performance for each bin. We can visually observe in Fig. 5a that absolute estimation error is larger for pixels in steeper slopes, i.e height estimation for a large number of pixels in flat areas yielded MAE close to 3.0 m, while pixels in very steep terrains yield MAE more than 4.0 m. An potential improvement in the model, using this result, could be with a modified network architecture that incorporates this additional information of terrain's slope. On the other hand, we observe in Fig. 5b minor correlation between the estimation error (<0.2 m) and the aspect of a pixel, thus not further investigating it.

Table 2. Object-wise comparison results of convolutional encoder-decoder neural network with Boutsoukis et al. [2] in 4 classes quantization. Different objects that belong to our testing dataset have been selected than the compared study, based on the same delineation of the same region.

Method	Number of class objects	Object-based accuracy (%)	Area-based accuracy(%)
Boutsoukis et al. [2]	Large: 90	91.11	91.39
	Medium.: 1671	80.73	
	Small: 2006	66.55	
ConvEnc-Dec	2604	88.02	90.71
ConvEnc-DecSkip	2604	**91.40**	**94.10**

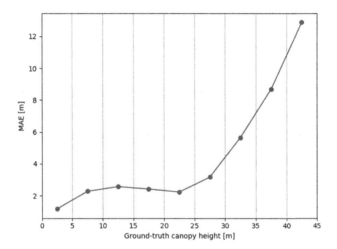

Fig. 4. Mean absolute errors for different groups of ground-truth height values using 'ConvEnc-DecSkip' method.

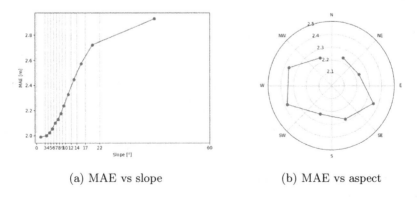

(a) MAE vs slope   (b) MAE vs aspect

Fig. 5. Correlation of mean absolute error (MAE) of canopy height estimation with slope and aspect of pixel location ('ConvEnc-DecSkip' model).

5 Discussion

Visual inspection of the estimated maps using 'ConvEnc-DecSkip' model suggest that information contained in multi-spectral spaceborne imagery can be accurately mapped in the height space. However, limitations in accuracy, can be observed visually especially in larger heights, mainly due to aleatoric uncertainty (inherent data ambiguity) as well as epistemic uncertainty (model insufficiency). The above mentioned uncertainties cannot be quantified directly from the presented model. The outcome observations of the visual inspection are further confirmed in the qualitative assessment.

Qualitative evaluation in pixel-wise manner yielded MAE of 2.29 which outperforms results in Gabon (MAE = 4.3) and approaches the results in Switzerland (MAE = 1.7), based on the state-of-the-art study of Lang et al. [7]. However, RMSE (=3.15)values of 'ConvEnc-DecSkip' outperforms both experiments in [7] (RMSE = 3.4 in Switzerland - RMSE = 5.6 in Gabon), which indicates that our model yields fewer highly erroneous estimations. In the object-wise comparison with the work of Boutsoukis et al. [2] that incorporates feature extraction with that are classified with support vector machines, the convolutional encoder-decoder network has performed better, although the given GSD of the spaceborne imagery (10 m) is significantly lower than of the airborne orthomosaic (40 cm).

The present work is considered to be a preliminary step for future work, since it investigates the performance of a basic deep learning approach to the canopy height estimation problem. This can be further used in comparison results with more sophisticated methods to allow for timely and robust estimation of height map in practical applications. Our focus lies on the exploitation of temporal correlations in sequential spaceborne imagery, apart from the spatial and spectral features that single-shot images allow.

Two major challenges that need to be addressed in future work include the time invariance of the a developed model (how well can it be used on past or future satellite images), as well as it's location invariance (how well can it be used or what need to be calibrated for different geographic positions). For this purpose further input and ground truth dataset should be used for testing purposes in different geographic positions and time.

Furthermore, fusion techniques of heterogeneous data sources can increase the accuracy of estimation. For example, the valuable measurements of the GEDI mission, if fused with the higher-resolution Sentinel-2 images can potentially improve performance of the present work. Finally, an important, but unexplored aspect of canopy height mapping, is the estimation of uncertainty in the output. A confidence value in each pixel of the estimated height map is necessary for a robust and practical estimation framework.

Acknowledgements. This study has been partially funded and supported by the European Union's Horizon 2020 innovation program under Grant Agreement No. 820852, e-shape (https://e-shape.eu/). LIDAR data was granted by the Cross-border cooperation programme Czech Republic–Bavaria Free State ETC goal 2014–2020, the Interreg V project No. 99 "Přeshraniční mapovani lesnich ekosystemů – cesta ke společnemu managementu NP Šumava a NP Bavorsky les /Grenzuberschreitende Kartierung der Waldokosysteme – Weg zum gemeinsamen Management in NP Sumava und NP Bayerischen Wald". We acknowledge the support of the "Data Pool Initiative for the Bohemian Forest Ecosystem" data-sharing initiative of the Bavarian Forest National Park.

References

1. Badrinarayanan, V., Handa, A., Cipolla, R.: SegNet: A Deep Convolutional Encoder-Decoder Architecture for Robust Semantic Pixel-Wise Labelling, May 2015. http://arxiv.org/abs/1505.07293
2. Boutsoukis, C., Manakos, I., Heurich, M., Delopoulos, A.: Canopy height estimation from single multispectral 2D airborne imagery using texture analysis and machine learning in structurally rich temperate forests. Remote Sens. **11**(23) (2019). https://doi.org/10.3390/rs11232853. www.mdpi.com/journal/remotesensing
3. Cailleret, M., Heurich, M., Bugmann, H.: Reduction in browsing intensity may not compensate climate change effects on tree species composition in the Bavarian Forest National Park. For. Ecol. Manag. **328**, 179–192 (2014)
4. Dubayah, R., et al.: The global ecosystem dynamics investigation: high-resolution laser ranging of the earth's forests and topography. Sci. Remote Sens. **1**, 100002 (2020). https://doi.org/10.1016/j.srs.2020.100002
5. Goetz, S., Steinberg, D., Dubayah, R., Blair, B.: Laser remote sensing of canopy habitat heterogeneity as a predictor of bird species richness in an eastern temperate forest, USA. Remote Sens. Environ. **108**(3), 254–263 (2007). https://doi.org/10.1016/j.rse.2006.11.016
6. Kaiser, Ł., Gomez, A.N., Chollet, F.: Depthwise separable convolutions for neural machine translation. In: 6th International Conference on Learning Representations, ICLR 2018 - Conference Track Proceedings. International Conference on Learning Representations, ICLR (2018)
7. Lang, N., Schindler, K., Wegner, J.D.: Country-wide high-resolution vegetation height mapping with Sentinel-2. Remote Sens. Environ. **233** (2019). https://doi.org/10.1016/j.rse.2019.111347. https://arxiv.org/abs/1904.13270
8. Mitchell, A.L., Rosenqvist, A., Mora, B.: Current remote sensing approaches to monitoring forest degradation in support of countries measurement, reporting and verification (MRV) systems for REDD+. Carbon Balance Manag. **12**(1), 1–22 (2017). https://doi.org/10.1186/s13021-017-0078-9
9. Petrou, Z.I., Tarantino, C., Adamo, M., Blonda, P., Petrou, M.: Estimation of vegetation height through satellite image texture analysis. In: ISPRS - International Archives of the Photogrammetry, Remote Sensing and Spatial Information Sciences XXXIX-B8, pp. 321–326 (2012). https://doi.org/10.5194/isprsarchives-xxxix-b8-321-2012. http://www.academia.edu/download/43548529/isprsarchives-XXXIX-B8-321-2012.pdf
10. Petrou, Z.I., Manakos, I., Stathaki, T., Mucher, C.A., Adamo, M.: Discrimination of vegetation height categories with passive satellite sensor imagery using texture analysis. IEEE J. Sel. Top. Appl. Earth Obs. Remote Sens. **8**(4), 1442–1455 (2015). https://doi.org/10.1109/JSTARS.2015.2409131. https://ieeexplore.ieee.org/abstract/document/7061969/

11. Ronneberger, O., Fischer, P., Brox, T.: U-Net: convolutional networks for biomedical image segmentation. In: Navab, N., Hornegger, J., Wells, W.M., Frangi, A.F. (eds.) MICCAI 2015. LNCS, vol. 9351, pp. 234–241. Springer, Cham (2015). https://doi.org/10.1007/978-3-319-24574-4_28

12. Rußwurm, M., Koerner, M.: Multi-temporal land cover classification with sequential recurrent encoders. ISPRS Int. J. Geo-Inf. **7**(4), 129 (2018). https://doi.org/10.3390/ijgi7040129. http://www.mdpi.com/2220-9964/7/4/129

13. Silveyra Gonzalez, R., Latifi, H., Weinacker, H., Dees, M., Koch, B., Heurich, M.: Integrating LiDAR and high-resolution imagery for object-based mapping of forest habitats in a heterogeneous temperate forest landscape. Int. J. Remote Sens. **39**(23), 8859–8884 (2018). https://doi.org/10.1080/01431161.2018.1500071. https://www.tandfonline.com/action/journalInformation?journalCode=tres20

14. Wang, X., Ouyang, S., Sun, O.J., Fang, J.: Forest biomass patterns across northeast China are strongly shaped by forest height. Forest Ecol. Manag. **293**, 149–160 (2013). https://doi.org/10.1016/j.foreco.2013.01.001

Implementation of a Random Forest Classifier to Examine Wildfire Predictive Modelling in Greece Using Diachronically Collected Fire Occurrence and Fire Mapping Data

Alexis Apostolakis[✉], Stella Girtsou, Charalampos Kontoes, Ioannis Papoutsis, and Michalis Tsoutsos

National Observatory of Athens, Athens, Greece
{alex.apostolakis,sgirtsou,kontoes,ipapoutsis,mtsoutsos}@noa.gr

Abstract. Forest fires cause severe damages in ecosystems, human lives and infrastructure globally. This situation tends to get worse in the next decades due to climate change and the expected increase in the length and severity of the fire season. Thus, the ability to develop a method that reliably models the risk of fire occurrence is an important step towards preventing, confronting and limiting the disaster. Different approaches building upon Machine Learning (ML) methods for predicting wildfires and deriving a better understanding of fires' regimes have been devised. This study demonstrates the development of a Random Forest (RF) classifier to predict "fire"/"non fire" classes in Greece. For this a prototype and representative for the Mediterranean ecosystem database of validated fires and fire related features has been created. The database is populated with data (e.g. Earth Observation derived biophysical parameters and daily collected climatic and weather data) for a period of nine years (2010–2018). Spatially it refers to grid cells of 500 m wide where Active Fires (AF) and Burned Areas/Burn Scars (BSM) were reported during that period. By using feature ranking techniques as Chi-squared and Spearman correlations the study showcases the most significant wildfire triggering variables. It also highlights the extent by which the database and selected features scheme can be used to successfully train a RF classifier for deriving "fire"/"non-fire" predictions over the country of Greece in the prospect of generating a dynamic fire risk system for daily assessments.

Keywords: Fire/non-fire classification · Machine Learning (ML) · Random Forest (RF) · Burned Scar Mapping (BSM) · EFFIS · FIRMS

1 Introduction

Forest fires affect severely the natural and rural ecosystems, human lives, critical infrastructures and assets in general. Last year's bush fires in New South Wales of Australia burned about 1.65 million hectares and claimed the lives of 6 people [1]. In the Amazonian forested ecosystems, the Brazilian Space Agency has reported during August 2019 a significant increase in fire occurrences of the order of 83% compared to the same

© Springer Nature Switzerland AG 2021
J. Lokoč et al. (Eds.): MMM 2021, LNCS 12573, pp. 318–329, 2021.
https://doi.org/10.1007/978-3-030-67835-7_27

period of the previous year. Furthermore, the extreme drought and heat wave events during the summer seasons of 2017 and 2018 have resulted in extraordinary fire events that affected severely the Mediterranean ecosystems [2]. The problem becomes considerably important if we account for the climate change scenarios which suggest substantial warming and increase of heat waves, drought and dry spell events across the entire Mediterranean [3] in the future years. In this regard, the access to validated information on the spatiotemporal patterns of wildfire behavior, from past fire occurrences and fire triggering factors, are of great importance for the implementation of the proper disaster risk mitigation policies.

In this context, a number of fire danger rating and prevention systems have been developed in Europe ingesting weather, topography, fuel, and fire ignition data [4, 5]. As stated in [6, 7] the models which have been used for predicting the fire ignition susceptibility in an area are classified into three groups; theoretical (or physics-based), semi-empirical, and empirical models. Theoretical and semi-empirical models are entirely or partly based on equations that describe the physics of the related to the fire ignition physical phenomena like fluid mechanics, combustion and heat transfer. On the other hand the empirical models are purely based on the statistical correlations between data extracted from historical fire records and their related environmental, biophysical, morphological, fuel, and climatic/weather data. These historical data can be regarded as vectors of variables which are known or believed to influence the ignition of a fire, and provide the essential input knowledge for the model to identify which are the areas and the time periods at high risk. Empirical models, gaining knowledge from long lasting histories of fire events and their triggering parameters have used either statistical correlations or ML methods as in the present study [7, 8].

ML algorithms learn directly from the data and develop their own internal model without being necessary to provide any expert knowledge or simulate precisely the physical parameters that feed the model running. Moreover, ML models detect and automatically formulate the complex mathematical relations that exist between the diverse input parameters and this is an important advantage over the physical-based models where the mathematics of those relations should be known in advance [8].

Regarding the parameters that affect the presence of wildfires, several studies referenced in [9] and [10], show that vegetation proxies as NDVI, topography (altitude, slope, aspect), soil moisture, fuel, and meteorological data are considerably influencing factors for fire risk and fire ignition.

Aiming to advance further the relevant research, this work provides the foundations of a data driven model that covers the Greece's national needs for predicting "fire"/"non fire" prone areas at the enhanced spatial resolution of 500 m. A prototype feature database was generated and organized in the form of a datacube structure, using a nine-year lasting record (2010–2018) of multi-source and multi-sensor data such as EO based essential environmental and ecosystem related indexes, meteo records, fuel classes, morphological features with diverse spatial and temporal resolutions. Furthermore, a Random Forest (RF) algorithm [8], has been parameterized and successfully trained to produce a satisfactory prediction of "fire"/"non fire" class mapping on a daily basis over Greece.

2 Study Area - Training Data Set

2.1 Study Area

The area of interest covers the Greece's territory (131.957 km^2) located in the southeast of the Mediterranean climatic zone, with mild and rainy winters, warm and dry summers and extended periods of sunshine throughout most of the year [11].

A major part of the country, up to 58.8% of the total surface, represents low altitude areas (0–500m) which are prone to fire ignition [10]. The topography and the dominant north winds in combination with the vegetation types in the central and southern parts of Greece are between the prime drivers for fire ignition during the summer period [10]. Last but not least the vegetation cover that makes Greece particularly prone to fire hazard and fire risk such as coniferous and mixed forests, sclerophyllous vegetation, natural grasslands, transitional woodlands, semi natural and pasture areas correspond to approximately 72% of the total surface of the country (source: Copernicus CORINE Land Cover 2018 of Greece) [12]. It is worth noting that Greece consists the typical case of the Mediterranean ecosystem in regard to fire risk and expected damages similar to the ones of France, Italy, Portugal, Spain, and Turkey, with the most severe events being associated with strong winds at lower altitudinal zones (<1000 m) during hot dry summer periods [12], and extended heat waves that tend to be longer and drier due to the climate change [13].

The official annual statistics [14, 15] confirm that the average number of fires and the corresponding burned areas in Greece have increased by a factor of 4 during the last decades. This makes obvious that wildfires continue to constitute the major threat for the environment and the society in Greece, often with significant cost in human lives, as in the extreme case of the fire in Mati on July 23, 2018, which claimed the lives of 102 people. This highlights the need for setting and put in operation reliable fire management systems providing knowledge on the contemporary fire regimes and supporting the preparedness and emergency response procedures.

2.2 Data Resources

Forest Fire Inventory. A reliable forest fire inventory is vital for the prediction of wildfires regimes in an area, given that new fire occurrences in the same location are used to happen under similar weather and environmental conditions. For this, an exhaustive forest fire inventory of fire occurrences and burn scar maps was compiled by exploiting diachronic data generated by the FireHub system of BEYOND [16], as well as the NASA FIRMS and the European Forest Fire Information System (EFFIS/JRC) data.

To be noted that the Centre of Earth Observation Research and Satellite Remote Sensing BEYOND of NOA runs the operational system FireHub that offers (a) the so called "Diachronic Burn Scar Mapping service" providing polygons of Burned Areas at high spatial resolution (10 m–30 m) for more than 35 years which cover the entire country (the Diachronic NOA-BSM product), and (b) the "Early Detection and Real Time Fire Monitoring Service" that is systematically archiving for the last 10 years the daily observations of Active Fire (AF) at the spatial resolution of 500 × 500 m (the AF product) [17]. The database was created by applying the proper masks to exclude areas which are unlikely for fire ignition (e.g. grid cells laid on urban, agriculture and water areas) [18] (Fig. 1).

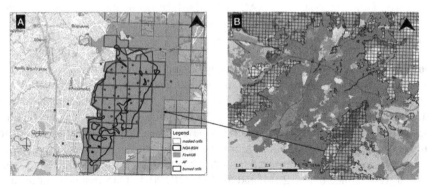

Fig. 1. (A) Blue cells validated as true positives (burned) from NOA-BSM (black polygon), AF (red points) and FireHub spotted active fires (orange cells). (B) Map shows how urban areas where excluded from the current study. (Color figure online)

The process for labeling the grid cells as "fire"/"non-fire" in order to create the training and validation dataset for the prediction algorithm (see Sect. 3), was initially based on FireHub evidences. At this stage, every grid cell intersecting with a BSM-NOA polygon and the corresponding AF detections had been labeled as fire cell. In a following step any remaining AF evidences were checked spatially and temporally against the EFFIS and FIRMS datasets. This resulted in a new set of "fire" cells which were also added in the database and used for the analysis. In conclusion, this process has returned a set of 12.978 "fire" cells which had been reported by independent observations of the three operating systems as FireHub, EFFIS, and FIRMS corresponding to the period 2010–2018. Each fire cell was assigned a unique fire event ID and the corresponding date of fire occurrence.

Additionally, an equivalent dataset of 12.585 "non-fire" cells spanning the same period (2010–2018) was created. The latter dataset was generated through a simple random selection spatially expanding over the entire Greece. It is worth-noting that the "non-fire" cells are recognized as areas where there is no presence of a recent fire outbreak. In practice and as explained in [19] these "non-fire" samples are representing pseudo-absence data because it is undefined if the conditions were in favor of a fire outbreak or not at their locations.

Meteorological Data. The meteorological data were derived from ERA5-Land, a reanalysis dataset providing hourly temporal resolution and native spatial resolution at 9 km. We obtained temperature, wind and precipitation datasets for a total of eight years (2010–2018) and computed the following parameters: maximum temperature, minimum temperature, mean temperature, dominant wind direction, maximum wind speed of the dominant direction, maximum wind speed, wind direction of the maximum wind speed, accumulated precipitation of the past 7 days.

EO Vegetation Data. Several studies consider NDVI as being of paramount importance in wildfires modelling [17]. In this study, the vegetation condition is represented by using the NDVI as proxy (Normalized Differential Vegetation Index) parameter, the latter being produced on 8-day intervals from NASA's MODIS dataset at 500 m spatial resolution. A number of 504 MODIS images were downloaded, stored and processed in order to calculate the NDVI data for the study period.

Topographic and Land Use Data. Topographic and Digital Elevation Model derived parameters are important factors for fire susceptibility prediction and mapping [10]. For the purposes of the study the Copernicus EU-DEM v1.1 at 25 m spatial resolution was used. This DEM was processed for deriving the morphological related features such as slope, aspect and curvature. Previous studies have shown that these features influence significantly the probability for fire occurrence [10]. Moreover, the land cover category assigned to each fire cell was retrieved from the CORINE 2012, for the cells representing areas that were burned before 2014, and the CORINE 2018 if the area was burned after 2014. Moreover, grid cells containing multiple land cover classes, were assigned the class with the highest fire proneness according to [10].

Table 1. Products and features extracted for the training of the model

Product	Source	Spatial resolution	Temporal resolution	Features
Wind (u-comp, v-comp)	ERA5-Land	9 km	Hourly	Dominant dir., Max speed of the dom. Dir., Max wind speed, Wind dir of the max speed
Temperature	ERA5- Land	9 km	Hourly	Max temperature, Min temperature, Mean temperature
Precipitation	ERA5- Land	9 km	Hourly	7-day accumulated precipitation
NDVI	NASA -MODIS	500 m	8-day	NDVI
DEM	Copernicus	25 m	Static	DEM, slope, aspect, curvature
Corine	Copernicus	100 m	6-year	CLC
"fire"/"non-fire" cells	NOA, EFFIS, NASA	500 m	Daily	

3   Methodological Approach and Implementation

3.1   Data Archiving and Modelling

The previous section makes obvious that the feature database has incorporated dynamic (weather/environmental) and less dynamic or static (DEM, land cover) data in multiple spatiotemporal resolutions. The multisource and multi type character of the data, together with its high spatiotemporal diversity and big volume, rendered the file storage and the manipulation of the data layers a challenging process. For this, an innovative earth observation datacube architecture was employed [20] for storing and managing the data and also for extracting and exploiting the features to be analyzed. This datacube is a one of its kind asset; it contains analysis ready spatio-temporal data in a structured and easily accessible architecture, providing a single access point to key environmental parameters that serve as wildfire drivers.

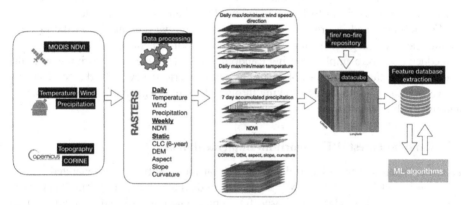

Fig. 2. System's architecture and processing blocks

Figure 2 shows the main blocks of the system's architecture and the corresponding processing steps. It encompasses access to the data, data downloading and manipulation, data ingestion and processing, information layer creation, feature space creation, ML algorithm training, implementation and validation. The Data processing block refers to a multitude of actions for the conversion of raw and/or archived FIRMS, EFFIS, ERA5, FireHub and Copernicus data so as to create the analysis ready data. Actually, this processing block has been designed to invoke a suite of properly developed Python scripts running over the datacube offering flexibility, scalability, reproducibility, transferability and capability of the system to respond to any scale, from local to national.

3.2   Feature Ranking

Guided by the need to validate the quality and adequacy of the feature database for addressing the "fire"/"no fire" (fire presence/absence) classification problem, it was necessary to apply feature ranking and examine the order of correlation between the independent variables, namely the wildfire related features of Table 1, with the dependent variable which was set by two distinct classes: the "fire" class and the "non-fire" class. For this, filter, wrapper, and embedded techniques for feature selection (and ranking) have been applied [21].

Table 1 shows the selected factors influencing fire prediction, the corresponding data sources and the features extracted and used as input to the "fire"/"non-fire" classification problem.

Firstly, for evaluating the categorical features (Corine Land Cover, dominant wind direction and wind direction of the maximum wind speed) the chi-squared filter method based on the Chi-squared statistics test was used [22]. For the remaining of the features (numerical), a correlation filter method based on Spearman's rank correlation coefficient has been used for evaluating the dependencies of the various feature combinations. It is worth noting that the Spearman's correlation has been selected as it can detect linear and non-linear monotonic relations [23] but also the correlations between the ordinal data. Actually the resulted number is ranging in the set $[-1, 1]$ indicating stronger correlation when the number is closer to 1 or -1, and weaker when the number is near to zero.

Furthermore, and in order to allow for a comparative ranking for all the features (numerical and categorical), three more feature ranking methods of the wrapper and embedded type were employed using Random Forest. The selected algorithms were the Sequential Feature Selection [24] using as score metric the Area Under the learning Curve (AUC) [25], the feature ranking through the measurement of the node impurity, and finally the permutation importance [22].

3.3 Random Forest (RF) Algorithm Implementation

As a final step and in order to demonstrate that the quality of the generated feature database is suited for letting a ML algorithm to distinguish between "fire" and "non-fire" classes, a RF model was chosen for implementation [26] and evaluation of its classification output. RF is ideal for this purpose because it is one of the most used and tested algorithms for predicting fire occurrence, susceptibility and risk [8]. Moreover RF is the most common representative of the family of decision trees ensembles which are commonly used in all sorts of classification problems and it can handle directly categorical data like wind direction and Corine Land Cover.

Many tests of the RF classifier were performed and a wide range of results was produced. The experiments with the highest scores were those where the shuffling method was used to create the training and test dataset. Actually, using shuffling provided the training set with indicative "fire" cells from almost every burned area. Considering that burned scars are small regions with very similar prevailing conditions, shuffling makes it easier for the algorithm to recognize fire cells and reduces the model's ability to adapt properly to new, previously unseen data.

In a further analysis step, and in order to increase the difficulty by simulating entirely unknown conditions for the test dataset, the shuffling process was left out. The target was to avoid neighboring "fire"/"non fire" cells, from the same day, and the same fire event, to be included both in the training and test datasets. Thus, following the k-fold cross validation methodology [26, 27] the feature space was split into 10 "folds" under the rule that the cell samples of a specific day cannot be distributed in more than one folds. This approach effectively allowed us to assess the model's capacity to generalize better, in the context of an operational forecasting application.

Furthermore, for tuning the RF model's internal parameters (decision trees number, trees depth, number of features in each tree) a random search hyper-parameterization process was performed [28, 29]. The metric that was selected for maximization was the recall [30] for the fire class, because the target was to achieve better performance in fire class prediction rather than in the "non-fire" class. The results that are presented in the next section indicate a significant learning potential of the RF algorithm using as input the specific feature database.

4   Results

4.1   The Spearman's Correlation

Figure 3 shows in a heat map the returned Spearman's rank correlation coefficient considering all possible pairs of numeric features.

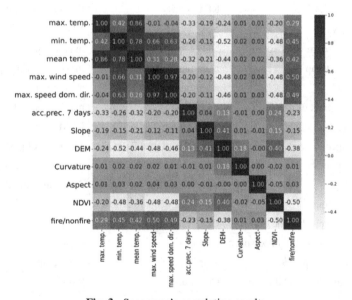

Fig. 3. Spearman's correlation results

Notable but not surprising the observations on this heat map indicate that: (1) the wind speed, the temperature, the NDVI and the DEM elevation are ranked as more meaningful and influencing features for "fire"/"non fire" class prediction, (2) The DEM elevation and the accumulated rainfall have inverse correlations with the dependent variable "fire"/"non-fire", (3) The two wind speed and the three temperature features are highly correlated one another.

4.2 The Chi-Squared Tests

The p-values of the chi-squared tests for the categorical inputs (Corine Land Cover, dominant wind direction and wind direction of the maximum wind speed) were extremely small ($<10^{-100}$) so the standard null hypothesis that these variables are independent was rejected beyond doubt. Consequently, it was showed that all the categorical variables are dependent with a high statistical significance with the "fire"/"non-fire" variable.

4.3 Comparative Feature Ranking

The ranking results from Sequential Feature Selection (SFS), RF impurity and the permutation importance are shown in Table 2; one can deduce that NDVI, wind speed and temperature variables are ranked between the first three places.

Table 2. Feature Ranking returned by Sequential Feature Selection, RF impurity and Permutation importance

Ranking	SFS AUC	RF impurity	Permutation importance
1	NDVI	max. speed of the dom. dir	NDVI
2	maximum wind speed	NDVI	max. temp
3	max. temp	max. temp	maximum wind speed
4	CLC	maximum wind speed	mean temp
5	acc. prec. of past 7 days	min. temp	max. speed of the dom. dir
6	DEM	mean temp	DEM
7	Aspect	acc. prec. of past 7 days	acc. prec. of past 7 days
8	Curvature	DEM	wind dir. of the max. speed
9	Slope	wind dir. of the max. speed	dominant wind direction
10	min. temp	dominant wind direction	CLC
11	max. speed of the dom. dir	CLC	Aspect
12	wind dir. of the max. speed	Aspect	Curvature
13	mean temp	Slope	min. temp
14	dominant wind direction	Curvature	Slope

4.4 The RF Classifier Results

The metrics of precision, recall, and f1-score returned from the application of RF are shown in Table 3.

The first two rows indicate classification results based on training and test data that were randomly produced at the rates of 90% and 10% respectively of the feature dataset while using the shuffling process. In contrast, the two last rows indicate the mean values

Table 3. RF metrics results

Training/Test set	Class	Precision	Recall	F1-Score
Training/Test with shuffling	Fire	0.94	0.92	0.93
	Non fire	0.92	0.94	0.93
k-fold cross validation (folds include entire days)	Fire	0.75	0.77	0.76
	Non fire	0.77	0.75	0.76

of the k-fold cross validation metrics without shuffling. The significance from invoking shuffling in the classification results is obvious.

In the first case (with shuffling) the results are quite satisfactory. The model is capable to distinguish quite well the "fire" from the "non-fire" class. In the case of the k-fold cross validation without shuffling, where training and test cells belong to different fire events, the mean recall value for the class fire remains still quite high as it was targeted. The difference and lower performance results in the k-fold cross validation case without invoking shuffling was expected as explained in 3.3.

5   Conclusion

This study showcased the creation of a prototype feature database that has proven useful for understanding the fire regime and predicting the "fire"/"non fire" classes in Greece. Moreover the study showed that this national scale feature database has been a reliable input for ML modelling by training successfully a RF algorithm, in order to distinguish between the two categories. As shown the parameters which are highly significant in the prediction have been *NDVI*, *maximum wind speed* and *maximum temperature* followed by the *DEM derived parameters* and *accumulated precipitation of the past 7 days*. The promising results which have been returned from the application of the RF classifier guide our steps towards building an enhanced fire risk model through invoking, testing, merging and validating results from different data driven ML approaches. Moreover, in order to address the daily requirements for emergency response there is need to translate the algorithm's "fire"/"non-fire" predictions to a number of fire risk classes. In this regard, it is necessary to conduct additional tests that examine the performances of various ML algorithms e.g. Artificial Neural Networks (ANN), Convolutional NN (CNN) or other ensemble algorithms like boosting trees. Furthermore, additional EO derived data (i.e. LST, soil moisture, proximity classes) could be considered for integration in the model in order to enhance the classification performance and introduce new aspects of the expected fire occurrences in Greece's ecosystem. This latter is an ongoing research study at the premises of the BEYOND Center of Excellence and subject of validation by the Greek Fire Brigade authority using fire data from the fire season of 2020.

Finally, since an ML Model can easily be transferred between feature databases with same features, a fire risk ML model developed for Greece could operate on a feature database with data for another location. Especially if this area has similar ecosystem properties like for example Mediterranean countries, the performance of the model could be similar. The challenge in this case would be the creation of harmonized data features with the ones used in the original development. Even though the level of performance of the ML model on another dataset cannot be "a priori" guaranteed, it is definitely worth to test the potential of transferring a developed for fire prediction ML model for Greece to another country or larger area.

Acknowledgements. This paper has been supported by using data and resources from the following Projects funded from EC and the Greek Government - Ministry of Development & Investments : (1) FRAMEWORK SERVICE CONTRACT FOR COPERNICUS EMERGENCY MANAGEMENT SERVICE RISK AND RECOVERY MAPPING- The European Forest Fire Information System (EFFIS) JRC/IPR/2014/G.2/0012/OC; (2) FRAMEWORK SERVICE CONTRACT FOR COPERNICUS EMERGENCY MANAGEMENT SERVICE RISK AND RECOVERY MAPPING - Program Call for tender JRC/IPR/2014/G.2/0012/OC; and (3) CLIMPACT: Flagship Initiative for Climate Change and its Impact by the Hellenic Network of Agencies for Climate Impact Mitigation and Adaptation.

References

1. The Copernicus Emergency Management Service Monitors Impact of Fires in Australia | Copernicus Emergency Management Service. https://emergency.copernicus.eu/mapping/ems/copernicus-emergency-management-service-monitors-impact-fires-autralia. Accessed 31 July 2020
2. European Commission: JRC Tecnical Report Forest Fires in Europe, Middle East and North Africa 2018 (2018)
3. Castellari, S., Kurnik, B.: Climate change, impacts and vulnerability in Europe 2016, no. 1. (2017)
4. Fares, S., et al.: Characterizing potential wildland fire fuel in live vegetation in the Mediterranean region. **74**, 1 (2017). https://doi.org/10.1007/s13595-016-0599-5
5. Lambert, J., Drenou, C., Denux, J.-P., Balent, G., Cheret, V.: Monitoring forest decline through remote sensing time series analysis. GISci. Remote Sens. **50**(4), 437–457 (2013). https://doi.org/10.1080/15481603.2013.820070
6. Pastor, E., Zárate, L., Planas, E., Arnaldos, J.: Mathematical models and calculation systems for the study of wildland fire behaviour. Progress Energy Combust. Sci. **29**(2), 139–153 (2003). https://doi.org/10.1016/S0360-1285(03)00017-0.
7. Hong, H., Tsangaratos, P., Ilia, I., Liu, J., Zhu, A.X., Xu, C.: Applying genetic algorithms to set the optimal combination of forest fire related variables and model forest fire susceptibility based on data mining models. The case of Dayu County, China. Sci. Total Environ. **630**, 1044–1056 (2018). https://doi.org/10.1016/j.scitotenv.2018.02.278
8. Jain, P., Coogan, S.C.P., Subramanian, S.G., Crowley, M., Taylor, S., Flannigan, M.D.: A review of machine learning applications in wildfire science and management (2020)
9. Pourtaghi, Z.S., Pourghasemi, H.R., Aretano, R., Semeraro, T.: Investigation of general indicators influencing on forest fire and its susceptibility modeling using different data mining techniques. Ecol. Indic. **64**, 72–84 (2016). https://doi.org/10.1016/j.ecolind.2015.12.030
10. Kontoes, C., Keramitsoglou, I., Papoutsis, I., Sifakis, N., Xofis, P.: National scale operational mapping of burnt areas as a tool for the better understanding of contemporary wildfire patterns and regimes. Sensors **13**(8), 11146–11166 (2013). https://doi.org/10.3390/s130811146
11. ΕΜΥ: Εθνική Μετεωρολογική Υπηρεσία. https://www.emy.gr/emy/el/. Accessed 31 July 2020

12. EEA: State of the environment report (SOER) No 1/2010 : The European environment: State and outlook 2010. Synthesis (2010)
13. Rivera, A., Bravo, C., Buob, G.: Climate Change and Land Ice (2017)
14. Kailidis, D., Karanikola, P.: Forest Fires 1900–2000. Giahoudi Press, Thessaloniki (2004)
15. Forest Fires in Europe 2006 | EU Science Hub. https://ec.europa.eu/jrc/en/publication/eur-sci entific-and-technical-research-reports/forest-fires-europe-2006. Accessed 13 Aug 2020
16. Beyond Centre of Excellence for EO based monitoring of Natural Disasters. https://www.bey ond-eocenter.eu/. Accessed 31 July 2020
17. Kontoes, C., Papoutsis, I., Themistocles, H., Ieronymidi, E., Keramitsoglou, I.: Remote Sensing Techniques for Forest Fire Disaster Management: The FireHub Operational Platform, Book Chapter No. 6, Integrating Scale in Remote Sensing and GIS (2017)
18. SEVIRI Monitor - NOA GIS. https://195.251.203.238/seviri/. Accessed 31 July 2020
19. Massada, A.B., Syphard, A.D., Stewart, S.I., Radeloff, V.C.: Wildfire ignition-distribution modelling: a comparative study in the Huron-Manistee National Forest, Michigan, USA (2013). https://doi.org/10.1071/WF11178
20. Killough, B.: Overview of the open data cube initiative. In: International Geoscience and Remote Sensing Symposium (IGARSS), vol. 2018-July, pp. 8629–8632 (2018). https://doi.org/10.1109/IGARSS.2018.8517694
21. Chandrashekar, G., Sahin, F.: A survey on feature selection methods. Comput. Electr. Eng. **40**(1), 16–28 (2014). https://doi.org/10.1016/j.compeleceng.2013.11.024
22. Bommert, A., Sun, X., Bischl, B., Rahnenführer, J., Lang, M.: Benchmark for filter methods for feature selection in high-dimensional classification data. Comput. Stat. Data Anal. **143**(September) (2020). https://doi.org/10.1016/j.csda.2019.106839
23. Dodge, Y.: The Concise Encyclopedia of Statistics, p. 502. Springer, Heidelberg (2010)
24. Feelders, A., Verkooijen, W.: On the Statistical Comparison of Inductive Learning Methods (1996)
25. Bradley, A.P.: The use of the area under the ROC curve in the evaluation of machine learning algorithms. Pattern Recogn. **30**(7), 1145–1159 (1997). https://doi.org/10.1016/S0031-320 3(96)00142-2
26. Ho, T.K.: Random decision forests. In: Proceedings of the International Conference on Document Analysis and Recognition, ICDAR, vol. 1, pp. 278–282 (1995). https://doi.org/10.1109/ICDAR.1995.598994
27. Stone, M.: Cross-Validatory Choice and Assessment of Statistical Predictions (1974)
28. Tonini, M., D'andrea, M., Biondi, G., Esposti, S.D., Trucchia, A., Fiorucci, P.: A machine learning-based approach for wildfire susceptibility mapping. The case study of the Liguria region in Italy. Geoscience **10**(3), 18 (2020). https://doi.org/10.3390/geosciences10030105
29. Bergstra, J., Ca, J.B., Ca, Y.B.: Random Search for Hyper-Parameter Optimization Yoshua Bengio (2012)
30. Kent, A., Berry, M.M., Luehrs, F.U., Perry, J.W.: Machine literature searching VIII. Operational criteria for designing information retrieval systems. Am. Doc. **6**(2), 93–101 (1955). https://doi.org/10.1002/asi.5090060209

Mobile eHealth Platform for Home Monitoring of Bipolar Disorder

Joan Codina-Filbà[1]([⊠])(iD), Sergio Escalera[2,4]([⊠])(iD), Joan Escudero[3]([⊠]),
Coen Antens[2]([⊠]), Pau Buch-Cardona[2,4]([⊠])(iD), and Mireia Farrús[1,4]([⊠])(iD)

[1] Universitat Pompeu Fabra, Barcelona, Spain
{joan.codina,mireia.farrus}@upf.edu
[2] Computer Vision Center, UAB, Cerdanyola del Vallès, Spain
{sergio.escalera,coen,pbuch}@cvc.uab.cat
[3] Pulso Ediciones SL, Sant Cugat del Vallès, Spain
j.escudero@pulso.com
[4] Universitat de Barcelona, Barcelona, Spain

Abstract. People suffering Bipolar Disorder (BD) experiment changes in mood status having depressive or manic episodes with normal periods in the middle. BD is a chronic disease with a high level of non-adherence to medication that needs a continuous monitoring of patients to detect when they relapse in an episode, so that physicians can take care of them. Here we present MoodRecord, an easy-to-use, non-intrusive, multilingual, robust and scalable platform suitable for home monitoring patients with BD, that allows physicians and relatives to track the patient state and get alarms when abnormalities occur.

MoodRecord takes advantage of the capabilities of smartphones as a communication and recording device to do a continuous monitoring of patients. It automatically records user activity, and asks the user to answer some questions or to record himself in video, according to a predefined plan designed by physicians. The video is analysed, recognising the mood status from images and bipolar assessment scores are extracted from speech parameters. The data obtained from the different sources are merged periodically to observe if a relapse may start and if so, raise the corresponding alarm. The application got a positive evaluation in a pilot with users from three different countries. During the pilot, the predictions of the voice and image modules showed a coherent correlation with the diagnosis performed by clinicians.

Keywords: Bipolar disorder · eHealth · Mobile monitoring · Data fusion

1 Introduction

Bipolarity —or bipolar disorder— is one of the most common forms of mental illness, which is characterised by episodes combining euphoria and depression

J. Lokoč et al. (Eds.): MMM 2021, LNCS 12573, pp. 330–341, 2021.
https://doi.org/10.1007/978-3-030-67835-7_28

phases, becoming usually a persistent disorder with a prevalence in the population of about 1% [25]. During bipolarity episodes, different aspects of the patient life become altered. Sleeping hours, for instance, increase in depression or decrease in manic states while sociability and sexual interest behave inversely. The detection of incipient changes on mood (prodromes) when patients are euthymic (normal status), allow clinicians to do psychological treatments designed for relapse prevention in conjunction with mood stabilisers [4]. Prodromes are defined as "any cognitive, behavioural and affective signs or symptoms that may make patients think they are entering an early stage of an episode" [16]. While objectively a patient may be able to detect prodromes [17], a high rating is not able to seek treatment at an early stage of future relapse [15].

The treatment of bipolarity faces several difficulties. On the one hand, the high level of non-adherence to medication; some patients, because they like the positive feelings of the 'highs' they have in manic episodes; others, because the idea of an indefinite treatment discourages them [17]. Another problem is that, once started, both mania and depression episodes produce a positive feedback that fuels themselves: at the early stages of mania, the increased sociability and less need to sleep arouses more social contacts and sleep disorders. The first stages of depression create a loss of interest in people, disrupting social contact, reducing activity and increasing laziness and sleeping time. These difficulties are reasons to make a close follow-up of the subject in order to detect the prodromes and tackle them. In addition, the chronic condition and the time changing mood with intermittent episodes of mania or depression, with normal episodes in between, makes the bipolar disorder a good candidate to develop a home monitoring tool to observe changes in patients behaviour and mood status to raise the corresponding alarms when mood destabilises in any sense.

The current health care environment demands an objective assessment of the patient state. While in many branches of medicine this can be performed using the corresponding biomarkers, this is not the case of psychiatry where the evaluation is based on an interview with the patient. An objective measure of the patient state is crucial to perform an accurate monitoring of the patient evolution and to interchange information between clinicians. Rating scales (RS) allow to obtain these objective measurements in a quantitative way. Many RSs have been validated to be reliable —different evaluators will produce consistent results for a given subject— and are able to produce a valid diagnosis. A RS is composed of different items to evaluate with a corresponding rubric indicating how to measure it. The combination of the scores of the items gives the final score, which becomes the objective measurement of the patient's state. There are self-administered scales thought to be filled by the user herself and clinician-administered scales filled by physicians.

The evaluation of the patients suffering BD is based on two different RSs: one to account for the severity of depression and the other for mania. The most commonly used scales are the Hamilton Depression Rating Scale (HDRS) [14], and the Young Mania Rating Scale (YMRS) [31]. Both scales require clinicians to evaluate and rate different aspects of the patient related to prodromes. Although the scales measure opposite states, they assess alterations in similar items (but in different sense), e.g. changes in physical activity, sleep, speech or sexual interest.

Automated and objective measures of these prodromes would simplify and shorten clinicians task to detect the outbreak of an episode. Also, the alarms raised by the system allow clinicians, relatives, or caregivers in general to take a proactive action to start the treatment before the user asks for it.

The current paper presents the MoodRecord system in the framework of the NYMPHA-MD Project [12]. The MoodRecord system identifies signs of deviations in mood based on the fusion of data collected using a smartphone, allowing early interventions. The different data sources captured by the MoodRecord application have been chosen to cover most of the items in the RSs. In this light, the paper describes the methodology used to extract mood with a special focus on video and voice modules as they are the most complex ones.

The paper is structured as follows. Section 2 briefly overviews related applications on automatic mood detection and related works on detection of mood from video and voice. Section 3 describes the MoodRecord mobile application created for patient continuous supervision in the framework of NYMPHA-MD, detailing the implementation of the video and voice modules. Section 4 contains the description of the pilot implementation and evaluation; and finally, Sect. 5 sketches the final conclusions and foreseen future work.

2   Related Work

Automatic home monitoring for any disorder, illness or impairment will be effective if combined with suitable self-monitoring platforms. Recently, monitoring platforms for computers and smartphones have been released (see [13,26] for an overview on the use of smartphones for the detection of medical disorders). The main drawback most of these systems present is the inability of collecting objective data from the behaviour of the patients. Moreover, they usually do not provide any feedback from healthcare providers. Some recent projects such as PSYCHE[1] or MONARCA[2] [13] are specifically related to the detection and treatment of BD. Although they in general put a step forward in the automatic monitoring of bipolar disorder, their main drawbacks are the high level of intrusiveness for patients due to a large number of sensors required, and the inability to target the caregiver's role.

Mood Detection from Voice. Speech is a powerful identifier for the detection of bipolar disorder, as it has been shown in several works [11,19,21]. Moreover, voice is a non-intrusive and ubiquitous sensor. The automatic detection of mental disorders —or other disorders that can be reflected in language— can be performed either by means of context-dependent features or acoustic-dependent

---

[1] www.psyche-project.org.

[2] www.monarca-project.eu.

features [3,20,29]. While the former ones require the use of language transcriptions, the latter make use of solely acoustic features extracted from speech. Specifically, the acoustic-based systems rely on the extraction of speech-based features regardless of the language, which makes it a language-independent system. The acoustic features can be, among others, spectral features, voice quality characteristics, or prosodic features extracted in their acoustic form.

Other studies report the usefulness of prosodic features to detect emotional states such as depression and mania. In [24], for example, several features such as pitch rate, jitter, shimmer and harmonic to noise ratio indices were shown to be indicators for patients with voice disability in contrast to healthy people —especially pitch rate—. In [27], jitter, shimmer and first and second formant frequencies differed significantly in depressed speech. In a similar way, [23] analyses the chaotic behaviour of vocal sounds in patients with depression, and [32] proposes a new feature value based on the nonlinear Teager energy operator to classify speech under stressed conditions. Other voice quality characteristic are further studied in [22] to detect depression and post-traumatic stress disorder, and speech pressure has also been reported as indicator for mania states [2].

Mood Detection from Video. The automatic analysis of perceived mood estimation through the analysis of human facial expressions (FE) refers to a set of methodologies and studies that relate artificial intelligence with behavioural sciences and neurology. Faces carry a wealth of social information, including information about identity, emotional and motivational cues, lip speech, and focus of attention as indicated by eye gaze, all of which are important for successful communication. Human emotion studies have been mainly described by either a categorical or a dimensional approach. The categorical approach, implemented in this study, maps any perceived emotion into a closed group of predefined FE. Among the existing categorical approaches for FE, Ekman's set of 7 universal expressions + neutral state [8,9] is one of the most used ones (Fig. 1). This set includes expressions that provide additional information to the non-invasive evaluation of depression and mania scales in BD. For a wider comprehension on Automatic Facial Expression Recognition (AFER) systems, and its current state-of-the-art methodology for accurate recognition based on deep learning approaches, the reader is referred to [5].

Fig. 1. 7 Ekman core facial expressions + neutral state. Image from [18].

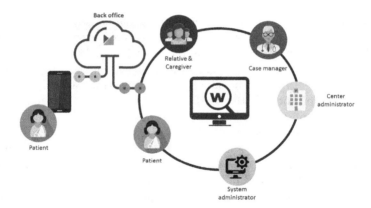

Fig. 2. Diagram of the MoodRecord system architecture showing its main parts (Front End/App, Back End and Back Office/Web) and potential users.

3   The MoodRecord Application

The *NYMPHA-MD Project: Next Generation Mobile Platforms for HeAlth, in Mental Disorders*[3] is an FP7 European project that launched a Pre-Commercial Procurement (PCP) bid to look for solutions on continuous monitoring of bipolar disorder. PULSO Ediciones S.L. led a team with the Universitat Pompeu Fabra and the Center for Computer Vision, in charge of the speech and face analysis modules, respectively. The team participated in the bid with the MoodRecord system, that passed satisfactorily all the eliminating rounds of the PCP (definition, prototype, and pilot).

The MoodRecord application developed according to the technical specifications for NYMPHA-MD project aims to provide a new way to manage patients diagnosed with bipolar disorder, developing a system that keeps an important role to informal caregivers. The system performs patient monitoring to detect and predict changes in patient's mood, maximising the treatment adherence by means of personalised data and feedback under the supervision of clinicians. The MoodRecord system consists of a web interface and a mobile application, both managed from a back-office server. The application is intended to be used by patients suffering BD (primary users) under clinician prescription. The website has been developed to provide clinicians with all data from patients, collected by the app, to simplify and improve their follow up. Patients, relatives and caregivers can access the web using a reduced set of functionalities. Figure 2 shows a diagram of the Mood Record schematic architecture, with the different stakeholders.

The MoodRecord App runs on Android and IOS systems. Figure 3a shows the sidebar menu of the MoodRecord application, with the different actions the

---

[3] https://cordis.europa.eu/project/id/610462.

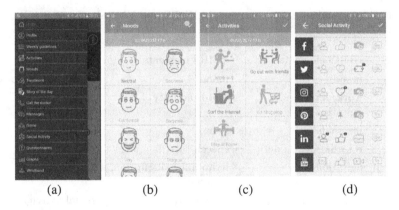

(a)   (b)   (c)   (d)

Fig. 3. Screenshots of the MoodRecord application. From left to right: a) main menu, b) mood selection screen, c) activity recording screen and d) social activity screen.

user can perform. The application manages a weekly guideline defined by the doctor, with a set actions and questionnaires that the patient must perform. The different actions available for the patient are the following:

- Mood: multiple choice mood selector (Fig. 3b).
- Activities: to report the activities performed during the day (Fig. 3c).
- Story of the day: to self-record their daily experiences, using the frontal camera for later video and audio analysis.
- Treatment: to report compliance to any treatment prescription.
- Social activity: to indicate daily activity on social networks (Fig. 3d).
- Questionnaires: to fill out different self-administered RSs available, either planned by doctors or when the system predicts some risk of suicide: Barratt's Impulsivity Scale, Beck Hopelessness Scale, Buss-Durkee Hostility Inventory, Generic health status scale: EQ-5D-5L, Perceived Stress Scale, Patient Health Questionnaire 15 and Personal and Social Performance.

The MoodRecord App includes a game that the user can play at any moment. The system monitors data about the time spent in the game and telemetry (velocity, intensity, attempts, etc.). The application also monitors the application's usage to measure how the patient adheres to the plan. Finally, the application is bound to an activity wristband (FitBit®) to record sleep times and physical activity. The data collected is highly correlated to the different items measured by HDRS and YMRS scales to evaluate bipolarity, and tries to cover the whole evaluation range. Table 1 shows this relationship between the data collected by the App and items in both scales.

The MoodRecord application integrates and reports the data extracted from the different modules and combines the different sources of information to estimate risk of mania or depression. Clinicians use the scales to evaluate patients during visits. Such evaluations are used by MoodRecord to recognize the normal states of the user. In the case where 30% of the previous week's registers,

Table 1. Correspondence between items in HDRS and YMRS scales and values measured by MoodRecord

Feature	HDRS elements	YMRS elements	MoodRecord
Mood	mood	mood	mood, video, voice
Activity	work/activities, agitation	motor activity	activity record, wristband
Sleep	insomnia	sleep	wristband
Speech	retardation, anxiety	speech rate, language	voice
Behaviour	agitation	irritability, disruption, aggressive	voice, game
Sex	genital symptoms	sexual interest	social activity
Insight	insight, suicide, guilt feelings	insight, paranoid ideas	self-administered scales
Other	somatic symptoms, loss of weight, hypochondria	appearance	

obtained or computed from any of the different sources, are outside these normality records, the system raises an alarm. The system will also raise an alarm if the scores of the self-administered scales overcome the threshold of "normal values". If the system detects a severe depression risk (indicated by a higher number of registers with a high deviation from the normality values) the system will automatically ask the user to answer the BHS questionnaire, to detect suicide tendencies and raise a severe alarm.

3.1   Mood Detection from Voice

For the NYMPHA-MD project hereby presented, we developed a Praat-based [1] module for prosodic feature extraction in order to detect mania, depression and normal states. The features extracted in our module are: mean fundamental frequency (F0), maximum value of F0, minimum F0, range of F0, first formant frequency (F1), second formant frequency (F2), relative jitter, absolute jitter, relative shimmer, absolute shimmer [10], noise-to-harmonic ratio (NHR), harmonic-to-noise ratio (HNR), intensity, number of pauses, speech rate, articulation rate, and average syllable duration. The last four features were computed without using any textual transcriptions by means of the algorithm described in [6].

A set of regression models is trained to identify YMRS and HDRS values. Regression is based on the combination of different models: (a) models that estimate the global HDRS and YMRS values, and (b) models that estimate some HDRS/YMRS items related to speech and language.

For each model, two different versions are combined into a hybrid model: (1) a general model to be used to classify any new patient once her normal status is defined, and (2) a personalized version for each patient. The hybrid system uses the general models until the personalized models outperform the general one. To obtain the personal model, the system uses the data filled by doctors during interviews. After each interview, models are retrained and check for performance.

General models have been trained with data provided by the NYMPHA-MD consortium during the second phase of the project and data collected during the pilot. This dataset is composed of 30 valid recordings with the corresponding evaluations by doctors. Although the dataset is small and most of the recordings were done with users in an euthymic state, the results obtained using support vector regression with a radial kernel showed that, while HDRS estimation achieved the best result by using low-level voice quality and formant features, the best result when estimating YMRS was achieved by using the whole combination of voice quality, formant, and prosodic features [11]. This can be explained by the fact that mania state is more expressive in terms of speech, and thus the need of prosodic information is much more relevant.

3.2 Mood Detection from Video

The NYMPHA-MD project includes an AFER system following Ekman's categorical proposal on emotional analysis over a universal reduced set of FEs, by classifying any FE within an 8-class categorical classifier. Such a system could benefit from existing validated public datasets. The solution adopted is applied over each individual frame independently, and uses a deep learning scheme that tackles: (a) *face detection*, by means of a Viola-Jones algorithm [28], which works with intensity fluctuation in order to capture faces trained in a near-frontal position from images, shown to be very fast and with a good performance for MoodRecord scenario, and (b) *classification problem*, by means of a standard deep neural network (Fig. 4) that will get the resulting face detected as a RGB input image to convey to a probability vector containing the confidence for each of the 8 possible FEs mentioned before. The predicted expression will be the one with major relevance. The predictions for each frame of the video add a temporal dimension to the analysis. Histograms of frequency can be drawn to compute regular or deviation patterns of apparent mood.

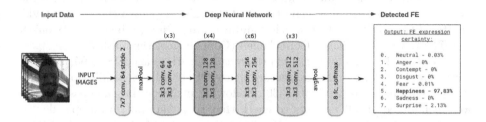

Fig. 4. Adapted ResNet50 neural network as a FE classifier.

Fig. 5. Visual results from the perceived emotion analysis from video.

Convolutional neural networks are very robust to image variability such as ethnicity, gender and age, hence performing well under unseen data. The model used (standard ResNet50 architecture, Fig. 4) has been fine tuned over an existing pretrained model (ImageNet [7]). After selecting the valid images for training, the final set was composed of 8832 images split into the eight categories. Since the dataset was unbalanced towards the neutral expression, additional data was included by manually scraping images from the internet. The dataset was divided in 70/20/10% (train/validation/test) splits. Once trained, the complete face analysis pipeline (face detector + facial expression classifier) was tested over real unseen data. Figure 5 shows a visual performance over a few testing examples: the system is capable of detecting near-frontal face area of interest (red bounding box), which is later passed to the facial expression perceived mood module, that outputs the perceived mood (top left corner from each of the samples shown in Fig. 5). Under semi-controlled environments (limited range of head pose angles, illumination and image quality) the system is capable of detecting the different FEs from a subject with over 90% accuracy in most of the classes (see confusion matrix in Fig. 6).

4   Pilot Testing

During the last phase of the NYMPHA-MD project, MoodRecord system was used on a pilot to measure the usability and feasibility of the application. The full pilot was run from January 2018 to March 2018, but the patients follow-up took 4 weeks, and it was conducted by the three institutions in charge of the project: the Mental Health Department of the Azienda Provinciale per i Servizi Sanitari (Trento, Italy), Corporació Sanitària Parc Taulí (Sabadell, Catalonia), and The Psychiatric Center Copenhagen (Copenhagen, Denmark).

The study includes 30 patients (10 on each location) aged 18 to 65 years old with a bipolar disorder diagnosis according to ICD-10 using Schedules for Clinical Assessments in Neuropsychiatry [30]. Following inclusion, baseline assessments were conducted on all patients to check that they were not in an episode. To adjust and personalise the system, each participant had to be in a normal state to start the monitoring, thus the system has a personalised "reference status" to compare with. The study was designed and powered to investigate differences in symptoms and functioning due to using a smartphone-based monitoring system.

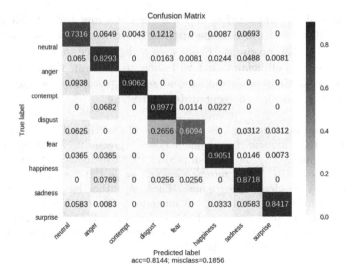

Fig. 6. Confusion matrix from test data.

Once a day the patients should enter and evaluate mood, activity level, enter a story of the day, whether medicine was intake and activity level on social media. The main focus of the study was to measure the system's usability and feasibility performance. The users rated the system, according to the USE and CSUQ scales with positive evaluations (average 4.4 out of 7). Even there where significant differences between users from Denmark and the ones from Italy or Spain that showed a similar criteria. Interestingly, audio and visual estimation provided consistent, correlated, and regular recognition patterns of mood for the patients, as verified by the physicians and possibly because of the no existence of episodes during the pilot.

Nowadays there is an undergoing process to fulfil the administrative steps required to get the CE Medical Certification (class type I or IIa) of the system. A prerequisite to be installed on different hospitals, that have already shown their interest in the product.

5   Conclusions and Future Work

We have presented MoodRecord, a system to monitor patients with Bipolar Disorder for early detection of relapses. The system includes an app for smartphones that collects data of different modes, which include the analysis patient's voice and face, to report the mood status so that clinicians can take the appropriate actions before the patient worsens.

The system is fully operative and has been tested in a pilot use case where patients have approved its usability. The vision system has shown a good performance in classification of facial expressions with external data and correlation with patient reports during the pilot. The voice analysis has shown consistent results with testing data using quality, formant and prosodic features.

The future of the application depends on obtaining the CE Medical Certificate that will allow to install the application in hospitals (there are already some interested in the product). This usage will produce more data from users and evaluations from clinicians, that will allow to fine tune the voice and image models to increase their performance and reduce possible bias under different kinds of population underrepresented during testing.

Acknowledgements. This work is part of the MYMPHA-MD project, which has been funded by the European Union under Grant Agreement N° 610462. It has also been partially supported by the Spanish project PID2019-105093GB-I00 (MINECO/-FEDER, UE) and CERCA Programme/Generalitat de Catalunya.), and by ICREA under the ICREA Academia programme. The last author has been funded by the Agencia Estatal de Investigación (AEI), Ministerio de Ciencia, Innovación y Universidades and the Fondo Social Europeo (FSE) under grant RYC-2015-17239 (AEI/FSE, UE). The authors would like to thank Ivan Latorre for his technical support and Giorgia Cistola for her help on the data preparation.

References

1. Boersma, P.: Praat: doing phonetics by computer [computer program] (2020). http://www.praat.org/
2. Carlson, G.A., Goodwin, F.K.: The stages of mania: a longitudinal analysis of the manic episode. Arch. Gen. Psychiatry **28**(2), 221–228 (1973)
3. Chien, Y.W., Hong, S.Y., Cheah, W.T., Yao, L.H., Chang, Y.L., Fu, L.C.: An automatic assessment system for Alzheimer's disease based on speech using feature sequence generator and recurrent neural network. Sci. Rep. **9**(1), 1–10 (2019)
4. Colom, F., Lam, D.: Psychoeducation: improving outcomes in bipolar disorder. Eur. Psychiatry **20**(5), 359–364 (2005)
5. Corneanu, C., Oliu, M., Cohn, J., Escalera, S.: Survey on RGB, 3D, thermal, and multimodal approaches for facial expression recognition: history, trends, and affect-related applications. IEEE Trans. Pattern Anal. Mach. Intell. **38**, 1 (2016)
6. De Jong, N.H., Wempe, T.: Praat script to detect syllable nuclei and measure speech rate automatically. Behav. Res. Methods **41**(2), 385–390 (2009)
7. Deng, J., Dong, W., Socher, R., Li, L.J., Li, K., Fei-Fei, L.: ImageNet: a large-scale hierarchical image database. In: CVPR 2009 (2009)
8. Ekman, P.: Universals and Cultural Differences in Facial Expressions of Emotions. University of Nebraska Press, Lincoln (1971)
9. Ekman, P., Oster, H.: Facial expressions of emotion. Annu. Rev. Psychol. **30**(1), 527–554 (1979)
10. Farrús, M., Hernando, J., Ejarque, P.: Jitter and shimmer measurements for speaker recognition. In: Proceedings of the Interspeech (2007)
11. Farrús, M., Codina-Filbà, J., Escudero, J.: Acoustic and prosodic information for home monitoring of bipolar disorder. Health Inf. J. (2021)
12. Faurholt-Jepsen, M.: The NYMPHA-MD project: next generation mobile platforms for health, in mental disorders. Eur. Psychiatry **41**(S1), S23–S23 (2017)
13. Gravenhorst, F., et al.: Mobile phones as medical devices in mental disorder treatment: an overview. Pers. Ubiquit. Comput. **19**(2), 335–353 (2014). https://doi.org/10.1007/s00779-014-0829-5

14. Hamilton, M.: The hamilton rating scale for depression. In: Sartorius, N., Ban, T.A. (eds.) Assessment of Depression, pp. 143–152. Springer, Heidelberg (1986). https://doi.org/10.1007/978-3-642-70486-4_14
15. Joyce, P.R.: Illness behaviour and rehospitalization in bipolar affective disorder. Psychol. Med. 15(3), 521–525 (1985)
16. Lam, D., Wong, G.: Prodromes, coping strategies, insight and social functioning in bipolar affective disorders. Psychol. Med. 27(5), 1091–1100 (1997)
17. Lam, D., Wong, G.: Prodromes, coping strategies and psychological interventions in bipolar disorders. Clin. Psychol. Rev. 25(8), 1028–1042 (2005)
18. Langner, O., Dotsch, R., Bijlstra, G., Wigboldus, D., Hawk, S., Knippenberg, A.: Presentation and validation of the radboud face database. Cogn. Emot. 24, 1377–1388 (2010)
19. Maxhuni, A., Muñoz-Meléndez, A., Osmani, V., Perez, H., Mayora, O., Morales, E.F.: Classification of bipolar disorder episodes based on analysis of voice and motor activity of patients. Pervasive Mobile Comput. 31, 50–66 (2016)
20. Mota, N.B., et al.: Speech graphs provide a quantitative measure of thought disorder in psychosis. PLoS ONE 7(4), e34928 (2012)
21. Muaremi, A., Gravenhorst, F., Grünerbl, A., Arnrich, B., Tröster, G.: Assessing bipolar episodes using speech cues derived from phone calls. In: Cipresso, P., Matic, A., Lopez, G. (eds.) MindCare 2014. LNICST, vol. 100, pp. 103–114. Springer, Cham (2014). https://doi.org/10.1007/978-3-319-11564-1_11
22. Scherer, S., Stratou, G., Gratch, J., Morency, L.P.: Investigating voice quality as a speaker-independent indicator of depression and PTSD. In: Interspeech, pp. 847–851 (2013)
23. Shimizu, T., Furuse, N., Yamazaki, T., Ueta, Y., Sato, T., Nagata, S.: Chaos of vowel/a/in Japanese patients with depression: a preliminary study. J. Occup. Health 47(3), 267–269 (2005)
24. Shinohara, S., Nakamura, M., Mitsuyoshi, S., Tokuno, S., Omiya, Y., Hagiwara, N.: Voice disability index using pitch rate. In: IEEE EMBS Conference on Biomedical Engineering and Sciences (IECBES), pp. 557–560 (2016)
25. Strakowski, S.M., DelBello, M.P., Adler, C.M., Fleck, D.E.: Bipolar Disorder. Oxford University Press, USA (2020)
26. Torous, J., Powell, A.C.: Current research and trends in the use of smartphone applications for mood disorders. Internet Interv. 2(2), 169–173 (2015)
27. Vicsi, K., Sztahó, D., Kiss, G.: Examination of the sensitivity of acoustic-phonetic parameters of speech to depression. In: IEEE 3rd International Conference on Cognitive Infocommunications (CogInfoCom), pp. 511–515 (2012)
28. Viola, P., Jones, M.: Robust real-time object detection. Int. J. Comput. Vis. 4(34–47), 4 (2001)
29. Voleti, R., Woolridge, S., Liss, J.M., Milanovic, M., Bowie, C.R., Berisha, V.: Objective assessment of social skills using automated language analysis for identification of schizophrenia and bipolar disorder. arXiv preprint arXiv:1904.10622 (2019)
30. Wing, J.K., et al.: SCAN: schedules for clinical assessment in neuropsychiatry. Arch. Gen. Psychiatry 47(6), 589–593 (1990)
31. Young, R.C., Biggs, J.T., Ziegler, V.E., Meyer, D.A.: A rating scale for mania: reliability, validity and sensitivity. Br. J. Psychiatry 133(5), 429–435 (1978)
32. Zhou, G., Hansen, J.H., Kaiser, J.F.: Nonlinear feature based classification of speech under stress. IEEE Trans. Speech Audio Process. 9(3), 201–216 (2001)

Multimodal Sensor Data Analysis for Detection of Risk Situations of Fragile People in @home Environments

Thinhinane Yebda[1](✉)(iD), Jenny Benois-Pineau[1](✉)(iD), Marion Pech[2](✉),
Hélène Amieva[2](✉)(iD), Laura Middleton[3](✉)(iD), and Max Bergelt[3](✉)

[1] LaBRI, Université de Bordeaux, Talence, France
thinhinane.yebda@u-bordeaux.fr, jenny.benois-pineau@u-bordeaux
[2] BPH, Université de Bordeaux, Bordeaux, France
{marion.pech,helene.amieva}@u-bordeaux.fr
[3] Department of Kinesiology, University of Waterloo, Waterloo, Canada
{laura.middleton,max.begler}@uwaterloo.ca

Abstract. Multimedia (MM) nowadays often means "Multimodality". The target application area of MM technologies further extends to healthcare. Health parameters monitoring, context and situational recognition in ambient assisted living - all these applications require tailored solutions. We are interested in development of AI solutions for prevention of risk situations of fragile people living at home. This research requires a tight collaboration of IT researchers with psychologists and kinesiologists. In this paper we present a large collaborative project between such actors for developing future solutions of risk situations detection of fragile people. We report on definition of risk scenarios which have been simulated in the data collected with the developed Android application. Adapted annotation scenarios for sensory and visual data are elaborated. A pilot corpus recorded with healthy volunteers in everyday life situations is presented. Preliminary detection results on LSC dataset show the complexity of real-life recognition tasks.

Keywords: Wearable sensors · Artificial intelligence · Time series · Data collection · Healthcare applications

1 Introduction

Nowadays research in Multimedia (MM) moves more and more towards what we call "multimodality". We deploy our MM know-how [11] in the general multimodal context, be it methods for video analysis or multi-sensory data. The application target in MM research has also moved to the urgent societal problems.

This research is supported by french national ANRT grant: $n°2018/0364$, AAP 2019 "Digital Health Challenge" grant, ALLOCATION: SSESE1902GA and InflexSys project and the UBx Waterloo project. We thank master students of Computer Science for Business Administration curriculum for their help in experiments.

J. Lokoč et al. (Eds.): MMM 2021, LNCS 12573, pp. 342–353, 2021.
https://doi.org/10.1007/978-3-030-67835-7_29

Healthcare and ambient assisted living are amongst them. Aging of population is the main reason for this. Europe is one of geographic areas which combines demographic aging and gerontogrowth. This second phenomenon, defined by the increase in the number of elderly people, is manifested in France for instance by the 35.2% increase in the number of people over 65 between 1999 and 2014, while the number of people over 80 has increased by 78.9% [18].

Frailty in older adults refers to a clinical state of increased vulnerability that carries a high risk for poor health outcomes including falls, incident disability, hospitalization, and death [4,20]. Frailty is manifested both by biological (age-related) vulnerability to stressors and by a decrease of physiological reserves.

Numerous studies have been realized to determine the most frequent risk situations faced by frail people such as falls [25]. Today we are seeking for detection of much more complex and "semantic" risk situations in everyday life of such people. It requires collection of multimodal data, both sensory and visual. It also requires a tight collaboration of researchers of different disciplines: IT, psychology, kinesiology. We conduct a joint study between University, a Company involving IT researchers and psychologists in Europe and - researchers from Kinesiology department of a University in Northern America. In this paper, we present ongoing research in the detection of risk situations for risk prevention of frail people.

The remainder of the paper is organized as follows: in Sect. 2 we shortly present related work, in Sect. 3 we discuss complex real-life risk definition scenarios and report on preliminary results on publicly available LSC dataset. Section 4 introduces data collection protocols and data annotation scenarios. Despite we are very far form concluding the work, we will present some preliminary conclusions of our experience in Sect. 5 and will outline the perspectives of the work.

2   State of the Art

In recent years, wearable technologies and the Internet of Things (IoT)-based applications aiming at supporting independent living of the elderly have considerably progressed. A frequent outcome targeted by such technologies is fall detection, which also constitutes one of the major causes of institutionalization [24]. Therefore, wearable sensors (WS) are being developed with the aim of becoming effective tools for prevention, early detection of falls, and monitoring general activities of daily living in older adults. A progress has been made possible thanks to the development of remote data collection methods and technologies [21]. Regarding the applications for monitoring activities of daily living and of independent living with elderly, several studies have shown the efficiency of WS and IoT systems from a technological point of view [2,7,8]. For remote monitoring, smartphone and smartwatch-based applications are frequently used for old adults [2,13,14]. The main advantage of these devices is that they are non-invasive and comfortable for users.

The subject being of a strong impact for the society, the research in this field is intensive and its industrial implementations such as Apple watch are

popular. We give a short overview of proposed solutions below. For fall detection
and prevention different types of WS and IoT-based applications have been
developed for older adults' care at home [14,16,17,19,26,28,29]. The authors
of [17] propose a smart and connected home health monitoring system. It is
composed of hardware such as sensors, a wearable sensor with an alarm button,
a gateway and of software for data collection. This technology is based on deep
learning and hidden Markov models with sensor orientation calibration [10,22].
In [16,19] a waist-worn fall detector based on an Attitude and Heading Reference
system and barometric sensor is designed. This system showed 100% of sensibility
for fall detection in diverse studies.

"Tagcare" is proposed in [29]. This monitoring system prototype has shown
a high accuracy (98%) for movement detection and sudden falling. Furthermore, wrist-worn wearables have been added to improve fall detection rates.
[14,28]. Finally, a fall detection system has been developed in an indoor environment, based on IoT and classifiers of "Ensemble-Random Forest" (RF) family.
It showed a success rate of above 94% for accuracy, sensitivity and specificity in
detecting three types of falls (forward, backward and lateral falls) and activities
of daily living (walking, stairs climbing, and sitting) [26]. As may be seen, various technological solutions reveal to be potentially interesting tools to support
independent living at home, and the recent results for the prevention of falls are
promising. Nevertheless, challenges and barriers to the wider adoption of WS
and IoT applications still exist [1]. While fall detection is quite well studied,
detection of more complex risks remains a challenge. In [27] "semantic risks"
were introduced which are in the focus of interest of psychologists and kinesiologists today. Their detection requires visual data collection together with sensor
measures. To this extent the recording of the data is similar to "lifelog". Such
lifelogs [3] have been used mainly for dairy generation in different studies, e.g. of
patients suffering from Alzheimer disease [10] for recognition of familiar scenes
[23]. The methodology nowadays consists of recognition of different situations
accordingly to designed taxonomy by Deep Neural Networks classifiers. In the
following section we will discuss the reality in risk definition scenarios and show
that the real-life data are difficult for detection of risk situations.

3   Risk Definition Scenarios and Preliminary Risk Detection on LSC Dataset

In this section we will discuss the complexity of risk scenarios in a real world
study and report on preliminary results which were obtained on general purpose
LSC dataset.

3.1   Risk Definition Scenarios

In [27] different semantic risk situations were reported such as the risk of fraud
and the risk of domestic accidents. The difficulty of real world studies with

frail subjects is that semantic risk situations are both complex and "complementary". It means that if we denote by A a real life situation which yields a risk R: $A \rightarrow R$ we can define risk annotation and risk detection scenarios in a lifelog-type data. Nevertheless, studying the most general population of subjects taking medications psychologists also define risk situations in the form $\bar{A} \rightarrow R$. This total ambiguity in definition can steel be overcome as the absence of A is considered for the all period of daily data recording. Hence, we have to consider two scenarios of risks.

- (1) $A \rightarrow R$ - "immediate" risk situations
- (2) $\bar{A} \rightarrow R$ - "long term" risks. If during a specified recording period the event A was not detected then the subject is in the risky situation R.

In this work for the first category of risk situations we can consider the same taxonomy as in previous research such as risks of fall, of domestic accidents, of fraud [27]. For the second category of risks we work with: (1) non drinking water during one daily recording which means risk of dehydration and (2) non taking medication risk of a health damage.

Speaking about the first category of risks, we can not conclude the person is in a risky situation each time the risky environment can be observed. For instance if the person enters into the kitchen this does not mean that each time he/she is in a risky situation. *We consider the subject condition risky if he/she is in unstable physiological condition (sensors) in a risky environment(visual).* An example of this is the increase of the Electrodermal Activity (EDA) measure and heart-rate activity in the kitchen. Here we can consider that domestic accident risk is of high probability.

3.2 Detecting Risk Situations Using Sensor Signals on a LSC Dataset

In the previous work [27] a method for annotation of risk situations and detection of them using only visual component of LSC dataset was proposed. Accordingly to the "risk situation model" we presented above, we need to detect the risk situation also in sensory signals. For this purpose we applied an LSTM network.

Long-Short Term Memory (LSTM) neural networks are widely used for time series classification as they build very long term links during learning [9]. They retain the relevant information from previous entries and use this information to modify the current output. We applied a standard LSTM architecture for risk detection on sensor signals.

LSTM Architecture. The basic LSTM unit is composed of cells with an input gate, output gate and forget gate. The particularity of LSTMs consists in the way in which the hidden state h is processed: in the case of simple RNNs, the processing of the recurrence, is ensured by a *tanh* function. For LSTMs, this processing is replaced by a "memory cell". The LSTM cell is characterized by a central node, containing the internal state (or memory) of the cell, and a number of "gates" of mentioned three categories. These gates are used to manage the sequential information storage (input and forget gates) and the role of the internal state in the production of each output (output gate). By closing the input gate, for example, new events are less taken into account. LSTMs use the concept of gating to overcome the vanishing an exploding gradient problem [5].

Experiments and Results. Thus experiments were conducted to recognize semantic risk situations on a part of Biometrics Data and Human Activity Data from signals part of LSC dataset [6]. Heart rate, galvanic skin response, skin temperature and the number of steps were used. We have selected four semantic risk classes: fall, risk of fall, risk of fraud and risk of domestic accidents and a no-risk - "Other" class. The risk situations were annotated using the image data only as reported in [27]. In sensor signals some data are missing during recording due to sensor failure. The lacking data were annotated as NaN (Not a number). The statistics of risks, no risk and NaN are presented in the Table 1.

Table 1. Table of the distribution of data between different classes

Risk situations	Number of risks situations	Percentage
Risk of fraud	120	0.22%
Risk of fall	347	0.63%
Risk of domestic accident	406	0.73%
Fall	2181	3.96%
Other	21152	38.43%
NaN	30826	56.01%
Total	55032	100%

The LSTM model was trained on these data after the NaN data have been removed. The optimization algorithm was Adam [15] with a fixed learning rate of 0.001 chosen experimentally. The framework used was Keras (https://keras.io/). The accuracy and Loss curves during training are illustrated in Fig. 1. In spite of a satisfactory accuracy of 80% the detailed analysis of these results showed that this accuracy is totally biased by the over-representation of the "Other" class. Indeed, for all the classes except "Other", the metrics of Recall and Precision are almost 0. The classes are not balanced. Indeed coming back to the Table 1 one can see that e.g. the risk of fall represents only 0.22% of all data, while "Other" class presence is of 38.43%. With such an unbalanced distribution it is useless to apply any data augmentation technique. Indeed, in our experiments, adding Gaussian noise to the measures of signals in risk situations for data augmentation did not bring improvement. Thus, coming back to our model of a risk situation of first type, see Sect. 3.1 it is clear that a general LSC dataset recording scenario is not adapted for sufficient data collection for risk situations. Hence, we have defined a new protocol for collection of multi-sensory data. We present it in the following section.

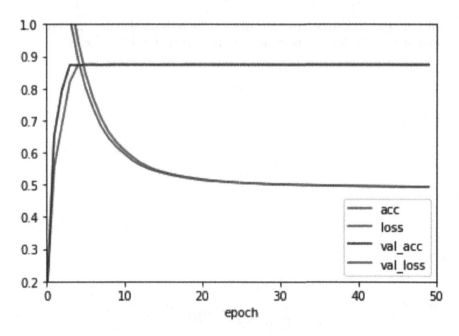

Fig. 1. LSTM curves obtained at LSC dataset.

4   Data Collection for Risk Prevention

In this section we will present the elaborated data recording protocol and resulting multimodal corpus "BIRDS".

4.1   Data Recording Protocol

We have set up a mirror experiment between an University in Europe and an University in Northen America for the recording of Corpus, we called BIRDS (Bio Immersive Risk Detection System): the same devices are used connected to developed Android application. First, a bracelet developed by Empatica, the E4 which is a medically-graded wearable device is used. It offers real-time physiological data acquisition, enabling researchers to conduct an in-depth analysis and visualization of the data. This bracelet is equipped with an accelerometer, an Electrodermal Activity (EDA) sensor, a PPG sensor, a Infrared Thermopile sensor and a bluetooth transmitter. In addition, a chest-worn wearable device MetaMotionR offers real-time and continuous monitoring of motion of the person. It offers the measures of different sensors, such as accelerometer, gyroscope, magnetometer. The set of sensors is completed with a camera. The latter is directly connected to a phone considered as the main controller of the whole device. The camera records 3 s of video every 10 s 10 fps frame rate.

The device was worn for thirty days by two healthy volunteers. The first volunteer was twenty-six-years old and wore the device for twenty-one days. The second volunteer was sixty-two years old and wore the device for nine days. The diagram of recordings with their duration is presented in Fig. 2 below.

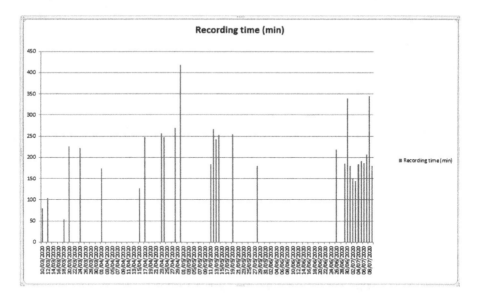

Fig. 2. Recording time diagram of the BIRDS corpus

Taking into account that "fall" is quite a well-studied situation with already available commercial solutions, and is a rare event, we work only with three "immediate risks": risk of falling, the risk of domestic accident, the risk of fraud. For long-term risks two situations are considered: medication and drinking water. To enrich the risk situations of first type accordingly to the model: context and person's condition, the healthy volunteers simulated "at risk" conditions by associating emotions such as fear, or stress.

We defined a risk of falling when the volunteer was in kneeling position or when he took stairs or was in the bathroom.

The risk of domestic accidents corresponds to the situation in the kitchen when the volunteer was cooking, took knives, etc.

The risk of intrusion/fraud was defined as a situation when the volunteer was with a person near a door.

The medication was identified when the volunteer simulated a drug intake. For the young volunteer it was a simulated situation. For the old volunteer it was a real-world situation.

For the risk of dehydration recording (second scenario), we asked the volunteer to periodically take a glass of water. The volunteers were required to write a short recording dairy approximately indicating the time instant when the subject puts the devices on himself, adjusts them, records, simulates various situations where each risk of first type or situation from our taxonomy is reported e.g., 13:45 entering into the kitchen (risk of domestic accident).

4.2 BIRDS Corpus Description

In its current state BIRDS corpus contains recordings of 30 days of the overall volume of 6316 min. The recording time per day varies from nearly two hours up to five hours. The overall and mean duration of recordings together with volumes of the data are given in the Table 4. The audio data are not used for privacy issues but are remain available. Figures 3 and 4 illustrate the content of the two datasets LSC and BIRDS. In Fig. 3 images from LSC dataset [6] are presented and in Fig. 4 video frames from different scenes of our corpus BIRDS are displayed. The two datasets were recorded with different camera positions. The LSC dataset is recorded with chest-worn camera, while BIRDS Dataset - with shoulder-worn camera. This position was set as proposed by ergotherapists in previous research[12], specifically adapted to the elderly subjects. The video frames in Fig. 4 depict various risk situations. One can notice that BIRDS corpus is more realistic with regard to the home environment of an elderly person. The corpus BIRDS will be made publicly available subject to fulfilling legal procedures accordingly to GDPR (https://gdpr-info.eu/) (Table 2).

Table 2. Description of the corpus BIRDS.

	Total sum	Mean	Standard deviation
Recording time (Min)	6316	210.53	76.47
Global data volume (MB)	50861	1695.36	4414.50
Sensor data volume (MB)	101.82	50.91	12.63
Video data volume (Mo)	23314.94	777.16	483.86
Sound sensor data volume (Mo)	200.84	6.69	2.50

Fig. 3. Examples of images from dataset LSC [6].

As it can be seen from the Table 3, the quantity of risk situations is not yet sufficient to do a significant training of the neural network and will have to be extended in the future recording sessions. The scenario of risk recognition has to be adapted to the environment of each subject.

Fig. 4. Data of corpus BIRDS.

Table 3. Risk situations in the corpus BIRDS

Risk situations	Number of risks situations
Risk of fall	26
Risk of domestic accident	64
Risk of fraud	4
Drug intake	18
Hydration by water intake	60

Table 4. Description of the corpus BIRDS.

	Total sum	Mean	Standard deviation
Recording time (Min)	6316	210.53	76.47
Global data volume (MB)	50861	1695.36	4414.50
Sensor data volume (MB)	101.82	50.91	12.63
Video data volume (Mo)	23314.94	777.16	483.86
Sound sensor data volume (Mo)	200.84	6.69	2.50

4.3   Corpus Annotation

A multimodal data annotation platform has been developed to annotate data from the BIRDS corpus Fig. 5. The data from the BIRDS dataset are annotated using recording diary that the subject has filled-in during the recording sessions in order to have ground truth about the data and the risk situations that have occurred. Contrarily to one of the first annotation protocols purely "by image" [27], in the case of immediate risk situations, psychologists focus first on the anomalies of physiological signals such as EDA and heart-rate and localize them in time with the help of the elastic window. The corresponding video clips are then visualized for the annotator to take his/her decision to which category belongs the anomaly. Note that with such a protocol, the long-term risks cannot be annotated. To identify medication and water drinking actions, the annotator has to visualize all video clips circa time instant reported by the subject in his recording dairy.

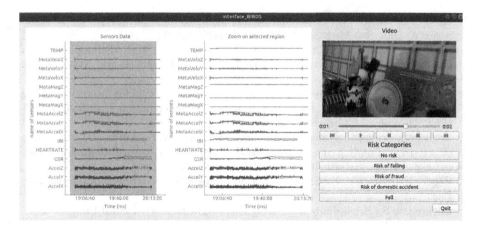

Fig. 5. Interface Screenshot for multimodal data annotation.

5   Conclusion

Hence in this paper we have presented a multi-disciplinary, multi-partner research on risk detection for fragile subjects using multimodal data recorded by wearable sensors. This research is a step ahead with regard to previous work on general-purpose lifelog data. We have extended the risk situation taxonomy for fragile persons, elaborated a two component - visual and multi-sensory data - model of risks. Recording scenario and protocol were elaborated as well and a new challenging corpus of data has been recorded and presented. We have shown, that detection of risk situations on the time series from wearable sensors with LSTM networks still remains a complex and open problem. Furthermore, the taxonomy of real-world risk situation comprises two types: immediate risks and long-term risks. While annotation approaches have to be different, the detection model has to remain in the same framework of temporal Neural Networks. In the follow -up of the work we will have to elaborate an efficient model for risk situation detection in case of missing data from signals and the fusion model: from visual and sensory components as well. Finally corpus enrichment is a continuous process which has to be conducted all along the project.

References

1. Baig, M.M., Afifi, S., GholamHosseini, H., Mirza, F.: A systematic review of wearable sensors and IoT-based monitoring applications for older adults-a focus on ageing population and independent living. J. Med. Syst. **43**(8), 233 (2019)
2. Bellagente, P., et al.: Remote and non-invasive monitoring of elderly in a smart city context. In: 2018 IEEE Sensors Applications Symposium (SAS), pp. 1–6. IEEE (2018)
3. Doherty, A.R., et al.: Experiences of aiding autobiographical memory using the sensecam. Hum.-Comput. Interact. **27**(1–2), 151–174 (2012)

4. Fried, L.P., et al.: Frailty in older adults: evidence for a phenotype. J. Gerontol. Series A: Biol. Sci. Med. Sci. **56**(3), M146–M157 (2001)

5. Greff, K., Srivastava, R.K., Koutník, J., Steunebrink, B.R., Schmidhuber, J.: LSTM: a search space odyssey. IEEE Trans. Neural Netw. Learn. Syst. **28**(10), 2222–2232 (2016)

6. Gurrin, C., Schoe, K., Joho, H., Munzer, B.: A test collection for interactive lifelog retrieval. In: MMM 2019, the 25th International Conference on MultiMedia Modeling, Thessaloniki, Greece (2019)

7. Hegde, N., Bries, M., Swibas, T., Melanson, E., Sazonov, E.: Automatic recognition of activities of daily living utilizing insole-based and wrist-worn wearable sensors. IEEE J. Biomed. Health Inf. **22**(4), 979–988 (2017)

8. Hegde, N., Sazonov, E.: Smartstep: a fully integrated, low-power insole monitor. Electronics **3**(2), 381–397 (2014)

9. Hochreiter, S., Schmidhuber, J.: Long short-term memory. Neural Comput. **9**(8), 1735–1780 (1997)

10. Hopper, L., Newman, E., Joyce, R., Smeaton, A., Irving, K.: Dementia ambient care: home-based monitoring and enablement of people with dementia. In: 9th Panhellenic Conference on Alzheimer's Disease and 1st Mediterranean on Neurodegenerative Diseases Conference, PICAD 2015, pp. 1–4 (2015)

11. Ionescu, B., Benois-Pineau, J., Piatrik, T., Quénot, G. (eds.): Fusion in Computer Vision - Understanding Complex Visual Content. ACVPR. Springer, Cham (2014). https://doi.org/10.1007/978-3-319-05696-8

12. Karaman, S., et al.: Hierarchical hidden Markov model in detecting activities of daily living in wearable videos for studies of dementia. Multim. Tools Appl. **69**(3), 743–771 (2014)

13. Kheirkhahan, M., et al.: A smartwatch-based framework for real-time and online assessment and mobility monitoring. J. Biomed. Inform. **89**, 29–40 (2019)

14. Khojasteh, S.B., Villar, J.R., Chira, C., González, V.M., De la Cal, E.: Improving fall detection using an on-wrist wearable accelerometer. Sensors **18**(5), 1350 (2018)

15. Kingma, D.P., Ba, J.: Adam: a method for stochastic optimization. CoRR abs/1412.6980 (2015)

16. Lin, Y., Wang, C., Wang, J., Dou, Z.: A novel dynamic spectrum access framework based on reinforcement learning for cognitive radio sensor networks. Sensors **16**(10), 1675 (2016)

17. Maimoon, L., et al.: SilverLink: developing an international smart and connected home monitoring system for senior care. In: Xing, C., Zhang, Y., Liang, Y. (eds.) ICSH 2016. LNCS, vol. 10219, pp. 65–77. Springer, Cham (2017). https://doi.org/10.1007/978-3-319-59858-1_7

18. Parant, A.: Blanchet mickaël, 2017, atlas des séniors et du grand âge en france. 100 cartes et graphiques pour analyser et comprendre, rennes, presses de l'ehesp, 120 p. Population **73**(4), 852–853 (2018)

19. Pierleoni, P., Belli, A., Palma, L., Pellegrini, M., Pernini, L., Valenti, S.: A high reliability wearable device for elderly fall detection. IEEE Sens. J. **15**(8), 4544–4553 (2015)

20. Rockwood, K., Mitnitski, A.: Frailty in relation to the accumulation of deficits. J. Gerontol. Ser. A: Biol. Sci. Med. Sci. **62**(7), 722–727 (2007)

21. Sabesan, S., Sankar, R.: Improving long-term management of epilepsy using a wearable multimodal seizure detection system. Epilepsy Behav. **46**, 56–57 (2015)

22. Seo, D., Yoo, B., Ko, H.: Data-driven smart home system for elderly people based on web technologies. In: Streitz, N., Markopoulos, P. (eds.) DAPI 2016. LNCS, vol. 9749, pp. 122–131. Springer, Cham (2016). https://doi.org/10.1007/978-3-319-39862-4_12

23. Talavera, E., Petkov, N., Radeva, P.: Towards unsupervised familiar scene recognition in egocentric videos. CoRR abs/1905.04093 (2019)

24. Tinetti, M.E., Kumar, C.: The patient who falls: "it's always a trade-off". JAMA 303(3), 258–266 (2010)

25. Vemulapalli, S.S., et al.: GIS-based spatial and temporal analysis of aging-involved accidents: a case study of three counties in Florida. Appl. Spatial Anal. Policy 10(4), 537–563 (2016). https://doi.org/10.1007/s12061-016-9192-4

26. Yacchirema, D., de Puga, J.S., Palau, C., Esteve, M.: Fall detection system for elderly people using IoT and ensemble machine learning algorithm. Pers. Ubiquit. Comput. 23(5–6), 801–817 (2019)

27. Yebda, T., Benois-Pineau, J., Pech, M., Amièva, H., Gurrin, C.: Detection of semantic risk situations in lifelog data for improving life of frail people. In: Proceedings of the 2020 International Conference on Multimedia Retrieval, pp. 402–406 (2020)

28. Yuan, J., Tan, K.K., Lee, T.H., Koh, G.C.H.: Power-efficient interrupt-driven algorithms for fall detection and classification of activities of daily living. IEEE Sens. J. 15(3), 1377–1387 (2014)

29. Zhu, L., Wang, R., Wang, Z., Yang, H.: TagCare: using RFIDs to monitor the status of the elderly living alone. IEEE Access 5, 11364–11373 (2017)

Towards the Development of a Trustworthy Chatbot for Mental Health Applications

Matthias Kraus[✉], Philip Seldschopf, and Wolfgang Minker

Ulm University, Ulm, Germany
{matthias.kraus,philip.seldschopf,wolfgang.minker}@uni-ulm.de

Abstract. Research on conversational chatbots for mental health applications is an emerging topic. Current work focuses primarily on the usability and acceptance of such systems. However, the human-computer trust relationship is often overlooked, even though being highly important for the acceptance of chatbots in a clinical environment. This paper presents the creation and evaluation of a trustworthy agent using relational and proactive dialogue. A pilot study with non-clinical subjects showed that a relational strategy using empathetic reactions and small-talk failed to foster human-computer trust. However, changing the initiative to be more proactive seems to be welcomed as it is perceived more reliable and understandable by users.

Keywords: Chatbot · Trustworthy agent · Proactivity · Mental health

1 Introduction

Mental disorders have a global impact on health and major social consequences affecting an estimated 792 million people worldwide [16]. However, the availability of mental health services and resources are scarce, especially in low and middle-income countries. Even in high-income countries such as Germany, there is a lack of psychiatrists and psychologists with a workforce of 62.75 per 100,000 population [33], while every tenth adult in Germany has current depressive symptoms [7]. Additionally, mental health issues often go untreated as people are unsure of the thresholds for treatment or are afraid of societal stigma and discrimination [3]. In order to enhance the accessibility of mental health care and to promote well-being, several public projects have been launched. For example, the EU-funded project *Mental health monitoring through interactive conversations*[1] (MENHIR) deals with researching and developing conversational technologies to promote mental health and assist people with mental ill health (depression and anxiety) to manage their conditions. The overall objective of the project is the creation of a chatbot (bot) that provides 24 - hour personalised social support, mental health coping strategies and education, and symptom and mood

---

[1] Ref. no.823907, https://menhir-project.eu.

© Springer Nature Switzerland AG 2021
J. Lokoč et al. (Eds.): MMM 2021, LNCS 12573, pp. 354–366, 2021.
https://doi.org/10.1007/978-3-030-67835-7_30

management for the identification of patterns that are indicative of relapse and recurrence. Using chatbots in e-health is an emerging topic over the last years and has resulted in various applications ranging from knowledge-based information agents to assist caregivers [36] to digital assistants in an intelligent operating room [30]. In order to be fully accepted and applicable in a delicate domain like health care, the design of trustworthy interaction strategies is inevitable. In the field of health care chatbots, this is a highly understudied topic as most research focuses on user satisfaction and technical acceptance but omits considering the human-computer trust (HCT) relationship [21]. Therefore, a chatbot incorporating relational dialogue strategies for establishing a trustworthy interaction is presented in this work. The chatbot has been created using the open source framework Rasa Stack (https://rasa.com). Relational dialogue strategies were implemented in the form of small talk and empathetic reactions to user input. Empathetic reactions and small talk are supposed to deepen relationships and to foster trust between agents and users [4,6] Additionally, an active version of the chatbot capable of sending push messages to users and taking the initiative during the dialogue was created. Recent studies have shown that active system actions, if provided appropriate, have a positive effect on the user's acceptance and trust towards the system [2,19,20]. In a pilot study, it was investigated how the relational strategies as well as the initiative of the chatbot affect the HCT relationship compared to a baseline variant. The study was conducted with subjects having no history of mental health treatment in order to create a control group for further experiments with subjects possessing mild mental disorders.

2   Related Work

2.1   Chatbots and Mental Health

A chatbot is generally an computer application that is able to engage in a natural dialogue with a human being [28]. The main purpose of a chatbot is to assist users in completing a task via goal-oriented dialogue. Compared to digital assistants such as Alexa or Siri, a chatbot is usually not tied to a device. Conversations with a chatbot take place in a chat program (e.g. Messenger like Telegram or Whatsapp) and are therefore specific to chat platforms. Thus, the communication of a chatbot is mainly conducted using written language [21].

Current research of chatbots in mental health suggest that the psychiatric use of chatbots is favourable, as it promotes self psycho-education and adherence [41]. Thus, the application of conversational chatbots in the mental health domain has been emerging over the last decade. For example Ly et al. [25] created the automated chatbot SHIM which was used to assess the effectiveness of strategies applied in positive psychology and cognitive behaviour therapy (CBT). Therefore, the bot was capable of providing empathetic responses on the user's mood and tailored content based on the user's previous inputs. In a small user study with students Shim was compared to a baseline and yielded significant effects on well-being and perceived stress. Additionally, it was found that users had high engagement in using the app. In the work of Fitzpatrick et al. [13], the conversational agent WOEBOT was evaluated in a study to determine the feasibility, acceptability, and preliminary efficacy of delivering a self-help program for college students who self-identify as having symptoms of anxiety and depression. Therefore, the chatbot used strategies being found in CBT. After a 2-week-trial, Woebot was perceived to have a high overall satisfaction and appears to be a feasible, engaging, and effective way to deliver psychological therapy. Denecke et al. [12] investigated the usability of a mental health chatbot SERMO for regulating emotions. Here, the emotions were directly recognised from the conversation using a lexicon-based approach. During the interaction, SERMO asks for the current mood, runs dialogues to retrieve information on a current event that impacted on the user as well as the emotion associated with the event. Based on this information it suggests suited activites and exercises. The results of the usability test showed that SERMO worked well for task completion and was easy to use. However, the interaction with SERMO was not perceived motivating or stimulating. Contrary to mentioned related work, the trustworthiness of an empathetic chatbot is observed in this paper. Therefore, we provide a short introduction on the concept of trust in the next section.

2.2   Human-Computer Trust

Trust is generally understood as the expected reliability of another party and the willingness to accept a remaining uncertainty [39]. The greater the willingness, the greater the confidence in that party. In the context of human-machine interaction the definition of Lee and See [23] describes trust in dealing with automation. Trust refers to the perceived degree of reliability of an automated system, taking into account the existing uncertainty. More precisely, this means that a high degree of perceived trust goes hand in hand with a high perceived reliability of the system [29]. Trust is a dynamic variable, which after an initial manifestation increasingly depends on the reliability of a system [22]. During the initial phase, the expectation conformity of a system plays a decisive role. Not only the direct influence of trust is relevant for the interaction with a system, but also the consequence on the perceived competence, the predictability about the system's behaviour as well as the reliability and acceptance towards a system [32]. Reliable systems not only lead to increased trust, but can also lead to an overestimation of the objective reliability of the system [40]. Low confidence leads to non-use, while excessive confidence, also known as overtrust, can lead to an increased risk when using the system [31]. To the best of our knowledge, the concept of trust has yet to be included into the domain of conversational mental health agents. Therefore, possible trust-building mechanisms for building chatbots are investigated in this work.

For measuring the trustworthiness of the developed chatbot, the Trust in Automated Systems scale [17] was used, where subjects could agree or disagree with statements about the system's impression. Sub-components of trust were measured using the HCT-model by Madsen and Gregor [26]. This hierarchical model is built on five basic components of trust: Personal attachment and faith form the bases for affect-based trust while perceived understandability, perceived technical competence, and perceived reliability constitute the bases for cognition-based trust. Affect-based trust refers to a long-term human-computer relationship, being established through frequent interactions with a system. Conversely, cognition-based trust refers to a more short-termed trust. Here, mostly the functionality and usability of a system are of importance.

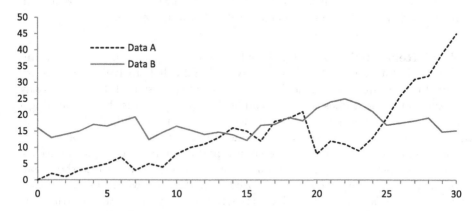

Fig. 1. Example of the PANAS-SF dialogue with the Rasa chatbot.

3   Development of Trustworthy Agents

3.1   Scenario

In this paper the implementation and evaluation of a prototypical mental health chatbot for mood and symptom monitoring is presented. Therefore, the main task of the chatbot is to interact with the user for having a daily mood check-in. This can then be used either by a future version of the bot or a psychiatrist to suggest activities or to provide feedback. For the daily check-in, 12 items of the Positive and Negative Affect Schedule (PANAS-SF) [42] questionnaire were used. This scale consists of a number of words that describe different feelings and emotions, e.g. interested, guilty, or active. Users can indicate to what extent (not at all to extremely) they have this feeling or emotion (see Fig. 1). Additionally, users had the opportunity to write freely about experiences of their day and their feelings. For rendering the chatbot trustworthy, an empathetic version capable of conducting small-talk is implemented. Additionally, the degree of initiative of the chatbot is manipulated using push notifications and by changing the dialogue initiative for a more proactive dialogue. The interaction design is described in detail in the following section.

3.2 Dialogue Design

The dialogue with the chatbot is initiated by the user with a simple greeting, which is reciprocated by the system. Afterwards, the bot initiates the daily check-in dialogue. However, the general goal is to build a chatbot that knows how to win people's trust to ensure improved usage. Therefore, the basic functionality of the chatbot's mood tracking dialogue is extended by adding relational dialogue strategies and a variation of the chatbot's initiative. The role of relational dialogue and the initiative is explained in the following sections.

Role of Relational Dialogue. People use a variety of types of social languages, including small talk and empathy, to build collaborative, trusting interpersonal relationships. In particular, the two constructs small talk and empathy have shown in past studies that they can lead to increased trust by establishing a long-term social-emotional relationship with their users [5]. Empathy is the mental process by which a person tries to understand the statements, behaviours or feelings of another person, from the counterpart's perspective or preconditions. The term "empathy" is not used uniformly in psychology. In this work, the social-psychological meaning of empathy is used [24]. Previous work has shown that digital emphatic agents are perceived as more caring, sympathetic and trustworthy than agents without emphatic abilities [6]. Above all, effective answers that correspond better to the situation of another than one's own should serve as the main instrument for inducing empathy [14]. The relational bot showed different emphatic reactions in different situations. During the daily check-in, for example, the bot repeatedly shows his appreciation and understanding for the user during very personal topics or provides an appropriate reaction to a negative mood on the day of the check-in, e.g *"Thank you! I really appreciate you talking to me about this."* or *"I know the questions are not always easy to answer but you are doing great.".* People use small talk, to establish interpersonal collaborative trusting relationships [9]. Research in the field of conversational agents has shown that it is not enough to limit conversations between agents and people to task-oriented topics [4]. The results suggest that small talk supports deepening relationships and building trust between virtual agents and users. It is inevitable that the agent will be able to establish a close relationship with the human interaction partner, especially in the long-term confrontation [27]. Therefore, the developed chatbot is able to deal with topics such as music preferences, about his person, how he feels today if he has any plans for today and some other topics like the weather. For an example, see the conversation depicted in Fig. 3.

Role of Initiative. When and how a chatbot interacts is an important factor in user studies on verbal behaviour, as previous work in human-robot interaction showed that it can shape the personality as well as the level of trust [1]. A high degree of initiative can also be perceived as proactive behaviour [2]. Proactivity is a term that is widely used in the domain of organisational psychology and management [34,35]. According to its definition, proactive behaviour is about taking control, anticipating and preventing problematic situations rather than reacting. Recent studies have shown that a medium-level of proactivity can be beneficial to the HCT relationship [20,37]. From the findings, it can be deduced that users are more likely to trust chatbots if they use an appropriate level of initiative. The initiative of the robot is manipulated in the form of push notification that intelligent remind users of their daily check-in, as well as dialogue control of the chatbot. Here, the chatbot has the initiative throughout the dialogue, e.g. starts small-talk directly, changes topics automatically. A comparison of the different variants can be seen in Fig. 2.

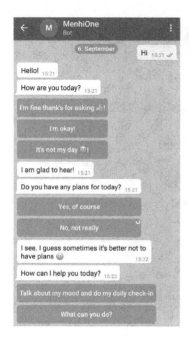
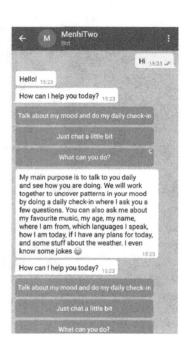

Fig. 2. Comparison of the proactive (left) to the non-proactive dialogue (right).

3.3 System Overview

The chatbots were implemented using the open-source framework Rasa Stack. This was due to the free availability and privacy issues, as personal data was not shared with external services. Rasa is a framework for creating conversational AI. The framework consists of two modules, one for dialogue control and one for natural language understanding (NLU). Rasa Core is a dialogue management system that is designed for using machine-learning in order to train a dialogue policy instead of a finite-state approach. The chatbot can learn through interactive learning by utilising so-called stories. A story is a representation of a dialogue between a user and the chatbot, converted into a specific format where user inputs are expressed as corresponding intents. The responses of the chatbot are expressed as corresponding action names. Rasa NLU is a statistically-driven NLU service for intent classification, response retrieval and entity extraction. When Rasa receives a message from the user, it attempts to predict the intent and extract the entities present in the message. This part is handled by Rasa NLU. Once the user's intention has been identified, the Rasa stack performs a specific action. In the example, visualised in Fig. 3, the intent of the user's utterance "Fine" would be "express_mood_positive". Then Rasa tries to predict what to do next. This decision is made taking into account several factors and is made by the Rasa Core unit. In the example, Rasa shows an empathetic reaction.

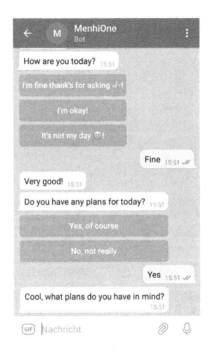

Fig. 3. Example dialogue with the Rasa chatbot. The system initiates the dialogue and then progresses to small-talk with the user.

It also predicted the next action that the model should perform - to continue with small-talk and to ask the user about his current plans. The more sample data Rasa has, the more likely it is that the right decision will be made. The model presented was trained with several stories and numerous example utterances for training the NLU. Telegram is used as platform for the interaction (see Fig. 3). Rasa offers extensions to be easily implemented in a Telegram chatbot. In addition, Telegram is used widely and can be reached via mobile phone, tablet and computer, which makes it much easier for test subjects to use it. A virtual cloud server was used to provide the chatbot. Rasa offers the possibility to run a HTTP server that handles requests using a trained Rasa model. Since a local server was used, the additional software Nginx [38] was installed to ensure the connection to Telegram. The Rasa NLU server was set behind an Nginx reverse proxy, where Nginx handles the SSL and then proxies to Rasa over HTTP. So-called cronjobs were used to send the test persons a daily push message. Under many operating systems there is the so-called Cron-System (Cron-Daemon), which makes it possible to execute automated tasks (jobs) at special times [11]. To send a message to a user, his Telegram user ID and Bot ID were required. Then the text of the message and the desired time of the cronjob could be determined.

4   Pilot Study

4.1   Study Setup

The baseline study was realized in a 2*2 between-subject experimental design. The dialogue strategy (relational, non-relational) and the initiative conditions (proactive, non-proactive) were implemented as two-step factors in four individual chatbots (e.g. relational-proactive, non-relational-proactive, ...). This results in four experimental groups in total. Participants had to interact with one of the chatbots on three consecutive days. As a cover up, they were told to test a novel chatbot on Telegram. Thereby their emotional state is checked in the form of a daily check-in. The dependent variables (trust, competence, understandability, reliability, personal attachment, faith, usability [8]) were collected at two different measuring points (after the first and last interaction with the chatbot) using 7-point Likert scales. The mean values were used for the evaluation for a more robust result. Validated psychological scales were used for testing the dependent variables. In order to rule out confounding variables the participants' technical affinity [18], and the predisposition to trust [29] were recorded prior to the experiment in combination with demographic data. In addition, the subjects had the opportunity to note any problems, impressions or irregularities in the conversations at the end of the two intermediate test questionnaires. A total of three conversations were carried out per respondent, which took place on three consecutive days. Any effects of the relational dialogue strategies should thus show their effect. Subjects were recruited at the University. Prerequisite for the participation was the possession of a mobile phone with the messaging program Telegram, as well as a fluent knowledge of the English language. As an incentive, the participants were promised a € 10 amazon voucher. All test persons were informed about the scientific purpose, as well as about the anonymous use of their data. A total of 41 (26 female) people took part in the experiment. However, 5 persons had to be excluded due to insufficient language knowledge (minimum B2). Subjects were between 19 and 30 years old (M = 24.83, SD = 2.93). Most of the subjects were psychology students or had an academic degree. They were randomly distributed to each study group.

Table 1. Descriptive statistics of the measured dependent variables with reference to the dialogue strategies (means and standard errors). All variables measured on 7-point Likert scales.

	Dialogue strategy			
	Relational	Non-relational	Proactive	Non-proactive
Trust	4.63 (.20)	4.69 (.21)	4.89 (.21)	4.41 (.21)
Reliability	4.06 (.27)	4.25 (.29)	4.58 (.29)	3.74 (.29)
Competence	3.62 (.30)	3.63 (.31)	3.84 (.31)	3.42 (.32)
Understandability	4.28 (.26)	4.43 (.27)	4.71 (.27)	4.00 (.27)
PA	1.90 (.20)	2.41 (.21)	2.24 (.21)	2.06 (.21)
Faith	2.08 (.23)	2.12 (.25)	2.19 (.25)	2.01 (.25)
Usability	2.63 (.09)	2.51 (.09)	2.61 (.09)	2.52 (.10)

4.2 Results

For an exploratory data analysis, a multivariate Analysis of Variance (MANOVA) was conducted for guaranteeing no significant confounding variables and for testing the significance of the relational and initiative strategies. Confounding group differences for the study conditions could be ruled out as we found no significant differences except regarding the initiative of the chatbot. Subjects who interacted with the non-proactive version of the chatbot had an almost significantly higher technological affinity as compared to the proactive group ($t(34) = -2.01$, $p = .053$). Therefore, this variable was used as a covariate to make up for noisy data when considering the initiative. However, there were no interaction effects between relational strategies and the initiative of the system (all p-values > 0.05). For further investigations, the effects of relational strategy and initiative were investigated separately. Therefore, we paired the individual samples and evaluated two study groups in each case: relational vs. non-relational and proactive vs. non-proactive. The results can be found in Table 1. One notable result was found when considering the relational strategies. Personal Attachment (PA) was rated higher for the non-relational strategy ($F(1, 31) = 3.14$, $p = .086$). Regarding the initiative of the chatbot two notable results were found. First, subjects rated the reliability of the proactive chatbot higher compared to the non-proactive version ($F(1, 31) = 3.92$, $p = .057$). Additionally, the understandability of the proactive chatbot was rated higher ($F(1, 31) = 3.26$, $p = .081$). For the other dependent variables no significant or notable results were found.

4.3 Perspectives of Trustworthy Agents for Mental Health Applications

The results of this baseline study towards the integration of trustworthy dialogue strategies in mental health application provided several indications which should be considered in future work. It was found that subjects were personally less attached to the chatbot capable of relational dialogue strategies. The construct of personal attachment to the system used in this study is comprised of: liking, i.e meaning that the user finds using the system agreeable and it suits their taste, as well as loving, i.e. that the user has a strong preference for the system, is partial to using it and has an attachment to it [26]. Hence, the inclusion of small talk and empathetic responses had not the intended effect of forming a trusted bond to the user. In fact, the opposite is the case. This seems rather strange at first sight, but could be explained that the interaction deals with very personal and sensitive content. Therefore, subjects seemed to be careful to open themselves to the chatbot. Privacy issues concerning the use of chatbots is an emerging topic and needs to be considered [15]. Another explanation could be that a kind of uncanny valley [10] was created, and the empathetic behaviour of the agent did not match the participants expectations of a Telegram chatbot. Hence, people could be more attached to the non-relational version, as they are supposed to be more used to such system behaviour. This forms an interesting topic when studying the effects of an empathetic agent on people with mental disorders, as they have more motivation to use the chatbot and hence have different expectations towards the system. Moreover, it was found that a proactive chatbot was perceived as more reliable and predictable. Proactive behaviour, like in our case, push notifications and more dialogue control by the system, seem to have an positive effect on trust and perceived competence as well. This is in line with other research [19,20,37] indicating that an appropriate level of proactivity can foster HCT with regard to competency and reliability. In organizational management, it was also shown that proactive behaviour can be positively related to perceived competency [34]. In summary, proactive behaviour may also be applicable in mental health applications, but further investigations are necessary. As a limitation of this work, the rather low usability of the chatbot needs to be addressed. This may have occurred, due to the rather rigid dialogue capabilities of the chatbots that are centered around the daily check-in dialogue. Hence, people seemed to have gotten bored and annoyed by the system after three days of usage. In order to avoid this in future studies, more dialogue topics need to be integrated.

5   Conclusion

The integration of trustworthy dialogue strategies in conversational agents for mental health applications is a highly understudied topic. In this work, a first step towards this goal is presented by testing the effects of relational strategies and initiative on the human-computer relationship. The results of a pilot study showed some interesting insights for extending the development of the created mental health chatbot. The capability of small and empathetic reactions seemed to have failed their purpose in establishing a bond to the user. Hence, other mechanisms for trust formation need to be considered. Contrary, proactive behaviour of a chatbot could be welcomed by user, as it is perceived as more reliable and understandable. Based on this pilot study with non-clinical subjects, further evaluations are planned involving people with slight mental disorders such as anxiety or mild depression. In doing so, a foundation for the development of trustworthy agents is intended to be created.

Acknowledgements. This research has received funding from the European Union's Horizon 2020 research and innovation programme under grant agreement No 823907 (MENHIR: Mental health monitoring through interactive conversations https://menhir-project.eu).

References

1. Aoyama, K., Shimomura, H.: Real world speech interaction with a humanoid robot on a layered robot behavior control architecture. In: Proceedings of the 2005 IEEE International Conference on Robotics and Automation, pp. 3814–3819. IEEE (2005)
2. Baraglia, J., Cakmak, M., Nagai, Y., Rao, R., Asada, M.: Initiative in robot assistance during collaborative task execution. In: The Eleventh ACM/IEEE International Conference on Human Robot Interaction, pp. 67–74. IEEE Press (2016)
3. Bhugra, D.: Attitudes towards mental illness: a review of the literature. Acta Psychiatr. Scand. **80**(1), 1–12 (1989)
4. Bickmore, T., Cassell, J.: Small talk and conversational storytelling in embodied conversational interface agents. In: AAAI Fall Symposium on Narrative Intelligence, pp. 87–92 (1999)
5. Bickmore, T.W., Picard, R.W.: Establishing and maintaining long-term human-computer relationships. ACM Trans. Comput.-Hum. Interact. (TOCHI) **12**(2), 293–327 (2005)
6. Brave, S., Nass, C., Hutchinson, K.: Computers that care: investigating the effects of orientation of emotion exhibited by an embodied computer agent. Int. J. Hum Comput Stud. **62**(2), 161–178 (2005)
7. Bretschneider, J., Kuhnert, R., Hapke, U.: Depressive symptoms among adults in Germany. J. Health Monit. **2**, 77–83 (2017)
8. Brooke, J.: SUS: a "quick and dirty' usability. In: Usability Evaluation in Industry, p. 189 (1996)
9. Cassell, J., Bickmore, T.: Negotiated collusion: modeling social language and its relationship effects in intelligent agents. User Model. User-Adapted Interact. **13**(1–2), 89–132 (2002)
10. Ciechanowski, L., Przegalinska, A., Magnuski, M., Gloor, P.: In the shades of the uncanny valley: an experimental study of human-chatbot interaction. Future Gener. Comput. Syst. **92**, 539–548 (2019)
11. Davidovi, Š., Guliani, K.: Reliable cron across the planet. Queue **13**(3), 30–39 (2015)
12. Denecke, K., Vaaheesan, S., Arulnathan, A.: A mental health chatbot for regulating emotions (SERMO)-concept and usability test. IEEE Trans. Emerg. Top. Comput. **1**, 1 (2020). https://doi.org/10.1109/TETC.2020.2974478
13. Fitzpatrick, K.K., Darcy, A., Vierhile, M.: Delivering cognitive behavior therapy to young adults with symptoms of depression and anxiety using a fully automated conversational agent (woebot): a randomized controlled trial. JMIR Mental Health **4**(2), e19 (2017)

14. Hoffman, M.L.: Empathy and Moral Development: Implications for Caring and Justice. Cambridge University Press, Cambridge (2001)

15. Ischen, C., Araujo, T., Voorveld, H., van Noort, G., Smit, E.: Privacy concerns in chatbot interactions. In: Følstad, A., Araujo, T., Papadopoulos, S., Law, E.L.-C., Granmo, O.-C., Luger, E., Brandtzaeg, P.B. (eds.) CONVERSATIONS 2019. LNCS, vol. 11970, pp. 34–48. Springer, Cham (2020). https://doi.org/10.1007/978-3-030-39540-7_3

16. James, S.L., et al.: Global, regional, and national incidence, prevalence, and years lived with disability for 354 diseases and injuries for 195 countries and territories, 1990–2017: a systematic analysis for the global burden of disease study 2017. Lancet **392**(10159), 1789–1858 (2018)

17. Jian, J.Y., Bisantz, A.M., Drury, C.G.: Foundations for an empirically determined scale of trust in automated systems. Int. J. Cogn. Ergon. **4**(1), 53–71 (2000)

18. Karrer, K., Glaser, C., Clemens, C., Bruder, C.: Technikaffinität erfassen-der fragebogen TA-EG. Der Mensch im Mittelpunkt technischer Systeme **8**, 196–201 (2009)

19. Kraus, M., Fischbach, F., Jansen, P., Minker, W.: A comparison of explicit and implicit proactive dialogue strategies for conversational recommendation. In: Proceedings of the 12th Language Resources and Evaluation Conference, pp. 429–435 (2020)

20. Kraus, M., Wagner, N., Minker, W.: Effects of proactive dialogue strategies on human-computer trust. In: Proceedings of the 28th ACM Conference on User Modeling, Adaptation and Personalization, pp. 107–116 (2020)

21. Laranjo, L., et al.: Conversational agents in healthcare: a systematic review. J. Am. Med. Inform. Assoc. **25**(9), 1248–1258 (2018)

22. Lee, J., Moray, N.: Trust, control strategies and allocation of function in human-machine systems. Ergonomics **35**(10), 1243–1270 (1992)

23. Lee, J.D., See, K.A.: Trust in automation: designing for appropriate reliance. Hum. Factors **46**(1), 50–80 (2004)

24. Linden, M., Hautzinger, M.: Verhaltenstherapiemanual, vol. 8. Springer, Heidelberg (2008)

25. Ly, K.H., Ly, A.M., Andersson, G.: A fully automated conversational agent for promoting mental well-being: a pilot RCT using mixed methods. Internet Interv. **10**, 39–46 (2017)

26. Madsen, M., Gregor, S.: Measuring human-computer trust. In: 11th Australasian Conference on Information Systems, vol. 53, pp. 6–8. Citeseer (2000)

27. Mattar, N., Wachsmuth, I.: Small talk is more than chit-chat. In: Glimm, B., Krüger, A. (eds.) KI 2012. LNCS (LNAI), vol. 7526, pp. 119–130. Springer, Heidelberg (2012). https://doi.org/10.1007/978-3-642-33347-7_11

28. McTear, M., Callejas, Z., Griol, D.: The Conversational Interface, vol. 6. Springer, Cham (2016). https://doi.org/10.1007/978-3-319-32967-3

29. Merritt, S.M., Heimbaugh, H., LaChapell, J., Lee, D.: I trust it, but i don't know why: effects of implicit attitudes toward automation on trust in an automated system. Hum. Factors **55**(3), 520–534 (2013)

30. Miehle, J., Ostler, D., Gerstenlauer, N., Minker, W.: The next step: intelligent digital assistance for clinical operating rooms. Innov. Surg. Sci. **2**(3), 159–161 (2017)

31. Moray, N., Inagaki, T., Itoh, M.: Adaptive automation, trust, and self-confidence in fault management of time-critical tasks. J. Exp. Psychol. Appl. **6**(1), 44 (2000)

32. Muir, B.M.: Trust in automation: Part i. theoretical issues in the study of trust and human intervention in automated systems. Ergonomics **37**(11), 1905–1922 (1994)

33. Organization, W.H.: Mental Health Atlas 2017. World Health Organization (2018)

34. Parker, S.K., Bindl, U.K., Strauss, K.: Making things happen: a model of proactive motivation. J. Manag. **36**(4), 827–856 (2010)
35. Parker, S.K., Wang, Y., Liao, J.: When is proactivity wise? A review of factors that influence the individual outcomes of proactive behavior. Ann. Rev. Organ. Psychol. Organ. Behav. **6**, 221–248 (2019)
36. Pragst, L., Ultes, S., Kraus, M., Minker, W.: Adaptive dialogue management in the kristina project for multicultural health care applications. In: Proceedings of the 19thWorkshop on the Semantics and Pragmatics of Dialogue (SEMDIAL), pp. 202–203 (2015)
37. Rau, P.L.P., Li, Y., Liu, J.: Effects of a social robot's autonomy and group orientation on human decision-making. Adv. Hum.-Comput. Interact. **2013**, 11 (2013)
38. Reese, W.: Nginx: the high-performance web server and reverse proxy. Linux J. **2008**(173) (2008)
39. Rotter, J.B.: Interpersonal trust, trustworthiness, and gullibility. Am. Psychol. **35**(1), 1 (1980)
40. Stanton, N.A., Walker, G.H., Young, M.S., Kazi, T., Salmon, P.M.: Changing drivers' minds: the evaluation of an advanced driver coaching system. Ergonomics **50**(8), 1209–1234 (2007)
41. Vaidyam, A.N., Wisniewski, H., Halamka, J.D., Kashavan, M.S., Torous, J.B.: Chatbots and conversational agents in mental health: a review of the psychiatric landscape. Can. J. Psychiatry **64**(7), 456–464 (2019)
42. Watson, D., Clark, L.A., Tellegen, A.: Development and validation of brief measures of positive and negative affect: the PANAS scales. J. Pers. Soc. Psychol. **54**(6), 1063 (1988)

Fusion of Multimodal Sensor Data for Effective Human Action Recognition in the Service of Medical Platforms

Panagiotis Giannakeris$^{(\boxtimes)}$ ⓘ, Athina Tsanousaⓘ, Thanasis Mavropoulosⓘ, Georgios Meditskosⓘ, Konstantinos Ioannidisⓘ, Stefanos Vrochidisⓘ, and Ioannis Kompatsiarisⓘ

Centre for Research and Technology Hellas, Information Technologies Institute, Thessaloniki, Greece
{giannakeris,atsan,mavrathan,gmeditsk,kioannid,stefanos,ikom}@iti.gr

Abstract. In what has arguably been one of the most troubling periods of recent medical history, with a global pandemic emphasising the importance of staying healthy, innovative tools that shelter patient well-being gain momentum. In that view, a framework is proposed that leverages multimodal data, namely inertial and depth sensor-originating data, can be integrated in health care-oriented platforms, and tackles the crucial task of human action recognition (HAR). To analyse person movement and consequently assess the patient's condition, an effective methodology is presented that is two-fold: initially, Kinect-based action representations are constructed from handcrafted 3DHOG depth features and the descriptive power of a Fisher encoding scheme. This is complemented by wearable sensor data analysis, using time domain features and then boosted by exploring fusion strategies of minimum expense. Finally, an extended experimental process reveals competitive results in a well-known benchmark dataset and indicates the applicability of our methodology for HAR.

Keywords: Action recognition · Sensor fusion · Depth sensors · Wearable sensors

1 Introduction

Considering the biological and psychological challenges that contemporary, urban mainly, settings pose for many people who are used to leading fast-paced but sedentary lives, it becomes apparent that maintaining a healthy lifestyle comprising mental and physical activities, as well as adequate rest is of paramount importance. Attaining the correct balance of activities is a task that greatly benefits from the latest advances in technologies such as pervasive sensors, artificial intelligence, human and health monitoring and assistive living [9]. They aid in the efficient logging of sleep/activity data [17] and thus the effective organisation of people's routines via reminders/motivation actions and suggestions [24].

© Springer Nature Switzerland AG 2021
J. Lokoč et al. (Eds.): MMM 2021, LNCS 12573, pp. 367–378, 2021.
https://doi.org/10.1007/978-3-030-67835-7_31

Particularly in unconventional circumstances, such as the present Covid-19 era, that people need to apply socially distancing criteria in all of their activities, often having to cope with the unavailability of experts, physical activity self-assessment via sensor-based methods is crucial.

Specifically in the field of medicine, data analysis coming from small, low cost, high performance sensors has been providing researchers the tools to develop efficient and versatile methods of assisting patients, in order to improve their lifestyle. People in need of monitoring tend to be more autonomous and less attached to their caretakers, when having access to personalised activity information. Knowing that reliable mechanisms, such as automatic push notifications in case of patient fall, are in place to ensure timely intervention, it provides obvious benefits to both their physical state, and mental state and sense of self-sufficiency. Passive patient monitoring is an incontrovertible area of application of the abovementioned systems, where patients with mental diseases like dementia can be supervised to avoid or prevent potentially hazardous circumstances.

In the present work, focus is placed on monitoring certain well-defined actions/human movements, usually pertaining to a rehabilitation scenario, by fusing inertial and depth sensor data, since the technique has proven to provide excellent results, while the required training data are easily obtainable. To this end, manually crafted features are extracted first, from depth and sensor modalities and are adapted for our HAR framework. Then, multimodal data analysis is evaluated, with particular attention to fusion mechanisms of minimal expense. Specifically, several classification algorithms on inertial and visual sensors were applied, both separately and with two different fusion strategies, in order to recognise 27 human actions of the UTD-MHAD multimodal dataset [5]. The contribution of the paper could be summarised in the following:

– The methodology of Fisher encoding with 3DHOG depth features is adapted for HAR and evaluated. Time domain features based on inertial sensors are also evaluated.
– Two inexpensive fusion strategies (one feature-level and one decision-level) are deployed and performance comparisons are made between the two as well as the separate modalities.
– Extensive evaluation of several classifiers is performed with numerous evaluation protocols on a well-known multimodal HAR benchmark dataset.

Corresponding analysis results could be integrated into unified multi-user-oriented medical platforms, servicing both patients [23] and caretakers[1].

2   Related Work

Human action recognition in the context of Ambient Assisted Living (AAL) is facilitated by a variety of sensors, which may include inertial, range and magnetic sensors, depth and RGB cameras and even atypical modality type sensors, such

---

[1] https://www.gatekeeper-project.eu/.

as electrocardiogram ones [22]. The multitude of existing sensor technologies is supplemented by respective analysis methodologies. Diverse studies elaborate on modern machine learning HAR approaches, such as the one found in [25] or [31] that focuses on deep learning, state-of-the-art techniques. Moreover, in [11] distinct neural networks are exploited for depth and inertial sensing before decision-level fusion is performed. However, to leverage the performance improvement of deep learning, large amounts of training data and computational resources are often required.

Kinect revolutionised the field by providing an easily accessible and affordable tool, capable of synchronised skeleton, depth and RGB data provision without the need for additional post-processing. Since its introduction in the consumer market, researchers wholeheartedly embraced it and exploited its capabilities to present novel methods of tackling HAR [8,19,21,30]. Despite the justified attention it gathered and the promising results, concerns are expressed regarding installation/setup complexity and computational efficiency [18]. In addition, privacy issues due to the RGB data are raised nowadays more often than before. As a consequence, many studies focus primarily on depth and skeleton information to deal with data anonymisation requirements.

A common denominator when talking about inertial sensing, is the use of accelerometers and gyroscopes, and depending on the field of application [1,10], they may be complemented by more specialised sensors, such as magnetometers or barometric altimeters. Applications and trends favourable to inertial sensing are illustrated in [2], which also includes details on the history of devices and predictions on future directions. An in-depth view of the most important features and technologies, coupled with significant drawbacks governing typical gyroscope and accelerometer outputs is provided in [26].

Since certain real life challenges are impossible to be tackled just by one modality, approaches that combine the two have also been tested with promising results and helped overcome certain otherwise insurmountable issues [4,12,32]. Three main fusion directions exist that apply to most HAR approaches and each is performed at a different workflow step [6]: (a) data-level fusion, (b) feature-level fusion, and (c) decision-level fusion. Data-level fusion corresponds to the concatenation of raw data as they are directly collected from the respective sensors. Feature-level fusion (early fusion) is performed after features have been extracted from raw data and entails fusion of retrieved feature sets. Lastly, decision-level fusion (late fusion) combines the results of individual sensors after the classification has been completed.

Depending on the problem, different fusion mechanisms and theories have been attempted, such as exploitation of Hidden Markov Models (HMM) for hand gesture recognition [20] to tackle different modality synchronisation issues or the Dempster-Shafer theory for late (decision-level) fusion for action recognition in [3]. The former methodology [20] reported individual recognition sensing accuracy of 84% (Kinect) and 88% (inertial), while the concatenated model achieved accuracy of 93%. In the latter [3], early (feature-level) fusion is achieved by merging each sensor's individually extracted feature sets (first represented as vectors and then normalised) before the classification process is activated. Reported scores varied between 2–23% compared to the individual ones. Similar improvements are exhibited in [33] when the authors combine ear-worn sensors and RGB-D (Red, Green, Blue and Depth) to perform walking analysis. Moreover, an ensemble of binary one-vs-all neural network classifiers is explored in [13] to improve indoor human action recognition robustness, which once trained, is able to be effortlessly embedded on portable devices. Furthermore, a task that benefits greatly (2–8% improvement) from sensor data fusion is identified in [16], which describes an approach that leverages an SVM classifier and combines depth maps with accelerometer data to perform fall detection.

3   Methodology

3.1   Inertial Sensors

One wearable inertial sensor was used to record human actions in UTD-MHAD dataset [5], which is used in this work. The sensor provided recordings of acceleration, angular velocity and magnetic strength. To perform the analysis on the inertial sensor signals, the features suggested in [14], a paper that conducts experiments on the same dataset, were extracted. Firstly the magnitude of the raw signals of accelerometers and gyroscopes was calculated, using the formula in Eq. 1, where a stands for the signal values of each axis. For the preprocessing stage, a moving window average for each 3 rows of data was applied. Following, three features were extracted from the filtered signal vectors of each axis and of the calculated magnitude. More specifically, the mean of each vector (Eq. 2), the average of the absolute first difference of each signal vector a (Eq. 3), as well as the average of the corresponding second difference of the signal vectors a (Eq. 4) were calculated. Analysis was performed on accelerometer and gyroscope signals, as well as on their concatenated features.

$$a_{mag} = \sqrt{a_x^2 + a_y^2 + a_z^2} \tag{1}$$

$$mean = \frac{1}{N} \sum a(n) \tag{2}$$

$$mean_{fd} = \frac{1}{N} \sum |a(n) - a(n-1)| \tag{3}$$

$$mean_{sd} = \frac{1}{N} \sum |a(n+1) - 2a(n) + a(n-1)| \tag{4}$$

3.2   Depth Sensors

Local Features. In order to extract features from depth videos, the well-established efficiency of the HOG descriptor (Histograms of Oriented Gradients) was leveraged. The process was performed on 3D volumes, as in [14], to capture spatio-temporal features that encode the actor's body shape and limp movements that happen when an action is performed. The **3DHOG** descriptors are calculated based on the gradient magnitude responses in the horizontal and vertical directions of frames. Next, the responses are aggregated over spatio-temporal blocks of pixels. A histogram of gradient responses, quantised into 8 bins (8 orientations), is constructed for each block and the responses of all pixels in that block are assigned linearly into neighboring bins. Finally, the histograms of a neighborhood of blocks are concatenated together to form a local 3DHOG descriptor. Our method is different in that aspect from the approach of [28] or [14], and does not result in 3D chunks of perfectly neighboring blocks. Instead, in order to speed up the calculations, strided sampling was applied, where a fixed number of pixels are skipped before the next block is taken. The blocks were chosen to have a size of 15×15 pixels in space, and 20 frames in time, as in [14]. The 3D chunks are created with the concatenation of 3×3 blocks in space and 2 blocks in time, and the stride parameter is set to 5 pixels on all directions. Therefore, each chunk is compiled by 18 histograms ($3 \times 3 \times 2$ blocks), resulting in a 144-dimensional 3DHOG descriptor. Finally, the local 3DHOGs are L1-normalised and reduced to half their size (70 components) using PCA.

Action Representation. The local 3DHOG descriptor's dimensionality depends on the choices for the spatial and temporal dimensions of the concatenation chunks and is fixed in a given setting (144 reduced to 70 after PCA). However, the number of local 3DHOG descriptors extracted in a sequence can be arbitrary and is determined by the duration of each video, which is not the same for every sequence in the dataset. Thus, we ought to apply a method that will allow us to aggregate the set of collected local 3DHOGs to a final fixed size meaningful representation for each sequence.

In order to build the final descriptors, Fisher encoding is applied, which is proven to be a more efficient and powerful method to synthesise action representations compared to other bag-of-words techniques [7,28,29]. First, a visual vocabulary based on the most prominent visual clues of the whole depth sequence is built. The computation of the most discriminating samples is performed by applying unsupervised clustering (Gaussian Mixture Model (GMM)) in the shallow representation hyperspace, as formed by the feature collection of each depth sequence.

Let $\{\mu_j, \Sigma_j, \pi_j; j \in R^L\}$ be the set of parameters for L Gaussian models, with μ_j, Σ_j and π_j standing respectively for the mean, the covariance and the prior probability weights of the jth Gaussian. Assuming that the D-dimensional 3DHOG descriptor is represented as $\overline{x}_i \in R^D; i = \{1, \ldots, N\}$, with N denoting the total number of descriptors, Fisher encoding is then built upon the first and second order statistics:

$$f_{1j} = \frac{1}{N\sqrt{\pi_j}} \sum_{i=1}^{N} q_{ij}\sigma_j^{-1}(\overline{x}_i - \overline{\mu}_j)$$

$$f_{2j} = \frac{1}{N\sqrt{2\pi_j}} \sum_{i=1}^{N} q_{ij}[\frac{(\overline{x}_i - \overline{\mu}_j)^2}{\sigma_j^2} - 1]$$

(5)

where q_{ij} is the Gaussian soft assignment of descriptor x_i to the jth Gaussian and is given by:

$$q_{ij} = \frac{exp[-\frac{1}{2}(x_i - \mu_j)^T\Sigma_j^{-1}(x_i - \mu_j)]}{\sum_{t=1}^{L} exp[-\frac{1}{2}(x_i - \mu_t)^T\Sigma_j^{-1}(x_i - \mu_t)]}$$

(6)

Distances, as calculated by Eq. 5, are next concatenated to form the resulting Fisher vector, $F_X = [f_{11}, f_{21}, \ldots, f_{1L}, f_{2L}]$. Finally, L2 and power normalisation is applied to all Fisher vectors.

3.3   Sensor Fusion

For the fusion of depth and inertial sensors, both early and late fusion schemes were deployed. Accelerometer and gyroscope features were combined with the features extracted from the depth videos. In order to combine the heterogeneous sources at feature level (early fusion), the sensor data were first L2-normalised and then concatenated with the Fisher vectors. To perform late fusion, the probability vectors of the predicted classes were combined by averaging: using the same classifier, the probabilities obtained from inertial and depth modalities were averaged and the class with the highest averaged probability was assigned to each test case. The amount of actions included in the dataset would not favour other forms of late fusion, like weighted late fusion, that compute weights based on the classification metrics of each class. The additional cost of fusing the modalities is low, given that concatenation and averaging calculations are simple as well as highly paralellizable.

4   Experiments and Results

4.1   Dataset and Evaluation Description

The evaluation of our methods was performed on a well-known public multimodal dataset for action recognition, **UTD-MHAD** [5]. This dataset provides captured data for 27 different types of actions, carried out by 8 subjects (4 female, 4 male), performing 1 to 4 trials for each action. The set contains in total 861 samples. Please refer to [5] for a detailed description and the full class list. This is a challenging dataset because it contains a high number of classes with substantial variability. Specifically, only about 30 samples correspond to each class on average.

Table 1. Performance of inertial sensors.

	Sbj Generic			Sbj Specific			Cross Sbj		
	Acc+Gyro	Acc	Gyro	Acc+Gyro	Acc	Gyro	Acc+Gyro	Acc	Gyro
LDA	0.787	0.665	0.659	0.806	0.806	0.890	0.786	0.609	0.637
kNN	0.476	0.515	0.476	0.806	0.876	0.806	0.437	0.500	0.437
NB	0.597	0.499	0.506	0.527	0.424	0.384	0.586	0.474	0.523
RF	0.693	0.569	0.592	0.913	0.813	0.828	0.656	0.527	0.567
LSVM	0.618	0.469	0.602	0.802	0.846	0.802	0.574	0.460	0.574
KSVM	0.318	0.418	0.336	0.602	0.792	0.599	0.346	0.451	0.351

In our effort to comply with all the evaluation scenarios that have been previously proposed for this dataset, we conduct our experiments based on three different evaluation protocols: a) *subject-generic* protocol, where each subject was used once as a test set. b) The *subject-specific* protocol, where each subject was examined separately. For each subject, two of the trials constitute the training set and the other two trials form the test set. c) The cross-subject protocol, where the models are trained on half of the subjects (1, 3, 5, 7) and tested on the other half (2, 4, 6, 8). The respective results refer to the average accuracy of all rounds of experiments. The classification algorithms evaluated in this work are: Linear Discriminant Analysis (LDA), k-Nearest Neighbours with 1 neighbour (k-NN), Naive Bayes (NB), Random Forests (RF), Linear Support Vector Machine (LSVM) and Kernel SVM (KSVM) with quadratic kernel. We also experimented with a higher number of neighbours for the k-NN classifier, but, the accuracy dropped significantly, mainly because the training set is small relative to the high number of classes.

4.2   Inertial Sensor Performance Analysis

The recordings of the wearable inertial sensor were tested for their performance together and separately. As seen in Table 1, which presents the accuracy levels of all experiments of the three evaluation scenarios, we cannot draw conclusions on which scheme performs best, as it seems that this varies depending on the classifier. In case of the subject specific evaluation scenario, the combination of accelerometer and gyroscope performs better. This is not the case though in the other two evaluation scenarios, where there are classifiers that produce better results using the readings of the one sensor only. Such observations are usually reported in relevant studies, where there is always present heterogeneity caused by different subjects, different sampling frequencies or even different placement of sensors. Another reason would be the number of actions recorded in the current dataset. Regarding the performance of the classification algorithms, LDA and RF produced the best accuracy levels. The experiments reproduced from the baseline paper [14] did not yield the same results, probably because of a misconception in the description of the evaluation or feature extraction steps.

Table 2. 8-fold cross validation using various GMM vocabulary sizes.

	4 words	8 words	16 words	32 words	64 words
LDA	0.886	0.959	0.973	**0.979**	0.962
kNN	0.926	0.957	0.968	**0.980**	0.979
NB	0.792	0.828	**0.853**	0.851	0.838
RF	0.921	0.938	0.956	**0.965**	0.954
LSVM	0.902	0.956	0.976	**0.990**	0.986
KSVM	0.011	0.008	0.008	**0.015**	0.005

Table 3. Performance of depth sensor.

	Sbj Generic	Sbj Specific	Cross Sbj
LDA	**0.856**	0.860	0.781
kNN	0.572	0.993	0.458
NB	0.796	0.670	0.681
RF	0.826	0.984	**0.809**
LSVM	0.779	**0.998**	0.747
KSVM	0.502	0.970	0.433

4.3 Depth Sensor Performance Analysis

To infer the optimal number of Gaussians of the GMM clustering, that is, the number of visual words of the vocabulary, an initial experiment was conducted, using 8-fold cross validation on the entire dataset with random splits. The values for the size of the codebook that were tester are: 4, 8, 16, 32 and 64 words. Table 2 shows the results. Nearly all the classifiers achieve their peak performance with 32 GMM words, therefore, the sweat spot is roughly around this value and is used in all further experiments. Table 3 shows the performance of the depth sensor for every classifier in every evaluation protocol. It can be seen that in general, LDA, Random Forests and Linear SVM perform consistently better than the others in all the tests. Moreover, the method performs better in the subject specific protocol, where there are no unseen subjects in the test set.

4.4 Sensor Fusion Performance Analysis

Figure 1 shows a comparison of the fusion approaches with the individual modalities for each evaluation protocol. In most cases the early fusion scheme performs better, or at least equal, to both the inertial and depth modalities and the late fusion scheme. This conclusion holds true for the majority of the classifiers in all tests. On the contrary, there are cases where late fusion performs worse than the separate modalities. In general, we can safely conclude that early fusion is the most appropriate technique, irrespective of the classifier.

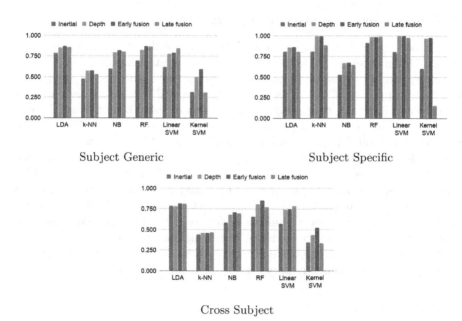

Fig. 1. Performance comparison of individual modalities with early and late fusion.

Table 4. Comparison with state-of-the-art. I = Inertial, D = Depth, I+D = Fusion.

Work	Modality	Sbj Generic	Sbj Specific	Cross Sbj	8-fold cv
Chen et al. 2015 [5]	I	–	–	0.661	–
	D	–	–	0.672	–
	I+D	–	–	0.791	–
Elmadany et al. 2015 [15]	D	–	–	0.734	–
Chen et al. 2015 [4]	I	0.764	0.883	–	–
	D	0.747	0.851	–	–
	I+D	**0.915**	0.972	–	–
Zhang et al. 2017 [34]	D	–	–	0.844	–
Ehatisham et al. 2019 [14]	I	–	–	–	0.916
	D	–	–	–	0.815
	I+D	–	–	–	0.970
Dawar et al. 2019 [12]	I	–	–	0.815	–
	D	–	–	0.759	–
	I+D	–	–	**0.892**	–
Weiyao et al. 2019 [32]	D	–	–	0.887	–
Sidor et al. 2020 [27]	D	0.886	0.993	–	–
Ours	I	0.787	0.913	0.786	0.904
	D	0.856	0.998	0.809	0.990
	I+D	0.873	**0.998**	0.853	**0.997**

4.5 Comparison with State-of-the-Art

Table 4 shows a detailed comparison with the state-of-the-art works in the same dataset. Our method's results are taken from the best performing classifier on the corresponding evaluation protocol and for each one of the inertial, depth and early fusion approaches. For other works, the reported results are presented on the corresponding field, depending on what protocols have been followed. It can be seen, that our method outperforms all other works on the subject specific and 8-fold cross validation protocols. Regarding the subject generic evaluation, our early fusion technique is surpassed by the decision-level fusion of [4], despite the fact that the separate modalities in our methodology perform better. This is an indication that more sophisticated fusion may boost our results in the case of unseen subjects. Regarding the cross subject evaluation, which is the most popular protocol, our fusion technique is surpassed by the deep learning-based fusion of [12], but our depth modality scores higher. Still, our method's early fusion scheme achieves competitive accuracy (down by a factor of 0.04) without data augmentation which is required in [12] to train deep CNNs.

5 Conclusions

In this work we have presented an effective methodology for human action recognition, based on fusion of inertial and depth data. Regarding the depth sensors, the 3DHOG and Fisher encoding methodology can produce discriminative features of actions and even compete with deep learning approaches, particularly for actions of previously seen subjects. The dataset used for this work consisted of many subjects and recorded actions. This heterogeneity seems to have affected the results of the inertial sensors' analysis. However, any discriminative information in the features can be exploited with a simple and inexpensive early fusion, as the results suggest. LDA, Random Forests and Linear SVMs are the best choices for HAR classification using these features. Overall, there is still room for improvement regarding actions of unseen subjects, which would require robustness to arbitrary physical dimensions or specific movement patterns of subjects.

Acknowledgment. This research has been financed by the European Regional Development Fund of the European Union and Greek national funds through the Operational Program Competitiveness, Entrepreneurship and Innovation, under the call RESEARCH - CREATE - INNOVATE (T1EDK-00686) and the EC funded project GATEKEEPER (H2020-857223).

References

1. Avci, A., Bosch, S., Marin-Perianu, M., Marin-Perianu, R., Havinga, P.: Activity recognition using inertial sensing for healthcare, wellbeing and sports applications: a survey. In: 23th International Conference on Architecture of Computing Systems 2010, pp. 1–10. VDE (2010)

2. Benser, E.T.: Trends in inertial sensors and applications. In: 2015 IEEE International Symposium on Inertial Sensors and Systems (ISISS) Proceedings, pp. 1–4 (2015)

3. Chen, C., Jafari, R., Kehtarnavaz, N.: Improving human action recognition using fusion of depth camera and inertial sensors. IEEE Trans. Hum.-Mach. Syst. **45**(1), 51–61 (2015)

4. Chen, C., Jafari, R., Kehtarnavaz, N.: A real-time human action recognition system using depth and inertial sensor fusion. IEEE Sens. J. **16**(3), 773–781 (2015)

5. Chen, C., Jafari, R., Kehtarnavaz, N.: UTD-MHAD: a multimodal dataset for human action recognition utilizing a depth camera and a wearable inertial sensor. In: 2015 IEEE International Conference on Image Processing (ICIP), pp. 168–172. IEEE (2015)

6. Chen, C., Jafari, R., Kehtarnavaz, N.: A survey of depth and inertial sensor fusion for human action recognition. Multimedia Tools Appl. **76**(3), 4405–4425 (2015). https://doi.org/10.1007/s11042-015-3177-1

7. Chen, C., Liu, M., Zhang, B., Han, J., Jiang, J., Liu, H.: 3D action recognition using multi-temporal depth motion maps and fisher vector. In: IJCAI, pp. 3331–3337 (2016)

8. Chen, L., Wei, H., Ferryman, J.: A survey of human motion analysis using depth imagery. Pattern Recogn. Lett. **34**(15), 1995–2006 (2013)

9. Chen, Y., Le, D., Yumak, Z., Pu, P.: EHR: a sensing technology readiness model for lifestyle changes. Mob. Netw. Appl. **22**(3), 478–492 (2017)

10. Collin, J., Davidson, P., Kirkko-Jaakkola, M., Leppäkoski, H.: Inertial sensors and their applications. In: Bhattacharyya, S.S., Deprettere, E.F., Leupers, R., Takala, J. (eds.) Handbook of Signal Processing Systems, pp. 51–85. Springer, Cham (2019). https://doi.org/10.1007/978-3-319-91734-4_2

11. Dawar, N., Ostadabbas, S., Kehtarnavaz, N.: Data augmentation in deep learning-based fusion of depth and inertial sensing for action recognition. IEEE Sens. Lett. **3**(1), 1–4 (2019)

12. Dawar, N., Ostadabbas, S., Kehtarnavaz, N.: Data augmentation in deep learning-based fusion of depth and inertial sensing for action recognition. IEEE Sens. Lett. **3**(1), 1–4 (2018)

13. Delachaux, B., Rebetez, J., Perez-Uribe, A., Satizábal Mejia, H.F.: Indoor activity recognition by combining one-vs.-all neural network classifiers exploiting wearable and depth sensors. In: Rojas, I., Joya, G., Cabestany, J. (eds.) IWANN 2013. LNCS, vol. 7903, pp. 216–223. Springer, Heidelberg (2013). https://doi.org/10.1007/978-3-642-38682-4_25

14. Ehatisham-Ul-Haq, M., et al.: Robust human activity recognition using multimodal feature-level fusion. IEEE Access **7**, 60736–60751 (2019)

15. Elmadany, N.E.D., He, Y., Guan, L.: Human action recognition using hybrid centroid canonical correlation analysis. In: 2015 IEEE International Symposium on Multimedia (ISM), pp. 205–210. IEEE (2015)

16. Kwolek, B., Kepski, M.: Human fall detection on embedded platform using depth maps and wireless accelerometer. Comput. Methods Programs Biomed. **117**(3), 489–501 (2014)

17. Lane, N.D., et al.: Bewell: sensing sleep, physical activities and social interactions to promote wellbeing. Mob. Netw. Appl. **19**(3), 345–359 (2014)

18. Lara, O.D., Labrador, M.A.: A survey on human activity recognition using wearable sensors. IEEE Commun. Surv. Tutor. **15**(3), 1192–1209 (2012)

19. Li, W., Zhang, Z., Liu, Z.: Action recognition based on a bag of 3D points. In: 2010 IEEE Computer Society Conference on Computer Vision and Pattern Recognition-Workshops, pp. 9–14. IEEE (2010)
20. Liu, K., Chen, C., Jafari, R., Kehtarnavaz, N.: Fusion of inertial and depth sensor data for robust hand gesture recognition. IEEE Sens. J. **14**(6), 1898–1903 (2014)
21. Liu, L., Shao, L.: Learning discriminative representations from RGB-D video data. In: Twenty-Third International Joint Conference on Artificial Intelligence (2013)
22. Masum, A.K.M., Bahadur, E.H., Shan-A-Alahi, A., Uz Zaman Chowdhury, M.A., Uddin, M.R., Al Noman, A.: Human activity recognition using accelerometer, gyroscope and magnetometer sensors: deep neural network approaches. In: 2019 10th International Conference on Computing, Communication and Networking Technologies (ICCCNT), pp. 1–6 (2019)
23. Mavropoulos, T., et al.: A smart dialogue-competent monitoring framework supporting people in rehabilitation. In: Proceedings of the 12th ACM International Conference on PErvasive Technologies Related to Assistive Environments, pp. 499–508 (2019)
24. Munson, S.A., Consolvo, S.: Exploring goal-setting, rewards, self-monitoring, and sharing to motivate physical activity. In: 2012 6th International Conference on Pervasive Computing Technologies for Healthcare (PervasiveHealth) and Workshops, pp. 25–32. IEEE (2012)
25. Ramasamy Ramamurthy, S., Roy, N.: Recent trends in machine learning for human activity recognition-a survey. Wiley Interdisc. Rev. Data Mining Knowl. Discov. **8**(4), e1254 (2018)
26. Shaeffer, D.K.: Mems inertial sensors: a tutorial overview. IEEE Commun. Mag. **51**(4), 100–109 (2013)
27. Sidor, K., Wysocki, M.: Recognition of human activities using depth maps and the viewpoint feature histogram descriptor. Sensors **20**(10), 2940 (2020)
28. Uijlings, J.R., Duta, I.C., Rostamzadeh, N., Sebe, N.: Realtime video classification using dense HOF/HOG. In: Proceedings of International Conference on Multimedia Retrieval, pp. 145–152 (2014)
29. Wang, H., Schmid, C.: Action recognition with improved trajectories. In: Proceedings of the IEEE International Conference on Computer Vision (ICCV), December 2013
30. Wang, J., Liu, Z., Wu, Y., Yuan, J.: Mining actionlet ensemble for action recognition with depth cameras. In: 2012 IEEE Conference on Computer Vision and Pattern Recognition, pp. 1290–1297. IEEE (2012)
31. Wang, J., Chen, Y., Hao, S., Peng, X., Hu, L.: Deep learning for sensor-based activity recognition: a survey. Pattern Recogn. Lett. **119**, 3–11 (2019)
32. Weiyao, X., Muqing, W., Min, Z., Yifeng, L., Bo, L., Ting, X.: Human action recognition using multilevel depth motion maps. IEEE Access **7**, 41811–41822 (2019)
33. Wong, C., McKeague, S., Correa, J., Liu, J., Yang, G.Z.: Enhanced classification of abnormal gait using BSN and depth. In: 2012 Ninth International Conference on Wearable and Implantable Body Sensor Networks, pp. 166–171. IEEE (2012)
34. Zhang, B., Yang, Y., Chen, C., Yang, L., Han, J., Shao, L.: Action recognition using 3D histograms of texture and a multi-class boosting classifier. IEEE Trans. Image Process. **26**(10), 4648–4660 (2017)

SpotifyGraph: Visualisation of User's Preferences in Music

Pavel Gajdusek and Ladislav Peska[✉]

SIRET Research Group, Department of Software Engineering,
Faculty of Mathematics and Physics, Charles University, Prague, Czech Republic
gajdusek.pavel@gmail.com, peska@ksi.mff.cuni.cz

Abstract. Many music streaming portals recommend lists of songs to the users. These recommendations are often results of black-box algorithms (from the user's perspective). However, irrelevant recommendations without the proper justification may considerably hinder the user's trust. Moreover, user profiles in music streaming services tend to be very large, consisting of hundreds of artists and thousands of tracks. So, not only the recommendation procedure details are hidden for the user, but he/she often lacks a sufficient knowledge about the source data the recommendations are derived from. In order to cope with these challenges, we propose SpotifyGraph application. The application aims on a comprehensible visualization of the relations within the Spotify user's profile and therefore improve understandability of provided recommendations.

1 Introduction

Recommender systems (RS) recently stormed many diverse domains throughout the internet and most users are confronted with some kind of recommendation on a daily basis. One of the main application domains for RS are music streaming services, such as Spotify. However, the recommendations given to users are usually based on some black-box algorithms and the users might not understand, why the songs or artist were recommended to them. Sometimes, this attitude works well, e.g. for rather obvious recommendations, or if some explanation strategy is employed. However, the situation becomes challenging for more picky users with strict preferences, for users with diverse interests in many music genres or styles, or if the provided explanation does not contain features relevant for the user.

Furthermore, given the fast consumption of objects (i.e., it requires only a few minutes to play a song), user profiles may easily grow to hundreds of artists or thousands of played tracks. Often, users does not have sufficient knowledge of what their profile may contain. Therefore, not only the recommending algorithms are black-box, but user's profiles, i.e. source data of the recommending algorithms, becomes partially black-box as well.

Therefore, it would be desirable to provide to users a comprehensible visualization of their profiles to help them understand at least the source data of RS, or which part of their profile contributes to given recommendations the most.

© Springer Nature Switzerland AG 2021
J. Lokoč et al. (Eds.): MMM 2021, LNCS 12573, pp. 379–384, 2021.
https://doi.org/10.1007/978-3-030-67835-7_32

In this demo paper, we propose a SpotifyGraph application aiming to provide an interactive graph-based visualization of Spotify user's profile. The application specifically aims on disclosing the inner structure of the underlined data as understood by Spotify. The applied visualization paradigm is essentially a hierarchical force-directed graph with several explanation strategies. The application also provides recommendations of additional artists related to the particular segment of user's profile. One of the additional SpotifyGraph's features is the ability to export/import other user's graphs and therefore it can be also used as a convenient form of communicating and sharing other user's music tastes.

In a broader sense, the goal of SpotifyGraph is to enable users to realize, how connected their world of music is and let them enjoy the moment of exploring the inner structure of it.

Spotify API is utilized to collect the necessary data. Apart from users, there are three main types of entities in Spotify API: *Artists*, *Albums* and *Tracks*. In SpotifyGraph, we primarily focused on visualizing artists connected to the current user. This option was selected for several reasons. First, given the cardinality of artists as compared to albums or tracks, artists are the most compact entity to represent and therefore allows reasonably uncluttered visualization and some performance improvements as well. Also, artist entities subjectively contains most of the user-understandable metadata, such as genres, images, popularity or related artists. However, in order to give users more complete information (i.e., for cases where the recommendation is not obvious) a list of (playable) artist's tracks should be provided as well. Overall, SpotifyGraph utilizes following data: *user's saved tracks and corresponding artists, user's followed artists, related artists for all artists listed in the previous steps, metadata for all collected artists* and finally *top tracks per artist* and respective metadata for them.

There are several projects that aim on music visualisation, e.g. geMsearch [3], Graphsify [2], music-map.com, or MusicMapp [1]. However, these application often focus on a different task (search results visualization [2,3], music as whole [1]), focus on different entities to visualize or do not incorporate the concept of a user profile. Only a small fraction of related work focus on the artists-level visualizations or complete user profile visualization, extension and sharing. Providing users a comprehensible and interactive visualization of their profiles seems like a unique goal among the related work.

2   SpotifyGraph Application

As was already mentioned, SpotifyGraph provides a graph-based visualization of user's music profile on the artists granularity level. So, let us first introduce two key concepts: artist's weight w_a and artist-to-artist edge's weight $e_{i,j}$. Weight of artist a in the context of current user u is defined as

$$w_a = w_f * f_{u,a} + w_l * \sum_{\forall t \in a} l_{u,t} + w_p * p_a \qquad (1)$$

where $f_{u,a}$ is a follow relation, $l_{u,t}$ is a like relation for a track t and p_a is an overall popularity of artist a. Hyperparameters w_f, w_l and w_p were kept static

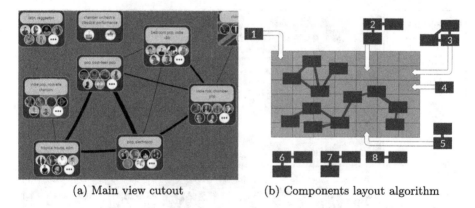

(a) Main view cutout   (b) Components layout algorithm

Fig. 1. Main view: (a) selection of the main view plane depicting the layout of artist's clusters. Users are allowed to move individual nodes, zoom into a part of the graph or select some of the clusters. (b) Small components layout algorithm illustration. Empty cells around the visualized content are detected and fitting small components are placed there. Remaining small components are displayed below the main part of the graph.

for the demonstration, but we plan to either auto-tune them or let users to set them manually in the future work. For artists i and j, their edge weight score is

$$e_{i,j} = r_{i,j} + r_{j,i} + w_g * |G_i \cap G_j| \tag{2}$$

where r is the related artists relation, G_i is a set of genres for artist i and w_g hyperparameter was kept at 0.2.

Note that for many regular Spotify users, the size of their profile may easily grow to several hundreds of entities to display. This would result into highly cluttered visualization incomprehensible for the user. Therefore, we aimed on visualizing aggregated information for clusters of artists first (i.e. the *Main view* - see Fig. 1a) and the content of a single selected cluster later (i.e. the *Single cluster view*, see Fig. 2).

2.1 Main View

Clustering and Layout. As a first task in the visualization pipeline, we need to construct the clusters of similar authors. In order to do that, some graph clustering technique is applied on each connected component of artists (w.r.t. edges defined in Eq. 2). Upon some preliminary experiments, we utilized a modified spectral clustering [6] for this demo. The clustering is performed recurrently, while we maintain adaptive maximal and minimal cluster sizes and merge too small clusters with the closest ones. The edge weight between arbitrary clusters C, $\bar{C}$ is calculated as the sum of their interconnecting edges, i.e. $e_{C,\bar{C}} = \sum_{\forall(i \in C, j \in \bar{C})} e_{i,j}$. We utilized a force-directed layout, namely Fruchterman-Reingold algorithm [4] to visualize the clusters. However, due to the size of individual nodes, there was a considerable volume of node overlaps in

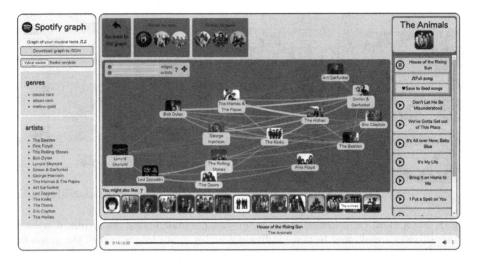

Fig. 2. Single cluster view of the SpotifyGraph application. Graph of artists from the selected cluster is depicted in the central area. Left panel provides some aggregated information about the current cluster. Top panel contains the most similar clusters to the current one. Below the artists graph, there is the list of top-k recommended unknown artists similar to this cluster. Upon clicking on recommendation, the related artists are highlighted in yellow. The right panel contains top songs for the selected artist. These can be played or set as liked. (Color figure online)

the visualization. Therefore, the layout is further processed via a growing tree overlaps removal [5].

Finally, we observed that user profiles often contain a larger volume of small components with 1 to 3 clusters and the layout algorithm often left a lot of free space around the corners. This results into an inefficiency, while displaying larger volumes of nodes (i.e., the layout significantly exceeds initially visible area). Therefore, we utilized a fixed layout for small components with 1–3 clusters and a custom placement strategy for them as illustrated on Fig. 1b.

Cluster Content Visualization. For each cluster of artists, several explanation strategies are utilized. First, names and images of top-k artists (based on Eq. 1) are depicted. This visually significant information (e.g., a well-known logo or a portrait) may give users a quick feedback on the cluster's content. Second, some important music style features (e.g., language of the lyrics) may be connected with the artists' country of origin, which can be parsed from genres metadata for some artists.[1] Therefore, if at least a 40% of cluster's artists share the same country of origin, it is displayed as a flag on the node's background. Examples of prevalently French and UK clusters are depicted on Fig. 1a.

---

[1] Generally, this information is often absent for artists from English-speaking countries but mostly present for the others.

Finally, the sequence of cluster's captions is generated from artist's genres via Algorithm 1. The algorithm aims to provide the most representative genres (or any other metadata) w.r.t. their cardinality, but omit too similar ones (e.g., if the "classical rock" genre is already in the caption, the "rock" genre would not provide any novel information and should be omitted). Note that the condition on line 7 could be exchanged e.g. with similarity of genre embeddings. However, we opted for this simple yet easily controllable solution due to a very plain vocabulary utilized in the genre metadata. For each cluster, first k captions are displayed. The exact volume depends on the level of zoom-in (i.e., more description is provided upon zooming into a part of the graph).

Algorithm 1: Cluster's caption generation

---

1 $G \leftarrow$ all artists' genres with removed country names
2 $G_f \leftarrow$ select unique $g \in G$; sort them by cardinality
3 $C \leftarrow$ [] (constructed caption)
4 $W_u \leftarrow$ [] (list of words already in caption, with repetition)
5 **foreach** g *in* G_f **do**
6     $W_g \longleftarrow$ g.split(' ')
7   **if** $|w \in W_u : w \in W_g| < 2$ **and** $|W_g \cap W_u| < |W_g|$ **then**
8        $C.append(g)$
9        $W_u.append(words)$
10 **return** C

2.2 Single Cluster View

If a user decides to inspect details of some cluster, the single cluster view is displayed (see Fig. 2). The same layout technique as in the main graph is utilized to display individual artists. Furthermore, user has a chance to remove less important nodes or edges via a slider on the left and make the visualization less cluttered.

Apart from this main plane, several panels with additional information are displayed. Among the most notable features is the recommendation component on the bottom. Depicted recommendations represent top-k unknown artists most similar to the current cluster C. The score for each artist $\hat{r}_j$ is calculated as $\hat{r}_j = \sum_{\forall i \in C} w_i * r_{i,j}$, where $r_{i,j}$ is the related artists' relation provided by Spotify API and w_i is the weight of artist i as defined in Eq. 1. Furthermore, in order to better explain the source of the recommendations, all related artists are highlighted upon the selection of a recommended artist j. This is also illustrated on Fig. 2, where the link between recommended *The Animals* and already known *The Doors*, *The Rolling Stones* and *The Kinks* is displayed. Analogical visualization could be utilized in Spotify as an additional explanation strategy for provided recommendations. Furthermore, artist's inspection panel (right) contains top tracks of the respective artist, which can be played or liked, so the user's profile can be enhanced by newly found songs.

3   Conclusions and Future Work

In this demo paper, we presented SpotifyGraph application for visualization of user's preferences in music. The application aims on a comprehensible interactive visualization of the underlined connections in Spotify user's profile in the form of force-directed graphs. It further provides recommendations targeted to a specific subset of user's profile, options to extend his/her Spotify profile and possibility to share his/her profile visualizations with others.

There are multiple future work directions including both application improvements and research opportunities. We plan to experiment with additional recommending strategies, such as purely content-based and hybrid recommendations, calibration of recommendations [7] or utilization of external recommending services. Cluster visualizations, especially caption generation may be improved as well. We plan to focus on the interplay between genre's cluster-wise support, its uniqueness among other clusters, information compactness (i.e. description length) and similarity to already selected content. We would also like to evaluate various layout pipelines and their effect on user's cognitive tasks. Also, some extensions towards tune-ability of the application can be made (e.g., edge weights calculation or clustering parameters). The main challenge here would be to find a comprehensible GUI for users, or to learn the hyperparameter values based on their behavior. Finally, the application can be also extended by some graph fusion tools enabling users to compare multiple profiles and extend their own in the process.

Acknowledgements. This paper was supported by Czech Science Foundation (GAČR) project 19-22071Y, and by Charles University grant SVV-260588. Source codes are available from gitlab.com/gajdusep/spotifygraph, where *testfiles* folder contains several examples of exported graphs. Live demo is available from gajdusep.github.io/spotifygraph.

References

1. Baig, M.H., Varghese, J.R., Wang, Z.: MusicMapp: a deep learning based solution for music exploration and visual interaction. In: MM 2018, pp. 1253–1255. ACM, New York (2018)
2. Bernardi, A., Zytek, A.: Graphsify: an interactive music exploration tool (2018). https://pdfs.semanticscholar.org/bcbc/88bf251622f08d36fca84ce1a269a44fd948.pdf
3. Esswein, C., Schedl, M., Zangerle, E.: geMsearch: personalized explorative music search. In: IUI Workshops, vol. 2068. CEUR-WS (2018)
4. Fruchterman, T.M.J., Reingold, E.M.: Graph drawing by force-directed placement. Softw. Pract. Exp. **21**(11), 1129–1164 (1991)
5. Nachmanson, L., Nocaj, A., Bereg, S., Zhang, L., Holroyd, A.: Node overlap removal by growing a tree. J. Graph Algorithms Appl. **21**(5), 857–872 (2017)
6. Ng, A.Y., Jordan, M.I., Weiss, Y.: On spectral clustering: analysis and an algorithm. In: NIPS 2001, pp. 849–856. MIT Press, Cambridge (2001)
7. Peška, L., Balcar, Š.: Fuzzy D'Hondt's algorithm for on-line recommendations aggregation. In: ORSUM@RecSys 2019, vol. 109, pp. 2–11. PMLR, 19 September 2019

A System for Interactive Multimedia Retrieval Evaluations

Luca Rossetto[2(✉)] [iD], Ralph Gasser[1] [iD], Loris Sauter[1] [iD],
Abraham Bernstein[2] [iD], and Heiko Schuldt[1] [iD]

[1] Department of Mathematics and Computer Science, University of Basel,
Basel, Switzerland
{ralph.gasser,loris.sauter,heiko.schuldt}@unibas.ch
[2] Department of Informatics, University of Zurich, Zurich, Switzerland
{rossetto,bernstein}@ifi.uzh.ch

Abstract. The evaluation of the performance of interactive multimedia retrieval systems is a methodologically non-trivial endeavour and requires specialized infrastructure. Current evaluation campaigns have so far relied on a local setting, where all retrieval systems needed to be evaluated at the same physical location at the same time. This constraint does not only complicate the organization and coordination but also limits the number of systems which can reasonably be evaluated within a set time frame. Travel restrictions might further limit the possibility for such evaluations. To address these problems, evaluations need to be conducted in a (geographically) distributed setting, which was so far not possible due to the lack of supporting infrastructure. In this paper, we present the *Distributed Retrieval Evaluation Server (DRES)*, an open-source evaluation system to facilitate evaluation campaigns for interactive multimedia retrieval systems in both traditional on-site as well as fully distributed settings which has already proven effective in a competitive evaluation.

Keywords: Interactive multimedia retrieval · Retrieval evaluation

1 Introduction

Due to the continuous growth of multimedia collections in terms of their size and diversity, multimedia retrieval has evolved to a major discipline in the general field of "Big Data" research. Tools and techniques to efficiently store, manage, and search such data corpora have become more important, and a lot of research effort went into exploring techniques to extract features from media data, to store and manage large quantities of such data, and to efficiently index it so as to facilitate fast access even for collections beyond billions of entries [3,4,8].

Despite all these efforts, however, it has been shown repeatedly [7,9] that the task of finding a particular item in a large enough collection still is an interactive task that requires cooperation between a human actor and a system. This results

© Springer Nature Switzerland AG 2021
J. Lokoč et al. (Eds.): MMM 2021, LNCS 12573, pp. 385–390, 2021.
https://doi.org/10.1007/978-3-030-67835-7_33

in the more general setting of *interactive retrieval*, in which users leverage end-to-end retrieval systems to explore media collections and to satisfy a particular information need, by refining queries and browsing through result sets.

Evaluating the performance of such systems is a far more difficult and complex undertaking than evaluating the algorithms used by them. Firstly, the human operator plays an important role in the overall combined human-system performance since the translation of an information need into a query is an inherently manual problem. Secondly, the task itself is more complex as its solution requires a sequence of steps involving a combination of techniques. One way of handling this complexity is by conducting evaluation campaigns such as the *Video Browser Showdown (VBS)* [12] for videos or the *Lifelog Search Challenge (LSC)* [5] for multimodal lifelog data. In both campaigns, teams from around the world gather once per year to compare their retrieval systems in a series of tasks. Each task formulates a particular information need, e.g., by depicting an example or by describing the desired object. The teams then have a predefined amount of time to find the item in question and to submit it to the *evaluation server*. Finding the correct item quickly is rewarded with a higher score, whereas wrong submissions or taking a lot of time are penalized. This setting incentivizes participants to continuously refine their systems in all aspects. It can be attributed to the success of such campaigns, that the evaluation setting has changed and adapted over the years. As systems become better, tasks need to become more challenging. With the increasingly complex techniques employed in systems, it is also no longer sufficient to simply rank these by their performance during an evaluation. Instead, one has to collect sufficient data so as to be able to explain why any one system performed better than another for a certain type of task, which requires specialized logging infrastructure.

The contribution of this paper is a demo of the *Distributed Retrieval Evaluation Server (DRES)*[1] – a modular and extendable open source system that generalizes not only the aforementioned, interactive evaluation setting conceptually but also enables a user to setup and hold various retrieval evaluations. DRES comes with a standardized API for logging, which can be used to collect metrics regarding the performance of individual systems. Since evaluating interactive retrieval systems in an on-site setting may not always be feasible, DRES is explicitly designed to support such evaluations in a distributed setting, where participants can reside in different locations. DRES has already been successfully used in multiple distributed retrieval evaluations outside of the larger international campaigns and is scheduled to replace the previously used *VBS Server*[2] from LSC 2020 and VBS 2021 onward. Its flexible architecture also enables its use in other retrieval evaluation campaigns.

The remainder of this paper is structured as follows: Sect. 2 briefly surveys related work. Section 3 introduces some of the concepts, gives a system overview and motivates some of the design decisions behind DRES. Finally, Sect. 4 provides some conclusion and outlook on future work.

---

[1] https://dres.dev/.
[2] https://github.com/klschoef/vbsserver/.

2   Related Work

Evaluating multimedia retrieval solutions in a competitive, challenge-focused setting has been an established practice for many years. The first such evaluation campaigns—such as the TREC Video Track [13] which later turned into TRECVID [1] or ImageCLEF [2], both established in 2001 and 2003 respectively—were set up as non-interactive evaluations. More evaluation campaigns in the multimedia domain have been started over the years – many of them in the context of the MediaEval benchmarking initiative, which has been active since 2008. However, none of these challenges have so far been evaluated interactively.

An early example for an interactive retrieval evaluation campaign was VideOlympics [14] from 2008, which had tasks similar to TRECVID's Ad-Hoc Video Search but took place live in front of an audience. The Video Browser Showdown (VBS) [12] campaign was started in 2012 [11] and has since been held annually in conjunction with the International Conference on MultiMedia Modelling, making it the longest running interactive multimedia retrieval campaign to date, relying on an ever increasing video collection [10]. The tasks evaluated during VBS have undergone some changes over the years. As of 2020, there were three types of tasks: (1) a *Visual Known-Item Search* (Visual KIS) task, where participants have to find an unique video segment of roughly 20 seconds in length from within a pre-defined dataset, (2) a *Textual Known-Item Search* (Textual KIS) task where an unique video segment must be found based on a precise textual description, and (3) an *Ad-Hoc Video Search* (AVS) task, where participants are required to find as many video segments as possible that match a rough textual description. This last task type is similar to the challenge posed by VideOlympics, but employs human judges rather than a pre-determined ground truth, since increasing dataset sizes made exhaustive pre-labelling of the data impractical. All task types are solved by *experts*, which are usually the developers of the retrieval systems. Visual KIS and AVS tasks are additionally solved by *novices* who are selected from the conference audience and have no prior experience with any particular system, in order to assess the usability of the retrieval systems for non-specialists.

Inspired by the VBS, the Lifelog Search Challenge (LSC) started in 2018 [6] as a workshop at the ACM International Conference on Multimedia Retrieval, where it is since held annually [5]. The challenge is similar to the Textual KIS task, but uses lifelog data consisting of image sequences as a retrieval target, which were captured by a wearable camera and annotated with various meta-information.

3   DRES: System Overview

The following describes the inner workings of the system, its architecture and interaction models as well as certain considerations made during its design in order to support present and future requirements.

3.1   Capabilities

DRES is designed to meet the requirements posed by interactive retrieval campaigns both presently and in the foreseeable future and to be easily extendable should new requirements arise. An *evaluation* in DRES consists of multiple *retrieval tasks*. Each task is based on a defined *media collection*, which can contain any type of media such as images, videos or audio. In order to support a wide range of evaluation settings, *retrieval tasks* can be configured by an evaluation coordinator to meet any given need. Such tasks are composed of a set of *hints* and a set of *targets*. The *target* can either be predefined as one or many media objects, a temporal range within a media object, or not specified at all. In the latter case, submissions are forwarded to a judgement mechanism, to have the correctness of a submitted result determined by an external (human) judgement. The *hints* are presented to the participants during a task and they can consist of text, images, or videos as well as any combination thereof. Hints can be arranged on a timeline such that the displayed information changes over time.

The flexible data model of DRES enables evaluation coordinators to build tasks of various types, such as the aforementioned *Textual KIS* type. Such a task's hints are textual and typically, they come with three hints each starting 0, 60 and 120 seconds into the task, where each hint is replaced by the next one and the last hint remains active until the end of the task. The target of such a task is then simply a temporal range within a video item.

Further configurations enable evaluation coordinators to specify if a submission preview is to be shown while the task is still running or if a submission needs to specify a temporal range in addition to a media item. The modular and flexible design would enable evaluation coordinators to expand upon the currently known Textual KIS task by, e.g. adding a doodle of the description for the second half of the task duration as a visual aid. A user management component keeps track of all participants of an evaluation and ensures that all submitted solutions to a task are attributed to the correct team and member. A dedicated component collects interaction- as well as result-logs submitted by the participating systems [9] which can be used to gain additional insights into the system behaviours and search strategies.

3.2   Architecture

Architecturally, DRES can be divided into two primary components: a back-end and a front-end. The back-end manages the *evaluation configurations* and the individual *evaluations*, i.e., instances of a specific configuration, as well as the required multimedia collections and user data. It communicates with the users through an interactive shell as well as a RESTful API, which adheres to the OpenAPI standard. The RESTful API is primarily used by the front-end, however, there is a public API that can be used to submit results and report system metrics such as user/system interactions and excerpts of query results.

The front-end runs in a web-browser in order to minimize software requirements on the user's side. The primary purpose of the front-end is to present the tasks to the participants during an active evaluation and to let the organizers manage the sequence of tasks. It also provides functionality for evaluation and task configuration as well as the management of users and media collections.

The back-end is primarily structured around a component called *RunExecutor*, which coordinates an arbitrary number of *RunManagers* that in turn are responsible for an individual evaluation. Each RunManager is initialized with an evaluation configuration that specifies the tasks of the evaluation and information about the participants trying to solve them. This configuration serves as a sort of template for the individual instances (i.e., a run) of an actual evaluation, which are then managed by a RunManager. The system currently supports *synchronous* runs, meaning all participants get the same task at the same time. *Asynchronous* runs where participants solve the same tasks but not necessarily all at the same time are planned for future expansions.

3.3 Demonstration

During the demonstration, visitors will not only be able to see the participant facing side of the system, which they might already have seen during an evaluation such as VBS or LSC, but be able to experience the entire workflow from setup of an evaluation, configuration of tasks and running of the campaign itself.

4 Conclusion and Outlook

In this paper, we introduced DRES, the Distributed Retrieval Evaluation Server, an open-source system that can be used to setup and host evaluation sessions for interactive multimedia retrieval solutions for both on-site as well as distributed settings. Its flexible data model and modular architecture enables it to support all types of evaluation tasks currently in use by established evaluation campaigns such as VBS and LSC as well as further constellations, which might become relevant in the future. The support for distributed evaluation settings opens up new avenues to advance the state of interactive multimedia retrieval by eliminating the spatial restrictions as well as reducing the organizational and financial overhead of holding an evaluation in one common location. First experiences gathered with distributed evaluations are promising and we expect that such settings will become a powerful augmentation to the established, localized ones. Currently, DRES supports *synchronous* evaluations, where all participants solve the same task at the same time. In future versions, we aim to also support *asynchronous* evaluations, where participants can solve the same tasks independently of one other hence offering spatial as well as temporal distribution. This would further reduce the burden placed on participants, especially in larger distributed settings, which might stretch across multiple time zones.

Acknowledgements. This work was partly supported by the Hasler Foundation in the context of the project City-Stories (contract no. 17055).

References

1. Awad, G., et al.: Trecvid 2019: an evaluation campaign to benchmark video activity detection, video captioning and matching, and video search & retrieval. In: Proceedings of TRECVID 2019. NIST, USA (2019)
2. Clough, P., Sanderson, M.: The CLEF 2003 cross language image retrieval track. In: Peters, C., Gonzalo, J., Braschler, M., Kluck, M. (eds.) CLEF 2003. LNCS, vol. 3237, pp. 581–593. Springer, Heidelberg (2004). https://doi.org/10.1007/978-3-540-30222-3_56
3. Gasser, R., Rossetto, L., Heller, S., Schuldt, H.: Cottontail DB: an open source database system for multimedia retrieval and analysis. In: Proceedings of the 28th ACM International Conference on Multimedia (MM 2020). ACM, Seattle, October 2020
4. Giangreco, I., Schuldt, H.: Adam pro: database support for big multimedia retrieval. Datenbank-Spektrum 16(1), 17–26 (2016)
5. Gurrin, C., et al.: Introduction to the third annual lifelog search challenge (LSC'20). In: Proceedings of the 2020 International Conference on Multimedia Retrieval, pp. 584–585 (2020)
6. Gurrin, C., et al.: [Invited papers] comparing approaches to interactive lifelog search at the lifelog search challenge (LSC 2018). ITE Trans. Media Technol. Appl. 7(2), 46–59 (2019)
7. Lokoč, J., et al.: Interactive search or sequential browsing? A detailed analysis of the video browser showdown 2018. ACM Trans. Multimed. Comput. Commun. Appl. 15(1), 1–18 (2019)
8. Pouyanfar, S., Yang, Y., Chen, S.C., Shyu, M.L., Iyengar, S.S.: Multimedia big data analytics: a survey. ACM Comput. Surv. 51(1) (2018)
9. Rossetto, L., et al.: Interactive video retrieval in the age of deep learning – detailed evaluation of VBS 2019. IEEE Trans. Multimed. 23, 243–256 (2021). https://doi.org/10.1109/tmm.2020.2980944
10. Rossetto, L., Schuldt, H., Awad, G., Butt, A.A.: V3C – a research video collection. In: Kompatsiaris, I., Huet, B., Mezaris, V., Gurrin, C., Cheng, W.-H., Vrochidis, S. (eds.) MMM 2019. LNCS, vol. 11295, pp. 349–360. Springer, Cham (2019). https://doi.org/10.1007/978-3-030-05710-7_29
11. Schoeffmann, K.: A user-centric media retrieval competition: the video browser showdown 2012–2014. IEEE Multimed. 21(4), 8–13 (2014)
12. Schoeffmann, K.: Video browser showdown 2012–2019: a review. In: 2019 International Conference on Content-Based Multimedia Indexing (CBMI), pp. 1–4. IEEE (2019)
13. Smeaton, A.F., Over, P., Taban, R.: The TREC-2001 video track report. In: TREC (2001)
14. Snoek, C.G., Worring, M., de Rooij, O., van de Sande, K.E., Yan, R., Hauptmann, A.G.: VideOlympics: real-time evaluation of multimedia retrieval systems. IEEE Multimed. 15(1), 86–91 (2008)

SQL-Like Interpretable Interactive Video Search

Jiaxin Wu[1]([✉]), Phuong Anh Nguyen[1], Zhixin Ma[2], and Chong-Wah Ngo[2]

[1] Department of Computer Science, City University of Hong Kong,
Hong Kong, China
{jiaxin.wu,panguyen2-c}@my.cityu.edu.hk
[2] School of Computing and Information Systems, Singapore Management University,
Singapore, Singapore
zxma.2020@phdcs.smu.edu.sg, cwngo@smu.edu.sg

Abstract. Concept-free search, which embeds text and video signals in a joint space for retrieval, appears to be a new state-of-the-art. However, this new search paradigm suffers from two limitations. First, the search result is unpredictable and not interpretable. Second, the embedded features are in high-dimensional space hindering real-time indexing and search. In this paper, we present a new implementation of the Vireo video search system (Vireo-VSS), which employs a dual-task model to index each video segment with an embedding feature in a low dimension and a concept list for retrieval. The concept list serves as a reference to interpret its associated embedded feature. With these changes, a SQL-like querying interface is designed such that a user can specify the search content (subject, predicate, object) and constraint (logical condition) in a semi-structured way. The system will decompose the SQL-like query into multiple sub-queries depending on the constraint being specified. Each sub-query is translated into an embedding feature and a concept list for video retrieval. The search result is compiled by union or pruning of the search lists from multiple sub-queries. The SQL-like interface is also extended for temporal querying, by providing multiple SQL templates for users to specify the temporal evolution of a query.

Keywords: SQL-like interpretable search · Concept-free search · Concept-based search · Interactive video search · Video browser showdown

1 Introduction

Video Browser Showdown (VBS) is a live interactive video search held in every year [4,5,9]. It is a well-known benchmark to evaluate the video search system in the literature. This benchmarking activities include three tasks: visual known-item search, textual known-item search, and ad-hoc video search. Visual known-item search provides a visual content of the target video as the query,

---

The original version of this chapter was revised: the acknowledgement section has been corrected. Additionally, the affiliations of the third and last author and the e-mail address of the last author have been corrected. The correction to this chapter is available at https://doi.org/10.1007/978-3-030-67835-7_51

and the participants need to find the corresponding video clip from a large video collection within a short period of time. The textual known-item task textually describes the audio-visual content of a video as the query. In contrast, ad-hoc video search (AVS) does not assume the knowledge of a target video. The task is to search as many video clips as possible that meet the query description in text.

One of the key features in solving these tasks is to perform the cross-modal search by understanding of video semantics and user search attention [9]. With a user' input query, the system should manage to effectively and efficiently find highly relevant video clips at the top of the ranked list. Most of the previous participants in VBS applied concept-based methods as their text-to-video search models [6]. Concept-based methods rely on concept classifiers to index a bunch of concepts in the videos, and in the real scenario, users could retrieve videos by these concepts. Due to the success of deep learning, the accuracy of concept detection has improved tremendously, boosting the performance of interactive search [9]. Recently, concept-free methods which embed video and text in a joint space has shown their supreme performances in the text-to-video retrieval, and become the new state-of-the-art in AVS task [3,10].

However, as concept-free methods perform matching in a black-box manner, the result is not interpretable. For instance, a user might be frustrated for requiring to attempt different ways of expressing a text query in order to obtain a satisfying result. Furthermore, the current concept-free models [3,10] embed features in high-dimensional space. The high demand in memory consumption hinders real-time indexing and search.

To solve the aforementioned shortcomings, in this paper, we introduce a new version of our Vireo video search system (Vireo-VSS), which incorporates a dual-task model [10] for the text-to-video search. The dual-task model trains the concept-free method and concept-based method in an end-to-end deep network. The concept-based method decodes the embedding feature of the concept-free method into a list of concepts for interpretation. In the implementation, we perform dimensionality reduction of the model to allow it suitable for real-time application. Besides, a SQL-like querying interface is designed for users to formulate the query in an explicit way. Instead of providing one text box for inputting the whole query, we allow several kinds of text boxes for querying, e.g., subject, predicate, object text boxes. The motivation is to relieve the user from the trial-and-error way of querying, and instead to focus on expressing the object-of-interest and their relationship in a fill-in-the-blank manner. The interface also allows users to specify time, location, and logical constraints with ease, instead of formulating these constraints into a long sentence. According to what the users put in this interface, the system will generate one or multiple sub-queries, and input them to the dual-task model for search. The following sections describe the Vireo-VSS in details.

2   Dual-Task Model for Real-Time Interactive Search

Our Vireo-VSS employs a dual-task model [10] for cross-modal search. The dual-task model is comprised of two tasks that are learnt end-to-end with neural

Table 1. Performance comparison (mean xinfAP) when reducing the dimensionality of embedding features from 2,048 to 256 on TRECVid AVS datasets.

Datasets	IACC.3			V3C1	Mean
Query sets	tv16	tv17	tv18	tv19	
2,048-dimensional models					
Concept search	0.148	0.147	0.091	0.115	0.125
Embedding search	0.163	0.232	0.118	0.160	0.168
Fusion search	0.185	0.241	0.123	0.185	0.184
256-dimensional models					
Concept search	0.140	0.144	0.087	0.111	0.121
Embedding search	0.146	0.229	0.121	0.160	0.164
Fusion search	0.166	0.243	0.126	0.174	0.177

networks. The first task is the textual-visual embedding matching which aims to minimize the distances between matched video-text pairs in a joint space. The other task is the multi-label concept classification to recover semantic concepts from the visual embedding. Two tasks are trained simultaneously to ensure the semantic consistency such that the embedding feature can properly reflect the video content when being decoded.

The dual-task model provides three schemes for search: embedding search, concept search and fusion search. Embedding search is based on the visual-textual embedding matching task. By inputing a textual query q, the model measures the similarity between the embeddings of the query $\tau(q)$ and a video $\phi(v_i)$. A score is computed for each video based on their cosine similarity:

$$score_{embedding}(q, v_i) = sim(\tau(q), \phi(v_i)). \tag{1}$$

Concept search is based on the trained model in the multi-label concept classification task. In the testing stage, the trained dual-task model provides each test video v_i a predicted concept vector $\hat{y}(v_i) \in \mathbb{R}^{n+}$. The dimension n is equal to the number of concepts in the concept bank, and each value of $\hat{y}(v_i)$ gives the predicted probability of a concept appearing in the video v_i. Given the user's input query q, a vector $c_q \in \{0,1\}^n$ will be formed composing of concepts extracted from q. Then, a concept score will be computed:

$$score_{concept}(q, v_i) = sim(c_q, \hat{y}(v_i)). \tag{2}$$

The fusion search uses a linear function to combine the embedding and concept scores as:

$$score_{combined}(q, v_i) = \theta * score_{concept}(q, v_i) + (1 - \theta) * score_{embedding}(q, v_i). \tag{3}$$

The fusion weight $\theta \in [0, 1]$ can be defined by the user in the interactive search. A value of 0 or 1 corresponds to the pure concept-based or pure concept-free search respectively.

However, the high-dimensional embedding features in the original dual-task model [10] hinders its usage in real-time index and search. Thus, we develop a dual-task model in low dimensional space in this paper for interactive search. We change the output dimension of the embedding feature from 2,048 to 256. As a result, the memory consumption is significantly reduced from about 16 GB to 2 GB, and the search speed is improved to 0.3 seconds per query on a standard PC. The empirical results also verify that the dimension reduction only slightly degrades the performance. Table 1 shows the comparison results on the TRECVid AVS task. We test them on two benchmarks datasets [1, 2] across four query sets released in the years of 2016–2019. Although the dimensionality reduction brings some drops in the performances on most query sets, the drops only happen on a relatively low ratio of queries. Only 38 out of 120 queries happen to drop in both the embedding search and concept search. Most of them are complex queries which describing rich interaction between human and object. For example, the query "Find shots of a person holding, talking or blowing into a horn" has degraded from 0.246 to 0.031 on embedding search. A big drop also happens on the query "Find shots of a person holding, opening, closing or handing over a box". The performance degradation might due to the lower capacity of the model in low-dimensional space in encoding complex information. We also try the indexing method KGraph[1]. While the speed is improved to 0.1 seconds per query, the retrieval performance drops further.

3   The SQL-Like Interface

The SQL-like interface is presented in Fig. 1. Rather than providing one text box only for users to input the query, we allow users to manually specify the search content (subject, predicate, object) and constraint (logical condition) in a semi-structured way. For the constraints, in each text box, users can specify the "OR" relation between terms using commas, and "AND" relation using semicolons. The "NOT" text box is allocated for the "NOT" statement in the query. Two examples of how a user can express queries into these text boxes are illustrated in Fig. 1. The system will decompose the SQL-like query into multiple sub-queries based on the constraint being claimed. For example, the query in Fig. 1(a) is parsed into two sub-queries: "two people kissing" and "bride and groom". These sub-queries are separately fed into the dual-task model to retrieve two sets of similar videos. In this example, the top-rank videos retrieved by the first query ("two people kissing") will be downgraded to lower rank if they are also ranked high by the second query ("bride and groom"). When a query is short, e.g., "woman wearing glasses", the user can simply input the whole query in one of the text boxes while leaving the remaining boxes blank.

---

[1] https://github.com/aaalgo/kgraph.

Quantity:	two		Quantity:	
Subject:	people		Subject:	black musician wearing white shirt
Predicate:	kissing		Predicate:	talking; standing
Object:			Object:	
Time:			Time:	
Location:			Location:	NYC subway station
NOT:	bride; groom		NOT:	

| (a) | (b) |

Fig. 1. The SQL-like interface for interactive search. The figure shows two examples of how to express queries. (a) An ad-hoc query "Find shots of two people kissing who are not bride and groom"; (b) A textual known item search (KIS) query "A black musician standing in a NYC subway station and talking to people. He wears a white shirt". The ad-hoc query is decomposed into two sub-queries and the KIS query is translated into "black musician wearing white shirt talking and standing in NYC subway station".

4   The Vireo Video Search System

We integrate the dual-task model presented in Sect. 2 and the SQL interface presented in Sect. 3 into our video search system. In the end, the Vireo-VSS

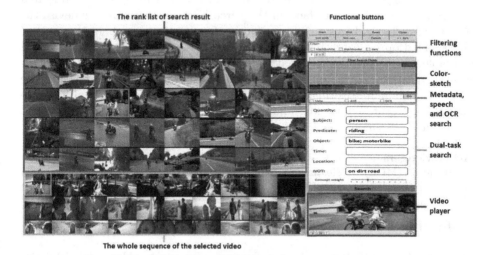

Fig. 2. Vireo-VSS provides multiple querying methods. In this example, a text query is combined with the color-sketch query to search for videos with "A person riding a bike or motorbike while not on a dirt road, and the resulting videos are constrained by the green color on the top-right corner" (Color figure online).

(shown in Fig. 2), which is going to be demonstrated in the VBS 2021, includes the following modules and functions:

○ *Query-by-sketch.* We keep using the simplified color-sketch retrieval model presented in [6] because of its promising performance in solving visual known-item search task.

○ *Query-by-text.* We provide two approaches for text-based search. First, a user can input a query into a text box to search for any text in the metadata, detected on-screen text, or video speech [6]. Second, a user can parse and input a query into the interface presented in Sect. 3 to search using our dual-task model.

○ *Query-by-example.* We utilize our approach in [7] which employs the nearest neighbor search for master-shot key-frames in the video dataset. Instead of using the feature extracted from CNN for matching, we use the embedding feature and the concept feature of the video segments extracted from our dual-task model.

○ *Temporal query.* As the temporal query is useful in solving textual known-item search query, we keep this function used in [8]. It is noted that this function is implemented for both sketch-based and text-based search, and it allows users to specify queries in time t and time $t + 1$.

○ *Filtering.* We provide three filtering functions to filter out black and white, black bordered, and dark video frames.

5   Conclusion

We have presented two new features, dual-task model for cross-modal search and SQL-like querying interface, in the Vireo-VSS. Furthermore, we devise the dual-task model by dimensionality reduction for real-time search, at the expense of a slight drop in search performance. To make search results predictable and tractable, we restrict the way that users formulate a query by filling in a SQL-like template. As the concept-free approach is insensitive to logical relation, the search system addresses this problem by automatically generating multiple sub-queries for feature embedding and search. We expect that these changes will make Vireo-VSS more capable of dealing with complex and verbose queries.

Acknowledgement. The research was partially supported by the Singapore Ministry of Education (MOE) Academic Research Fund (AcRF) Tier 1 grant and the National Natural Science Foundation of China (No. 61872256).

References

1. Awad, G., et al.: Trecvid 2016: evaluating video search, video event detection, localization, and hyperlinking. In: TRECVID 2016 Workshop (2016)
2. Berns, F., Rossetto, L., Schoeffmann, K., Beecks, C., Awad, G.: V3C1 dataset: an evaluation of content characteristics. In: ICMR, pp. 334–338 (2019)
3. Li, X., Xu, C., Yang, G., Chen, Z., Dong, J.: W2vv++: fully deep learning for ad-hoc video search. In: ACM MM (2019)

4. Lokoč, J., et al.: Interactive search or sequential browsing? A detailed analysis of the video browser showdown 2018. ACM TOMM **15**(1), 29:1–29:18 (2019)

5. Lokoč, J., Bailer, W., Schoeffmann, K., Muenzer, B., Awad, G.: On influential trends in interactive video retrieval: video browser showdown 2015–2017. IEEE TMM **20**(12), 3361–3376 (2018)

6. Nguyen, P.A., Lu, Y.-J., Zhang, H., Ngo, C.-W.: Enhanced VIREO KIS at VBS 2018. In: Schoeffmann, K., et al. (eds.) MMM 2018. LNCS, vol. 10705, pp. 407–412. Springer, Cham (2018). https://doi.org/10.1007/978-3-319-73600-6_42

7. Nguyen, P.A., Ngo, C.-W., Francis, D., Huet, B.: VIREO @ video browser showdown 2019. In: Kompatsiaris, I., Huet, B., Mezaris, V., Gurrin, C., Cheng, W.-H., Vrochidis, S. (eds.) MMM 2019. LNCS, vol. 11296, pp. 609–615. Springer, Cham (2019). https://doi.org/10.1007/978-3-030-05716-9_54

8. Nguyen, P.A., Wu, J., Ngo, C.-W., Francis, D., Huet, B.: VIREO @ video browser showdown 2020. In: Ro, Y.M., et al. (eds.) MMM 2020. LNCS, vol. 11962, pp. 772–777. Springer, Cham (2020). https://doi.org/10.1007/978-3-030-37734-2_68

9. Rossetto, L., et al.: Interactive video retrieval in the age of deep learning - detailed evaluation of VBS 2019. IEEE TMM 1 (2020)

10. Wu, J., Ngo, C.W.: Interpretable embedding for ad-hoc video search. In: ACM MM (2020)

VERGE in VBS 2021

Stelios Andreadis[✉], Anastasia Moumtzidou, Konstantinos Gkountakos,
Nick Pantelidis, Konstantinos Apostolidis, Damianos Galanopoulos,
Ilias Gialampoukidis, Stefanos Vrochidis, Vasileios Mezaris,
and Ioannis Kompatsiaris

Centre for Research and Technology Hellas, Information Technologies Institute,
Thessaloniki, Greece
{andreadisst,moumtzid,gountakos,pantelidisnikos,kapost,
dgalanop,heliasgj,stefanos,bmezaris,ikom}@iti.gr

Abstract. This paper presents VERGE, an interactive video search
engine that supports efficient browsing and searching into a collection of images or videos. The framework involves a variety of retrieval
approaches as well as reranking and fusion capabilities. A Web application enables users to create queries and view the results in a fast and
friendly manner.

1 Introduction

VERGE is an interactive video search engine that integrates several retrieval
modalities and provides users with a user interface (UI) for formulating different
types of queries and visualising the most relevant shots and videos. After a multi-
year participation in the Video Browser Showdown (VBS) competition [14], the
engine has been adjusted so as to support the Ad-Hoc Video Search (AVS) and
the Known Item Search Visual and Textual (KIS-V, KIS-T) tasks. This year two
new search modalities are introduced, i.e. Face Detection (Sect. 2.4) and Activity
Recognition (Sect. 2.6), while previously used methodologies are improved. In
addition, the latest version of the VERGE UI (Sect. 3) is presented, where fewer
search options enable the same assortment of retrieval modules, offering a more
compact and friendly usage.

2 Video Retrieval Framework

The VERGE framework involves a multitude of search modalities, implemented
as services, that can be used independently, fused or consecutively to rerank the
top results. Through a UI the users are able to readily create queries and view
the most relevant images or videos that match the criteria. A detailed description
of the integrated retrieval methodologies follows in the next subsections, while
the architecture of the framework is depicted in Fig. 1.

© Springer Nature Switzerland AG 2021
J. Lokoč et al. (Eds.): MMM 2021, LNCS 12573, pp. 398–404, 2021.
https://doi.org/10.1007/978-3-030-67835-7_35

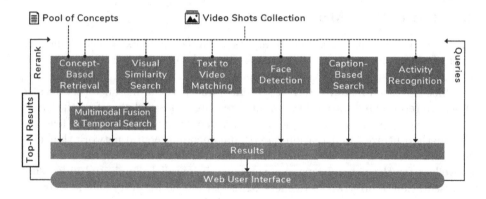

Fig. 1. The VERGE Framework

2.1 Visual Similarity Search

This module retrieves visually similar content using Deep Convolutional Neural Networks (DCNNs). These features are the output of the last pooling layer of the fine-tuned GoogleNet architecture presented in [13]. The dimension of the last pooling layer is 1024 and it is used as global image representation. Eventually, an IVFADC index database vector is created for fast binary indexing using these vectors and K-Nearest Neighbors are computed for the query image [7].

2.2 Concept-Based Retrieval

This module annotates each keyframe with a pool of concepts, which comprises 1000 ImageNet concepts, a selection of 300 concepts out of the 345 concepts of the TRECVID SIN task [12], 500 event-related concepts, 365 scene classification concepts, 580 object labels and 30 style-related concepts. To obtain the annotation scores for the 1000 ImageNet concepts, we used an ensemble method, averaging the concept scores from three pre-trained models that employ different DCNN architectures, namely the EfficientB3, EfficientB5 [15] and InceptionResNetV2. To obtain scores for the subset of 300 concepts from the TRECVID SIN task, we trained and employed two models based on the EfficientB1 and EfficientB3 architectures on the official SIN task dataset. For the event-related concepts we used the pre-trained model of EventNet [5]. Regarding the extraction of the scene-related concepts, we utilized the publicly available VGG16 model fine-tuned on the Places365 dataset [20]. Object detection scores were extracted using models pre-trained on the established MS COCO and Open Images V4 datasets, with 80 and 500 detectable objects, respectively. For the style-related concepts we employed the pre-trained models of [17]. Finally, to offer a cleaner representation of the concept-based annotations we employed various text similarity measures between all concepts' labels. After manual inspection of the text analysis results we formed groups of very similar concepts for which we create a common label and assign the max score of its members.

2.3   Text to Video Matching Module

The text to video matching module inputs a complex free-text query along with a set of video shots and returns a ranked list with the most relative video shots w.r.t. to the input textual query. For this, the method presented in [3] is utilized, in which a textual instance (e.g. a sentence) and a visual instance (i.e. a video shot) are represented into a new common feature space and therefore the direct comparison between free-text queries and video or image instances is feasible. The method utilizes an attention-based dual encoding neural network that uses two similar modules [1], each consisting of multi-level encoding for the video shot as well as for the natural language sentence, in parallel. For initial video shot representation, a pre-trained Resnet-152 model is used for every shot's keyframe whereas each word sentence is initially encoded as a bag-of-words vector. Then, both the sentence and the keyframe representations go through three different encoders (i.e. mean-pooling, attention-based [3], bi-GRU sequential model, and biGRU-CNN [9]). This multilevel encoding is used in order to project both text and video instances into a common feature space following the approach of [2]. When it comes to training data, two different datasets were combined, the TGIF [11] which contains approx. 100k short animated GIFs with one short description per each, and the MSR-VTT [19] consisting of 10k short video clips, each accompanied by 20 short descriptions.

2.4   Face Detection

This module is a specialization of object detection to human faces. For each shot, we extract the number of humans that appear. So, the user can easily distinguish the results of single-human or multi-human activities using as a searching parameter the number of persons. The selection of a face detector in contrast to a person detector is because in crowd-centred scenes the faces can be detected more efficiently as there are fewer occlusions. To address this, we select the implementation of BiFPN [16], a bidirectional feature pyramid network that allows easy and fast multi-scale feature fusion. The model is trained using Google Open Images [10] dataset, keeping only the defections of the "Human-face" class.

2.5   Video Captioning - Caption-Based Search

This module aims to generate for each shot a representative sentence/caption using words included in vocabulary, and thus the user can retrieve videos by simple text search. Video captioning approaches comprise two separate components: (i) a feature extractor that typically extracts the features of a video by sampling among the frames using a fixed number as a step, and (ii) an encoder-decoder that encodes the content and subsequently assigns it to words. To address this, an RNN-based neural network is used [4] based on [18] that takes into account the similarity of the words using semantic clusters. The model is pre-trained on MSR-VTT [19], a widely-known dataset in video captioning domain.

2.6   Activity Recognition

This module generates predictions of human-related activities for each shot, and thus, the user can filter the videos using activity-based keywords. A ranked list of 400 predefined human-related activities and the corresponding scores are generated using a deep learning-based approach. The model architecture is based on a 3D-Resnet Convolutional Neural Network (3D-CNN), similarly to [6] that encodes Spatio-Temporaly the input shots to human-related activities. In particular, the model architecture consists of a ResNet with 50 layers, the input video is fed in the form of: 112 [pixel] x 112 [pixel] x 3 [channel] x 16 [frame], and the model is pre-trained using Kinetics-400 dataset [8].

2.7   Multimodal Fusion and Temporal Search

This module fuses the results of two or more search modules, such as visual features (Sect. 2.1), visual concepts (Sect. 2.2) and color features, in a late fusion manner and retrieves similar shots with a two-step algorithm. The first step is the computation of a tensor L whose surfaces capture the similarity of the results between modality pairs, while the second involves the computation of the final ranked list. In the first step, the top-N results are retrieved for each modality, and thus, K lists are produced; one per modality. Each surface of the tensor is a 2D array with size $N \times N$ with values 0 or 1. In this sparse matrix, the value 1 is observed in the position (i, j) of the 2D surface if the i-th element of the first list coincides with the j-th element of the second list. The second step of the algorithm involves four stages to obtain the final result: (i) the extraction of a list from each 2D surface of L, (ii) the bi-modal ranking of the retrieved results, (iii) the merging of the rankings, and (iv) the duplicate removal for obtaining the final list. We further re-rank the top-N retrieved shots, by considering adjacent keyframes as temporally close, so as to perform temporal search.

3   VERGE User Interface and Interaction Modes

The VERGE UI integrates the above modalities in a friendly and efficient way. Aiming to be a compact and easy-to-use tool, this year the users are provided with fewer options in the UI, but without compromising the variety of alternative retrieval capabilities (e.g. one input field for all keyword-based searches).

As seen in Fig. 2, VERGE consists of three main components: (i) the dashboard menu on the left, (ii) the results panel that spans most of the screen, and (iii) the filmstrip on the bottom. The dashboard menu starts with a countdown timer that shows the remaining time for submission during VBS, a back button to restore previous results, and a switch button to select between obtaining new results and reranking. Then, it continues with four search options. The first option is a text input field where the user can type a free-text query and retrieve the most relevant video shots (Sect. 2.3). The second option offers a long list of concepts (Sects. 2.2, 2.6), supporting autocomplete search and multiple selection.

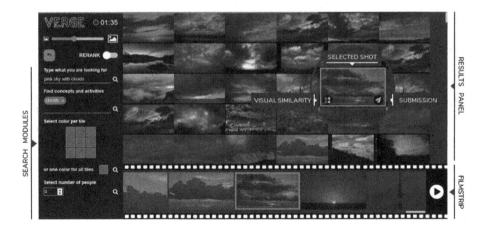

Fig. 2. The VERGE User Interface (Color figure online)

The next option can bring shots of a selected color; the user is able to pick a color from a pop-up palette either for the complete image or for certain parts of the image by coloring a 3 × 3 grid. The last option in the menu allows the user to retrieve shots where a specified number of people appear in them (Sect. 2.4). All the results are displayed in the central panel as single shots in a grid view or as groups of shots (videos). Hovering over a shot reveals two additional functionalities: getting visually similar images (Sect. 2.1) and submitting a shot to the contest. Clicking on it updates the filmstrip with the frames of the video it belongs to, while the button on the right can play the video.

To demonstrate the capabilities of VERGE in the VBS contest, we present some usage scenarios that tackle different types of queries. For a KIS-V query that shows a pink, cloudy sky, the user can select the concept "clouds" and rerank the results by color (Fig. 2). For a KIS-T query that reads "playing the drum in a subway station", the user can type the sentence in the free text search or alternatively combine the concepts "subway station/platform" and "drum". Lastly, the AVS query that asks for shots of a single kid smiling can be addressed with the concept "child", then a reranking by selecting one person to appear and, when a matching image appears, visual similarity can bring more relative results.

4   Future Work

The usability and the effectiveness of the retrieval methodologies as well as the user interface will be evaluated during VBS 2021 and will identify the direction of future algorithms and implementations in VERGE.

Acknowledgements. This work has been supported by the EU's Horizon 2020 research and innovation programme under grant agreements H2020-825079 Mind-Spaces, H2020-779962 V4Design, H2020-780656 ReTV, and H2020-832921 MIRROR.

References

1. Dong, J., Li, X., Xu, C., Ji, S., He, Y., et al.: Dual encoding for zero-example video retrieval. In: Proceedings of IEEE Conference on CVPR 2019, pp. 9346–9355 (2019)
2. Faghri, F., Fleet, D.J., et al.: VSE++: improving visual-semantic embeddings with hard negatives. In: Proceedings of the British Machine Vision Conference (BMVC) (2018)
3. Galanopoulos, D., Mezaris, V.: Attention mechanisms, signal encodings and fusion strategies for improved ad-hoc video search with dual encoding networks. In: Proceedings of the ACM International Conference on Multimedia Retrieval, (ICMR 2020). ACM (2020)
4. Gkountakos, K., Dimou, A., Papadopoulos, G.T., Daras, P.: Incorporating textual similarity in video captioning schemes. In: 2019 IEEE International Conference on Engineering, Technology and Innovation (ICE/ITMC), pp. 1–6. IEEE (2019)
5. Ye, G., Li, Y., Xu, H., et al.: EventNet: a large scale structured concept library for complex event detection in video. In: Proceedings of the ACM MM (2015)
6. Hara, K., et al.: Can spatiotemporal 3D CNNs retrace the history of 2D CNNs and imagenet? In: Proceedings of the IEEE Conference on Computer Vision and Pattern Recognition (CVPR) (2018)
7. Jegou, H., et al.: Product quantization for nearest neighbor search. IEEE Trans. Pattern Anal. Mach. Intell. **33**(1), 117–128 (2010)
8. Kay, W., et al.: The kinetics human action video dataset. arXiv preprint arXiv:1705.06950 (2017)
9. Kim, Y.: Convolutional neural networks for sentence classification. arXiv preprint arXiv:1408.5882 (2014)
10. Krasin, I., Duerig, T., Alldrin, N., Ferrari, V., Abu-El-Haija, S., et al.: OpenImages: a public dataset for large-scale multi-label and multi-class image classification (2017). https://storage.googleapis.com/openimages/web/index.html
11. Li, Y., Song, Y., Cao, L., Tetreault, J., et al.: TGIF: a new dataset and benchmark on animated GIF description. In: Proceedings of IEEE CVPR 2016 (2016)
12. Markatopoulou, F., Moumtzidou, A., Galanopoulos, D., et al.: ITI-CERTH participation in TRECVID 2017. In: Proceedings of the TRECVID 2017 Workshop, USA (2017)
13. Pittaras, N., Markatopoulou, F., Mezaris, V., Patras, I.: Comparison of fine-tuning and extension strategies for deep convolutional neural networks. In: Amsaleg, L., Guðmundsson, G.Þ., Gurrin, C., Jónsson, B.Þ., Satoh, S. (eds.) MMM 2017. LNCS, vol. 10132, pp. 102–114. Springer, Cham (2017). https://doi.org/10.1007/978-3-319-51811-4_9
14. Schoeffmann, K.: Video browser showdown 2012–2019: a review. In: 2019 International Conference on Content-Based Multimedia Indexing (CBMI), pp. 1–4. IEEE (2019)
15. Tan, M., Le, Q.V.: EfficientNet: rethinking model scaling for convolutional neural networks. arXiv preprint arXiv:1905.11946 (2019)
16. Tan, M., Pang, R., Le, Q.V.: EfficientDet: scalable and efficient object detection. In: Proceedings of the IEEE Conference on Computer Vision and Pattern Recognition (2020)
17. Tan, W.R., Chan, C.S., Aguirre, H.E., Tanaka, K.: Ceci n'est pas une pipe: a deep convolutional network for fine-art paintings classification. In: 2016 IEEE ICIP, pp. 3703–3707. IEEE (2016)

18. Venugopalan, S., Rohrbach, M., Donahue, J., et al.: Sequence to sequence-video to text. In: Proceedings of the IEEE ICCV, pp. 4534–4542 (2015)
19. Xu, J., Mei, T., Yao, T., Rui, Y.: MSR-VTT: a large video description dataset for bridging video and language. In: The IEEE Conference on CVPR, June 2016
20. Zhou, B., Lapedriza, A., et al.: Places: a 10 million image database for scene recognition. IEEE Trans. PAMI **40**(6), 1452–1464 (2017)

NoShot Video Browser at VBS2021

Christof Karisch$^{(\boxtimes)}$ (iD), Andreas Leibetseder (iD), and Klaus Schoeffmann

Institute of Information Technology (ITEC), Klagenfurt University, Klagenfurt,
Austria
chkarisch@edu.aau.at, {aleibets,ks}@itec.aau.at

Abstract. We present our NoShot Video Browser, which has been successfully used at the last Video Browser Showdown competition VBS2020
at the MMM2020. NoShot is given its name due to the fact, that it neither makes use of any kind of shot detection nor utilize the VBS master
shots. Instead videos are split into frames with a time distance of one second. The biggest strength of the system lies in its feature "time cache",
which shows results with the best confidence in a range of seconds.

Keywords: Video retrieval · Interactive video search · Video analysis

1 Introduction

As a participating system for the 9th edition (VBS2020) of the annual interactive
video retrieval challenge known as the Video Browser Showdown [4,5,8,10], the
NoShot Video Browser provides a simple and easy-to-use web UI for searching
in big video collections, such as the competition's currently employed V3C1
dataset [1,9] including approx. 1 000 h of video. The system has a very low query
and user interaction response time and can be used on every modern browser.

It specifically was developed at the AAU Klagenfurt in order to provide a
simpler UI approach as the already existing, feature-rich diveXplore [2,3,6,11,12]
system. Yet possessing by far fewer features, NoShot, however, shines through its
clean easy-to-use UI, which especially has many advantages for the challenge's
novice session: for determining a final team score the VBS traditionally features
two sessions – one for expert users, where usually the systems are used by their
developers and one for novice users, where novice users not familiar with the
systems compete against each other. In the latter case, therefore, it is paramount
to provide self-explaining, simple interfaces in order to maximize the success rate
of a system.

In addition to basic UI improvements and code refactorings, new features
such as a novel brightness filter is added for improving search results. Through
extensive testing and optimizations, We expect the system to perform well in
VBS2021.

© Springer Nature Switzerland AG 2021
J. Lokoč et al. (Eds.): MMM 2021, LNCS 12573, pp. 405–409, 2021.
https://doi.org/10.1007/978-3-030-67835-7_36

2   NoShot Video Browser

The NoShot system has already been used for the Video Browser Showdown 2020. It achieved the 7th place. As the system has been used in a competition for the first time, we were able to test the features of our system.

Every video in the dataset is split into keyframes with one second of time distance. NoShot uses YOLO 9k [7] to analyze every of those frames and to label them with up to five of the 9000 classes. The classes are organized into a hierarchy, so we additionally store the parent class id. This step takes a lot of time and processing power and is therefore done beforehand. The classification data is loaded into an Apache Solr[1] database. Apache Solr processes a search query with the VBS2020 data in under 100 milliseconds. It provides automatic caching and pagination, which further speeds up search queries. A search API written in Node.js[2] gets search requests of the frontend and builds a Solr search query. This query is sent to the Solr server via a http json request. The front end is written in javaScript and uses Vue.js[3] as the UI framework. The whole NoShot system is able to process a search request in a fraction of a second.

3   Time Cache

A core feature of the NoShot Video Browser is its time cache. This feature allows the system to show only the best keyframes of a certain time period measured in seconds. The best keyframes are found by sorting by the confidence returned from the YOLO 9000 CNN. In a preprocessing step the best keyframes for all categories are cached in Apache Solr. Furthermore, we create multiple caches (for 30, 60 and 180 s). By caching the results in Apache Solr, we ensure, that every search query has a low response time. At the VBS2020, especially the longest cache (180 s time cache) was useful, as we could successfully remove duplicate results. If a certain scene is shown for a long time only one frame per category is displayed. As this frame has a high confidence for that category, the detected object is often shown fully on the frame.

4   The NoShot GUI

The front end of the NoShot Video Browser is shown in Fig. 1. To enable the user to switch between several search views, the front end provides a tab view. The search field under the tab view is designed big to help the user to start the search query fast. After hitting the return key, a new search window appears. In this window the search keyword can be changed. While the user is typing, the search window refreshes. To help the user to find the best fitting of all 9000 classes, we show a autocomplete tool. The search window uses pagination for

---

[1] https://lucene.apache.org/solr/.

[2] https://nodejs.org.

[3] https://vuejs.org.

big result lists. The cache input sets the length of the time cache (1, 30, 60 or 180 s). The range input shows the frames before and after each result keyframe. The Video view is accessed via an overlay window on top of the tab view and provides video playback functionality. It is opened by holding the control key while clicking on a frame. The video playback position is set to the frame clicked. By holding the shift key while clicking on a frame, the video is opened in the video summary window. In this window the frames of a video are shown. With the cache input the video summary window can be set to only show a fraction of the video frames.

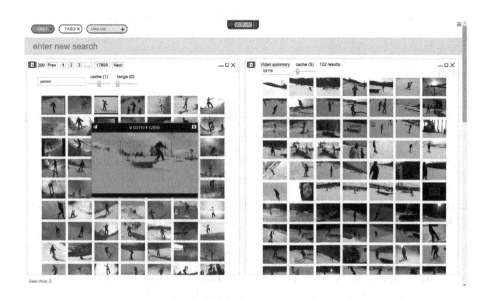

Fig. 1. The NoShot Video Browser

5   Improvements

In comparison with the last version of NoShot, we add a new feature called brightness filter. We add a preprocessing step to analyze the brightness of all keyframes in our database. The brightness filter separates the frame in several sections of the same size. The result of this analysis is stored in the Apache Solr database. The user of the system gets a new input field to specify a section, where retrieved result keyframes should be bright. The system filters all keyframes according to the selected section. Brightness is defined relative to the other sections of the same frame, thus for some frames the brightest regions could appear light gray, while for other frames they appear white. To improve the usability, we include the adjacent areas of a keyframe in the brightness search.

6 Conclusion

We introduce our NoShot Video Browser for participating in VBS2021, which is a web-technologies based interactive video search tool. We are confident, that our time cache feature is beneficial for users of the system. We further improved the system, by adding a brightness search filter. By combining the keyword search with the time cache and the brightness filter, the user can reduce the amount of results significantly. The most important factor for the new feature is its fast response time.

Acknowledgments. This work was funded by the FWF Austrian Science Fund under grant P 32010-N38.

References

1. Berns, F., Rossetto, L., Schoeffmann, K., Beecks, C., Awad, G.: V3c1 dataset: an evaluation of content characteristics. In: Proceedings of the 2019 on International Conference on Multimedia Retrieval, pp. 334–338. ACM (2019)
2. Leibetseder, A., Kletz, S., Schoeffmann, K.: Sketch-based similarity search for collaborative feature maps. In: Schoeffmann, K., et al. (eds.) MMM 2018. LNCS, vol. 10705, pp. 425–430. Springer, Cham (2018). https://doi.org/10.1007/978-3-319-73600-6_45
3. Leibetseder, A., Münzer, B., Primus, J., Kletz, S., Schoeffmann, K.: diveXplore 4.0: the ITEC deep interactive video exploration system at VBS2020. In: Ro, Y.M., et al. (eds.) MMM 2020. LNCS, vol. 11962, pp. 753–759. Springer, Cham (2020). https://doi.org/10.1007/978-3-030-37734-2_65
4. Lokoc, J., Bailer, W., Schoeffmann, K., Muenzer, B., Awad, G.: On influential trends in interactive video retrieval: video browser showdown 2015–2017. IEEE Trans. Multimedia 1 (2018). https://doi.org/10.1109/TMM.2018.2830110
5. Lokoč, J., et al.: Interactive search or sequential browsing? A detailed analysis of the video browser showdown 2018. ACM Trans. Multimedia Comput. Commun. Appl. **15**(1), 29:1–29:18 (2019). https://doi.org/10.1145/3295663
6. Primus, M.J., Münzer, B., Leibetseder, A., Schoeffmann, K.: The ITEC collaborative video search system at the video browser showdown 2018. In: Schoeffmann, K., et al. (eds.) MMM 2018. LNCS, vol. 10705, pp. 438–443. Springer, Cham (2018). https://doi.org/10.1007/978-3-319-73600-6_47
7. Redmon, J., Farhadi, A.: Yolo9000: better, faster, stronger. In: Proceedings of the IEEE Conference on Computer Vision and Pattern Recognition (CVPR), July 2017
8. Rossetto, L., et al.:Interactive video retrieval in the age of deep learning-detailed evaluation of VBS 2019. IEEE Trans. Multimedia (2020)
9. Rossetto, L., Schuldt, H., Awad, G., Butt, A.A.: V3C – a research video collection. In: Kompatsiaris, I., Huet, B., Mezaris, V., Gurrin, C., Cheng, W.-H., Vrochidis, S. (eds.) MMM 2019. LNCS, vol. 11295, pp. 349–360. Springer, Cham (2019). https://doi.org/10.1007/978-3-030-05710-7_29
10. Schoeffmann, K.: A user-centric media retrieval competition: the video browser showdown 2012–2014. MultiMedia, IEEE **21**(4), 8–13 (2014). https://doi.org/10.1109/MMUL.2014.56

11. Schoeffmann, K., Münzer, B., Leibetseder, A., Primus, J., Kletz, S.: Autopiloting feature maps: the deep interactive video exploration (diveXplore) system at VBS2019. In: Kompatsiaris, I., Huet, B., Mezaris, V., Gurrin, C., Cheng, W.-H., Vrochidis, S. (eds.) MMM 2019. LNCS, vol. 11296, pp. 585–590. Springer, Cham (2019). https://doi.org/10.1007/978-3-030-05716-9_50
12. Schoeffmann, K., et al.: Collaborative feature maps for interactive video search. In: Amsaleg, L., Gumundsson, G., Gurrin, C., Jonsson, B., Satoh, S. (eds.) MMM 2017. LNCS, vol. 10133, pp. 457–462. Springer, Heidelberg (2017). https://doi.org/10.1007/978-3-319-51814-5_41

Exquisitor at the Video Browser Showdown 2021: Relationships Between Semantic Classifiers

Omar Shahbaz Khan[1(✉)], Björn Þór Jónsson[1], Mathias Larsen[1],
Liam Poulsen[1], Dennis C. Koelma[2], Stevan Rudinac[2], Marcel Worring[2],
and Jan Zahálka[3]

[1] IT University of Copenhagen, Copenhagen, Denmark
omsh@itu.dk
[2] University of Amsterdam, Amsterdam, Netherlands
[3] Czech Technical University in Prague, Prague, Czech Republic

Abstract. Exquisitor is a scalable media exploration system based on interactive learning, which first took part in VBS in 2020. This paper presents an extension to Exquisitor, which supports operations on semantic classifiers to solve VBS tasks with temporal constraints. We outline the approach and present preliminary results, which indicate the potential of the approach.

Keywords: Interactive learning · Video browsing · Temporal relations

1 Introduction

The Video Browser Showdown (VBS), now in its 10th anniversary edition, has emerged as an important vehicle for the evolution of the multimedia field [5]. During VBS, researchers are given a series of never-before-seen task descriptions, based on a collection of 7,475 video clips [9], and asked to interactively retrieve either one specific video segment or multiple relevant segments, depending on the task type. VBS allows researchers working on media exploration and search tools to apply their techniques in a realistic setting and better understand the pros and cons of both the underlying techniques and the interfaces. The lessons learned during the competition can then inspire new methods and further research. In addition, the competitive setting makes for an exciting event where the ranking of systems can also give hints to their usability and applicability.

Exquisitor, a prototype media exploration system based on interactive learning, took part for the first time in VBS 2020, where it placed 5th out of 11 systems [2]. The goal of Exquisitor, as applied to VBS, is to build a semantic classifier for the information need represented in each task, and use that classifier—along with metadata filters and a video timeline explorer—to solve the task. Exquisitor uses the video segmentation supplied with the VBS collection and represents each video segment independently by semantic features

© Springer Nature Switzerland AG 2021
J. Lokoč et al. (Eds.): MMM 2021, LNCS 12573, pp. 410–416, 2021.
https://doi.org/10.1007/978-3-030-67835-7_37

derived from its keyframe. When building the semantic classifier, Exquisitor suggests keyframes to the user and asks for feedback on those suggestions. Once the user spots a potentially relevant keyframe, the video explorer can then be used to explore the actual content and internal structure of the full video clip.

For many VBS tasks, the task description applies to more than one video segment, often focusing on different semantic concepts in different segments, and sometimes providing an explicit temporal relationship. Unsurprisingly, therefore, all the strongest VBS competitors provide temporal queries as a major technique [4,6,7,10]. Since video segmentation tends to split the video by semantic concepts, a classifier built to find one segment may not find the other, and the system should provide support to utilise the relationship between concepts in video segments.

In this paper, we present a new version of Exquisitor, where the major extension is the support for utilising relationships between semantic classifiers. While each semantic classifier is developed in the same manner as before, using independent video segments, the results of two semantic classifiers can now be combined in various ways, with an optional temporal relationship specification. In this paper we briefly outline the method and interface for combining two semantic models and show how two models combined could be used to solve two VBS 2020 tasks, one of which the team failed to solve during the competition. We will present and evaluate the methods in more detail in a later publication.

2   Exquisitor

Exquisitor is a user relevance feedback approach capable of handling large scale collections in real time [3,8]. The Exquisitor system used for VBS consists of three parts: (1) a web-based user interface for receiving and judging video suggestions; (2) an interactive learning server, which receives user judgments and produces a new round of suggestions; and (3) a web server which serves videos and video thumbnails. Due to the computational efficiency of the system, all three components can run locally on a laptop.

Exquisitor Server: Exquisitor is fueled by a semantic model that combines interactive multimodal learning with cluster-based indexing. Each keyframe in each modality is represented by an efficient representation containing the most important semantic features, compressed using an index-based method [11]. This representation is further clustered using a cluster-based indexing approach [1]. When building a semantic classifier C, a linear SVM classifier is iteratively refined based on user interactions (positive and negative examples). In each round of interaction, the resulting separating hyperplane forms k-farthest neighbour queries posed to the cluster-based indexes. Finally, late fusion is performed on the retrieved results, to produce the 25 top-ranked results to suggest to the user.

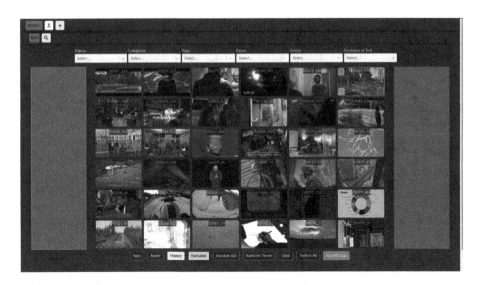

Fig. 1. Exquisitor's interface for building semantic classifiers. See text for details. (Color figure online)

Exquisitor Interface: The interface for building classifiers is shown in Fig. 1. By hovering over a keyframe, the user can choose to view the video, submit it to the VBS server, label it as a positive/negative example, or mark it as seen. Using the 'next' button, the user can also mark all videos as seen and get a full screen of new videos based on the current semantic classifier. Positive (green column) and negative (red column) examples are immediately used to update the model.

Interactive Learning and VBS: The tasks in VBS have three different flavours: Textual Known-Item-Search (KIS) tasks present a gradually evolving text description matching a short video segment; Visual KIS tasks show the video clip sought; and Ad-hoc Video Search (AVS) tasks ask for all segments matching a description. In these tasks, the aim of interactive learning is to create a classifier that is good enough to bring the correct answer(s) to the screen. For KIS tasks, a submitted result is considered as a positive example; once the correct result has been submitted the task is complete. For AVS tasks the process is identical, except that all videos on screen can be submitted at once using a special button, and the process only ends once time has expired.

3   Operations on Semantic Classifier Rankings

To ground the presentation, consider the two textual KIS tasks in Table 1, both of which have a temporal component. Task T_1 was solved by 6 teams during VBS 2020, and was generally considered a difficult task. There are many videos with bridesmaids and brides and grooms, respectively, but in this particular video they do not co-occur in a keyframe during the segment that was considered a solution to the task, and hence we failed to solve this task. Task T_6, on the other hand, was the only text-based task solved by all teams. The Exquisitor team solved it efficiently during the competition by building a classifier for elevators, since (a) the elevator and the bike co-exist in the same keyframe and (b) elevators are rare, so the keyframe is quickly suggested for inspection. Note, however, that since there are many examples of bikes in the collection, but most of them outdoors, building a classifier for bikes is not a productive method to solve T_6.

Table 1. Two example textual KIS tasks from VBS 2020.

Task	Description
T_1	Seven bridesmaids in turquoise dresses walking down a street, and three still images of the bride and couple. The bridesmaids walk on the sidewalk towards the camera. The photos of the couple and bride are taken in a park
T_6	Red elevator doors opening, a bike leans inside, doors closing and reopening, bike is gone. Zoom-in on bike, zoom-out from empty elevator. The bike is silver, the text "ATOMZ' is visible

Classifier Ranking Operations: The rankings obtained by two semantic classifiers, C_1 and C_2, can be combined with a keyframe relationship operation, C_1 *op* C_2, where $op \in \{\cap, \cup, \setminus, \dot{-}\}$. Furthermore, a temporal constraint can optionally be added, which requires either a maximum distance between keyframes (*within* <*frames*>) or a minimum distance (*after* <*frames*>). The result of the classifier ranking operation is a list of videos satisfying both the relationship constraint and optional temporal constraint. Each video is represented by a list of keyframes, annotated by the classifier(s) they appear in, and the videos are ranked by an average score based on the accumulated rank of their scenes from each classifier and the total number of scenes.

As an example, consider solving task T_6 by intersection of rankings produced by semantic classifiers for bikes and elevators. A video would be returned as an answer only if both classifiers return a scene from that video. Since the task description indicates that the two elements should be close to each other, a temporal constraint of *within* 1, for example, would avoid videos where bikes and elevators are far apart.

User Interface: Figure 2 shows the interface for classifier ranking operations. As the figure shows, the result of the merge is a list of the 10 top-ranked videos, where each video is represented by three colour-coded keyframes. Yellow keyframes are from C_1 and blue from C_2, while keyframes appearing in both classifiers are shown as green. The interface shows the highest ranked frame of each colour; if no keyframe appears in both classifiers, the third frame is the

second highest frame from one classifier. Additionally, summary information on
the number of keyframes in the video and classifiers is shown to the left of the
keyframes.

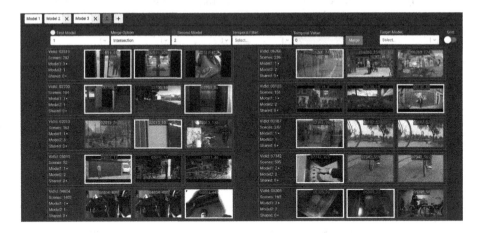

Fig. 2. Exquisitor's interface for semantic classifier operations. See text for details.

Evaluation: To evaluate the usefulness of classifier ranking operations, we
attempted to solve the two tasks of Table 1, both by building a single classifier
and by building two classifiers and intersecting their rankings. These experiments
were carried out in a calm setting, with no time limit, unlike the competitive
environment of VBS. Furthermore, for this evaluation, the entire task text was
considered. To estimate the user workload, we counted the number of interactions with the system, where an interaction is any action taken by the user, such
as labelling a keyframe as a positive example or changing to a different interface
component. We chose to stop after around 75 interactions; once we reached this
limit, we considered the task to be unsolved.

Table 2. Effectiveness experiment results

Task	Models	Interactions	Solved
T_1	'bridesmaid'	76	No
	'bride'	78	No
	'bridesmaid' ∩ 'bride'	60	Yes
T_6	'elevator'	8	Yes
	'bike'	75	No
	'elevator' ∩ 'bike'	15	Yes

Table 2 summarises the results for the two tasks. Consider first T_1, a difficult
task which was not solved by Exquisitor during the competition. As the table

shows, a simple intersection of the results produced by two classifiers could solve the task. Now consider task T_6, which was significantly easier. Table 2 shows that a single classifier on 'elevator' is the fastest approach to solve this task, due to the composition of the collection; this was fortunately the approach taken during the competition. Had we chosen to focus on 'bike' instead, however, the results suggest we would have failed to solve the task. Building rough classifiers for each concept and intersecting their rankings, however, is also an efficient method to solve the task; since the order in which the models are built does not matter the method is robust.

4 Conclusions

We have outlined an extension to the Exquisitor system, supporting operations on semantic classifiers to solve VBS tasks with temporal constraints. Our preliminary results indicate that this new approach has significant potential, and we look forward to testing the approach in the competitive setting.

Acknowledgments. This work was supported by a PhD grant from the IT University of Copenhagen and by the European Regional Development Fund (project Robotics for Industry 4.0, CZ.02.1.01/0.0/0.0/15 003/0000470).

References

1. Guðmundsson, G.Þ., Jónsson, B.Þ., Amsaleg, L.: A large-scale performance study of cluster-based high-dimensional indexing. In: Proceedings of International Workshop on Very-large-scale Multimedia Corpus, Mining and Retrieval (VLS-MCM). Firenze, Italy (2010)
2. Jónsson, B.Þ., Khan, O.S., Koelma, D.C., Rudinac, S., Worring, M., Zahálka, J.: Exquisitor at the video browser showdown 2020. In: Ro, Y.M., et al. (eds.) MMM 2020. LNCS, vol. 11962, pp. 796–802. Springer, Cham (2020). https://doi.org/10.1007/978-3-030-37734-2_72
3. Khan, O.S., et al.: Interactive learning for multimedia at large. In: Jose, J.M., et al. (eds.) ECIR 2020. LNCS, vol. 12035, pp. 495–510. Springer, Cham (2020). https://doi.org/10.1007/978-3-030-45439-5_33
4. Kratochvíl, M., Veselý, P., Mejzlík, F., Lokoč, J.: SOM-hunter: video browsing with relevance-to-SOM feedback loop. In: Ro, Y.M., et al. (eds.) MMM 2020. LNCS, vol. 11962, pp. 790–795. Springer, Cham (2020). https://doi.org/10.1007/978-3-030-37734-2_71
5. Lokoč, J., et al.: Interactive search or sequential browsing? A detailed analysis of the Video Browser Showdown 2018. ACM TOMM **15**(1) (2019)
6. Lokoč, J., Kovalčík, G., Souček, T.: VIRET at video browser showdown 2020. In: Ro, Y.M., Cheng, W.-H., Kim, J., Chu, W.-T., Cui, P., Choi, J.-W., Hu, M.-C., De Neve, W. (eds.) MMM 2020. LNCS, vol. 11962, pp. 784–789. Springer, Cham (2020). https://doi.org/10.1007/978-3-030-37734-2_70
7. Nguyen, P.A., Wu, J., Ngo, C.-W., Francis, D., Huet, B.: VIREO @ video browser showdown 2020. In: Ro, Y.M., et al. (eds.) MMM 2020. LNCS, vol. 11962, pp. 772–777. Springer, Cham (2020). https://doi.org/10.1007/978-3-030-37734-2_68

8. Ragnarsdóttir, H., et al.: Exquisitor: breaking the interaction barrier for exploration of 100 million images. In: Proceedings of the ACM Multimedia. Nice, France (2019)

9. Rossetto, L., Schuldt, H., Awad, G., Butt, A.A.: V3C - a research video collection. In: Kompatsiaris, I., Huet, B., Mezaris, V., Gurrin, C., Cheng, WH., Vrochidis, S. (eds.) Proceedings of MultiMedia Modeling (MMM). Thessaloniki, Greece (2019)

10. Sauter, L., Amiri Parian, M., Gasser, R., Heller, S., Rossetto, L., Schuldt, H.: Combining Boolean and multimedia retrieval in vitrivr for large-scale video search. In: RO, Y.M., et al. (eds.) MMM 2020. LNCS, vol. 11962, pp. 760–765. Springer, Cham (2020). https://doi.org/10.1007/978-3-030-37734-2_66

11. Zahálka, J., Rudinac, S., Jónsson, B.Þ., Koelma, D.C., Worring, M.: Blackthorn: large-scale interactive multimodal learning. IEEE TMM **20**(3) (2018)

VideoGraph – Towards Using Knowledge Graphs for Interactive Video Retrieval

Luca Rossetto[1]([✉])[iD], Matthias Baumgartner[1][iD], Narges Ashena[1][iD],
Florian Ruosch[1][iD], Romana Pernisch[1][iD], Lucien Heitz[1,2][iD],
and Abraham Bernstein[1][iD]

[1] Department of Informatics, University of Zurich, Zurich, Switzerland
{rossetto,baumgartner,ashena,ruosch,pernisch,heitz,bernstein}@ifi.uzh.ch
[2] Digital Society Initiative, University of Zurich, Zurich, Switzerland

Abstract. Video is a very expressive medium, able to capture a wide variety of information in different ways. While there have been many advances in the recent past, which enable the annotation of semantic concepts as well as individual objects within video, their larger context has so far not extensively been used for the purpose of retrieval. In this paper, we introduce the first iteration of VideoGraph, a knowledge graph-based video retrieval system. VideoGraph combines information extracted from multiple video modalities with external knowledge bases to produce a semantically enriched representation of the content in a video collection, which can then be retrieved using graph traversal. For the 2021 Video Browser Showdown, we show the first proof-of-concept of such a graph-based video retrieval approach.

Keywords: Interactive video retrieval · Knowledge-graphs · Multi-modal graphs

1 Introduction

Video is inherently able to capture various information in diverse ways. With increasing advances in machine learning-based content analysis techniques, it becomes increasingly possible to extract and annotate a large amount of this information. While much progress has been made toward said means of information extraction, the integration of such information in the context of multimedia data and for purposes of organization or retrieval of multimedia documents remains understudied. For this edition of the Video Browser Showdown [11], we introduce *VideoGraph*, a Knowledge Graph based video retrieval prototype. Based on similar approaches introduced in *LifeGraph* [9,10] at the Lifelog Search Challenge 2020 [5], VideoGraph uses graph exploration techniques to query a graph composed of information extracted from the challenge dataset [2,14] combined with general knowledge bases [15] which contain general information about the world. This combination enables querying for richer concepts and situations as those which can be detected directly using currently available methods.

© Springer Nature Switzerland AG 2021
J. Lokoč et al. (Eds.): MMM 2021, LNCS 12573, pp. 417–422, 2021.
https://doi.org/10.1007/978-3-030-67835-7_38

In the remainder of this paper, we first describe the methods used to construct the graph in Sect. 2 before outlining the methods used for querying it in Sect. 3. Finally, Sect. 4 offers some outlook and concluding remarks.

2   VideoGraph Construction

To construct the graph, we make use of information which is directly part of the video dataset [2,14], information which can be extracted from the videos by various means as well as external knowledge bases which provide additional context. The following provides an overview of all these data sources and how they contribute to the graph.

2.1   Wikidata

Wikidata [15], a sister project to the Wikipedia, is *"a free and open knowledge base that can be read and edited by both humans and machines"*.[1] It stores structured data in a graph form which is continuously expanded by a large, international community of volunteers. Since Wikidata can be seen as a general knowledge base without any particular restriction in topics, we use it as a backbone for the construction of the semantic relations between all the concepts extracted from the video dataset. Due to the large diversity in content found within the V3C dataset, we do not use any additional external knowledge bases. This is in contrast to LifeGraph, which also made use of the "Classification of Everyday Living" (COEL) [3].

2.2   Semantic Video Metadata

The videos in the dataset come already with some semantic metadata pulled directly from Vimeo. This includes user-defined tags, Vimeo categories, and textual descriptions. Categories and tags will be added to the graph as resources, therefore linking together videos with the same categories/tags, making their retrieval simpler. However, we will enrich the tags and categories by employing a simple link to Wikidata. This makes it possible, to not only query user defined tags and categories but also include similar concepts as defined within Wikidata.

Even though not all videos include descriptions, we extract entities from the latter where applicable. This adds semantic information where available. Unfortunately, this does not solve the problem for videos which lack descriptions and also do not have many tags or categories attached to them. For these, we have to rely heavily on other forms of information extraction, as described below.

To represent semantic information describing the video contents in the graph, we will rely on existing standards as much as possible [1].

---

[1] https://www.wikidata.org/wiki/Wikidata:Main_Page.

2.3 Textual Semantic Information from Video

Textual information in the videos is a rich source to extract *VideoGraph*'s entities from. We use textual information extracted from the V3C1 videos using Optic Character Recognition (OCR) and Automated Speech Recognition (ASR) technologies. We consider two sources for the textual information. First, the speech in the videos and, second, the visible text within the scenes. For the former, we make use of previously extracted, publically available data.[2]

The latter is also based on data already generated for [13]. Neither OCR nor ASR data are however perfect. There are missing values for videos with little to no dialogue or a language different than English. Also, the generated ASR/OCR data is noisy and does not perfectly reflect the speech/text in the videos.

We nevertheless apply entity extraction to both text sources in order to be able to link the contained information to the rest of the graph. In addition to the textual sources, we also use lower-level semantic information which resulted from prior analyses [2] of the dataset.[3]

2.4 Visual Semantic Information from Video

For the extraction of semantic information from the visual component of the videos, we primarily rely on previously generated semantic annotations [13] which were produced by the Google Cloud Vision API.[4] The detectors provided by the API were applied to the representative frame of every video segment in the dataset and for each produced non-localized semantic labels with unique ids. The ids used by the service were already known to Wikidata, so they could be easily mapped to the relevant QIDs. This mapping enables us to link the video segments via their contained semantic concepts not only to each other but also to other concepts, which may not be directly detectable.

2.5 Technical Video Metadata

In order to support an end-to-end retrieval process, the graph does not only need to contain semantic information describing the content of the videos, but also the technical information which is required to interact with the video files themselves, including filenames, codec and frame rate information, shot boundaries, etc. To capture this information, we use the Ontology for Media Resources [6] wherever applicable and make extensions where necessary. Since the graph does not only need to know about every video as a whole but also about every annotated shot, we use the Media Fragments Ontology [8] to represent every video shot from the dataset as a graph node.

---

[2] https://github.com/lucaro/V3C1-ASR.
[3] https://github.com/klschoef/V3C1Analysis.
[4] https://cloud.google.com/vision.

3 VideoGraph Exploration

Based on the graph described in Sect. 2, the proposed system supports several means of query construction and subsequent graph exploration methods.

3.1 Query Formulation

To retrieve videos, a user can select an arbitrary number of tags of which each corresponds to a node in the graph. Any selection of such tags then needs to be translated into a query against the graph. We follow the approach introduced in [10], where we fill a query template with the selected tags.

In this implementation, we aim to improve the retrieval of videos by adding negation. The user can select tags, which specifically should not be linked to the videos.

Additionally, we provide the user with a raw text input to search specifically within the transcribed audio and detected textual elements in the video. Since these are not tags, we use a different mechanism for querying.

Lastly, in contrast to [10] tags can have different origins. Therefore, we provide an interface, where the user can select the origin of a specific tag. However, this origin is optional.

3.2 Graph Exploration

Analogously to the approach described in [10], graph exploration will start at the nodes which have been specified in the query. The graph is then traversed with increasing depth from each start node until either a sufficiently large number of video segments or a maximum depth is reached. The results of the traversals from all start nodes are aggregated and the resulting video segments are scored by their inverse distance to the initial nodes, favoring segments with the shortest distance to the highest number of start nodes.

The full-text components of the query are evaluated independently of the graph since no semantic expansion is necessary there. The results of the two processes are independently sent to the user interface, where their relative weight can be adjusted dynamically.

3.3 Graph Extension

Content-based filtering techniques are employed to extend the existing graph. Using correlation-based similarity of video tags [7], additional edges between video nodes with similar tags are introduced. The similarity score for two tags is calculated based on their co-occurrence count in videos.

Furthermore, cosine distances between nodes are leveraged for additional graph extension. Vector cosine-based similarity is used to match video nodes that are alike [7]. The similarity score of videos is based on the assigned video tags they share. The tags create a projection space for finding neighboring nodes. The extended graph will account for neighbors discovered this way by introducing additional edges between the respective nodes.

3.4   User Interaction

Analogously to LifeGraph, the user interface is a modified version of *vitrivr-ng* [4], a browser-based component taken from the content-based multimedia retrieval stack *vitrivr* [12]. vitrivr-ng enables multiple forms of query formulation, of which only the text-based modalities are relevant for our application. Queries on VideoGraph consist of a collection of tags, which can either refer to semantic concepts which have been detected directly or indirectly, as well as free text. The means offered by vitrivr-ng for browsing the retrieved results as well as the UI-based re-ordering and filtering mechanisms are adopted without any major changes.

The UI also offers late-filtering functionality which enables a user to hide a subset of already retrieved results based on Boolean filter criteria which can be applied to metadata associated to the individual results. In order to make optimal use of this functionality, VideoGraph will expose various properties of the individual video shots, both semantic as well as technical, to this filtering mechanism.

4   Conclusion

In this paper, we introduced VideoGraph, our first attempt at representing the complex content of a video collection as a knowledge graph. Guided by insights gained from LifeGraph, which introduced knowledge graph-based exploration of lifelog data, VideoGraph is an initial proof-of-concept with the aim of evaluating the capabilities and limitations of such a graph representation in combination with reasonably simple graph traversal mechanisms for querying. Insights gained from this first evaluation of such a knowledge graph-based video retrieval approach should be able to inform future developments into more complex knowledge representations, such as graph-video co-embeddings.

Acknowledgement. This work was partially funded by the University of Zurich, the Digital Society Initiative, the Swiss Re Institute, and the Swiss National Science Foundation under contract number 200020_184994.

References

1. Arndt, R., Troncy, R., Staab, S., Hardman, L.: COMM: a core ontology for multimedia annotation. In: Staab, S., Studer, R. (eds.) Handbook on Ontologies. IHIS, pp. 403–421. Springer, Heidelberg (2009). https://doi.org/10.1007/978-3-540-92673-3_18

2. Berns, F., Rossetto, L., Schoeffmann, K., Beecks, C., Awad, G.: V3c1 dataset: an evaluation of content characteristics. In: Proceedings of the 2019 on International Conference on Multimedia Retrieval, pp. 334–338 (2019)

3. Bruton, P., Langford, J., Reed, M., Snelling, D.: Classification of everyday living version 1.0 (2019). https://docs.oasis-open.org/coel/COEL/v1.0/os/COEL-v1.0-os.html. Accessed 23 Jan 2019

4. Gasser, R., Rossetto, L., Schuldt, H.: Towards an all-purpose content-based multimedia information retrieval system. arXiv preprint arXiv:1902.03878 (2019)
5. Gurrin, C., et al.: An introduction to the third annual lifelog search challenge, LSC 2020. In: ICMR 2020, The 2020 International Conference on Multimedia Retrieval. ACM, Dublin (2020)
6. Lee, W., et al.: Ontology for media resources 1.0. W3C Recommendation, **9** (2012). https://www.w3.org/TR/mediaont-10/
7. Manjula, R., Chilambuchelvan, A.: Content based filtering techniques in recommendation systems using user preferences. Int. J. Innov. Eng. Technol. **7**(4), 149–154 (2016)
8. Mannens, E., et al.: A URI-based approach for addressing fragments of media resources on the web. Multimed. Tools Appl. **59**, 691–715 (2010). https://doi.org/10.1007/s11042-010-0683-z
9. Rossetto, L., Baumgartner, M., Ashena, N., Ruosch, F., Pernischova, R., Bernstein, A.: A knowledge graph-based system for retrieval of lifelog data. In: International Semantic Web Conference, Proceedings of the ISWC 2020 Demos and Industry Tracks, pp. 223–228, no. 2721. CEUR-WS (2020). http://ceur-ws.org/Vol-2721/paper557.pdf
10. Rossetto, L., Baumgartner, M., Ashena, N., Ruosch, F., Pernischová, R., Bernstein, A.: LifeGraph: a knowledge graph for lifelogs. In: Proceedings of the Third Annual Workshop on Lifelog Search Challenge, pp. 13–17 (2020)
11. Rossetto, L., et al.: Interactive video retrieval in the age of deep learning-detailed evaluation of VBS 2019. IEEE Trans. Multimed. **23**, 243–256 (2020)
12. Rossetto, L., Giangreco, I., Tanase, C., Schuldt, H.: vitrivr: a flexible retrieval stack supporting multiple query modes for searching in multimedia collections. In: Proceedings of the 24th ACM International Conference on Multimedia, pp. 1183–1186 (2016)
13. Rossetto, L., Amiri Parian, M., Gasser, R., Giangreco, I., Heller, S., Schuldt, H.: Deep learning-based concept detection in vitrivr. In: Kompatsiaris, I., Huet, B., Mezaris, V., Gurrin, C., Cheng, W.-H., Vrochidis, S. (eds.) MMM 2019. LNCS, vol. 11296, pp. 616–621. Springer, Cham (2019). https://doi.org/10.1007/978-3-030-05716-9_55
14. Rossetto, L., Schuldt, H., Awad, G., Butt, A.A.: V3C – a research video collection. In: Kompatsiaris, I., Huet, B., Mezaris, V., Gurrin, C., Cheng, W.-H., Vrochidis, S. (eds.) MMM 2019. LNCS, vol. 11295, pp. 349–360. Springer, Cham (2019). https://doi.org/10.1007/978-3-030-05710-7_29
15. Vrandečić, D., Krötzsch, M.: Wikidata: a free collaborative knowledgebase. Commun. ACM **57**(10), 78–85 (2014)

IVIST: Interactive Video Search Tool in VBS 2021

Yoonho Lee[✉], Heeju Choi, Sungjune Park, and Yong Man Ro

Image and Video Systems Lab, School of Electrical Engineering, KAIST, Daejeon, South Korea
{sml0399benbm,02heeju,sungjune-p,ymro}@kaist.ac.kr

Abstract. This paper presents a new version of the Interactive VIdeo Search Tool (IVIST), a video retrieval tool, for the participation of the Video Browser Showdown (VBS) 2021. In the previous IVIST (VBS 2020), there were core functions to search for videos practically, such as object detection, scene-text recognition, and dominant-color finding. Including core functions, we newly supplement other helpful functions to deal with finding videos more effectively: action recognition, place recognition, and description searching methods. These features are expected to enable a more detailed search, especially for human motion and background description which cannot be covered by the previous IVIST system. Furthermore, the user interface has been enhanced in a more user-friendly way. With these enhanced functions, a new version of IVIST can be practical and widely-used for actual users.

Keywords: Video Browser Showdown (VBS) · Interactive video retrieval · Action and place recognition

1 Introduction

Video Browser Showdown (VBS) competition aims to resolve two main tasks within a time limit (5–7 min): Known-Item Search (KIS) and Ad-hoc Video Search (AVS). For KIS part, users try to find an exact single instance that fits with a presented visual presentation (visual KIS) or fits with a text sentence presented by the moderator (textual KIS). For AVS part, users search as many as possible scenes that correspond to a given general description [2, 14]. As one of the participants in VBS 2020, IVIST [1] is a video retrieval system searching candidates for a scene that users want to find among a set of video collections. It was built with various useful functionalities, for example, object detection, scene-text detection, dominant-color finding, and so on.

In this paper, we present a newly enhanced version of IVIST. Experiences from VBS 2020 drove an improvement of the previous IVIST with the integration of new functions. The new version has been updated in two ways. First, for the user interface, more intuitive and user-friendly usage can be available, for example, the size of the thumbnail has been increased for users to find out whether it is a desirable video at once. Second, practical

---

Y. Lee and H. Choi—Both authors have equally contributed.

© Springer Nature Switzerland AG 2021
J. Lokoč et al. (Eds.): MMM 2021, LNCS 12573, pp. 423–428, 2021.
https://doi.org/10.1007/978-3-030-67835-7_39

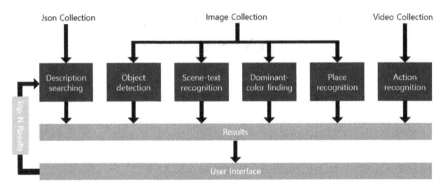

Fig. 1. IVIST system architecture

searching functions have been integrated, for example, action and place recognition, and description searching functions. Action recognition allows video sets to be divided into more specified types depending on which actions are conducted in the videos. On the other hand, place recognition complements a limitation of searching backgrounds. Description searching could be an effective method to deal with textual KIS and AVS tasks using pre-categorized phrases [13].

For the following sections, this paper is organized as follows: an overall architecture of the system (Sect. 2), explanations of main searching functions (Sect. 3), and conclusion (Sect. 4).

2   Overall Architecture of IVIST

IVIST uses ReactJS, Flask, and MongoDB for the front-end, back-end, and database, respectively. The front-end, back-end, and database are so closely connected that they can interact with each other. The process of our system is straight forward and intuitive. We first perform our analysis on the given dataset and build metadata which supports the categories (e.g., action – running or jumping, place – mountain or city) we made. Provided metadata are stored in the MongoDB database. When users type in keywords or categories with 'AND'/'OR' options on the front-end user-interface, the front-end delivers the query to the back-end server. Then, the back-end server accesses the database, receives the corresponding result data sets, and then, passes it back to the front-end. After that, the front-end displays probable results on the screen. Therefore, users can check the results of a query via thumbnails and play videos checking whether the results are what they want or not.

The new version has been updated in two ways. For the user interface, more intuitive usage has been available. The size of the thumbnail has been increased, so users can find demanding thumbnail faster than before. Then, the video can be played from the corresponding part when the thumbnail is clicked. Moreover, it has been user-friendly designed so that the user can operate various functions very easily.

Functions are well defined on the left black box and users can add queries from top to the bottom and start search. For the main function side, object detection, scene-text detection, dominant-color finding, action and place recognition and description searching

Fig. 2. Results from scene-text recognition with keyword "coffee".

can be used for search. Those searching functionalities can be combined to create various types of queries or re-rank the top results. The detailed principle of functions will be covered in the next section, while the general architecture of IVIST can be seen in Fig. 1.

As shown in the simulation results in Fig. 2, it is acquired by scene-text query with the word 'coffee'. Given thumbnails are scenes related to the coffee.

3   Main Functions in IVIST

This section provides an overview of functional improvements. One major focus was enhancements of the main functions in the previous version of IVIST (Sect. 3.1). Furthermore, three other functions have been newly developed for more various query formation in IVIST: action recognition (Sect. 3.2), place recognition (Sect. 3.3), and description searching (Sect. 3.4).

3.1   Existing Capabilities

We decided to use object detection, scene-text detection, and searching through dominant color functionalities for IVIST in VBS 2020. On top of it, we improve it in the aspect of the performance of object detection, because the performance of object detection models decreases when they encounter low-light images or videos. In order to cope with this problem, we used EnlightenGan [15] which lightens images so that object detection models can identify objects in the video with a low-light condition. To train EnlightenGan, we used several datasets [16–19] following Jiang et al. [15]. IVIST lightens low-light images using EnlightenGan and detects an object in the resulting image using HTC [20], the object detection model used by IVIST in VBS 2020.

3.2 Action Recognition

When people want to find or describe a video, they can use information about the activity in the video. This functionality is especially useful when people are trying to search the video with little information related to objects. For example, if a user wants to find a video showing a person enjoying surfing, it is not easy to find the video using object detection with object key words 'surfboard', 'person', or 'sea', as many videos are containing those objects. There can be a video that shows a shop and a man selling surfboard. In this case, searching by action information 'surfing' can narrow the range of candidates.

The main action recognition model used for IVIST is 3D ResNet-200 [3] which turned out to be one of the most accurate 3D CNN models. To adopt 3D ResNet to recognize action in the videos, we need a large amount of data to provide reliable performance. We pre-trained 3D ResNet-200 with Kinetics-700 and Moments in Time (MiT) [4, 5] and used UCF-101, HMDB-51, and ActivityNet [6–8] to boost its performances following Kataoka et al. [9].

3.3 Place Recognition

People sometimes want to find videos of some places. However, natural or remote places, such as a mountain or an ocean, are difficult to find with object detection or action recognition as those scenes usually do not include people or detectable objects. Place recognition improves the quality of video retrieval by overcoming this limitation.

We selected two popular deep neural networks, VGG-16 and ResNet-152 [10, 11] to enable place recognition. These two networks were adopted to recognize the places in the videos. While the VGG-based model shows higher accuracy in determining the most probable category, the ResNet-based model has a good performance on finding the top 5 probable categories as shown in Zhou et al. [12]. So, we used both models, the ResNet-based model to select the 3 most probable candidates, and the VGG-based model to find the most probable candidate. We decided to use a Places365 dataset [12]. The problem of the Places365 dataset is that there are too many similar and specific categories like 'mountain', 'mountain path', and 'valley' which makes people difficult to find the intended scene. To handle this problem, we grouped similar categories so that users can select the most similar ones.

3.4 Description Searching

In textual KIS and AVS, text can also be an intuitive and quick method to find desirable results. For example, suppose that users try to find scenes with 'people riding a bike'. In this case, users can pick a keyword, 'ride'. By using the keyword, users can make a query to find a video description containing the word 'ride'. Then, a video with the description, 'many riders from Paris and Argentina came to the middle of Paris to ride together', can be found by the query.

For implementation, we adopted given json text files which are describing the video scenes [13] with corresponding frame information. Categories are previously made: people, dog, bike, park, ride, etc. Each category contains different words with similar meanings (e.g., they, them, people, audience) so that the user can find scenes containing similar words with one keyword. Then, we searched components of category within the given entire json texts and match the video with a category if we could find words. Finally, we stored the video information of json text files into each matching category. As a result, when the user makes a query text with the title of a category, frame information of the category group is returned as output.

4 Conclusion

This paper introduces a new version of IVIST, focusing on the improvements of toolkits and user-interface. The modification on the user-interface enhances a task-solving capability, and also a user-friendly interface and customizable search options allow a user to create a flexible query that fits better with the situation. Moreover, existing functions (e.g., object detection, scene-text detection, dominant-color finding) have been improved to get desirable results. The system also supports newly added functions such as action and place recognition and description searching functions. However, there are rooms for improvements in the future, and we plan to enhance the system with various simulations.

References

1. Park, S., Song, J., Park, M., Ro, Y.M.: IVIST: interactive video search tool in VBS 2020. In: Ro, Y.M., et al. (eds.) MMM 2020. LNCS, vol. 11962, pp. 809–814. Springer, Cham (2020). https://doi.org/10.1007/978-3-030-37734-2_74
2. Cobârzan, C., et al.: Interactive video search tools: a detailed analysis of the video browser showdown 2015. Multimed. Tools Appl. **76**(4), 5539–5571 (2016). https://doi.org/10.1007/s11042-016-3661-2
3. Hara, K., Kataoka, H., Satoh, Y.: Can spatiotemporal 3D CNNS retrace the history of 2D CNNS and imagenet? In: Proceedings of the IEEE Conference on Computer Vision and Pattern Recognition, pp. 6546–6555 (2018)
4. Carreira, J., Noland, E., Hillier, C., Zisserman, A.: A short note on the kinetics-700 human action dataset (2019). arXiv preprint arXiv:1907.06987
5. Abu-El-Haija, S., et al.: Youtube-8m: a large-scale video classification benchmark (2016). arXiv preprint arXiv:1609.08675
6. Soomro, K., Zamir, A.R., Shah, M.: UCF101: a dataset of 101 human actions classes from videos in the wild (2012). arXiv preprint arXiv:1212.0402
7. Kuehne, H., Jhuang, H., Garrote, E., Poggio, T., Serre, T.: HMDB: a large video database for human motion recognition. In 2011 International Conference on Computer Vision, pp. 2556–2563. IEEE (2011)
8. Caba Heilbron, F., Escorcia, V., Ghanem, B., Carlos Niebles, J.: ActivityNet: a large-scale video benchmark for human activity understanding. In: Proceedings of the IEEE Conference on Computer Vision and Pattern Recognition, pp. 961–970 (2015)
9. Kataoka, H., Wakamiya, T., Hara, K., Satoh, Y.: Would mega-scale datasets further enhance spatiotemporal 3D CNNs? (2020). arXiv preprint arXiv:2004.04968
10. Simonyan, K., Zisserman, A.: Very deep convolutional networks for large-scale image recognition (2014). arXiv preprint arXiv:1409.1556
11. He, K., Zhang, X., Ren, S., Sun, J.: Deep residual learning for image recognition. In: Proceedings of the IEEE Conference on Computer Vision and Pattern Recognition, pp. 770–778 (2016)

12. Zhou, B., Lapedriza, A., Khosla, A., Oliva, A., Torralba, A.: Places: a 10 million image database for scene recognition. IEEE Trans. Pattern Anal. Mach. Intell. **40**(6), 1452–1464 (2017)
13. Rossetto, L., Schuldt, H., Awad, G., Butt, Asad A.: V3C – a research video collection. In: Kompatsiaris, I., Huet, B., Mezaris, V., Gurrin, C., Cheng, W.-H., Vrochidis, S. (eds.) MMM 2019. LNCS, vol. 11295, pp. 349–360. Springer, Cham (2019). https://doi.org/10.1007/978-3-030-05710-7_29
14. Lokoč, J., et al.: Interactive search or sequential browsing? A detailed analysis of the video browser showdown 2018. ACM Trans. Multimed. Comput. Commun. Appl. (TOMM) **15**(1), 1–18 (2019)
15. Jiang, Y., et al.: EnlightenGAN: deep light enhancement without paired supervision (2019). arXiv preprint arXiv:1906.06972
16. Wei, C., Wang, W., Yang, W., Liu, J.: Deep retinex decomposition for low-light enhancement (2018). arXiv preprint arXiv:1808.04560
17. Kalantari, N.K., Ramamoorthi, R.: Deep high dynamic range imaging of dynamic scenes. ACM Trans. Graph. **36**(4), 144-1 (2017)
18. Cai, J., Gu, S., Zhang, L.: Learning a deep single image contrast enhancer from multi-exposure images. IEEE Trans. Image Process. **27**(4), 2049–2062 (2018)
19. Dang-Nguyen, D.T., Pasquini, C., Conotter, V., Boato, G.: Raise: a raw images dataset for digital image forensics. In: Proceedings of the 6th ACM Multimedia Systems Conference, pp. 219–224 (2015)
20. Chen, K., et al.: Hybrid task cascade for instance segmentation. In: Proceedings of the IEEE Conference on Computer Vision and Pattern Recognition, pp. 4974–4983 (2019)

Video Search with Collage Queries

Jakub Lokoč, Jana Bátoryová, Dominik Smrž, and Marek Dobranský[(✉)]

SIRET Research Group,
Department of Software Engineering, Faculty of Mathematics and Physics,
Charles University, Prague, Czech Republic
`lokoc@ksi.mff.cuni.cz, marekdobr@gmail.com`

Abstract. Nowadays, popular web search portals enable users to find available images corresponding to a provided free-form text description. With such sources of example images, a suitable composition/collage of images can be constructed as an appropriate visual query input to a known-item search system. In this paper, we investigate a querying approach enabling users to search videos with a multi-query consisting of positioned example images, so-called collage query, depicting expected objects in a searched scene. The approach relies on images from external search engines, partitioning of preselected representative video frames, relevance scoring based on deep features extracted from images/frames, and is currently integrated into the open-source version of the SOMHunter system providing additional browsing capabilities.

Keywords: Interactive video retrieval · Deep features · Multi-query retrieval

1 Introduction

Effective search initialization is an important step to solve a known-item search (KIS) task, especially for searched items residing somewhere in an extensive multimedia collection (e.g., [14]). Hence, many KIS systems rely on various automatic annotation approaches to enable text/keyword search or support querying by drawn sketches or provided example images [6,7,12]. This paper focuses on search initialization with a set of example images organized as a collage on a canvas. Several KIS systems already tested a "collage query" interface at the Video Browser Showdown (VBS) to enable users to search for object classes in regions detected by deep object detection networks [1,11,15]. A semantic sketch approach targeting image pixels of an object class was also presented by the vitrivr system [13] at VBS. However, to the best of our knowledge, the vocabulary of supported classes was limited for the tested collage approaches, and thus users could not address all their search needs.

A narrow vocabulary supported by a KIS system can be complemented with example images from external web search engines that provide effective free-form text search. For example, at VBS 2019 the VIRET system [8,9] tested this option,

© Springer Nature Switzerland AG 2021
J. Lokoč et al. (Eds.): MMM 2021, LNCS 12573, pp. 429–434, 2021.
https://doi.org/10.1007/978-3-030-67835-7_40

enabling convenient drag and drop interface for external images. However, the tested implementation did not support the positioning of the images and so only a global similarity between a query image and a video frame was involved.

In this paper, we present a component that enables users to interactively edit a collage query comprising available images collected at an external search engine supporting effective free-form text search (e.g., Google Images). Considering representative video frames divided into local regions, the ranking model is based on a similarity matching strategy for two sets of images (collage images and video frame sub-regions). In our implementation, each (sub-)image is represented by a deep feature $r \in \mathbb{R}^n$, and the similarity of two representations is modeled with the cosine similarity.

Using this approach, users are not limited by a deep object detector trained for a fixed set of classes and can construct rich collages of searched objects. On the other hand, the users can still face problems with expressing their needs at the external search engine, it takes more time to construct a collage query, and the system inherently relies on external engine availability and a network connection. Nevertheless, we find this approach intuitive and simple enough to represent an interesting baseline for the Video Browser Showdown.

The collage query component is integrated into the SOMHunter system [2, 3] that provides additional functionality necessary for video browsing and communication with the VBS server. However, to properly benchmark the querying approach based on collage queries, we do not consider the text query component provided with SOMHunter. In other words, to initialize the search, users have to start only with a collage query, or its temporal variant [5]. Figure 1 shows an illustration of the user interface.

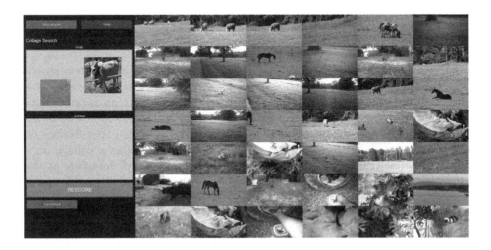

Fig. 1. Illustration of a user interface with temporal collage queries.

2   System Overview

For VBS 2021, we design a system prototype on top of the SOMHunter engine [3] providing useful features for the Video Browser Showdown competition. In order to search videos, the same set $\mathbb{F}$ of selected representative video frames is utilized. However, instead of the built-in text search mode, we implement a querying panel supporting temporal collage query creation/manipulation. On demand, the collage query is passed to a collage query processing component to rank frames $f \in \mathbb{F}$ and the resulting top-ranked frames are returned to SOMHunter's back end.

The involved components are depicted in Fig. 2, showing the collage query component as an external module. The expected search workflow starts at an external search engine, where users find an appropriate set of images. The images are placed on the collage canvas in SOMHunter's front end, where users can easily edit (i.e., move, resize, remove) the images. Once a collage query is ready and users press the re-score button, the collage query component receives the query and extracts deep features from the images. Currently, selected 128-dimensional features (based on a variant of the W2VV++ model [4,5,10]) used by the previous version of SOMHunter are considered. To support searching in image sub-regions, database video frames (i.e., the selected representative frames) are divided to a set of regions with overlaps. An example of frame partitioning is depicted in Fig. 3. Please note that there are many ways to define regions in images and the setting can be changed in the future. Therefore, the canvas supports an arbitrary position of collage images. Currently, the intersection over

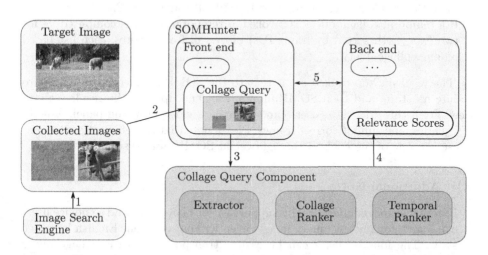

Fig. 2. 1. Users search for available images on the internet. 2. Users design a collage query. 3. Collage Query Component receives a collage query. 4. The result of relevance scoring is returned to SOMHunter. 5. SOMHunter updates the front end with new results.

union (IoU) is used to select an appropriate frame sub-region for each collage image.

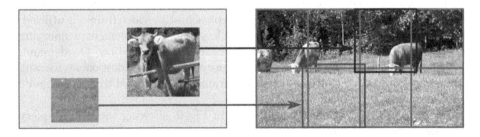

Fig. 3. Mapping of a collage query to a database video frame divided to 8 sub-regions.

The query processing method consists of the following steps:

- A deep feature representation $r_q \in \mathbb{R}^n$ of each collage image q is extracted online with an involved deep network.
- For each collage image q and a database frame $f \in \mathbb{F}$, the best matching frame sub-region f_{bmq} is selected (based on IoU, see Fig. 3); the similarity score $s_{f,q} = \sigma(r_{f_{bmq}}, r_q)$ is computed, where $r_{f_{bmq}} \in \mathbb{R}^n$ is a precomputed deep representation of f_{bmq}.
- Frame scores $s_{f,q}$ for all collage images q are aggregated to the overall score s_f of the frame. Scores s_f are used to rank all frames $f \in \mathbb{F}$.
- If a temporal collage query is invoked, a temporal fusion is computed [5] for two lists (both sorted by frame IDs) of overall frame scores, each list based on one collage query.

The resulting top-ranked set is sent to the SOMHunter core component to update its state, and then SOMHunter's front end updates the result set grid view. From this grid view, users can still open a video browsing panel, issue k-NN queries, and submit correct frames. Please note that this efficient component re-use was one of the goals of the lightweight SOMHunter system [3].

2.1 Additional Notes

- External search engines usually support various languages and so novice users can search for suitable query images even without excellent English language skills. The found images can be easily "transported" to the collage query canvas by making use of clipboard.
- The collage query component can be integrated as an external module if a different platform (e.g., for involved deep networks) is required.

3   Conclusions

This paper presents a component for SOMHunter that enables users to query a multimedia database with collage queries. The main features of the ranking model are presented as well as the basic variant of the user interface.

Acknowledgment. This paper has been supported by Czech Science Foundation (GAČR) project 19-22071Y and by Charles University grant SVV-260588.

References

1. Amato, G., et al.: The VISIONE video search system: exploiting off-the-shelf text search engines for large-scale video retrieval (2020)
2. Kratochvíl, M., Veselý, P., Mejzlík, F., Lokoč, J.: SOM-hunter: video browsing with relevance-to-SOM feedback loop. In: Ro, Y.M., et al. (eds.) MMM 2020, Part II. LNCS, vol. 11962, pp. 790–795. Springer, Cham (2020). https://doi.org/10.1007/978-3-030-37734-2_71
3. Kratochvíl, M., Mejzlýk, F., Veselí, P., Souček, T., Lokoč, J.: SOMHunter: lightweight video search system with SOM-guided relevance feedback. In: Proceedings of the 28th ACM International Conference on Multimedia, MM 2020. ACM (2020, in press)
4. Li, X., Xu, C., Yang, G., Chen, Z., Dong, J.: W2VV++: fully deep learning for ad-hoc video search. In: Proceedings of the 27th ACM International Conference on Multimedia, MM 2019, Nice, France, 21–25 October 2019, pp. 1786–1794 (2019). https://doi.org/10.1145/3343031.3350906
5. Lokoč, J., et al.: A W2VV++ case study with automated and interactive text-to-video retrieval. In: Proceedings of the 28th ACM International Conference on Multimedia, MM 2020. Association for Computing Machinery, New York (2020). https://doi.org/10.1145/3394171.3414002
6. Lokoč, J., Bailer, W., Schoeffmann, K., Münzer, B., Awad, G.: On influential trends in interactive video retrieval: video browser showdown 2015–2017. IEEE Trans. Multimed. **20**(12), 3361–3376 (2018). https://doi.org/10.1109/TMM.2018.2830110
7. Lokoč, J., et al.: Interactive search or sequential browsing? A detailed analysis of the video browser showdown 2018. ACM Trans. Multimed. Comput. Commun. Appl. **15**(1), 291–2918 (2019). https://doi.org/10.1145/3295663. http://doi.acm.org/10.1145/3295663
8. Lokoč, J., Kovalčík, G., Souček, T., Moravec, J., Čech, P.: A framework for effective known-item search in video. In: In Proceedings of the 27th ACM International Conference on Multimedia (MM 2019), 21–25 October 2019, Nice, France, pp. 1–9 (2019). https://doi.org/10.1145/3343031.3351046
9. Lokoč, J., Kovalčík, G., Souček, T., Moravec, J., Čech, P.: VIRET: a video retrieval tool for interactive known-item search. In: Proceedings of the 2019 on International Conference on Multimedia Retrieval, ICMR 2019, pp. 177–181. ACM, New York (2019). https://doi.org/10.1145/3323873.3325034. http://doi.acm.org/10.1145/3323873.3325034
10. Mettes, P., Koelma, D.C., Snoek, C.G.M.: Shuffled imagenet banks for video event detection and search. ACM Trans. Multimed. Comput. Commun. Appl. (TOMM) **16**(2), 1–21 (2020)

11. Nguyen, P.A., Wu, J., Ngo, C.-W., Francis, D., Huet, B.: VIREO @ video browser showdown 2020. In: Ro, Y.M., et al. (eds.) MMM 2020. LNCS, vol. 11962, pp. 772–777. Springer, Cham (2020). https://doi.org/10.1007/978-3-030-37734-2_68

12. Rossetto, L., et al.: Interactive video retrieval in the age of deep learning - detailed evaluation of VBS 2019. IEEE Trans. Multimed. (2020)

13. Rossetto, L., Parian, M.A., Gasser, R., Giangreco, I., Heller, S., Schuldt, H.: Deep learning-based concept detection in vitrivr. In: MultiMedia Modeling - 25th International Conference, MMM 2019, Thessaloniki, Greece, 8–11 January 2019, Proceedings, Part II, pp. 616–621 (2019). https://doi.org/10.1007/978-3-030-05716-9_55

14. Rossetto, L., Schuldt, H., Awad, G., Butt, A.A.: V3C – a research video collection. In: Kompatsiaris, I., Huet, B., Mezaris, V., Gurrin, C., Cheng, W.-H., Vrochidis, S. (eds.) MMM 2019, Part I. LNCS, vol. 11295, pp. 349–360. Springer, Cham (2019). https://doi.org/10.1007/978-3-030-05710-7_29

15. Truong, T.-D., et al.: Video search based on semantic extraction and locally regional object proposal. In: Schoeffmann, K., et al. (eds.) MMM 2018, Part II. LNCS, vol. 10705, pp. 451–456. Springer, Cham (2018). https://doi.org/10.1007/978-3-319-73600-6_49

Towards Explainable Interactive Multi-modal Video Retrieval with Vitrivr

Silvan Heller(✉) , Ralph Gasser , Cristina Illi , Maurizio Pasquinelli ,
Loris Sauter , Florian Spiess , and Heiko Schuldt

Department of Mathematics and Computer Science, University of Basel,
Basel, Switzerland
{silvan.heller,ralph.gasser,cristina.illi,maurizio.pasquinelli,
loris.sauter,florian.spiess,heiko.schuldt}@unibas.ch

Abstract. This paper presents the most recent iteration of the vitrivr multimedia retrieval system for its participation in the Video Browser Showdown (VBS) 2021. Building on existing functionality for interactive multi-modal retrieval, we overhaul query formulation and results presentation for queries which specify temporal context, extend our database with index structures for similarity search and present experimental functionality aimed at improving the explainability of results with the objective of better supporting users in the selection of results and the provision of relevance feedback.

Keywords: Video browser showdown · Interactive video retrieval · Content-based retrieval · Explainability

1 Introduction

The Video Browser Showdown (VBS) [17] is a major evaluation campaign for *interactive video retrieval* and celebrates its 10th anniversary in 2021. The vitrivr system and its predecessors have been long-running participants to VBS [19], utilising different methods from sketch-based queries to deep-learning methods for concept detection, and text retrieval for OCR and ASR in recent years [20,22].

vitrivr always had a focus on multi-modal multimedia retrieval, enabling users to mix and match different modalities to find a particular item in a collection. Being a general-purpose retrieval system, vitrivr has found many applications ranging, for example lifelog search [6,16].

In this paper, we present the vitrivr system as envisioned to participate at VBS'21, including its changes to temporal queries, newly added index structures and improvements towards the explainability of results. We have used the time since the last VBS to make improvements to query formulation and our retrieval model [8], to take another look at relevance feedback and results presentation [7], and to enhance our open source database, which is described in more detail in [3]. The backend and database layer are also used by another system participating at VBS, vitrivr-VR [23].

© Springer Nature Switzerland AG 2021
J. Lokoč et al. (Eds.): MMM 2021, LNCS 12573, pp. 435–440, 2021.
https://doi.org/10.1007/978-3-030-67835-7_41

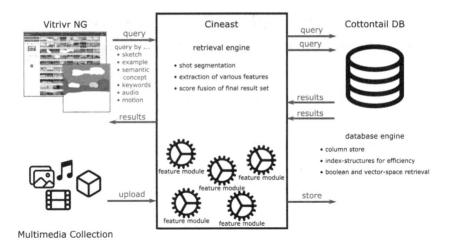

Fig. 1. Architecture overview of vitrivr [6]

The remainder of this paper is structured as follows: Sect. 2 gives an overview of the vitrivr system, Sect. 3 discusses our new approach to temporal scoring, Sect. 4 introduces the index structures we will be using for similarity search, Sect. 5 gives examples of the envisioned user-facing explanations of results and Sect. 6 concludes.

2   vitrivr

vitrivr[1] is an open source, content-based multimedia retrieval stack with explicit support for different modalities, such as images, audio, video, and 3D models. As such, vitrivr serves as a platform for a wide range of multimedia research activities such as gesture retrieval, search and exploration in VR and cultural heritage applications. The stack covers all aspects of multimedia retrieval, namely feature extraction and storage, content serving, query formulation and execution, and result presentation. In Fig. 1, we provide an architectural overview of vitrivr and its three main components: The storage layer Cottontail DB [3], the retrieval engine Cineast [18] and the presentation layer vitrivr-ng.

In vitrivr, many different kinds of features and data sources are used. We extract a variety of low-level visual features for color and texture, use more traditional features such as SURF [1], deep features [9] and textual information (ASR, OCR) for fulltext search [20]. For an in-depth discussion of the retrieval model unifying different types of features, we refer to [8].

While vitrivr has always been focused on *retrieval*, we have recently experimented with different presentation modes and relevance feedback for a stronger focus on *exploration*, which has shown promising results for Ad-Hoc Video Search (AVS)-style tasks [7].

---

[1] https://vitrivr.org/.

3   Temporal Querying

As discussed in [6,8], users of vitrivr can currently formulate temporal queries by combining multiple, independent queries. Other systems at VBS employ different approaches to the same problem [13,14]. In addition, users must specify a temporal bound wherein potential matches should occur, e.g., results of query A within 30 seconds of query B. The two queries are then processed independently in parallel by the retrieval engine. The results are combined by the front-end such that the temporal proximity of matching pairs in both result sets within the specified limits lead to a higher overall score.

Evaluations with different versions of vitrivr [7] and against other systems have shown that user experience and retrieval speed deteriorates quickly for large result sets. Furthermore, user studies have found that users almost never submit more than two queries jointly, even if they see longer portions of a video and the general feedback indicates that users find the current presentation as well as query formulation to be very unintuitive.

This feedback, combined with the proven effectiveness of temporal querying, has led to revisiting this functionality. Aggregating results by temporal closeness in the front-end suffers from inherent architectural limitations: Since Cineast only sends a limited number of results per query, some segments might be present in multiple result lists, but get lost due to the cutoff. For this iteration of Cineast, we therefore plan to compute matches for temporal queries already in the retrieval engine to address this issue and improve performance.

Additionally, we are experimenting with simpler ways to enable users to express temporal dependencies while preserving the multi-modality and expressiveness of the current front-end, for instance by using different ways of presenting matches for temporal queries.

4   Index Structures for Similarity Search

Already with the first shard of the V3C [21] dataset – V3C1 [2] – fitting all relevant data into main memory on commodity hardware is challenging. vitrivr uses Cottontail DB [3] as its database, which makes it well-positioned for future increase in dataset size. However, as these datasets become larger, linear nearest neighbor search (NNS) quickly becomes a bottleneck, especially in a time-critical and competitive setting such as VBS, where every second counts.

Promising techniques for approximate NNS that we currently consider are LSH-based methods [10], PQ [11] and the more recent ScANN [4] algorithm. Due to the curse of dimensionality, however, accelerating NNS with such indexes often comes at the cost of sacrificing either execution performance or accuracy. We will therefore focus on ways to optimize the trade-off between the two, e.g., by combining different index structures or by progressively improving result sets, as more accurate information becomes available after some time.

Fig. 2. Concept bounding boxes can be used to inspire the user when improving the query [5].

5   Towards the Explainability of Search Results

The topic of explainability has become increasingly important in recent years for two reasons: First, understanding the inner workings of machine learning algorithms, which are often considered to be black boxes, helps developers to improve their systems. Second, and often overlooked, by providing information on the retrieval process, users can improve their queries in an interactive and iterative way, which is exactly the goal of interactive retrieval. In contrast to relevance feedback which aims at improving results based on feedback by the user (as done in i.e. [12]), the objective of explaining results is to help users improve their query based on the information conveyed by the result.

Displaying explanations for different modalities from those employed by the user during their search can also inspire them to think of new ways of modifying their query. Especially so, if the user is stuck and does not get satisfying results with the current modality. These visualizations can then be used interactively to improve query formulation, for example, by performing more-like-this searches on specific objects or proposing concepts users had previously not considered to the query. As an example, consider Fig. 2, where detected concepts are highlighted with their bounding boxes [5,15].

vitrivr aims to explain results in two ways: First, on the level of an individual result by showing precisely why a feature was considered a result a match, e.g., by showing feature visualizations And second, on a result set level by giving users information about properties of all returned results such as co-occurring concepts and frequently occurring textual descriptions.

6   Conclusion

In this paper, we presented the current iteration of the vitrivr system participating at VBS 2021. Compared to previous participations, vitrivr will add improvements for temporal queries, provide additional index structures at database level, and improve the overall query process and enhance the user experience by explaining the retrieval results.

Acknowledgements. This work was partly supported by the Hasler Foundation in the context of the project City-Stories (contract no. 17055) and by the Swiss National Science Foundation, project Polypheny-DB (contract no. 200021_172763).

References

1. Bay, H., Ess, A., Tuytelaars, T., Van Gool, L.: Speeded-up robust features (SURF). Comput. Vis. Image Underst. **110**(3), 346–359 (2008)
2. Berns, F., Rossetto, L., Schoeffmann, K., Beecks, C., Awad, G.: V3C1 dataset: an evaluation of content characteristics. In: Proceedings of the International Conference on Multimedia Retrieval (2019)
3. Gasser, R., Rossetto, L., Heller, S., Schuldt, H.: Cottontail DB: an open source database system for multimedia retrieval and analysis. In: Proceedings of the 28th ACM International Conference on Multimedia (2020)
4. Guo, R., et al.: Accelerating large-scale inference with anisotropic vector quantization. arXiv preprint arXiv:1908.10396 (2020)
5. He, K., Gkioxari, G., Dollár, P., Girshick, R.: Mask R-CNN. In: Proceedings of the IEEE International Conference on Computer Vision (2017)
6. Heller, S., Parian, M.A., Gasser, R., Sauter, L., Schuldt, H.: Interactive lifelog retrieval with vitrivr. In: Proceedings of the Third ACM Workshop on Lifelog Search Challenge, LSC@ICMR 2020, Dublin, Ireland (2020)
7. Heller, S., Parian, M., Pasquinelli, M., Schuldt, H.: Vitrivr-explore: guided multimedia collection exploration for ad-hoc video search. In: Satoh, S., et al. (eds.) SISAP 2020. LNCS, vol. 12440, pp. 379–386. Springer, Cham (2020). https://doi.org/10.1007/978-3-030-60936-8_30
8. Heller, S., Sauter, L., Schuldt, H., Rossetto, L.: Multi-stage queries and temporal scoring in vitrivr. In: IEEE International Conference on Multimedia & Expo Workshops (ICMEW) (2020)
9. Howard, A.G., et al.: MobileNets: efficient convolutional neural networks for mobile vision applications. arXiv preprint arXiv:1704.04861 (2017)
10. Indyk, P., Motwani, R.: Approximate nearest neighbors: towards removing the curse of dimensionality. In: Proceedings of the Thirtieth Annual ACM Symposium on Theory of Computing (1998)
11. Jegou, H., Douze, M., Schmid, C.: Product quantization for nearest neighbor search. IEEE Trans. Pattern Anal. Mach. Intell. **33**(1), 117–128 (2010)
12. Jónsson, B.Þ., et al.: Exquisitor: interactive learning at large. arXiv preprint arXiv:1904.08689 (2019)
13. Lokoc, J., et al.: A W2VV++ case study with automated and interactive text-to-video retrieval. In: Proceedings of the 28 ACM International Conference on Multimedia (2020)

14. Nguyen, P.A., Wu, J., Ngo, C.-W., Francis, D., Huet, B.: VIREO @ video browser showdown 2020. In: Ro, Y.M., et al. (eds.) MMM 2020. LNCS, vol. 11962, pp. 772–777. Springer, Cham (2020). https://doi.org/10.1007/978-3-030-37734-2_68

15. Ren, S., He, K., Girshick, R.B., Sun, J.: Faster R-CNN: towards real-time object detection with region proposal networks. CoRR abs/1506.01497 (2015). http://arxiv.org/abs/1506.01497

16. Rossetto, L., Gasser, R., Heller, S., Amiri Parian, M., Schuldt, H.: Retrieval of structured and unstructured data with vitrivr. In: Proceedings of the ACM Workshop on Lifelog Search Challenge (2019)

17. Rossetto, L., et al.: Interactive video retrieval in the age of deep learning - detailed evaluation of VBS 2019. IEEE Trans. Multimed. (2020)

18. Rossetto, L., Giangreco, I., Heller, S., Tănase, C., Schuldt, H.: Searching in video collections using sketches and sample images – the cineast system. In: Tian, Q., Sebe, N., Qi, G.-J., Huet, B., Hong, R., Liu, X. (eds.) MMM 2016. LNCS, vol. 9517, pp. 336–341. Springer, Cham (2016). https://doi.org/10.1007/978-3-319-27674-8_30

19. Rossetto, L., et al.: IMOTION – searching for video sequences using multi-shot sketch queries. In: Tian, Q., Sebe, N., Qi, G.-J., Huet, B., Hong, R., Liu, X. (eds.) MMM 2016. LNCS, vol. 9517, pp. 377–382. Springer, Cham (2016). https://doi.org/10.1007/978-3-319-27674-8_36

20. Rossetto, L., Amiri Parian, M., Gasser, R., Giangreco, I., Heller, S., Schuldt, H.: Deep learning-based concept detection in vitrivr. In: Kompatsiaris, I., Huet, B., Mezaris, V., Gurrin, C., Cheng, W.-H., Vrochidis, S. (eds.) MMM 2019. LNCS, vol. 11296, pp. 616–621. Springer, Cham (2019). https://doi.org/10.1007/978-3-030-05716-9_55

21. Rossetto, L., Schuldt, H., Awad, G., Butt, A.A.: V3C – a research video collection. In: Kompatsiaris, I., Huet, B., Mezaris, V., Gurrin, C., Cheng, W.-H., Vrochidis, S. (eds.) MMM 2019. LNCS, vol. 11295, pp. 349–360. Springer, Cham (2019). https://doi.org/10.1007/978-3-030-05710-7_29

22. Sauter, L., Amiri Parian, M., Gasser, R., Heller, S., Rossetto, L., Schuldt, H.: Combining boolean and multimedia retrieval in vitrivr for large-scale video search. In: Ro, Y.M., et al. (eds.) MMM 2020. LNCS, vol. 11962, pp. 760–765. Springer, Cham (2020). https://doi.org/10.1007/978-3-030-37734-2_66

23. Spiess, F., Gasser, R., Heller, S., Rossetto, L., Sauter, L., Schuldt, H.: Competitive interactive video retrieval in virtual reality with vitrivr-VR. In: International Conference on Multimedia Modeling MMM, pp. 441–447. Springer, Heidelberg (2021)

Competitive Interactive Video Retrieval in Virtual Reality with vitrivr-VR

Florian Spiess[1]([✉]) [iD], Ralph Gasser[1] [iD], Silvan Heller[1] [iD], Luca Rossetto[2] [iD],
Loris Sauter[1] [iD], and Heiko Schuldt[1] [iD]

[1] Department of Mathematics and Computer Science,
University of Basel, Basel, Switzerland
{florian.spiess,ralph.gasser,silvan.heller,
loris.sauter,heiko.schuldt}@unibas.ch
[2] Department of Informatics,
University of Zurich, Zurich, Switzerland
rossetto@ifi.uzh.ch

Abstract. Virtual Reality (VR) has emerged and developed as a new modality to interact with multimedia data. In this paper, we present vitrivr-VR, a prototype of an interactive multimedia retrieval system in VR based on the open source full-stack multimedia retrieval system vitrivr. We have implemented query formulation tailored to VR: Users can use speech-to-text to search collections via text for concepts, OCR and ASR data as well as entire scene descriptions through a video-text co-embedding feature that embeds sentences and video sequences into the same feature space. Result presentation and relevance feedback in vitrivr-VR leverages the capabilities of virtual spaces.

Keywords: Video Browser Showdown · Virtual Reality · Interactive video retrieval

1 Introduction

The Video Browser Showdown (VBS) [14] is an annual interactive video search competition where video retrieval systems are evaluated competitively through a variety of search tasks on large video datasets.

Virtual reality (VR) technology has developed rapidly in recent years, and due to technological advances that have continued to make the hardware required more accessible, affordable, and comfortable, VR capable devices have become much more commonplace. VR has great potential when interacting with multimedia, especially in the context of retrieval, due to the ability of users to be immersed in a virtual space, which provides a near limitless presentation area. Additionally, hand and head tracking directly translates real world movement into movement within the virtual space, making spatial orientation intuitive.

We leverage components of the existing vitrivr system [16], which has participated at VBS for a number of years [4,18], and have built a new system,

© Springer Nature Switzerland AG 2021
J. Lokoč et al. (Eds.): MMM 2021, LNCS 12573, pp. 441–447, 2021.
https://doi.org/10.1007/978-3-030-67835-7_42

vitrivr-VR, with a virtual reality based interface to tackle the problem of efficient and intuitive interactive video retrieval in large video collections by exploiting the potential of multimedia retrieval in VR. vitrivr-VR has voice input for initial queries and uses a number of query refinement and relevance feedback tools engineered to enable intuitive and efficient video search in VR.

In the remainder of this paper we contextualize our work with respect to existing research in Sect. 2, present an overview of the vitrivr-VR architecture in Sect. 3, describe the query mechanisms used in Sect. 4, outline the interactive retrieval process in Sect. 5, and conclude in Sect. 6.

2   VR Multimedia Retrieval Interfaces

A few approaches to perform multimedia retrieval and exploration in VR have already been proposed and have shown the potential of this new modality [9], even in competitive settings such as the Lifelog Search Challenge (LSC) [1], where vitrivr has also been a participant in recent years [5]. While these existing systems have shown that multimedia retrieval and exploration is possible in VR, they primarily investigate how traditional 2D interfaces can be transferred into the VR space without significant changes to the mode of interaction and media presentation. In [1], for example, different methods of interacting with traditional 2D user interfaces in VR are explored, however, the user interfaces themselves are not adapted to the virtual space beyond their placement in the 3D space. Although these interfaces are immediately familiar from traditional 2D interfaces, they lose much of their efficiency in VR, especially for complex interactions, such as input of specific text.

Other approaches propose methods to make use of the space and freedom of movement available in VR. [3] introduces a system that enables 3D sketch-based model retrieval in VR through the fine-granular input possible with 3D tracked controllers. The retrieval results are displayed in a traditional 2D scrolling list. A system that explores result presentation in VR is described in [13], which proposes a simple approach to result presentation in virtual spaces by directly mapping feature similarity scores to the three spatial dimensions.

With vitrivr-VR we deliberately attempt to avoid traditional 2D user interfaces and presentation methods designed for 2D displays in favor of investigating the effectiveness of user interfaces designed with virtual spaces in mind.

3   System Overview

As the name suggests, vitrivr-VR shares some components with vitrivr, an open source multimedia retrieval stack.[1] It also uses Cottontail DB [2] as its storage layer and Cineast [15] as its retrieval engine. The presentation layer is a VR environment as opposed to the traditional web interface. Cottontail DB is a column store that allows combining Boolean retrieval and nearest neighbor search.

---

[1] https://vitrivr.org.

Cineast is responsible for both retrieval and feature extraction, supporting a multitude of features and media types.

To communicate with Cineast, vitrivr-VR uses the RESTful API provided through the OpenAPI specifications. vitrivr-VR is implemented in Unity[2] using the Unity XR Interaction Toolkit[3] to interface with the HTC Vive Pro headset through OpenVR.

The existing components from the vitrivr stack provide the core multimedia retrieval capabilities [7], while vitrivr-VR enables query formulation, query refinement, relevance feedback and browsing through its VR user interface. vitrivr-VR provides speech-to-text based text input for textual queries, query-by-example (QbE) through previously retrieved results and a number of query refinement and relevance feedback options.

4   Querying Mechanisms

vitrivr offers a multitude of feature descriptors, many of which focusing on various visual aspects of a video. Since the primary query formulation mechanism for vitrivr-VR is speech input which is automatically transcribed into text, we limit ourselves to the features which operate on textual input. These include previously introduced capabilities [17], such as full-text search in automatically transcribed dialog, visible text or automatically generated scene captions as well as tag-based search using a large number of individually detected instances of objects or semantic concepts visible in a scene. In addition, we use a video-text co-embedding mechanism inspired by [11], which is capable of capturing a richer semantic representation when compared to the individual tags while understanding a larger vocabulary than the already used captioning approach. The effectiveness of such methods has recently been demonstrated in [12].

We also make use of relevance feedback for non text-based querying, which is used by many systems in interactive video retrieval [6,8,10] and is also well suited for the result presentation and interaction mechanisms of vitrivr-VR.

5   Interactive Retrieval Process in VR

In vitrivr-VR, using the system to find the figurative needle in the haystack is an interactive process which can be categorised into three phases; *initial query, result organisation* and *refinement queries* A simplified interaction flow diagram of the process is shown in Fig. 1.

---

[2] https://unity.com.

[3] https://docs.unity3d.com/Packages/com.unity.xr.interaction.toolkit@0.9.

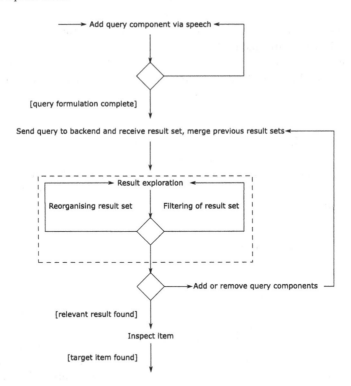

Fig. 1. Interaction flow diagram of vitrivr-VR

5.1   Initial Query

Recent VBS installments have shown that text-based retrieval, be it for a limited, known set of concepts or automatically generated scene captions, is a successful strategy for an initial query. However, typing words for text-based search is not very efficient in VR since the state of the art in finger-tracking for text input is not yet fast and stable. Hence, vitrivr-VR relies on speech-to-text[4] to formulate the initial query. In particular, vitrivr-VR supports traditional full-text search in OCR, ASR and scene caption data, dedicated concept search, as well as video-text co-embedding search as described in Sect. 4. The OCR and ASR data is the same as used in [17]. At the end of the initial query, a text-based similarity query is issued.

5.2   Result Organisation

In the second phase, during *result organisation*, users are able to re-arrange the results in order to effectively browse the result collection. One advantage of VR is the virtually unlimited space for result presentation, which we exploit in this

---

[4] https://docs.microsoft.com/en-us/windows/mixed-reality/voice-input-in-unity.

phase by initially displaying the result set spherically around the user, where each item's spatial context is indicative of its score. Media items can be pulled from the results display and manually positioned in space. Items 'pinned' in this way will return to their position in the results display once unpinned. A list of concepts sorted by their frequency within the result set is presented in close proximity to the user. Subsequently, users are enabled to spatially cluster the results along certain attributes such as concepts by dragging them from a list and positioning them in space, or remove items from the result set matching a certain filter criterion. The goal of this result organisation is to determine whether the searched item is already in the result set. Each item can be inspected to view the associated video segment, neighboring segments as well as additional information such as a list of concepts detected within the segment. From this view, items can also be submitted to the competition system. In case the target item is not yet in the result set, the third phase is launched.

5.3 Refinement Queries

From the previous phases, a set of query components produced a result set which was organised and filtered, and is spatially explorable within vitrivr-VR. In case a user did not yet succeed in finding the target item, our system provides multiple means to facilitate query refinement and additional queries: (i) query-by-example for result set expansion or reduction (ii) more-like-this queries, both visually and semantically (iii) relevance feedback. Result sets of these new queries might be seamlessly merged into the pre-existing result set without user interaction. Alternatively, taking advantage of the available space in VR, results of additional queries can be presented spatially separated from the original query in order to give users more control over the merging of the two result sets. Additionally, a query and its result set can be stashed away temporarily or discarded permanently, so the user can focus on other queries.

6 Conclusion

In this paper, we introduced vitrivr-VR, a virtual reality multimedia retrieval system integrated into the open source vitrivr stack. vitrivr-VR provides a VR interface for intuitive query formulation and results presentation in a virtual space. We expect video retrieval in VR to be a competitive, intuitive and, user-friendly alternative to traditional 2D interfaces and for our VBS participation to provide us with valuable insights into the effectiveness of VR based systems in competitive interactive video retrieval.

Acknowledgements. This work was partly supported by the Hasler Foundation in the context of the project City-Stories (contract no. 17055).

References

1. Duane, A., Þór Jónsson, B., Gurrin, C.: VRLE: lifelog interaction prototype in virtual reality: lifelog search challenge at ACM ICMR 2020. In: Proceedings of the Third Annual Workshop on Lifelog Search Challenge (2020)
2. Gasser, R., Rossetto, L., Heller, S., Schuldt, H.: Cottontail DB: an open source database system for multimedia retrieval and analysis. In: Proceedings of the 28th ACM International Conference on Multimedia (2020)
3. Giunchi, D., James, S., Steed, A.: 3D sketching for interactive model retrieval in virtual reality. In: Proceedings of the Joint Symposium on Computational Aesthetics and Sketch-Based Interfaces and Modeling and Non-Photorealistic Animation and Rendering (2018)
4. Heller, S., et al.: Towards explainable interactive multi-modal video retrieval with vitrivr. In: International Conference on Multimedia Modeling MMM (2021)
5. Heller, S., Parian, M.A., Gasser, R., Sauter, L., Schuldt, H.: Interactive lifelog retrieval with vitrivr. In: Proceedings of the Third ACM Workshop on Lifelog Search Challenge (2020)
6. Heller, S., Parian, M., Pasquinelli, M., Schuldt, H.: Vitrivr-explore: guided multimedia collection exploration for ad-hoc video search. In: Satoh, S., et al. (eds.) SISAP 2020. LNCS, vol. 12440, pp. 379–386. Springer, Cham (2020). https://doi.org/10.1007/978-3-030-60936-8_30
7. Heller, S., Sauter, L., Schuldt, H., Rossetto, L.: Multi-stage queries and temporal scoring in vitrivr. In: IEEE International Conference on Multimedia & Expo Workshops (2020)
8. Jónsson, B.Þ., Khan, O.S., Koelma, D.C., Rudinac, S., Worring, M., Zahálka, J.: Exquisitor at the video browser showdown 2020. In: Ro, Y., et al. (eds.) MMM 2020. LNCS, vol. 11962, pp. 796–802. Springer, Cham (2020). https://doi.org/10.1007/978-3-030-37734-2_72
9. Khanwalkar, S., Balakrishna, S., Jain, R.: Exploration of large image corpuses in virtual reality. In: Proceedings of the 24th ACM International Conference on Multimedia (2016)
10. Kratochvíl, M., Veselý, P., Mejzlík, F., Lokoč, J.: SOM-hunter: video browsing with relevance-to-SOM feedback loop. In: Ro, Y., et al. (eds.) MMM 2020. LNCS, vol. 11962, pp. 790–795. Springer, Cham (2020). https://doi.org/10.1007/978-3-030-37734-2_71
11. Li, X., Xu, C., Yang, G., Chen, Z., Dong, J.: W2VV++ fully deep learning for ad-hoc video search. In: Proceedings of the 27th ACM International Conference on Multimedia (2019)
12. Lokoč, J., et al.: A W2VV++ case study with automated and interactive text-to-video retrieval. In: ACM Multimedia (2020)
13. Nakazato, M., Huang, T.S.: 3D MARS: immersive virtual reality for content-based image retrieval. In: IEEE International Conference on Multimedia and Expo (2001)
14. Rossetto, L., et al.: Interactive video retrieval in the age of deep learning-detailed evaluation of VBS 2019. IEEE Trans. Multimed. **23**, 243–256 (2020)
15. Rossetto, L., Giangreco, I., Heller, S., Tănase, C., Schuldt, H.: Searching in video collections using sketches and sample images – the Cineast system. In: Tian, Q., Sebe, N., Qi, G.-J., Huet, B., Hong, R., Liu, X. (eds.) MMM 2016. LNCS, vol. 9517, pp. 336–341. Springer, Cham (2016). https://doi.org/10.1007/978-3-319-27674-8_30

16. Rossetto, L., Giangreco, I., Tanase, C., Schuldt, H.: Vitrivr: a flexible retrieval stack supporting multiple query modes for searching in multimedia collections. In: Proceedings of the 24th ACM International Conference on Multimedia (2016)
17. Rossetto, L., Parian, M.A., Gasser, R., Giangreco, I., Heller, S., Schuldt, H.: Deep learning-based concept detection in vitrivr. In: International Conference on Multimedia Modeling (2019)
18. Sauter, L., Parian, M.A., Gasser, R., Heller, S., Rossetto, L., Schuldt, H.: Combining Boolean and multimedia retrieval in vitrivr for large-scale video search. In: International Conference on Multimedia Modeling (2020)

An Interactive Video Search Tool: A Case Study Using the V3C1 Dataset

Abdullah Alfarrarjeh[2]($\boxtimes$), Jungwon Yoon[1], Seon Ho Kim[1], Amani Abu Jabal[2],
Akarsh Nagaraj[1], and Chinmayee Siddaramaiah[1]

[1] Integrated Media Systems Center,
University of Southern California, Los Angeles, USA
{jungwony,seonkim,akarshna,siddaram}@usc.edu
[2] Department of Computer Science,
German Jordanian University, Amman, Jordan
{abdullah.alfarrarjeh,amani.abujabal}@gju.edu.jo

Abstract. This paper presents a prototype of an interactive video
search tool for the preparation of MMM 2021 Video Browser Showdown (VBS). Our tool is tailored to enable searching for the public
V3C1 dataset associated with various analysis results including detected
objects, speech recognition, and visual features. It supports two types of
searches: text-based and visual-based. With a text-based search, the tool
enables users for querying videos using their textual descriptions, while
with a visual-based search, one provides a video example to search for
similar videos. Metadata extracted by recent state-of-the-art computer
vision algorithms for object detection and visual features are used for
accurate search. For an efficient search, the metadata are managed in
two database engines: Whoosh and PostgreSQL. The tool also enables
users to refine the search results by providing relevance feedback and
customizing the intermediate analysis of the query inputs.

Keywords: Interactive video search · Video Browser Showdown ·
Relevance feedback

1 Introduction

Due to the popularity of video applications (e.g., YouTube and Facebook) and
the proliferation of camera-enabled smartphone users, unprecedented number
of videos are being generated. Based on the "statista" website[1], the number
of uploaded videos per minute is 500 h of video as of May 2019. Given such
large-scale sets of collected videos, enabling efficient video search is essential.

The problem of video search has been investigated in different forms. One
approach is to enable users to search for a given query image. This approach
refers to a content-based video search or visual search, and various techniques

---

[1] https://www.statista.com/statistics/259477/hours-of-video-uploaded-to-youtube-
every-minute/.

© Springer Nature Switzerland AG 2021
J. Lokoč et al. (Eds.): MMM 2021, LNCS 12573, pp. 448–454, 2021.
https://doi.org/10.1007/978-3-030-67835-7_43

have been proposed to learn the visual content of videos then extract representative visual features (e.g., color histogram [7], edge histogram [10], and SIFT [9]). The other approach is to represent each video using a set of keywords tagged either manually or automatically. With the recent advancement in deep learning algorithm, learning the content of images and extracting relevant keywords is feasible. Then, the video search problem becomes analogous to document search; hence the state-of-the-art methods (e.g., inverted index file) for efficient document search can be utilized for video search, which is known as a text-based video search [13]. Since growing number of videos are tagged automatically with spatial metadata (e.g., GPS location, field-of-view [8], scene location [2]), the videos can be searched only using spatial criteria [8]. This approach is referred to as a spatial-based video search. Moreover, another approach uses both spatial and visual properties of videos in searching (e.g., spatial-visual search [1,3]).

In this paper, we propose a video search tool referred to as Interactive Video Search Tool (*IVS*). *IVS* supports different types of searches, including text-based search and visual search. To enable these types of searches, diverse metadata are being extracted using the state-of-the-art algorithms for object detection, speech recognition, character recognition, and visual features. The metadata are stored efficiently in two database engines (Whoosh and PostgreSQL). Moreover, the tool enables user interaction by providing feedback for the retrieved videos and initiates subsequent search according to the relevant feedback. Also, user intervention is allowed to customize the search by updating the query inputs to prioritize some of them (e.g., prioritizing a specific keyword among the query keywords in the case of text-based search). As a case study, *IVS* is used to search the V3C1 dataset composed of 7k videos. This dataset is already associated with several metadata by the provider; however, we extracted further metadata from the dataset as described in Subsect. 2.3.

2   Search Tool Architecture

Here, we describe the underlying components and the adopted built-in algorithms in our proposed search tool (*IVS*). Figure 1 shows the GUI for *IVS*.

Fig. 1. The GUI of the Web Tool

2.1 Video Dataset

The V3C1 dataset [4] is composed of 7,475 videos. The mean duration time of videos is 481 s. Each video is segmented into a set of short clips, and each short clip is represented by a keyframe. On average, the number of generated short clips per video is 145, and the mean duration time of clips is 3.2 s. The total number of generated clips for the whole video dataset is 1,082,657; therefore, the V3C1 dataset is represented by a set of 1,082,657 keyframes (i.e., images). Now, the video search task gets transformed into searching for images.

2.2 Query Tasks

Using the reference video dataset, $\mathcal{IVS}$ supports three types of queries: Visual Known Item Search (*Visual KIS*), Textual Known Item Search (*Textual KIS*), and Ad-hoc Video Search (*AVS*). Each of these query types aims at retrieving the keyframes that satisfy the query criteria.

The *Visual KIS* query enables a user to search within the reference video dataset for keyframes similar to the scenes depicted in a given 20-second clip. Meanwhile, with the *Textual KIS* query, the user provides a detailed text description instead of a video to search for keyframes capturing scenes that satisfy the query description. The *AVS* query is similar to the *Textual KIS* query as both of them are based on a textual description of the target scene. However, they are different in the number of the desired query results. In particular, the *Textual KIS* query aims to find the best keyframe that matches the query criteria while the *AVS* query expects to find a set of keyframes satisfying the target scene description.

2.3 Image Representation Metadata

Image search depends on the availability of rich image descriptors that depict various aspects of the image properties. Without loss of generality, $\mathcal{IVS}$ uses the following metadata as descriptors for the dataset.

- **Detected Objects:** One way of describing images is to specify the objects captured in images. For this sake, several deep learning methods for image recognition and object detection can be used. Each keyframe in the V3C1 video dataset is associated with a subset of the 1000 objects detectable by the NASnet convolutional neural network [14]. Given that each detected object is associated with a detection confidence score, we consider only the first ten objects with the highest confidence scores to reduce the impact of detection inaccuracy in searching. Furthermore, given that the NASnet algorithm annotates the detected object in a general and abstract manner (e.g., person), we have considered another state-of-the-art object detection algorithm (known as DenseCap [6]) which detects objects with more detailed annotations (e.g., a person playing baseball)[2]; hence, adding more contextual information about objects.

---

[2] When considering only nouns, the size of the vocabulary of DenseCap is 849 objects.

- **Faces:** The keyframes of the V3C1 dataset are augmented with the set of human faces that have been detected via the Apple Image Core API[3]. This API reports different labels that imply the number of the detected faces (i.e., no faces, one face, two faces, three faces, four faces, or many faces). Since the face detection implies a person's existence, $\mathcal{IVS}$ utilizes the metadata extracted from the face detection analysis as indicators for persons in the search approach.
- **Visual features:** Various approaches have been proposed for extracting visual features that depict the visual content of an image (e.g., color histogram [7], edge histogram [10], and SIFT [9]). The keyframes of the V3C1 dataset are associated with visual features consisted of color and edge histograms. In addition, we used one of the recent state-of-the-art approaches (referred to as regional maximum activations of convolutions (R-MAC) [12]), which utilizes a deep neural network for learning features from an image. R-MAC extracts a feature vector composed of 512 dimensions.
- **Detected Text:** Optical character recognition (OCR) enables extracting the written text that appears in certain portions of images (e.g., the text written on a traffic sign or a shop storefront captured in an image). Given that a subset of the keyframes of the V3C1 dataset is labeled with few or much text (based on the analysis results via the Apple Image Core API for recognizing text (See footnote 3)), we have extracted text from this subset using TesseractOCR [11]. Furthermore, since some short clips of the V3C1 dataset include speech, such clips have been analyzed by the Google Cloud Speech-to-Text API to extract spoken keywords to tag them with their corresponding keyframes. For both texts extracted by either OCR or speech recognition, it has been processed to filter out the non-English text and invalid English words using the Python language detection[4] and the part-of-speech tagging[5] libraries, respectively.
- **Dominant Color:** Each keyframe in the V3C1 dataset is labeled with the dominant color. The dominant color is either one of the colors: blue, cyan, gray, green, magenta, orange, red, violet, yellow, black/white, or undetermined.

2.4 Storage and Indexing

To enhance the search performance of $\mathcal{IVS}$, we have stored and indexed the metadata of the reference video dataset into two different database engines. First, the textual metadata (i.e., detected objects, faces, detected texts, and dominant colors) are stored into Whoosh which is a pure-python full-text search engine library. Using Whoosh, each keyframe is considered a document that is represented by keywords categorized into predefined categories (i.e., detected objects, faces, detected texts, and dominant colors). These keyframes and their corresponding keywords are organized in an inverted index file. After that, the

---

[3] https://developer.apple.com/documentation/coreimage/cifacefeature.

[4] https://pypi.org/project/langdetect/.

[5] https://www.nltk.org/book/ch05.html.

keyframes are retrieved based on the well-known text-based retrieval methods (e.g., TF-IDF and the Okapi BM25F ranking functions). Second, the visual features are stored in PostgreSQL. Since each keyframe is represented by a high-dimensional vector, PostgreSQL enables storing such high-dimensional data in a special data type known as "cube"[6] that supports organizing data using a Generalized Search Tree-based index (GiST)[7]; hence supporting efficient k-nearest neighbor (kNN) queries.

2.5   Searching Approach

Query Parsing and Preprocessing. For executing the text-based queries (i.e., *Textual KIS* and *AVS*), the textual description of a query needs to be interpreted. Thus, a natural language processing (NLP) library is used to interpret the query text to extract meaningful keywords and discard unnecessary ones (e.g., stop words and punctuation). In $\mathcal{IVS}$, an NLP library called *SpaCy* [5] is used. The extracted keywords are classified into adjectives, nouns, verbs, and numeric keywords. Thereafter, these words are mapped to predefined terms in the tool dictionary representing the objects, colors, and faces. Then, the mapped terms are used for searching the video dataset by utilizing the built indexes. For example, if the query includes the word "turquoise", it is mapped to the set of primary colors that constitute that specific target color (e.g., turquoise is mapped to blue and green) for utilizing the color metadata while searching. Also, if the query includes the word "a man", it is mapped into "one face" for utilizing the face metadata.

Regarding the visual-based queries (i.e., *Visual KIS*), the provided 20-seconds video is segmented to extract a set of keyframes. Then, the extracted keyframes (referred to as *query keyframes*) are processed to generate a visual feature descriptor per each. Moreover, each of the query keyframes is processed using object detection, speech recognition, and character recognition algorithms to identify its textual keywords. Both visual feature descriptors and textual keywords are used for searching the video dataset.

Retrieval and Ranking. For the textual queries, the keywords identified through the query parsing and preprocessing are used to search the textual metadata dataset utilizing the built indexes within the search tool. Given that the metadata is stored into different categories, each of the query keywords is mapped to a specific metadata category. At executing the search query, the text indexes corresponding to each metadata category are employed to retrieve the best keyframes associated with keywords matching the query keywords. Then, the retrieved keyframes are ranked using either the TF-IDF or BM25 scoring functions (based on user selection) supported by Whoosh. The query keywords are prioritized based on their corresponding categories. Based on our evaluation,

---

[6] https://www.postgresql.org/docs/10/cube.html.
[7] https://www.postgresql.org/docs/11/textsearch-indexes.html.

we found that detected object metadata are the most essential for finding relevant keyframes. Therefore, keywords related to the detected object category have a higher priority than the others; however, the other keywords are used to increase the priority of the retrieved keyframes matching both detected object keywords and other keywords[8]. In $\mathcal{IVS}$, the prioritizing mechanism of the search results is implemented using a Boolean query as outlined in Eq. 1.

$$
\begin{aligned}
TextQuery = \{keywords(category : DetectedObjects)\} \ OR \\
\{keywords(category : DetectedObjects) \ AND \ keywords(category : Others)\}
\end{aligned}
\tag{1}
$$

On the other hand, for the visual queries, for each query keyframe, two queries are executed: visual-based search using the extracted visual feature descriptors and text-based using the extracted textual keywords. Each search ranks the retrieved keyframes individually using some ranking functions. The text search ranks the keyframes using TF-IDF or BM25 while the visual search ranks them using a distance-based similarity function (either Euclidean or cosine distance based on user preference). Thereafter, the common keyframes retrieved from both visual-based and text-based search are kept, but their ranking scores are aggregated from both search mechanisms.

Interactive Search. $\mathcal{IVS}$ provides two mechanisms for interactive search. First, a user can help in updating the parsing results of the textual query. In particular, the user can delete or add some keywords. Moreover, the user can participate in emphasizing higher priority for some keywords. The high-priority keywords are combined with the "and" operator in the Boolean query composing the textual query (see Eq. 2). Second, the user can provide feedback for the retrieved results of a query by selecting the most relevant keyframes among the retrieved ones. Thereafter, a visual search is executed to retrieve similar reference keyframes to each of the selected keyframes.

$$
\begin{aligned}
TextQuery = (\{keywords(category : DetectedObjects)\} \ OR \\
\{keywords(category : DetectedObjects) \ AND \ keywords(category : Others)\}) \ AND \\
\{keywords(category : \star \ and \ emphasized)\}
\end{aligned}
\tag{2}
$$

3 Conclusion

This paper presents an interactive video search tool ($\mathcal{IVS}$). The $\mathcal{IVS}$ tool supports both textual and visual queries and aims at providing an efficient and accurate search utilizing various metadata types and hence different indexing techniques. In addition, the $\mathcal{IVS}$ tool enables getting users' feedback either by customizing the query or by relevance feedback to enhance the search results.

---

[8] If a keyframe does not match the detected objects keyword but matches the other query keywords, it is omitted from the retrieved keyframes.

Acknowledgment. This research has been supported in part by the USC Integrated Media Systems Center and unrestricted cash gifts from Oracle. The authors also acknowledge the USC Center for Advanced Research Computing (CARC) for providing computing resources for conducting some of the experiments. Also, thanks to Dr. Aiichiro Nakno for his help in using CARC.

References

1. Alfarrarjeh, A., et al.: Hybrid indexes for spatial-visual search. In: Thematic Workshops of ACM MM, pp. 75–83 (2017)
2. Alfarrarjeh, A., et al.: A data-centric approach for image scene localization. In: Big Data, pp. 594–603. IEEE (2018)
3. Alfarrarjeh, A., et al.: A class of R*-tree indexes for spatial-visual search of geotagged street images. In: ICDE, pp. 1990–1993. IEEE (2020)
4. Berns, F., et al.: V3C1 dataset: an evaluation of content characteristics. In: ICMR, pp. 334–338 (2019)
5. Explosion Inc.: spaCy (2020). https://spacy.io/
6. Johnson, J., et al.: DenseCap: fully convolutional localization networks for dense captioning. In: CVPR, pp. 4565–4574 (2016)
7. Kasutani, E., Yamada, A.: The MPEG-7 color layout descriptor: a compact image feature description for high-speed image/video segment retrieval. In: ICIP, vol. 1, pp. 674–677. IEEE (2001)
8. Kim, S.H., et al.: MediaQ: mobile multimedia management system. In: MMSys, pp. 224–235 (2014)
9. Lowe, D.: Distinctive image features from scale-invariant keypoints. IJCV **60**(2), 91–110 (2004). https://doi.org/10.1023/B:VISI.0000029664.99615.94
10. Nazir, A., et al.: Content based image retrieval system by using HSV color histogram, discrete wavelet transform and edge histogram descriptor. In: iCoMET, pp. 1–6. IEEE (2018)
11. Smith, R.: An overview of the tesseract OCR engine. In: ICDAR, vol. 2, pp. 629–633. IEEE (2007)
12. Tolias, G., et al.: Particular object retrieval with integral max-pooling of CNN activations. In: ICLR (2016)
13. Yee, K.P., et al.: Faceted metadata for image search and browsing. In: HCI, pp. 401–408 (2003)
14. Zoph, B., et al.: Learning transferable architectures for scalable image recognition. In: CVPR, pp. 8697–8710 (2018)

Less is More - diveXplore 5.0 at VBS 2021

Andreas Leibetseder[(⊠)] and Klaus Schoeffmann

Institute of Information Technology (ITEC), Klagenfurt University,
Klagenfurt, Austria
{aleibets,ks}@itec.aau.at

Abstract. As a longstanding participating system in the annual Video Browser Showdown (VBS2017-VBS2020) as well as in two iterations of the more recently established Lifelog Search Challenge (LSC2018-LSC2019), diveXplore is developed as a feature-rich *Deep Interactive Video Exploration* system. After its initial successful employment as a competitive tool at the challenges, its performance, however, declined as new features were introduced increasing its overall complexity. We mainly attribute this to the fact that many additions to the system needed to revolve around the system's core element – an interactive self-organizing browseable featuremap, which, as an integral component did not accommodate the addition of new features well. Therefore, counteracting said performance decline, the VBS 2021 version constitutes a completely rebuilt version 5.0, implemented from scratch with the aim of greatly reducing the system's complexity as well as keeping proven useful features in a modular manner.

Keywords: Video retrieval · Interactive video search · Video analysis

1 Introduction

The annual video retrieval challenge known as the Video Browser Showdown [7, 8,10,12] (VBS) as well as the newly developed Lifelog Search Challenge [3,4] (LSC) based on personal lifelog data are team competitions that involve quick interactive search of large multimodal databases. As such they include multiple tasks, such as Known-Item Search (KIS) along with Ad-hoc Video Search[1] (AVS) as well as two competition modes featuring experts and novices using the various competing systems. These conditions, coupled with a relatively tight per-task time-limit of merely a few minutes, in general imposes two requirements on a participating system: the need to be as sophisticated as possible in order to retrieve meaningful results, while at the same time being as simple to use as possible for accommodating novice users familiar with neither the tasks nor the systems.

The diveXplore system [5,6,9,13,14] continuously participated in the VBS challenge since its 5th iteration (VBS2017-VBS2020) and grew in complexity

---

[1] https://www-nlpir.nist.gov/projects/tv2018/Tasks/ad-hoc/.

© Springer Nature Switzerland AG 2021
J. Lokoč et al. (Eds.): MMM 2021, LNCS 12573, pp. 455–460, 2021.
https://doi.org/10.1007/978-3-030-67835-7_44

along with the size of the competition's gradually growing dataset [11], which currently comprises approximately 1000 hours of video [2]. Initially developed as a video exploration tool utilizing the concept of self-organizing feature maps [1], many new and exploratory features were added to the system for each VBS iteration, which led to an increasing deviation from its original core feature: video exploration through browsing. Consequently, the system became hard to maintain and complex to alter. Therefore, in order to competitively participate in VBS2021, diveXplore 5.0 is rebuilt from scratch focusing on the integration of successful features from the past challenges, while still offering video exploration features at a more modular level, preventing unnecessary feature dependencies. Figure 1 portrays the system's new architecture, technology and features. The remaining sections describe diveXplore in detail as well as highlight novel additions to the system.

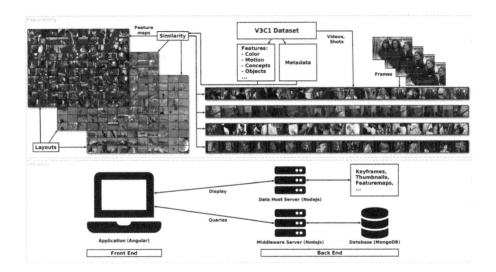

Fig. 1. Architecture and of *diveXplore* 5.0

2   diveXplore 5.0

As a veteran system at the VBS, diveXplore was first developed for VBS2017 [14] incorporating an approach similar to the winning system at VBS2016 [1] for its main interface: an interactive, collaborative and self-organizing featuremap. Users were able to organize this map according to several similarity features, such as deep concepts, color, texture and motion. Additionally, these features could be utilized for similarity search on a keyframe basis with results being shown in separate views. Over the course of the last VBS iterations, these views became more and more important, as new features were introduced, such as a shot-aware video player, storyboards, color filter and textual concept as well as hand-drawn sketch search. All of these components needed to be integrated in accordance with the ever-present featuremap, such that navigating shots or videos in any of the views required re-positioning the currently displayed featuremap or even loading a new one. This overall slowed system performance in addition to being very error-prone. Therefore, the focus of diveXplore 5.0 is to remove these strong dependencies and construct separate views that do not have to be updated during most interactions.

2.1   Architecture

As illustrated in Fig. 1, the system builds on several offline preprocessing steps in order to facilitate smooth interactive retrieval during the competition. First, the V3C1 dataset is split into custom shots as well as uniformly sampled frames using a one-second interval. This makes video retrieval very flexible since query results can be ordered on different levels of granularity. Subsequently, all included analyses can be started, which mostly include feature extraction utilizing hand-crafted as well as deep neural network approaches. The technologies used for creating the systems are web-based. The back end comprises two NodeJs[2] servers, where one is used for hosting keyframes, thumbnails and featuremaps, while the other server acts as a RESTful middleware between a client written in Angular[3] and a MongoDB[4] database.

2.2   Features

As mentioned, diveXplore 5.0 keeps the most useful features from previous VBS iterations, while adjusting them to accommodate aforementioned analyses on shot as well as frame level. Consequently, it still provides a textual concept search that allows for the search of different deep concepts such as objects, attributes, locations as well as metadata. Furthermore, the possibility for similarity search is maintained as well, especially since it proved invaluable for AVS tasks: a user can select a particular shot or frame and can retrieve similar items by choosing a similarity measure such as deep concepts, color, texture or motion features. In addition, the system includes two additional explorative views that borrow from previously integrated components, yet incorporate novel strategies. The following sections describe these new features in more detail – concept-based featuremaps and a video-based similarity filter.

Concept-Based Featuremaps. Figure 1 already indicates that diveXplore 5.0 keeps the concept of featuremaps, albeit in a different form: the large map incorporating the shots of all videos organized by different concepts is removed while maintaining and organizing smaller maps for individual concepts such as objects,

---

[2] https://nodejs.org.
[3] https://angular.io.
[4] https://www.mongodb.com.

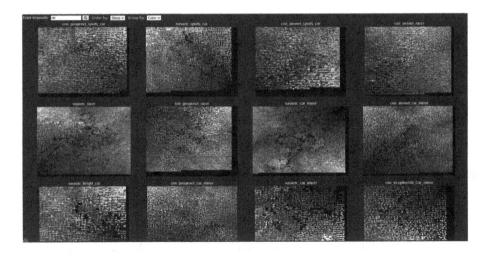

Fig. 2. Concept-based featuremaps.

which is illustrated in Fig. 2. By entering valid concepts, such as 'car' in the example, users can retrieve one or more featuremaps depending on how many incorporated deep nets include this particular concept. These smaller maps can subsequently be organized by utilizing all measures available for similarity search as well as recognition confidence and video affiliation. This feature is available for the multi-map overview shown in Fig. 2 as well as an individual fullscreen view after selecting an individual featuremap.

Video-Based Similarity Filter. The video-based similarity filter, shown in Fig. 3, is a new addition based on the system's previously integrated storyboard feature, which provided a user with small summaries of each individual video. The main difference to the former implementation is the addition of interaction through several combinable filters. The view as well makes use of deep concepts and similarity, but for comparing entire videos or sections of them. As an example, the figure shows a situation where full videos are ordered by their creation period, i.e. between 2008 and 2012. Similarly, a user can as well select one or multiple concepts such as 'person' and 'apple' and group video sections according to either the frequency or classification confidence of these terms. Since this view allows for a multitude of different combinations it will expectedly be subject to many improvements in future system updates.

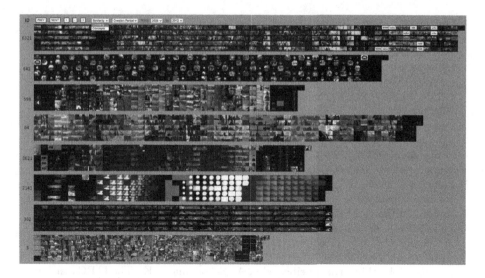

Fig. 3. Video-based similarity filter.

3   Conclusion

We present diveXplore 5.0 – a completely rebuilt version of a long-time competing system at the annual Video Browser Showdown. While keeping some proven useful features, such as concept and similarity search, the system includes two additional views for dataset exploration: concept-based featuremaps enabling the exploration and organization of individual concepts, as well as a video-based similarity filter, which allows for video- or segment-wide similarity search according to various combinable criteria. All features are integrated on a shot- as well as uniformly sampled frame-basis in order to increase the system's flexibility. We expect these compared to diveXplore 4.0 much reduced and more carefully integrated components to shape a competitive system for VBS2021.

Acknowledgments. This work was funded by the FWF Austrian Science Fund under grant P 32010-N38.

References

1. Barthel, K.U., Hezel, N., Mackowiak, R.: Navigating a graph of scenes for exploring large video collections. In: Tian, Q., Sebe, N., Qi, G.-J., Huet, B., Hong, R., Liu, X. (eds.) MMM 2016. LNCS, vol. 9517, pp. 418–423. Springer, Cham (2016). https://doi.org/10.1007/978-3-319-27674-8_43
2. Berns, F., Rossetto, L., Schoeffmann, K., Beecks, C., Awad, G.: V3C1 dataset: an evaluation of content characteristics. In: Proceedings of the 2019 on International Conference on Multimedia Retrieval, pp. 334–338. ACM (2019)
3. Gurrin, C., et al.: Introduction to the third annual lifelog search challenge (LSC'20). In: Proceedings of the 2020 International Conference on Multimedia Retrieval, pp. 584–585 (2020)

4. Gurrin, C., et al.: [Invited papers] comparing approaches to interactive lifelog search at the lifelog search challenge (LSC2018). ITE Trans. Media Technol. Appl. **7**(2), 46–59 (2019)
5. Leibetseder, A., Kletz, S., Schoeffmann, K.: Sketch-based similarity search for collaborative feature maps. In: Schoeffmann, K., et al. (eds.) MMM 2018. LNCS, vol. 10705, pp. 425–430. Springer, Cham (2018). https://doi.org/10.1007/978-3-319-73600-6_45
6. Leibetseder, A., Münzer, B., Primus, J., Kletz, S., Schoeffmann, K.: diveXplore 4.0: the ITEC deep interactive video exploration system at VBS2020. In: Ro, Y.M., et al. (eds.) MMM 2020. LNCS, vol. 11962, pp. 753–759. Springer, Cham (2020). https://doi.org/10.1007/978-3-030-37734-2_65
7. Lokoc, J., Bailer, W., Schoeffmann, K., Muenzer, B., Awad, G.: On influential trends in interactive video retrieval: video browser showdown 2015–2017. IEEE Trans. Multimed. 1 (2018). https://doi.org/10.1109/TMM.2018.2830110
8. Lokoč, J., et al.: Interactive search or sequential browsing? A detailed analysis of the video browser showdown 2018. ACM Trans. Multimed. Comput. Commun. Appl. **15**(1), 29:1–29:18 (2019). https://doi.org/10.1145/3295663
9. Primus, M.J., Münzer, B., Leibetseder, A., Schoeffmann, K.: The ITEC collaborative video search system at the video browser showdown 2018. In: Schoeffmann, K., et al. (eds.) MMM 2018. LNCS, vol. 10705, pp. 438–443. Springer, Cham (2018). https://doi.org/10.1007/978-3-319-73600-6_47
10. Rossetto, L., et al.: Interactive video retrieval in the age of deep learning – detailed evaluation of VBS 2019. IEEE Trans. Multimed. **23**, 243–256 (2021). https://doi.org/10.1109/TMM.2020.2980944
11. Rossetto, L., Schuldt, H., Awad, G., Butt, A.A.: V3C – a research video collection. In: Kompatsiaris, I., Huet, B., Mezaris, V., Gurrin, C., Cheng, W.-H., Vrochidis, S. (eds.) MMM 2019. LNCS, vol. 11295, pp. 349–360. Springer, Cham (2019). https://doi.org/10.1007/978-3-030-05710-7_29
12. Schoeffmann, K.: A user-centric media retrieval competition: the video browser showdown 2012–2014. IEEE MultiMed. **21**(4), 8–13 (2014). https://doi.org/10.1109/MMUL.2014.56
13. Schoeffmann, K., Münzer, B., Leibetseder, A., Primus, J., Kletz, S.: Autopiloting feature maps: the deep interactive video exploration (diveXplore) system at VBS2019. In: Kompatsiaris, I., Huet, B., Mezaris, V., Gurrin, C., Cheng, W.-H., Vrochidis, S. (eds.) MMM 2019. LNCS, vol. 11296, pp. 585–590. Springer, Cham (2019). https://doi.org/10.1007/978-3-030-05716-9_50
14. Schoeffmann, K., et al.: Collaborative feature maps for interactive video search. In: Amsaleg, L., Guðmundsson, G.Þ., Gurrin, C., Jónsson, B.Þ., Satoh, S. (eds.) MMM 2017. LNCS, vol. 10133, pp. 457–462. Springer, Cham (2017). https://doi.org/10.1007/978-3-319-51814-5_41

SOMHunter V2 at Video Browser Showdown 2021

Patrik Veselý$^{(\boxtimes)}$, František Mejzlík, and Jakub Lokoč

Department of Software Engineering, Faculty of Mathematics and Physics,
Charles University, Prague, Czech Republic
vesely-patrik@email.cz, frankmejzlik@gmail.com, lokoc@ksi.mff.cuni.cz

Abstract. This paper presents an enhanced version of an interactive video retrieval tool SOMHunter that won Video Browser Showdown 2020. The presented enhancements focus on improving text querying capabilities since the text search model plays a crucial part in successful searches. Hence, we introduce the ability to specify multiple text queries with further positional specification so users can better describe positional relationships of the objects. Moreover, a possibility to further specify text queries with an example image is introduced as well as consequent changes to the user interface of the tool.

Keywords: Interactive video retrieval · Self-organizing maps · Relevance feedback · Deep learning

1 Introduction

This year, Video Browser Showdown 2020 (VBS) [2,12,13,18,20,21] was, as usual, an exhibition of a variety of different approaches to interactive video retrieval with nine competing tools [1,5,7,9,10,14,17,19,22] from around the world. We introduced SOMHunter [7] there for the first time, employing three easy-to-use concepts – temporal text query search based on W2VV++ model [11] (employing advanced visual features [16]), optional display based on dynamic self-organizing map (SOM) [6] and a feedback mechanism [3] based on "liking" relevant frames. Participating for the first time, the tool managed to win first place solving 15 known-item search tasks (out of 22). Joining the VBS competition is a great opportunity to take a feedback and push systems forward. Shortly after, we started working on new features [15] for Lifelog Search Challange 2020 (LSC) [4]. We also released a light-weight web-oriented version of our tool [8] as an open-source project to help new teams considering participation at VBS or LSC events.

Taking a look at state-of-the-art VBS tools, all have one thing in common—each one of them implements text querying in some form. Additionally, VBS post-competition analysis suggests that textual models play a critical part in finding the target scene in our tool. In total 10 out of 15 tasks that were successfully solved, were submitted directly from a state employing only textual query (mostly the temporal one).

© Springer Nature Switzerland AG 2021
J. Lokoč et al. (Eds.): MMM 2021, LNCS 12573, pp. 461–466, 2021.
https://doi.org/10.1007/978-3-030-67835-7_45

This fact led us to focus on improving the text querying capabilities of SOMHunter. From our experience, a limited ability to describe positional relationships of entities present in frames is an issue. Therefore, we introduce a different approach to formulate a text query where users can provide multiple positioned descriptions arranged on a query canvas. While still being able to use a text query describing the frame as a whole, this brings new possibilities to exploit positions of searched objects (instead of position specification with a free-form text). The price for this feature is a multi-representation of selected frames, currently statically divided to overlapping regions. Since one text query can lead to a lot of somewhat relevant frames, we plan to introduce an intuitive text query vector relocation mechanism supporting construction of temporal queries (see Sect. 2.2).

While temporal and positional text queries based on W2VV++ model [11] are promising mechanisms, in many cases they are just not enough to solve a challenging task. As an example, during the last VBS 2020 competition, 5 out of 15 solved tasks were correctly submitted from the state using also the relevance feedback approach or the nearest neighbours search. With this in mind, we are aware that other search modes should be preserved in the tool.

2 Newly Included Text Querying Options

According to the VBS 2020 post-competition analysis we participated in, the text query model is truly a powerful element of the SOMHunter system. However, the employed bag-of-words (bow) variant of the W2VV++ model has limitations to describe object properties – the order of words does not matter in the model. For example, position relationships between objects in the frame cannot be effectively handled. In consequence, the query feature vector does not have to be situated in the desired part of the feature space, in the desired cluster of frame feature vectors. Hence, we introduce two extensions that try to tackle limitations of the bow variant.

2.1 Localized Text Queries

We introduce additional (optional) text queries describing a simple-shape area of the frame alongside the usual query describing the whole frame. A user simply places queries on the frame-shaped canvas, where each query separately describes the selected area. With positioned text queries, users can rely on exact position specification, and do not have to specify the position for the text search model.

To compute a relevance score for a frame, we consider a limited fixed number of frame subregions for which features are extracted in the pre-processing phase. Each positioned text query is mapped to a matching subregion and the relevance score is computed for corresponding feature vectors. Finally, an aggregated score for the whole frame is computed and used for the final ordering of all frames. Naturally, this approach multiplies the number of features to maintain for each frame, but as of now, this is not a limiting factor for the dataset of V3C1 representative frames (about 1M) and so it is a price we are willing to pay for more precise options to specify text queries.

2.2   Text Query Vector Relocation

A text query feature vector does not have to necessarily occur in the neighbourhood (or center) of searched vectors of relevant dataset frames due to limitations of a text embedding model or improper text query formulation. Therefore, multimedia search systems often provide a k nearest neighbours (kNN) search functionality to relocate the initial query vector in the feature space, towards the images of interest.

In a standard search scenario, this relocation takes part in a sequence of search steps starting with a text query and then followed by kNN browsing actions. However, if a user wants to search by a temporal text query, it might be cumbersome and confusing to use kNN search in the main display to relocate all parts of the temporal text query. To ease the initial query vector relocation, we propose a prompt panel showing a SOM-based grid of frames summarizing the neighbourhood of the text query feature vector. In this map, users can select a more appropriate example image that replaces the text query in the ranking model. The idea is illustrated in Fig. 1.

2.3   User Interface

From the start, the main goal of SOMHunter is to stay as simple as possible to use and with the regular addition of new features, it is getting harder and harder to implement these features into the UI.

Adding the possibility to construct multiple queries with its area specification, while still keeping the tool interface simple, can be a challenging task. Therefore, we take inspiration in a concept that has been already "battle-tested" in the VIRET tool [14]. We use a frame-shaped canvas where users can place

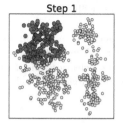 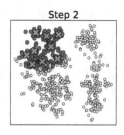 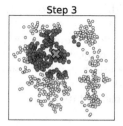

Fig. 1. Relocation of an initial text query vector (green cross). Gray circles illustrate frames vectors in feature vector space, where blue circles denote the top relevant ones to the query. Orange crosses in Step 2 show representative neighbours displayed in a prompt panel, while the effect of query relocation is illustrated in Step 3. (Color figure online)

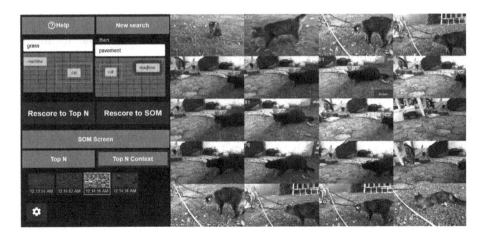

Fig. 2. The new user interface including also the component for constructing the introduced area text queries. This gives a user possibility to use multiple text queries that also capturing the positional relationships relative to the frame.

a text query and specify its area of the effect. With this approach, users can determine frame subareas related to the text queries. Using general region specification gives a convenient layer of abstraction that is independent on the actual implementation under the hood. Hence, it is possible to change the model implementation without changing the UI behaviour in the future (Fig. 2).

3   Conclusion

We presented an updated version of the interactive video retrieval system SOM-Hunter that extends the previous temporal query model with optional queries describing positional relationships of objects in frames. To improve and enrich the initial temporal text query formulation, we introduced an additional SOM-organized panel of frames providing an opportunity to move the text query embedding vector to another part of the joint feature vector space.

Acknowledgments. This paper has been supported by Czech Science Foundation (GAČR) project 19-22071Y and by Charles University grant SVV-260588.

We are grateful to Tomáš Souček for his help with frame selection and feature extraction. Also, we thank Miroslav Kratochvíl for his valuable feedback and help with the implementation.

References

1. Andreadis, S., et al.: VERGE in VBS 2020. In: Ro, Y.M., et al. (eds.) MMM 2020. LNCS, vol. 11962, pp. 778–783. Springer, Cham (2020). https://doi.org/10.1007/978-3-030-37734-2_69

2. Cobârzan, C., et al.: Interactive video search tools: a detailed analysis of the video browser showdown 2015. Multimed. Tools Appl. **76**(4), 5539–5571 (2016). https://doi.org/10.1007/s11042-016-3661-2
3. Cox, I.J., Miller, M.L., Minka, T.P., Papathomas, T.V., Yianilos, P.N.: The Bayesian image retrieval system, pichunter: theory, implementation, and psychophysical experiments. IEEE Trans. Image Process. **9**(1), 20–37 (2000)
4. Gurrin, C., et al.: Comparing approaches to interactive lifelog search at the lifelog search challenge (LSC2018). ITE Trans. Media Technol. Appl. **7**(2), 46–59 (2019)
5. Jónsson, B.Þ., Khan, O.S., Koelma, D.C., Rudinac, S., Worring, M., Zahálka, J.: Exquisitor at the video browser showdown 2020. In: Ro, Y.M., et al. (eds.) MMM 2020. LNCS, vol. 11962, pp. 796–802. Springer, Cham (2020). https://doi.org/10.1007/978-3-030-37734-2_72
6. Kohonen, T.: The self-organizing map. Neurocomputing **21**(1–3), 1–6 (1998)
7. Kratochvíl, M., Veselý, P., Mejzlík, F., Lokoč, J.: SOM-hunter: video browsing with relevance-to-SOM feedback loop. In: Ro, Y.M., et al. (eds.) MMM 2020. LNCS, vol. 11962, pp. 790–795. Springer, Cham (2020). https://doi.org/10.1007/978-3-030-37734-2_71
8. Kratochvíl, M., Mejzlík, F., Veselý, P., Souček, T., Lokoč, J.: SOMHunter: lightweight video search system with SOM-guided relevance feedback. In: Proceedings of the 28th ACM International Conference on Multimedia, MM 2020. ACM (2020, in press)
9. Le, N.-K., Nguyen, D.-H., Tran, M.-T.: An interactive video search platform for multi-modal retrieval with advanced concepts. In: Ro, Y.M., et al. (eds.) MMM 2020. LNCS, vol. 11962, pp. 766–771. Springer, Cham (2020). https://doi.org/10.1007/978-3-030-37734-2_67
10. Leibetseder, A., Münzer, B., Primus, J., Kletz, S., Schoeffmann, K.: diveXplore 4.0: the ITEC deep interactive video exploration system at VBS2020. In: Ro, Y.M., et al. (eds.) MMM 2020. LNCS, vol. 11962, pp. 753–759. Springer, Cham (2020). https://doi.org/10.1007/978-3-030-37734-2_65
11. Li, X., Xu, C., Yang, G., Chen, Z., Dong, J.: W2VV++ fully deep learning for ad-hoc video search. In: Proceedings of the 27th ACM International Conference on Multimedia, pp. 1786–1794 (2019)
12. Lokoč, J., Bailer, W., Schoeffmann, K., Münzer, B., Awad, G.: On influential trends in interactive video retrieval: video browser showdown 2015–2017. IEEE Trans. Multimed. **20**(12), 3361–3376 (2018)
13. Lokoč, J., et al.: Interactive search or sequential browsing? A detailed analysis of the video browser showdown 2018. ACM Trans. Multimed. Comput. Commun. Appl. (TOMM) **15**(1), 1–18 (2019)
14. Lokoč, J., Kovalčík, G., Souček, T.: VIRET at video browser showdown 2020. In: Ro, Y.M., et al. (eds.) MMM 2020. LNCS, vol. 11962, pp. 784–789. Springer, Cham (2020). https://doi.org/10.1007/978-3-030-37734-2_70
15. Mejzlík, F., Veselỳ, P., Kratochvíl, M., Souček, T., Lokoč, J.: Somhunter for lifelog search. In: Proceedings of the Third Annual Workshop on Lifelog Search Challenge, pp. 73–75 (2020)
16. Mettes, P., Koelma, D.C., Snoek, C.G.M.: Shuffled imagenet banks for video event detection and search. ACM Trans. Multimed. Comput. Commun. Appl. (TOMM) **16**(2), 1–21 (2020)
17. Nguyen, P.A., Wu, J., Ngo, C.-W., Francis, D., Huet, B.: VIREO @ video browser showdown 2020. In: Ro, Y.M., et al. (eds.) MMM 2020. LNCS, vol. 11962, pp. 772–777. Springer, Cham (2020). https://doi.org/10.1007/978-3-030-37734-2_68

18. Rossetto, L., et al.: Interactive video retrieval in the age of deep learning - detailed evaluation of VBS 2019. IEEE Trans. Multimed. (2020)
19. Sauter, L., Amiri Parian, M., Gasser, R., Heller, S., Rossetto, L., Schuldt, H.: Combining boolean and multimedia retrieval in vitrivr for large-scale video search. In: Ro, Y.M., et al. (eds.) MMM 2020. LNCS, vol. 11962, pp. 760–765. Springer, Cham (2020). https://doi.org/10.1007/978-3-030-37734-2_66
20. Schoeffmann, K.: A user-centric media retrieval competition: the video browser showdown 2012–2014. IEEE Multimed. **21**(4), 8–13 (2014)
21. Schoeffmann, K.: Video browser showdown 2012–2019: a review. In: 2019 International Conference on Content-Based Multimedia Indexing (CBMI), pp. 1–4. IEEE (2019)
22. Park, S., Song, J., Park, M., Ro, Y.M.: IVIST: interactive video search tool in VBS 2020. In: Ro, Y.M., et al. (eds.) MMM 2020. LNCS, vol. 11962, pp. 809–814. Springer, Cham (2020). https://doi.org/10.1007/978-3-030-37734-2_74

W2VV++ BERT Model at VBS 2021

Ladislav Peška$^{(\boxtimes)}$, Gregor Kovalčík, Tomáš Souček, Vít Škrhák,
and Jakub Lokoč

SIRET Research Group,
Department of Software Engineering Faculty of Mathematics and Physics,
Charles University, Prague, Czech Republic
{peska,lokoc}@ksi.mff.cuni.cz, gregor.kovalcik@gmail.com,
tomas.soucek1@gmail.com, vitek.skrhak@seznam.cz

Abstract. The W2VV++ model BoW variant integrated to VIRET and SOMHunter systems has proven its effectiveness in the previous Video Browser Showdown competition in 2020. As a next experimental interactive search prototype to benchmark, we consider a simple system relying on the more complex BERT variant of the W2VV++ model, accepting a rich text input. The input can be provided by keyboard or by speech processed by a third-party cloud service. The motivation for the more complex BERT variant is its good performance for rich text descriptions that can be provided for known-item search tasks. At the same time, users will be instructed to specify as rich text description about the searched scene as possible.

Keywords: Known-item search · Deep learning · Interactive video retrieval

1 Introduction

The Video Browser Showdown (VBS) [3,9,10] is a respected international initiative to benchmark interactive search systems with a set of challenging tasks over a shared dataset (currently the V3C1 dataset [15]). The last year, a performance boost was achieved with the W2VV++ model [7] (using [13]) that maps video frames and text descriptions to a joint feature space. In connection with temporal query fusion in systems like VIRET [11,12] or SOMHunter [6], the text search model was a very important factor behind many successfully solved tasks. Therefore, we plan to investigate a state-of-the-art free-form text search model for representative video frames and a variant of a temporal fusion, combined with a recommended search strategy where users provide a much longer text description of searched scenes.

Let us note that temporal models were also applied earlier on a sequence of sketches drawn by a user [2,17]. Recently at VBS 2020, vitrivr [16] and VIREO [14] tools also added a temporal search mode.

In order to enter the text input, a keyboard can be used, which is a preferred option by users with good keyboard writing skills. Another option is to use a voice

J. Lokoč et al. (Eds.): MMM 2021, LNCS 12573, pp. 467–472, 2021.
https://doi.org/10.1007/978-3-030-67835-7_46

input, which was recently introduced for the Lifelog Search Challenge 2020 by the Voxento system [1] relying on a Google Cloud Service. Voxento supports spoken queries as well as several simple commands like "stop recording" or "submit". In our prototype, we do not focus on commands as these actions can be issued by a simple interaction with a mouse click or a dedicated keyboard key. Users can just click a button and record a speech-based description. After its processing, the text can be still corrected manually by keyboard.

In order to rank database with respect to a longer text query for known-item search tasks, the key system component is the W2VV++ BERT variant as suggested in the recent W2VV++ case study [8]. According to the reported results with a smaller benchmark dataset focusing on known-item search tasks over a set of selected frames, the BERT variant outperformed the BoW version by a large margin. Hence, we believe that this more complex video-text joint embedding model could be suitable in a combination with rich descriptions that can be provided for observed scenes.

2   System Overview

The main feature of the tested prototype is searching by an existing powerful video-text search model. Another important feature is the generalization of 2-point temporal queries into n-point context-aware queries, which should be more suitable for continuous description of the scene provided by users. Overall, the system architecture is actually quite simple and consists of four components.

- User interface for text query input and result set browsing. This component can be created by making use of a few classes from the VIRET framework. Specifically, we re-use just the dataset of selected frames and the presentation grid. In addition, a text box for typing and editing sub-queries and control buttons (e.g., for speech recording) are included in the querying panel. As for the results visualization, the primary option would be to display one result (i.e., several key frames of the returned scene) per-line. Nonetheless, inspired by the performance of the vitrivr tool [16], we also plan to explore additional results aggregation, visualization, and browsing strategies.
- W2VV++ BERT model component [8], implemented in Python can run as a server application providing API for a text-to-vector transformation and subsequently a text-to-frame rating.
- Context-aware ranking model that provides the final ranking of objects based on the results of individual sub-queries.
- Third party services for optional sub-tasks. For example, if users rely on the speech input, an external service can be called to transform a recorded audio signal to text.

3   Context-Aware Query Ranker

One novel feature of the proposed tool is the context-aware ranker. This component is responsible for the final ranking of the results based on the per-frame

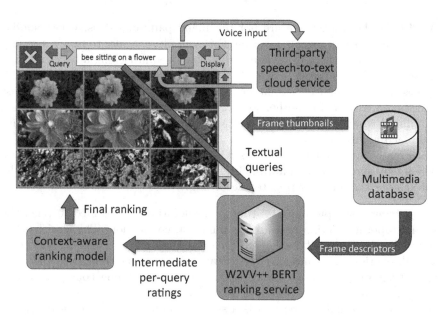

Fig. 1. System overview.

ratings of individual sub-queries provided by the W2VV++ BERT component. In the previous years, temporal queries have proven to be an important querying mechanism in both SOMHunter and VIRET tools [6,11]. Nonetheless, only 2-point temporal queries were utilized in VIRET and SOMHunter performed at most 3-point temporal query with a greedy search to aggregate individual sub-queries. Furthermore, the evaluation was performed on individual frames separately in both cases. The proposal of context-aware ranker is an alternative to the temporal queries. It relaxes the temporal ordering to mere temporal co-location context (i.e., all sub-queries must be reasonably answered in a certain temporal context window). The proposal of context-aware ranker is based on the following assumptions (Fig. 1):

- A1: Users can provide much longer or more detailed textual description of the remembered scene, especially if they are instructed to do so.
- A2: The overall description may span across the whole scene and no single frame can contain all information described in the text. Nonetheless, we should still be able to separate segments of the text (i.e., sub-queries) that correspond to a single frame. This can be done either explicitly by the user or implicitly e.g., via detecting sentence separators.
- A3: While providing a stream of text, it may be an additional burden for users to order the descriptions according to some timeline (i.e., to explicitly define what happened first, what next etc.). Also, several sub-queries may correspond to the same points on the timeline. However, we may be certain that all sub-queries correspond to a relatively short time-window of the searched scene.

- A4: In the datasets, there are not too many pairs of videos, where similar events happens in a reverse ordering. For example, consider to have a segment of a video, where there is a scene with a tractor approaching a field and then an aerial view of a forest. It is rather unlikely that there are too many other videos, where similar events happen in a different ordering, i.e., first a forest and then a tractor. In another words, we assume that if a temporal ordering is reduced to a mere temporal co-location of events, we would not introduce too much additional noise.[1]
- A5: Finally, we assume that users will often describe longer-lasting events, which may spread across multiple frames. Therefore, per-frame results to the sub-queries should be aggregated accordingly.

By putting assumptions A1–A4 into consideration, we decided to relax the original temporal ordering into a mere temporal co-location. This decision has a potential of both simplifying the user interface, allowing users to submit additional sub-queries easily and also considerably decrease the computational complexity of the query. Context-aware ranking utilizes a sliding context window, within which, we aggregate results of all sub-queries.

To be more formal, suppose $\mathcal{Q}$ is a set of sub-queries, $\mathcal{F}$ is a list of frames and $r_{i,j} \in (0,1)$ is the relevance of frame $f_i \in \mathcal{F}$ to the sub-query $q_j \in \mathcal{Q}$. In the current implementation, $r_{i,j}$ is provided by the W2VV++ BERT component, but any other frame-level relevance estimator can be used. Furthermore, suppose w is the size of the context window and $c_k \in \mathcal{C}$ is a list of consecutive frames from f_i to f_{i+w}, i.e., candidates for the searched scene. Candidates are derived from the list of frames iteratively by applying a certain step size.[2] Now, for any candidate list c_k, its rating is calculated as follows:

$$r_k = \alpha_1\left(\left[\alpha_2\left([r_{i,j}; f_i \in c_k]\right); q_j \in \mathcal{Q}\right]\right), \tag{1}$$

where α_2 and α_1 are a frame-wise aggregation of some sub-query and a query-wise aggregation respectively. Some plain aggregations such as *mean* or *min* for α_1 and *max* for α_2 can be utilized. Nonetheless, considering assumption A5, we also plan to evaluate some additional aggregations. Specifically, assuming that $r_{i,j}$ can be interpreted as a probability that the frame f_i answers the query q_j, α_2 can be defined as:

$$\alpha_{2,prob} = 1 - \prod_{\forall i \in c_k} (1 - r_{i,j}) \tag{2}$$

Alternatively, if $r_{i,j}$ is interpreted as a level of fulfillment, different variants of fuzzy logic T-conorms [5] may be utilized to aggregate the per-frame results.

In order to speed-up the α_2 aggregations across the list of candidates, we plan to utilize some sliding-window aggregation algorithms, e.g., [4].

---

[1] Note that we are aware that reverse events could be quite common within a single video as similar cuts may repeat frequently. Nonetheless, this would still reduce the task at hand to merely finding the right scene within a video.

[2] Note that padding is employed on the edges of individual videos.

Several variants are also plausible for α_1 aggregation. Here the main motivation is that some sub-queries may be incorrect (e.g., due to some incompatibilities in human vs. computer perception of the scene). In such cases, a few bad-performing sub-queries should not completely disqualify the candidate. On the other hand, candidates should provide reasonable results for as many queries as possible. Therefore, we plan to experiment with several metrics ranging from product to sum aggregations.

4   Conclusion

We present a simplified variant of the VIRET interactive video retrieval tool that tests a state-of-the-art text-video joint embedding model. The updated version should let users to specify a rich text query comprising different parts of the searched scene in an arbitrary order. The browsing experience is also improved with a history of searches.

Acknowledgements. This paper has been supported by the Charles University Grant Agency (GA UK) project number 1310920, by Czech Science Foundation (GAČR) project 19-22071Y and by Charles University grant SVV-260588.

References

1. Alateeq, A., Roantree, M., Gurrin, C.: Voxento: a prototype voice-controlled interactive search engine for lifelogs. In: Proceedings of the Third Annual Workshop on Lifelog Search Challenge, LSC 2020, pp. 77–81. ACM, New York (2020)
2. Blažek, A., Lokoč, J., Skopal, T.: Video retrieval with feature signature sketches. In: Traina, A.J.M., Traina, C., Cordeiro, R.L.F. (eds.) SISAP 2014. LNCS, vol. 8821, pp. 25–36. Springer, Cham (2014). https://doi.org/10.1007/978-3-319-11988-5_3
3. Cobârzan, C., et al.: Interactive video search tools: a detailed analysis of the video browser showdown 2015. Multimed. Tools Appl. **76**(4), 5539–5571 (2016). https://doi.org/10.1007/s11042-016-3661-2
4. Hirzel, M., Schneider, S., Tangwongsan, K.: Sliding-window aggregation algorithms: tutorial. In: Proceedings of the 11th ACM International Conference on Distributed and Event-based Systems, pp. 11–14. ACM (2017)
5. Klement, E.P., Mesiar, R., Pap, E.: Families of t-norms. In: Klement, E.P., Mesiar, R., Pap, E. (eds.) Triangular Norms, vol. 8, pp. 101–119. Springer, Dordrecht (2000). https://doi.org/10.1007/978-94-015-9540-7_4
6. Kratochvíl, M., Veselý, P., Mejzlík, F., Lokoč, J.: SOM-hunter: video browsing with relevance-to-SOM feedback loop. In: Ro, Y.M., et al. (eds.) MMM 2020. LNCS, vol. 11962, pp. 790–795. Springer, Cham (2020). https://doi.org/10.1007/978-3-030-37734-2_71
7. Li, X., Xu, C., Yang, G., Chen, Z., Dong, J.: W2VV++: fully deep learning for ad-hoc video search. In: Proceedings of the 27th ACM International Conference on Multimedia, MM 2019, Nice, France, 21–25 October 2019, pp. 1786–1794 (2019)
8. Lokoč, J., et al.: A W2VV++ case study with automated and interactive text-to-video retrieval. In: Proceedings of the 28th ACM International Conference on Multimedia, MM 2020. ACM, New York (2020)

9. Lokoč, J., Bailer, W., Schoeffmann, K., Münzer, B., Awad, G.: On influential trends in interactive video retrieval: video browser showdown 2015–2017. IEEE Trans. Multimed. **20**(12), 3361–3376 (2018)

10. Lokoč, J., et al.: Interactive search or sequential browsing? A detailed analysis of the video browser showdown 2018. ACM Trans. Multimed. Comput. Commun. Appl. **15**(1), 29:1–29:18 (2019)

11. Lokoč, J., Kovalčík, G., Souček, T., Moravec, J., Čech, P.: A framework for effective known-item search in video. In: Proceedings of the 27th ACM International Conference on Multimedia, MM 2019, pp. 1777–1785. ACM, New York (2019)

12. Lokoč, J., Kovalčík, G., Souček, T., Moravec, J., Čech, P.: VIRET: a video retrieval tool for interactive known-item search. In: Proceedings of the 2019 on International Conference on Multimedia Retrieval, ICMR 2019, pp. 177–181. ACM, New York (2019)

13. Mettes, P., Koelma, D.C., Snoek, C.G.M.: Shuffled imagenet banks for video event detection and search. ACM Trans. Multimed. Comput. Commun. Appl. (TOMM) **16**(2), 1–21 (2020)

14. Nguyen, P.A., Wu, J., Ngo, C.-W., Francis, D., Huet, B.: VIREO @ video browser showdown 2020. In: Ro, Y.M., et al. (eds.) MMM 2020. LNCS, vol. 11962, pp. 772–777. Springer, Cham (2020). https://doi.org/10.1007/978-3-030-37734-2_68

15. Rossetto, L., Schuldt, H., Awad, G., Butt, A.A.: V3C – a research video collection. In: Kompatsiaris, I., Huet, B., Mezaris, V., Gurrin, C., Cheng, W.-H., Vrochidis, S. (eds.) MMM 2019. LNCS, vol. 11295, pp. 349–360. Springer, Cham (2019). https://doi.org/10.1007/978-3-030-05710-7_29

16. Sauter, L., Amiri Parian, M., Gasser, R., Heller, S., Rossetto, L., Schuldt, H.: Combining boolean and multimedia retrieval in vitrivr for large-scale video search. In: Ro, Y.M., et al. (eds.) MMM 2020. LNCS, vol. 11962, pp. 760–765. Springer, Cham (2020). https://doi.org/10.1007/978-3-030-37734-2_66

17. Yuan, J., et al.: Video browser showdown by NUS. In: Schoeffmann, K., Merialdo, B., Hauptmann, A.G., Ngo, C.-W., Andreopoulos, Y., Breiteneder, C. (eds.) MMM 2012. LNCS, vol. 7131, pp. 642–645. Springer, Heidelberg (2012). https://doi.org/10.1007/978-3-642-27355-1_64

VISIONE at Video Browser Showdown 2021

Giuseppe Amato, Paolo Bolettieri, Fabrizio Falchi, Claudio Gennaro, Nicola Messina, Lucia Vadicamo(✉), and Claudio Vairo

ISTI-CNR, Via G. Moruzzi 1, 56124 Pisa, Italy
{giuseppe.amato,paolo.bolettieri,fabrizio.falchi,claudio.gennaro,
nicola.messina,lucia.vadicamo,claudio.vairo}@isti.cnr.it

Abstract. This paper presents the second release of VISIONE, a tool for effective video search on large-scale collections. It allows users to search for videos using textual descriptions, keywords, occurrence of objects and their spatial relationships, occurrence of colors and their spatial relationships, and image similarity. One of the main features of our system is that it employs specially designed textual encodings for indexing and searching video content using the mature and scalable Apache Lucene full-text search engine.

Keywords: Content-based video retrieval · Video search · Information search and retrieval · Surrogate text representation

1 Introduction

In the last decade, we have witnessed an exponential growth of multimedia content, mainly due to the pervasive use of cameras and social media. However, as visual data (e.g. video and images) are usually poorly annotated or not annotated at all, the use of scalable content-based retrieval systems and techniques for automatic visual analysis have become crucial to managing large visual archives. In this paper, we present a content-based video retrieval system, called VISIONE, which leverages various artificial intelligence techniques for automatic analysis of video keyframes in synergy with specially designed textual encoding of the visual content that facilitates the use of mature and scalable full-text search technologies for indexing and searching large-scale video collections.

A first release of VISIONE [1,6], which participated in the 2019 edition of the Video Browser Showdown (VBS) [11], is described in details in [2]. VBS is an international video search competition that is held annually since 2012 [13]. The V3C1 dataset [5], consisting of 7,475 videos, has been used in the competition since 2019. So far, three types of search tasks are considered in the competition: *Known-Item-Search (KIS)*, *textual KIS* and *Ad-hoc Video Search (AVS)*. The KIS task simulates the situation in which a user wants to find a particular video clip that he/she has watched before performing the search. The textual KIS concerns the case in which the user wants to find a particular video clip that

J. Lokoč et al. (Eds.): MMM 2021, LNCS 12573, pp. 473–478, 2021.
https://doi.org/10.1007/978-3-030-67835-7_47

he/she has never seen but of which a detailed textual description is provided. For the AVS task, instead, a general textual description is provided to the user who is asked to find as many video shots as possible that fit the given description.

One of the main limitations of the first version of VISIONE was the poor performance on textual KIS tasks. In facts, in our first participation in the VBS competition, the search in VISIONE was based only on object detection, colors, scene tags and visual similarity, and this proved to be not good enough to resolve textual KIS tasks in a reasonable time. To overcome this limitation, in this new release of our system, we integrated a retrieval module that allows searching for a target scene using natural language queries. Moreover, inspired by several systems that participated in previous editions of VBS, like [8,10,12], we introduced the possibility of performing a temporal search, where the user can describe two consecutive (or temporally close) keyframes of the same target video. Finally, several improvements have been made to the interface and in the selection of the best scoring functions used for ranking the results. All these novel aspects of our system are described in Sect. 3. The next section, instead, provides an overview of VISIONE and its functionalities.

2   VISIONE Video Search System

VISIONE provides several search functionalities in order to allows a user to search for a video by formulating textual or visual queries describing the content of a scene of a target video. In particular, it supports:

- *query by scene description*: the user can provide a textual description in natural language (e.g. "A tennis player serving a ball on the court");
- *query by keywords*: the user can specify keywords related to the target scene (e.g. "tennis, indoor, athlete, action");
- *query by object location*: the user can draw on a canvas simple diagrams to specify the objects that appear in a target scene and their spatial locations;
- *query by color location*: the user can specify some colors present in a target scene and their spatial locations;
- *query by visual example*: an image can be used as a query to retrieve video scenes that are visually similar to it. The image can be selected in the browsing interface as one of the results of a previous search iteration, or uploaded from URL/local file system.

Moreover, some filters are available to specify the aspect ratio of the target scene and if it is in color or in b/w. Figure 1 shows a screenshot of the search interface.

To support the above mentioned search functionalities, VISIONE exploits content analysis and artificial intelligence techniques to understand and represent the visual content of the video keyframes, including (i) a Transformer Encoder Reasoning Network [9] to extract relation-aware textual and visual features that enable our system to search images using textual descriptions; (ii) an image annotation engine [4] to extract scene tags; (iii) state-of-the-art object

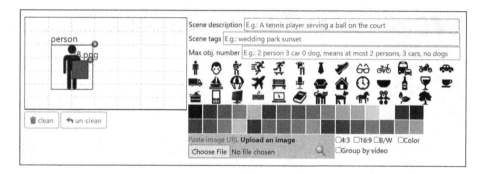

Fig. 1. Search interface. (Color figure online)

detectors to identify and localize objects in the video keyframes; (iv) spatial colors histograms to identify dominant colors and their locations: (v) the R-MAC [14] deep visual descriptors to support the similarity search functionality.

One of the main peculiarity of our system is that all the different descriptors extracted from the video keyframes (features, scene tags, colors/object classes and locations) as well as the queries formulated by the user through the search interface (e.g., keywords describing the target scene and/or diagrams depicting objects and colors locations) are encoded using specifically-designed textual representations (see [2] for the details). This choice allows us to exploit mature and scalable full-text search technologies for indexing and searching large-scale video database without the need to implement dedicated access methods. In particular, VISIONE relies on the Apache Lucene full-text search engine.

3   New VISIONE Functionalities for VBS 2021

This section provides an overview of the improvements performed to the system compared to the first release of VISIONE that participated in VBS 2019.

Query by Textual Description. To address the limitations of the previous version of VISIONE during the textual KIS, in this improved version we added an ad-hoc subsystem for searching keyframes using textual descriptions. Textual descriptions are full natural language sentences, usually between 5 to 50 words in length, describing a visual scene. For example, a valid textual description could be "*A tightly packed living room with a tv screen larger than the fireplace right beside it*". These textual descriptions can include objects details, expressed using their physical or semantic attributes, and they can specify the spatial or abstract relationships linking objects together.

This visual search using natural language descriptions as a query is achieved by using a recently developed deep neural network architecture, called Transformer Encoder Reasoning Network (TERN) [9], which is able to match images

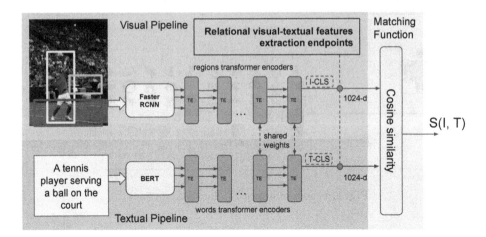

Fig. 2. Overview of the TERN architecture. Orange boxes are Transformer Encoder (TE) layers. Final TE layers share their weights for better stability during the training phase. (Color figure online)

and sentences in a highly-semantic common space. The core of the architecture is constituted of recently introduced deep relational modules called *transformer encoders* [15], which can spot out hidden intra-object relationships. In particular, in the visual pipeline, a stack of transformer encoders try to find links between image regions pre-extracted using a state-of-the-art object detector (Faster-RCNN); in the textual pipeline, using a pretrained BERT model plus another stack of transformer encoder layers, the model searches for relationships between sentence words. An overview of the architecture is shown in Fig. 2.

The extracted cross-modal features are normalized and in principle very similar to visual descriptors like RMAC [14]. Hence we indexed them using the same textual encoding that we already exploited to index the RMAC descriptors (see [2,3]).

Temporal Query. To support temporal queries, in the new version of our system, we have added a second canvas and associated input text boxes to the user interface, that allows users to simultaneously search for two keyframes that are temporally close in a video segment but that are different in the represented content. The search is executed by performing two queries to the index, each providing its own output results. The resulting keyframes, which belong to the same video and whose temporal distance is less than a given threshold δ, are then combined as pairs and shown in the result section of the interface. We use $\delta = 20$ s, however we plan to integrate the possibility to specify a different temporal threshold in the user interface. In this way, the user can exploit temporal relation between video keyframes when searching for a target video.

Improvements in the Searching Implementation. The search process in VISIONE relies on five search operations which implement the five search functionalities presented in Sect. 2. The results of these search operations are combined and ranked according to some text scoring functions (see [2] for more details). In the first implementation of VISIONE, we selected the text scoring function to be used for each search operation by performing some very preliminary tests and by (subjectively) estimating the performance of the system. Recently, in [2], we performed a more in-depth and objective analysis to select the best rankers combination for our system. In particular, we tested 64 different implementations of our system using all the queries output collected during the participation at the VBS2019 challenge in order to select the configuration that has the best performance in terms of effectiveness (i.e. how good is the system in returning at least one relevant result in the top positions of the result set). We used this newly established configuration in the new release of the system.

User Interface. Some improvements have also been made to the VISIONE user interface. We have integrated the possibility to search by similarity also using images uploaded from a URL or file system, as previously only images from the indexed data collection were allowed to be used as query examples. In addition, to boost the efficiency of our system during AVS tasks, we have added the possibility of selecting multiple images to be submitted as a response while automatically removing from the browsing interface all images that have already been submitted during the running AVS session.

4   Conclusion and Future Work

In this paper, we presented the second version of the VISIONE system, focusing on the new functionalities that we integrated in our system to better handle both KIS and AVS tasks. However, we plan to further improve our system in several ways, including exploiting video-text matching approaches (now the system uses only image-text matching), different color analysis techniques, more advanced techniques for organizing search results, and the use of textual speech and OCR annotations that are already provided by the VBS community. Moreover, we would like to integrate collaborative browsing and search functionalities. Finally, we are investigating the possibility of improving the bounding-box search tool by realizing a more precise match between user-defined rectangles during query and image bounding-boxes. The idea is to define a similarity function between two images based on the aggregation of the degree of overlap between the image bounding-boxes. Since there are several ways in which the bounding-boxes can be matched, the computation of this similarity defines an assignment problem, which can be solved in theory with the well-known Hungarian algorithm [7]. Given the complexity of the algorithm, this solution will presumably only be used to reorder the result-set of a query.

Acknowledgements. This work was partially funded by: AI4Media - A European Excellence Centre for Media, Society and Democracy (EC, H2020 n. 951911; AI4EU project (EC, H2020, n. 825619); AI4ChSites, CNR4C program (Tuscany POR FSE 2014–2020 CUP B15J19001040004).

References

1. Amato, G., et al.: VISIONE at VBS2019. In: Kompatsiaris, I., Huet, B., Mezaris, V., Gurrin, C., Cheng, W.-H., Vrochidis, S. (eds.) MMM 2019. LNCS, vol. 11296, pp. 591–596. Springer, Cham (2019). https://doi.org/10.1007/978-3-030-05716-9_51
2. Amato, G., et al.: The visione video search system: exploiting off-the-shelf text search engines for large-scale video retrieval. arXiv preprint arXiv:2008.02749 (2020)
3. Amato, G., Carrara, F., Falchi, F., Gennaro, C., Vadicamo, L.: Large-scale instance-level image retrieval. Inf. Process. Manage. **57**, 102100 (2019)
4. Amato, G., Falchi, F., Gennaro, C., Rabitti, F.: Searching and annotating 100M images with YFCC100M-HNfc6 and MI-file. In: Proceedings of the 15th International Workshop on Content-Based Multimedia Indexing, CBMI 2017, pp. 26:1–26:4. ACM (2017)
5. Berns, F., Rossetto, L., Schoeffmann, K., Beecks, C., Awad, G.: V3C1 dataset: an evaluation of content characteristics. In: Proceedings of the 2019 on International Conference on Multimedia Retrieval, ICMR 2019, pp. 334–338. Association for Computing Machinery (2019)
6. Bolettieri, P., et al.: An image retrieval system for video. In: Amato, G., Gennaro, C., Oria, V., Radovanović, M. (eds.) SISAP 2019. LNCS, vol. 11807, pp. 332–339. Springer, Cham (2019). https://doi.org/10.1007/978-3-030-32047-8_29
7. Kuhn, H.W.: The hungarian method for the assignment problem. Naval Res. Logist. Q. **2**(1–2), 83–97 (1955)
8. Lokoč, J., Kovalčík, G., Souček, T.: VIRET at video browser showdown 2020. In: Ro, Y.M., et al. (eds.) MMM 2020. LNCS, vol. 11962, pp. 784–789. Springer, Cham (2020). https://doi.org/10.1007/978-3-030-37734-2_70
9. Messina, N., Falchi, F., Esuli, A., Amato, G.: Transformer reasoning network for image-text matching and retrieval. In: International Conference on Pattern Recognition (ICPR) 2020 (2020, accepted)
10. Nguyen, P.A., Wu, J., Ngo, C.-W., Francis, D., Huet, B.: VIREO @ video browser showdown 2020. In: Ro, Y.M., et al. (eds.) MMM 2020. LNCS, vol. 11962, pp. 772–777. Springer, Cham (2020). https://doi.org/10.1007/978-3-030-37734-2_68
11. Rossetto, L., et al.: Interactive video retrieval in the age of deep learning - detailed evaluation of VBS 2019. IEEE Trans. Multimed. **23**, 243–256 (2021). https://doi.org/10.1109/TMM.2020.2980944
12. Sauter, L., Amiri Parian, M., Gasser, R., Heller, S., Rossetto, L., Schuldt, H.: Combining boolean and multimedia retrieval in vitrivr for large-scale video search. In: Ro, Y.M., et al. (eds.) MMM 2020. LNCS, vol. 11962, pp. 760–765. Springer, Cham (2020). https://doi.org/10.1007/978-3-030-37734-2_66
13. Schoeffmann, K.: Video browser showdown 2012–2019: a review. In: 2019 International Conference on Content-Based Multimedia Indexing (CBMI), pp. 1–4 (2019)
14. Tolias, G., Sicre, R., Jégou, H.: Particular object retrieval with integral max-pooling of CNN activations. arXiv preprint arXiv:1511.05879 (2015)
15. Vaswani, A., et al.: Attention is all you need. In: Advances in Neural Information Processing Systems, pp. 5998–6008 (2017)

IVOS - The ITEC Interactive Video Object Search System at VBS2021

Anja Ressmann$^{(\boxtimes)}$ and Klaus Schoeffmann

Institute of Information Technology (ITEC), Klagenfurt University,
Klagenfurt, Austria
anjare@edu.aau.at, ks@itec.aau.at

Abstract. We present IVOS, an interactive video content search system that allows for object-based search and filtering in video archives. The main idea behind is to use the result of recent object detection models to index all keyframes with a manageable set of object classes, and allow the user to filter by different characteristics, such as object name, object location, relative object size, object color, and combinations for different object classes – e.g., *"large person in white on the left, with a red tie"*. In addition to that, IVOS can also find segments with a specific number of objects of a particular class (e.g., *"many apples"* or *"two people"*) and supports similarity search, based on similar object occurrences.

Keywords: Content-based video retrieval · Interactive video search · Object detection · Deep learning

1 Introduction

The Video Browser Showdown (VBS) [5,13] is a live annual video content search competition, where international teams compete against each other in an equal setting (same environment/room, same dataset, same tasks, etc.). Many different approaches have been proposed over the years since the start of the VBS in 2012 [9], and it has become a solid platform for evolutionary evaluation of multimedia retrieval in the multimedia community [6,7].

In its 10th edition, VBS2021 challenges users of video content search systems to solve Known-Item Search (KIS) and Ad-hoc Video Search (AVS) tasks for the V3C1 [8] dataset within a particular time limit of typically five minutes per task (this is often decided during the competition, depending on the performance of the teams). The tasks are issued with different modalities – either visually or just by a textual description – and are separated over different sessions, where experts (i.e., the system developers) and novices (volunteers from the audience) need to solve these tasks. This requires not only an effective video content search system, but also an interface that is easy to use. Since the VBS is performed annually, search systems are typically evolutionary improved over the years, integrating experiences and lessons learned from previous competitions.

© Springer Nature Switzerland AG 2021
J. Lokoč et al. (Eds.): MMM 2021, LNCS 12573, pp. 479–483, 2021.
https://doi.org/10.1007/978-3-030-67835-7_48

For VBS2021 we introduce *IVOS*, the *ITEC Interactive Video Object Search system*, which allows to search and explore a video archive based on objects detected by state-of-the-art object detection models. While clearly inspired by the VISIONE system [1] at VBS2019, IVOS is designed according to our previous experience at several VBS competitions, where we used a rather complex search system (diveXplore 1.0–4.0) [3,10–12] that often overwhelmed users with too many available search features and too much presented information – this is true for novices, but also for experts, unfortunately. One particular challenging component was the text-based semantic search for concepts, which were detected in full frames by a CNN model trained for all 21,841 ImageNet concept classes. Here, users were often confused by the many available concepts and the rather low detection confidence. Thus, when challenged with specific tasks, users did not really know what they should actually search for – which was attributed to both the huge number of available concepts, and their specific English names, some of which were often not clear to some users due to the language barrier. However, also choosing the right query type among the many provided ones (map browsing, query-by-concept with different models, query-by-example, query-by-similarity, and query-by-sketch), and effectively switching between them, turned out to be rather puzzling for some users. These problems became especially obvious under the strong time pressure at the VBS (which is reinforced when other teams sit nearby and start to submit results). Therefore, IVOS follows a completely different strategy and provides only object-based exploratory search for a small and manageable set of 80 object classes, which can be selected by small icons, grouped by categories.

2   Object-Based Exploratory Search

The IVOS system uses the recently proposed YOLOv4 object detection network [2], trained for the 80 object classes of the MS COCO dataset [4], to detect objects in keyframes of the master-shot reference of V3C [8], which contains about one million keyframes. Each detected object is saved with its name, location, and size, and the dominant color of the object bounding-box is extracted.

This information is finally stored in a database and used for search in a Web interface, which is shown in Fig. 1. Here the user can search for keyframes that contain one or more objects, which can be further filtered by location, color, relative size, and detection confidence. This allows the user to perform explorative queries such as *"a person on the left"*, or *"a small car in blue at the bottom"* (see details in the next section).

The analysis system in the back uses the detected object information to combine/hyperlink similar keyframes (in terms of detected objects), which finally allows the user to quickly see other shots/keyframes that are similar to the current one. Additionally, statistics of object detections in keyframes are aggregated over the entire video and this aggregated information is used to interlink similar videos, which can also be explored by the user, as shown in Fig. 2b.

Although currently not implemented, until the actual VBS2021 competition, we plan to further extend the system by temporal search features, such as filtering segments for *"object A should appear before object B"* and *"object C is visible for 3s"*.

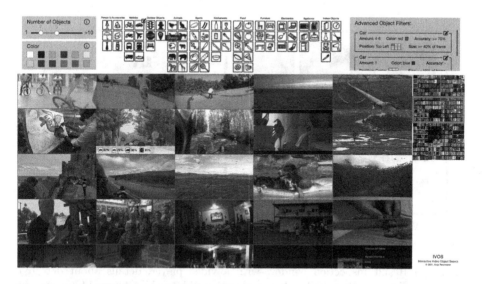

Fig. 1. The IVOS system allows exploratory search and filtering by objects and their specific characteristics.

3   The IVOS User Interface

The interface of IVOS (see Fig. 1) presents objects available for search as small icons that are grouped by category and act as query triggers when clicked. Depending on feedback from user tests, we might add a search bar that helps to select objects by their name.

The query results are presented below, together with a visual scrollbar, and can be inspected by simply hovering over it with the mouse. A score considering the confidence of the detected objects in the keyframes, as well as the quality with which advanced query filters are satisfied determines the order of the keyframes.

When clicking the thumbnail, a new dialog modal appears, which shows the corresponding video segment (i.e., a video player), details of the detected objects, and hyperlinks to other similar videos and keyframes/shots (see Fig. 2a).

Additional quick filter settings can be made in the top left corner, where the results can be filtered by a specific number of detected instances, and/or by specific colors of the keyframe/shot.

More advanced filter options for an object can be set by clicking the edit icon of the corresponding object icon. This opens a new dialog modal where the user is able to combine several object filters and set specific filter characteristics for each of them: (i) the number of detected instances, (ii) the color, (iii) the required object confidence, (iv) the location in the frame, as well as (v) the relative size to the frame. Temporal filters can be set here as well (see Fig. 2b). If multiple specifications for an object are set, keyframes must satisfy all of them.

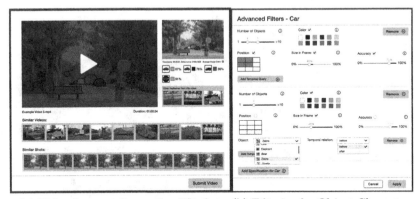

(a) Video Segment Inspection Window   (b) Filtering by Object Characteristics

Fig. 2. (a) Result inspection window with a video player, details about detected objects, and links to other similar shots and similar videos. (b) Advanced filter options that allow to create combined object queries.

4 Summary

The IVOS system was developed for simple but effective content search in video archives. It uses only 80 object classes, which are available as small icons that should be easily understandable and manageable for a user and can be combined with each other. The analysis of the content is based on state-of-the-art object detection models (YOLOv4 [2]) and further processing, which allows to extract the dominant color of an object, as well as to find other shots and videos that are similar to the current one.

Acknowledgments. This work was funded by the FWF Austrian Science Fund under grant P 32010-N38.

References

1. Amato, G., et al.: VISIONE at VBS2019. In: Kompatsiaris, I., Huet, B., Mezaris, V., Gurrin, C., Cheng, W.-H., Vrochidis, S. (eds.) MMM 2019. LNCS, vol. 11296, pp. 591–596. Springer, Cham (2019). https://doi.org/10.1007/978-3-030-05716-9_51

2. Bochkovskiy, A., Wang, C.Y., Liao, H.Y.M.: Yolov4: optimal speed and accuracy of object detection. arXiv preprint arXiv:2004.10934 (2020)
3. Leibetseder, A., Münzer, B., Primus, J., Kletz, S., Schoeffmann, K.: diveXplore 4.0: the ITEC deep interactive video exploration system at VBS2020. In: Ro, Y.M., et al. (eds.) MMM 2020. LNCS, vol. 11962, pp. 753–759. Springer, Cham (2020). https://doi.org/10.1007/978-3-030-37734-2_65
4. Lin, T.-Y., et al.: Microsoft COCO: common objects in context. In: Fleet, D., Pajdla, T., Schiele, B., Tuytelaars, T. (eds.) ECCV 2014. LNCS, vol. 8693, pp. 740–755. Springer, Cham (2014). https://doi.org/10.1007/978-3-319-10602-1_48
5. Lokoc, J., Bailer, W., Schoeffmann, K., Muenzer, B., Awad, G.: On influential trends in interactive video retrieval: video browser showdown 2015–2017. IEEE Trans. Multimed. 1 (2018). https://doi.org/10.1109/TMM.2018.2830110
6. Lokoč, J., et al.: Interactive search or sequential browsing? A detailed analysis of the video browser showdown 2018. ACM Trans. Multimed. Comput. Commun. Appl. 15(1), 29:1–29:18 (2019). https://doi.org/10.1145/3295663
7. Rossetto, L., et al.: Interactive video retrieval in the age of deep learning-detailed evaluation of VBS 2019. IEEE Trans. Multimed. 23, 243–256 (2020)
8. Rossetto, L., Schuldt, H., Awad, G., Butt, A.A.: V3C – a research video collection. In: Kompatsiaris, I., Huet, B., Mezaris, V., Gurrin, C., Cheng, W.-H., Vrochidis, S. (eds.) MMM 2019. LNCS, vol. 11295, pp. 349–360. Springer, Cham (2019). https://doi.org/10.1007/978-3-030-05710-7_29
9. Schoeffmann, K.: Video browser showdown 2012–2019: a review. In: 2019 International Conference on Content-Based Multimedia Indexing (CBMI), pp. 1–4. IEEE (2019)
10. Schoeffmann, K., Münzer, B., Leibetseder, A., Primus, J., Kletz, S.: Autopiloting feature maps: the deep interactive video exploration (diveXplore) system at VBS2019. In: Kompatsiaris, I., Huet, B., Mezaris, V., Gurrin, C., Cheng, W.-H., Vrochidis, S. (eds.) MMM 2019. LNCS, vol. 11296, pp. 585–590. Springer, Cham (2019). https://doi.org/10.1007/978-3-030-05716-9_50
11. Schoeffmann, K., Münzer, B., Primus, J., Leibetseder, A.: The divexplore system at the video browser showdown 2018-final notes. arXiv preprint arXiv:1804.01863 (2018)
12. Schoeffmann, K., Münzer, B., Primus, M.J., Kletz, S., Leibetseder, A.: How experts search different than novices-an evaluation of the divexplore video retrieval system at video browser showdown 2018. In: 2018 IEEE International Conference on Multimedia & Expo Workshops (ICMEW). IEEE (2018)
13. Schöffmann, K., Bailer, W.: Video browser showdown. ACM SIGMultimed. Records 4(2), 1–2 (2012)

Video Search with Sub-Image Keyword Transfer Using Existing Image Archives

Nico Hezel$^{(\boxtimes)}$, Konstantin Schall, Klaus Jung, and Kai Uwe Barthel

HTW Berlin, University of Applied Sciences - Visual Computing Group,
Wilhelminenhofstraße 75, 12459 Berlin, Germany
`hezel@htw-berlin.de`, `barthel@htw-berlin.de`
`http://visual-computing.com`

Abstract. This paper presents details of our frame-based Ad-hoc Video Search system with manually assisted querying that will be used for the Video Browser Showdown 2021 (VBS2021). The main contributions of our new system consist of an improved automatic keywording component, better visual feature vectors which have been fine-tuned for the task of image retrieval, and an improved visual presentation of the search results. Additionally, we use a more powerful joint textual/visual search engine based on Lucene, which can perform a search according to the temporal sequence of textual or visual properties of the video frames.

Keywords: Content-based video retrieval · Exploration · Image browsing · Visualization · CNNs · Automatic keywording

1 Introduction

At the Video Browser Showdown, participants use their interactive video browsing systems to find specific video clips as fast as possible. VBS2021 will use the V3C1 dataset in collaboration with NIST, i.e. **TRECVID 2020**, which consists of 7475 video files, amounting for 1000 hours of video content. As in the previous competitions [14] the VBS participants have to solve **Known-Item Search (KIS)** and **Ad-Hoc Video Search (AVS)** tasks that are issued as live presentation of scenes of interest, either as a visual 20 seconds clip, or as a textual description. For KIS the goal is to find the correct segment, for AVS as many suiting segments as possible have to be found.

Over the last years many deep learning-based improvements have led to better visual feature vectors [5], improved automatic keywording [1], and shared visual and textual feature embeddings [6]. However, finding specific video clips in huge collections still remains a very challenging task.

After participating in several VBS competitions, we have learned about the difficulties associated with this type of video search tasks. Typically, trying to find a specific video clip consists of an iterative sequence of search and browsing operations. Starting from an initial search, relevance feedback and search refinements are used to gradually improve the result. Especially for KIS, the

© Springer Nature Switzerland AG 2021
J. Lokoč et al. (Eds.): MMM 2021, LNCS 12573, pp. 484–489, 2021.
https://doi.org/10.1007/978-3-030-67835-7_49

main problem is to find a good initial image that can be used as a query candidate. Although the visual similarity search usually works well, a successful search requires an image that is very similar to the desired target image. When searching in large video collections, there may be many images that are similar to the search query, but not necessarily to the desired result. Since automatic keywording is still not accurate enough, another problem occurs when using text search. A vague text search leads to a huge amount of possible matching images, making it difficult to narrow down the results. A too specific text search could miss the right result because of the possible inaccuracy of the keywords.

In the past, the most successful systems [10] have been those that exploit the chronological order of different scenes in a video. For example, it is easier to find a clip that first shows a moving car and then the view of a castle by combining these two searches than by trying to search for the specific car only.

In a way, there was a certain randomness in how fast a video clip could be found or not. If a good initial query was available, the search could usually be narrowed down quickly. If such a query could not be found, a lot of time was spent to even start the specific search.

Systems that allowed to view many images simultaneously were at an advantage, as they helped to get a better overview of the result. This can be achieved by using (hierarchical) self-sorting maps [16] or image graphs [2].

2   Automatic Keywording

To address the problems mentioned above, this paper focuses primarily on an improved automatic keywording method. In principle, there are three different approaches to determine keywords for an unknown image.

Methods employing multilabel classification use networks to predict the probability of the presence of a certain object in an image. Known implementations [8] can distinguish between several thousand object classes. To ensure the result quality remains nearly the same for a higher number of categories, exponentially more training images are necessary. If images contain several objects, the classification accuracy usually suffers.

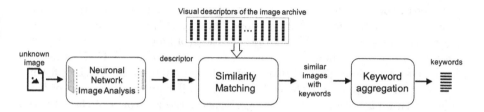

Fig. 1. Keywording of an unknown image using keywords from other images.

Object detection schemes address the recognition and the localization of multiple objects in an image [12]. These schemes work well if the number of object classes is limited, but their accuracy drops if the number of classes is increased.

(a) blue sea wood summer beach heaven vacation sun coast sky water nature recreation old landscape

(b) fence wood sea beach blue old wooden water sand summer texture vacation protection ocean

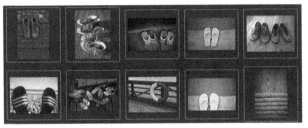

(c) travel summer water footwear vacation shoes leisure pair shoe foot blue relax beauty sandal

Fig. 2. Keywording of an entire image and two sub-images. On the right the ten most similar images are shown. The proposed keywords are shown below. a) Entire image: Only global concepts are described, the fence and the shoes are missing, b) sub-image (center row, right column): Details such as the wooden fence are recognized, c) sub-image (center row, center column): The shoes/sandals are recognized.

For example, the Open Images Dataset V6[1] uses only 600 different objects, which is still not enough for general video retrieval with arbitrary objects. Many important concepts (such as "sky", "wood" or "forest") are missing.

A third approach makes use of large collections of images that have previously been tagged. Figure 1 shows the principle of such a scheme. A CNN is used to determine a visual descriptor (feature vector) for an unknown image. This feature vector is then used to search the collection for similar images. The keywords for the unknown image are generated by aggregating the keywords of the retrieved images taking into account the respective feature vector similarities.

The advantage of such an approach is its ability to provide keywords from a much larger and/or more specific vocabulary than previous methods. The

---

[1] https://storage.googleapis.com/openimages/web/index.html.

accuracy of the system can easily be improved by increasing the keyword quality, the collection size or by improving the visual feature vectors. We have developed the online keywording system Akiwi[2], which is an example of such a system using 20 million labeled images.

Although reusing keywords from other images can be used effectively to label unknown images with keywords from a very extensive vocabulary, the quality is often still not good enough. Figure 2a shows the result for an example image using Akiwi. The overall situation and atmosphere of the scene is well captured, but important objects are not recognized. This could be due to the fact that no comparable scene might be available in the image collection. To circumvent this problem, we have decided to split the image into 3×3 overlapping sub-images that are labeled individually. The keywords of the sub-images are then combined as global keywords for the entire image. Figure 2b and c show the results for two sub-images. This time the important objects are identified. Dividing the images into nine sub-images significantly increases the computational complexity. Because of the large receptive field of the CNNs local feature vectors are strongly correlated, therefore predicting distinct keywords from them is difficult. Therefore an approach to decorrelate the local features vectors is needed to be able to predict keywords from subareas of the feature map directly.

3   Improved Image Feature Vectors

Aggregated convolutional activations have proven to form a highly representative feature vector [13]. Additionally, retrieval quality can be enhanced if these feature vectors are produced by CNNs which have been fine-tuned for the specific task of image retrieval [5].

We presented a simple, yet effective supervised aggregation method built on top of existing regional pooling approaches in [15]. In addition to regional max pooling, we calculate regional average activations of extracted feature maps. Parameters for each of the pooled regional feature vectors are learned to perform a weighted aggregation to a single global descriptor. These aggregation weights can automatically be learned while the chosen backbone CNN is fine-tuned. This form of parameterized global feature vector aggregation is further referred to as DARAC.

The feature vectors used in our new video retrieval system were generated by utilizing a ResNet152 [9] for activation map extraction in combination with a DARAC-head for global feature aggregation. The backbone was pretrained by classifying ImageNet [11] and the DARAC parameters were randomly initialized. Subsequently the resulting network was fine-tuned with a combination of the ImageNet and the Google Landmarks v2 dataset [17]. The Arcface loss [7], which enforces a larger inter-class margin compared to the commonly used softmax cross entropy, has been used during this fine-tuning procedure.

---

[2] http://www.akiwi.eu/.

4   Search Result Visualization

For displaying a search result, we again rely on visually arranging the images. In [4] we developed a projection quality measure to evaluate the quality of visual sorting. We used this measure to optimize a self-sorting map (SSM) [16] for speed and projection quality. A SSM has a complexity of $n \log(n)$ like a hierarchical Self-Organizing Map (SOM), but is faster in practice. We combined the advantages of SSM and SOM by keeping the swapping approach of the SSM but changing the target calculation. Instead of calculating the target as a constant mean vector for the entire block region, we use a sliding box filter which can be efficiently calculated using integral images. Additionally we continuously reduce the block size.

In earlier publications, we introduced the idea of graph-based image navigation and proposed an efficient algorithm for building hierarchical image similarity graphs [2]. For this VBS we again use real-time visual exploration of millions of images with an image graph. This allows us to instantly check particular images for similar related images without having to perform a full image search.

5   Search System

Our video search system needs several offline preparation steps. Initially we extract two frames per second from every video clip. Frames with nearly identical features vector are merged and therefore shared over several shots or even videos. Next for each frame 20 keywords are determined as described before. Chronological continuous frames describing the same visual content form a shot. Any search query will check the entire collection of frames but we only show a representative of a shot in our interface to reduce clutter.

The hierarchical image graph is build using a fusion of the visual and the keyword similarities. A detailed description how to build such a graph efficiently can be found in [3]. Building a graph of one million images takes one hour on a single core of a Xeon E3-1230 v3 CPU.

Our search system allows the temporal concatenation of different search components, which consist of search queries for keywords, visual feature vectors, color sketches or combinations thereof. A search always retrieves a larger set of result images, which is then condensed to 256 cluster centers to merge groups of very similar images. This number has proven to be large enough to cover enough variations of a search concept and at the same time not too large so that individual images in a visually sorted 16×16 grid are still perceivable at a glance. If the user zooms to one of them, related images are retrieved from the graph, which in turn are displayed visually sorted.

References

1. Asano, Y.M., Rupprecht, C., Vedaldi, A.: Self-labelling via simultaneous clustering and representation learning. In: ICLR. OpenReview.net (2020)

2. Barthel, K., Hezel, N., Schall, K., Jung, K.: Real-time visual navigation in huge image sets using similarity graphs. In: ACM Multimedia, pp. 2202–2204. ACM (2019)

3. Barthel, K.U., Hezel, N.: Visually exploring millions of images using image maps and graphs, chapter 11, pp. 289–315. John Wiley & Sons, Ltd. (2019)

4. Barthel, K.U., Hezel, N., Jung, K.: Visually browsing millions of images using image graphs. In: Proceedings of the 2017 ACM on International Conference on Multimedia Retrieval, ICMR 2017, pp. 475–479. Association for Computing Machinery, New York (2017)

5. Cao, B., Araujo, A., Sim, J.: Unifying deep local and global features for image search. arXiv pp. arXiv-2001 (2020)

6. Chen, Y.C., et al.: Uniter: Universal image-text representation learning. In: ECCV (2020)

7. Deng, J., Guo, J., Xue, N., Zafeiriou, S.: Arcface: additive angular margin loss for deep face recognition. In: Proceedings of the IEEE Conference on Computer Vision and Pattern Recognition, pp. 4690–4699 (2019)

8. Durand, T., Mehrasa, N., Mori, G.: Learning a deep convnet for multi-label classification with partial labels. In: Proceedings of the IEEE Conference on Computer Vision and Pattern Recognition, pp. 647–657 (2019)

9. He, K., Zhang, X., Ren, S., Sun, J.: Deep residual learning for image recognition. In: Proceedings of the IEEE Conference on Computer Vision and Pattern Recognition, pp. 770–778 (2016)

10. Kratochvíl, M., Veselỳ, P., Mejzlík, F., Lokoč, J.: Som-hunter: video browsing with relevance-to-som feedback loop. In: International Conference on Multimedia Modeling, pp. 790–795. Springer (2020)

11. Krizhevsky, A., Sutskever, I., Hinton, G.E.: Imagenet classification with deep convolutional neural networks. In: Advances in Neural Information Processing Systems, pp. 1097–1105 (2012)

12. Liu, L., et al.: Deep learning for generic object detection: a survey. Int. J. Comput. Vision $128(2)$, 261–318 (2020)

13. Radenović, F., Tolias, G., Chum, O.: Fine-tuning CNN image retrieval with no human annotation. IEEE Trans. Pattern Anal. Mach. Intell. $41(7)$, 1655–1668 (2018)

14. Rossetto, L., et al.: Interactive video retrieval in the age of deep learning - detailed evaluation of VBS 2019. IEEE Trans. Multimed. 23, 1 (2020)

15. Schall, K., Barthel, K.U., Hezel, N., Jung, K.: Deep aggregation of regional convolutional activations for content based image retrieval. In: 2019 IEEE 21st International Workshop on Multimedia Signal Processing (MMSP), pp. 1–6. IEEE (2019)

16. Strong, G., Gong, M.: Self-sorting map: an efficient algorithm for presenting multimedia data in structured layouts. IEEE Trans. Multimed. $16(4)$, 1045–1058 (2014)

17. Weyand, T., Araujo, A., Cao, B., Sim, J.: Google landmarks dataset v2-a large-scale benchmark for instance-level recognition and retrieval. In: Proceedings of the IEEE/CVF Conference on Computer Vision and Pattern Recognition, pp. 2575–2584 (2020)

A VR Interface for Browsing Visual Spaces at VBS2021

Ly-Duyen Tran[1]([⊠]), Manh-Duy Nguyen[1], Thao-Nhu Nguyen[1], Graham Healy[1], Annalina Caputo[1], Binh T. Nguyen[2,3], and Cathal Gurrin[1]

[1] Dubin City University, Dublin, Ireland
ly.tran2@mail.dcu.ie
[2] AISIA Research Lab, Ho Chi Minh, Vietnam
[3] Vietnam National University, Ho Chi Minh University of Science, Ho Chi Minh, Vietnam

Abstract. The Video Browser Showdown (VBS) is an annual competition in which each participant prepares an interactive video retrieval system and partakes in a live comparative evaluation at the annual MMM Conference. In this paper, we introduce EOLAS, which is a prototype video/image retrieval system incorporating a novel virtual reality (VR) interface. For VBS'21, EOLAS represented each keyframe of the collection by an embedded feature in a latent vector space, into which a query would also be projected to facilitate retrieval within a VR environment. A user could then explore the space and perform one of a number of filter operations to traverse the space and locate the correct result.

Keywords: Video browser showdown · Interactive retrieval · Virtual reality · Eolas

1 Introduction

As the volume of multimedia data increases, there is a need for valid experimental comparisons between competing approaches. The Video Browser Showdown (VBS) meets this need by providing a means of comparing interactive search systems using known-item search tasks over large video collections (partly) in front of a live audience [8]. In this work, we introduce an experimental VBS prototype developed by a DCU and HCMUS-based team participating for the first time. The experimental system called EOLAS is an interactive retrieval system that provides a novel Virtual Reality (VR) interface to large multimedia libraries. The system comprises two main components, the back-end embedding and clustering techniques for data storage/retrieval and the front-end voice and gesture-controlled VR interface for interaction.

This paper's contribution is in describing EOLAS, where users can explore the dataset and seek a specific video in a user-friendly virtual environment without the visual interface constraints imposed by using a desktop screen. Moreover, our voice control protocol allows users to avoid the laborious task of typing text

© Springer Nature Switzerland AG 2021
J. Lokoč et al. (Eds.): MMM 2021, LNCS 12573, pp. 490–495, 2021.
https://doi.org/10.1007/978-3-030-67835-7_50

when in VR. The embedding scheme, combined with a clustering step, also helps users get a better experience analyzing the dataset.

2   Related Systems

The VBS has witnessed many video retrieval systems that participated in various approaches during the previous nine years. This section briefly describes several top-performing teams in the competition last year to indicate the current state-of-the-art.

The main goal of SOM-Hunter [5], the top-performing system in 2020, was to combine intuitive text search and browsing with more advanced optional options for experts. The top-ranked frames for a query were displayed using a self-organizing map (SOM) to visualize the high-dimensional data. Additionally, users could choose relevant frames for relevance feedback and they could repeat this process until a suitable result was found, or they could start-over at any time with a new query. The runner-up and the third-placed systems were "VIRET" [6] and the "vitrivr" system [9], respectively. Both tools supported novel query generation methodologies, such as Query-by-Sketch, which facilitated users to draw a query and Query-by-Example, allowing a user to search for visually similar keyframes to a selected one. They still had the typical approach of using a textual query in which each image frame was annotated by its semantic concepts such as detected objects or scene text appearing within the frame. VIRET aimed their interface at novice users by supporting easier query re-formulation, while 'vitrivr" focused on their storage database structure to reduce searching time. A fourth system to note was Exquisitor [4], which was a notable and novel prototype based on large-scale interactive learning. It relies on chosen positive and negative examples of visualized keyframes to learn a simple relevance model. The retrieval approach is supported with an efficient index.

VR systems for interactive retrieval have not yet received significant research attention. However, one notable approach is the VR platform for multimodal Lifelog data [2], which was best placed in the 2018 Lifelog Search Challenge [3] by focusing on providing novice users with an immersive and easy to search and browse large archives. Eolas will build upon the user-friendly nature of the VR-lifelog tool [2] by supporting easy querying and implementing a state-of-the-art latent space index support targeted browsing of the collection, as shown to be effective by previous participants. The 360-degree view afforded by a VR platform can help users by providing a larger display area.

3   An Overview of Eolas

Eolas can be viewed in terms of the two prominent components: indexing the dataset in which we annotated all frames into a latent feature space and our unique user interaction support for the VR environment. The overview of Eolas can be depicted in Fig. 1.

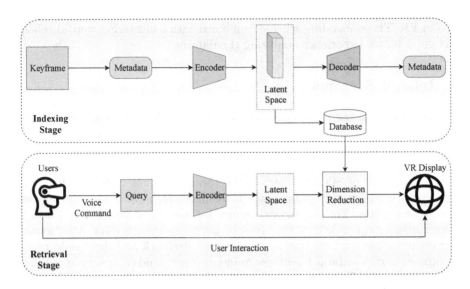

Fig. 1. EOLAS workflow. In the indexing stage, each frame was embedded into a latent vector space based on its metadata by training an Autoencoder network and then was stored in the index. Users wearing a VR device could input a semantic query by using our voice command protocol. The query was also converted into the same latent space as the database. All feature vectors were then transformed into 3D space for visualisation in the VR environment to find the result.

3.1 Source Data

In this task, we use the first part of the Vimeo Creative Commons Collection dataset[1] (V3C) called V3C1 [1], a wide-ranging video collection. V3C1 dataset includes 7,475 videos (1,000 h), categorized into different content categories ranging from Food, Fashion, Art to Instructional videos. The videos were segmented into shots which were represented by over a million keyframes. Additionally, we integrated various forms of provided metadata, such as shot captions, visual concepts, and text extracted from the shots as an input for the retrieval engine. We also amended additional metadata to support user filters, such as a face detector's output that enumerated the number of faces in the shot, a color histogram, audio detection, and camera motion vectors. The sources are presented in Table 1.

3.2 Search Engine

Since each keyframe was represented by the various forms of metadata presented above, we approached the video retrieval problem as a semantic challenge and every keyframe was annotated by combining the three primary sources of textual information (captions, concepts, and text).

---

[1] https://sigmm.hosting.acm.org/2019/07/06/the-v3c1-dataset-advancing-the-state-of-the-art-in-video-retrieval/.

Table 1. List of all metadata

Name	Definition
Caption	Text generated from Google Cloud Speech-to-Text API
Concept	Detection of objects in keyframes
Text	Detection of text in keyframes (OCR)
Color	Detection of dominant color in keyframes
Faces	Detection of faces in keyframes
Histogram	Histogram data on bitrate, resolution, duration and upload dates
Camera motion	Detection of static or moving cameras
Speech detector	Detection of music or silence in videos

Index Construction. The semantic data of a frame was first embedded into a feature space using the Glove model [7], followed by a concatenation step to be fused into a single vector. We then built an Autoencoder network in which an encoder module was used to create a lower-dimensional space to store representative features in the vector and the decoder component needed to ensure these encoded attributes could be reverted to its original value. It is noted that the Autoencoder network would be integrated with the RNN sequence model to cope with the issue of arbitrary length textual descriptions. After the training stage, a keyframe could be annotated by its meaningful features and stored in our index for fast retrieval. An extra supervised learning clustering technique was applied to discover similar groups of keyframes, which could be useful while users solve a task. This step was done once while indexing data and the groups were then stored in our index.

Retrieval. A textual query was also fed into the encoding module to be converted into the same vector space with keyframe embedded features in the retrieval stage. We then applied a dimensional reduction algorithm to visualize the data in the 3D virtual space, where the users' initial position was based on the query's projection into space. The video keyframes were also illustrated in this space through their 3D converted vector features. The ocular distances within this space indicated the similarity between the query and frames. Keyframes were clustered into groups from which several images would be shown. When in the virtual environment, users could select relevant keyframes. All images from their groups with the query's features were then transformed into a 1D line, which we called similarity path. This path could help users to focus on these keyframes rather than the entire dataset.

3.3   User Interaction

There are two phases within which the user interacts with the system. For each task, the user uses voice commands prompting the system to start a new search from the spoken query. Using ASR, the spoken inquiry is converted to text query (which is embedded in the latent space) and is used to bring the VR user to a starting cluster in the 3D space. In the second phase of interaction, the user can navigate the space to locate the required videos. The user has the option to select the most suitable videos and use a voice command to ask the system to take him/her to the closest video cluster to the selected videos. The videos are ranked and shown to the user in decreasing rank order on a similarity path. At any time, the user can (1) restart the search, (2) go back to the previous cluster, (3) go to any point in the similarity path on the minimap, or (4) choose more videos in order to rerank the similarity path (Fig. 2).

Fig. 2. A VR visual interface prototype

VR Visual Interface. Our proposed visual interface is designed to be accessible for novice users. The user interface is kept minimal, consisting of two parts: a minimap showing the current similarity path, which is always floating on the right of the user's head, and a saved section on the left side of the user. All videos in the 3D space are represented by tangible objects that the user can grab and move in the space or put in the saved section. The user can travel around the environment, either using voice commands to navigate the similarity path on the minimap or through a traditional VR teleportation system by choosing the nearby clusters visible from the current location.

4   Conclusions

This paper introduces EOLAS, which is an experimental video retrieval system built on the VR platform. By utilising the wide and surrounding view in the virtual environment, this tool can easily visualize the grouped videos in the 3D dimensional space based on their encoded features and provide users with an intuitive and novel access mechanism for large video archives.

Acknowledgments. This publication has emanated from research supported in party by research grants from Science Foundation Ireland under grant numbers 18/CRT/6223 and 18/CRT/6224.

References

1. Berns, F., Rossetto, L., Schoeffmann, K., Beecks, C., Awad, G.: V3C1 dataset: an evaluation of content characteristics. In: Proceedings of the 2019 on International Conference on Multimedia Retrieval, ICMR 2019, pp. 334–338. Association for Computing Machinery, New York (2019). https://doi.org/10.1145/3323873.3325051
2. Duane, A., Gurrin, C., Huerst, W.: Virtual reality lifelog explorer: lifelog search challenge at ACM ICMR 2018. In: Proceedings of the 2018 ACM Workshop on The Lifelog Search Challenge, pp. 20–23 (2018)
3. Gurrin, C., et al.: Comparing approaches to interactive lifelog search at the lifelog search challenge (LSC2018). ITE Trans. Media Technol. Appl. **7**(2), 46–59 (2019)
4. Jónsson, B.Þ., Khan, O.S., Koelma, D.C., Rudinac, S., Worring, M., Zahálka, J.: Exquisitor at the video browser showdown 2020. In: Ro, Y.M., et al. (eds.) MMM 2020. LNCS, vol. 11962, pp. 796–802. Springer, Cham (2020). https://doi.org/10.1007/978-3-030-37734-2_72
5. Kratochvíl, M., Veselý, P., Mejzlík, F., Lokoč, J.: SOM-hunter: video browsing with relevance-to-SOM feedback loop. In: Ro, Y.M., et al. (eds.) MMM 2020. LNCS, vol. 11962, pp. 790–795. Springer, Cham (2020). https://doi.org/10.1007/978-3-030-37734-2_71
6. Lokoč, J., Kovalčík, G., Souček, T.: VIRET at video browser showdown 2020. In: Ro, Y.M., et al. (eds.) MMM 2020, Part II. LNCS, vol. 11962, pp. 784–789. Springer, Cham (2020). https://doi.org/10.1007/978-3-030-37734-2_70
7. Pennington, J., Socher, R., Manning, C.D.: GloVe: global vectors for word representation. In: Empirical Methods in Natural Language Processing (EMNLP), pp. 1532–1543 (2014). http://www.aclweb.org/anthology/D14-1162
8. Rossetto, L., et al.: Interactive video retrieval in the age of deep learning - detailed evaluation of VBS 2019. IEEE Trans. Multimed. 1 (2020)
9. Sauter, L., Amiri Parian, M., Gasser, R., Heller, S., Rossetto, L., Schuldt, H.: Combining boolean and multimedia retrieval in vitrivr for large-scale video search. In: Ro, Y.M., et al. (eds.) MMM 2020. LNCS, vol. 11962, pp. 760–765. Springer, Cham (2020). https://doi.org/10.1007/978-3-030-37734-2_66

Correction to: SQL-Like Interpretable Interactive Video Search

Jiaxin Wu, Phuong Anh Nguyen, Zhixin Ma, and Chong-Wah Ngo

Correction to:
Chapter "SQL-Like Interpretable Interactive Video Search"
in: J. Lokoč et al. (Eds.): *MultiMedia Modeling*, LNCS 12573,
https://doi.org/10.1007/978-3-030-67835-7_34

The original version of the book was inadvertently published with an incorrect acknowledgement in chapter 34. The acknowledgement has been corrected and reads as follows:

Acknowledgement: The research was partially supported by the Singapore Ministry of Education (MOE) Academic Research Fund (AcRF) Tier 1 grant and the National Natural Science Foundation of China (No. 61872256).

The affiliation of the third author, Zhixin Ma, was incorrect. In the contribution it read "School of Information System," but correctly it should be "School of Computing and Information Systems".

The affiliation of the last author, Chong-Wah Ngo, was not correct. In the book it read "Department of Computer Science, City University of Hong Kong, Hong Kong, China". Instead, the correct affiliation is: "School of Computing and Information Systems, Singapore Management University, Singapore, Singapore".

Additionally, his e-mail address "cscwngo@cityu.edu.hk" was also incorrect. The correct e-mail address is: "cwngo@smu.edu.sg".

The chapter and the book have been updated with the changes.

The updated version of this chapter can be found at
https://doi.org/10.1007/978-3-030-67835-7_34

Author Index

Printed in the United States
by Baker & Taylor Publisher Services